THE BEHAVIOR OF TELEOST FISHES

THE BEHAVIOR
OF TELEOST FISHES

Edited by TONY J. PITCHER

THE JOHNS HOPKINS UNIVERSITY PRESS
Baltimore, Maryland

© 1986 Tony J. Pitcher
All rights reserved.
First published in 1986, by
The Johns Hopkins University Press
701 West 40th Street
Baltimore, Maryland 21211

Library of Congress Cataloging in Publication Data

Main entry under title:

The Behaviour of teleost fishes.

   Includes bibliographies and indexes.
   1. Osteichthys—Behavior.   2. Fishes—Behavior.
I. Pitcher, T.J.
QL6393.B44   1986   597'.50451   85–19887
ISBN 0-8018-3246-2

Printed and bound in Great Britain

CONTENTS

PART FOUR: APPLIED FISH BEHAVIOUR

To Susannah and Tamsin

CONTRIBUTORS

H. Bleckmann: Department of Neurosciences A-001, School of Medicine, University of California at San Diego, La Jolla, California, USA

P. Colgan: Department of Biology, Queens University, Kingston, Ontario, Canada K7L 3N6

G.J. FitzGerald: Département de Biologie, Université Laval, Cité Universitaire, Quebec, P.Q., Canada G1K 7P4

R.N. Gibson: Scottish Marine Biological Association, Dunstaffnage Marine Research Laboratory, PO Box 3, Oban, Argyll PA34 4AD, UK

D.M. Guthrie: Department of Zoology, Williamson Building, University of Manchester, Manchester M13 9PL, UK

M.R. Gross: Department of Biological Sciences, Simon Fraser University, Burnaby, British Columbia, Canada V5A 1S6

T.J. Hara: Freshwater Institute, 501 University Crescent, Winnipeg, Manitoba, Canada R3T 2N6

P.J.B. Hart: Department of Zoology, School of Biological Sciences, Adrian Building, University of Leicester, Leicester LE1 7RH, UK

A.D. Hawkins: Department of Agriculture and Fisheries Scotland, Marine Laboratory, PO Box 101, Victoria Road, Aberdeen AB9 8DB, UK

G.S. Helfman: Department of Zoology and Institute of Ecology, University of Georgia, Athens, Georgia 30602, USA

F.A. Huntingford: Department of Zoology, University of Glasgow, Glasgow G12 8QQ, UK

A.E. Magurran: School of Animal Biology, University College of North Wales, Bangor, Gwynedd LL57 2UW, UK

M. Milinski: Arbeitsgruppe für Verhaltensforschung, Abteilung für Biologie, Ruhr-Universität Bochum, Postfach 102148, D-4630 Bochum 1, Federal Republic of Germany

D.L.G. Noakes: Department of Zoology, College of Biological Sciences, University of Guelph, Guelph, Ontario, Canada N1G 2W1

K. O'Hara: Department of Zoology, University of Liverpool, Brownlow Street, PO Box 147, Liverpool L69 3BX, UK

J. Parzefall: Zoologisches Institut und Zoologisches Museum, Universität Hamburg, Martin-Luther-King-Platz 3, 2 Hamburg 13, Federal Republic of Germany

T.J. Pitcher: School of Animal Biology, University College of North Wales, Bangor, Gwynedd LL57 2UW, UK

R.C. Sargent: Department of Biological Sciences, Simon Fraser University, Burnaby, British Columbia, Canada V5A 1S6

G. Turner: School of Animal Biology, University College of North Wales, Bangor, Gwynedd LL57 2UW, UK

Contributors

C.S. Wardle: Department of Agriculture and Fisheries Scotland, Marine Laboratory, PO Box 101, Victoria Road, Aberdeen, AB9 8DB, UK

R.J. Wootton: Department of Zoology, University College, University of Wales, Aberystwyth, Dyfed SY23 3DA, UK

PREFACE

This book is about the behaviour of teleosts, a well-defined, highly successful, taxonomic group of vertebrate animals sharing a common body plan and forming the vast majority of living bony fishes. There are well over 22 000 living species of teleosts, including nearly all those of importance in commercial fisheries and aquaculture. Teleosts are represented in just about every conceivable aquatic environment from temporary desert pools to the deep ocean, from soda lakes to sub-zero Antarctic waters. Behaviour is the primary interface between these effective survival machines and their environment: behavioural plasticity is one of the keys to their success.

The study of animal behaviour has undergone revolutionary changes in the past decade under the dual impact of behavioural ecology and sociobiology. The modern body of theory provides quantitatively testable and experimentally accessible hypotheses. Much current work in animal behaviour has concentrated on birds and mammals, animals with ostensibly more complex structure, physiology and behavioural capacity, but there is a growing body of information about the behaviour of fishes. There is now increasing awareness that the same ecological and evolutionary rules govern teleost fish, and that their behaviour is not just a simplified version of that seen in birds and mammals. The details of fish behaviour intimately reflect unique and efficient adaptations to their three-dimensional aquatic environment.

The book aims to bring together accounts of the major functional topics in fish behaviour, reviewed in the light of current behavioural theory. Reviews at this level are intended to be of use to a broad spectrum of senior undergraduates, masters-level students and research workers interested in fish biology and fisheries. Each chapter commences with the fundamental principles often taken for granted in pure research monographs. It then sets out the evidence for the main conceptual basis of each topic, goes on to discuss current areas of controversy, and, finally, attempts to map out the contemporary research frontiers and speculate where new advances are likely to be made. Each chapter closes with a summary of its contents.

The book is divided into four sections, covering the bases of behaviour, sensory modalities, behavioural ecology and applied aspects of fish behaviour. Each section opens with a brief overview of its contents. I have tried to strike a balance between a comprehensive coverage of the field and those topics which I feel reflect the thrust of contemporary research activity in fish behaviour: unfortunately, there certainly was not room to review everything at this depth. Others might of course have made a different selection, and inevitably some will feel that there are omissions of topics such as the central nervous system, learning, the constraints of functional morphology, migration, and groups or habitats of particular interest. In many cases I can plead that review books have recently appeared in these areas and these are listed in the introductory sections where appropriate. But for a few omissions I have no real excuse other than personal choice.

I have attempted to maintain a high standard of contributions throughout the book, and many of the chapters have been independently refereed before revision and editing. Potential contributors were advised that manuscripts might not be accepted for inclusion, and indeed some contributions did fall by the wayside in that respect. I would like to thank the following for their help as referees: Professor P. Bateson FRS, Dr J.H.S. Blaxter, Dr J. Davenport, Dr T.R. Halliday, Dr R.N. Hughes, Dr F. Huntingford, Dr G.J.A. Kennedy, Dr J.R. Krebs FRS, Professor A. Manning, Professor J. Maynard-Smith FRS, Professor W.R.A. Muntz FRS, Dr M. Nelissen, Dr L. Partridge and Dr G.W. Potts. S.J. Pitcher gave invaluable help in compiling the index. Finally, I would like to thank Dr A. Magurran and Dr P.J.B. Hart for providing advice, support and encouragement throughout this book's gestation.

Tony J. Pitcher
Bangor

ACKNOWLEDGEMENTS FOR ILLUSTRATIONS

Permission to reproduce figures has been gratefully received from the following copyright owners:

Chapter 1: Figure 1, *Ecological Society of America*, Tempe, Arizona. Figure 2, *Paul Parey* Berlin. Figure 3, *The Society for the Study of Evolution*, Lawrence, Kansas. Figure 4, *Paul Parey*, Berlin. **Chapter 2:** Figure 1, *American Psycological Association*, Arlington, Virginia. Figure 2a, *American Psycological Association*, Arlington, Virginia. Figure 2b, *The Psychomonic Society of America*, Guadaloupe, Texas. Figure 2c, *American Psycological Association*, Arlington, Virginia. Figure 3a, *Yale University Press*, New Haven, Connecticut. Figure 3b, *E.J. Brill*, Leiden. Figure 3c, *E.J. Brill*, Leiden. Figure 4, *E.J. Brill*, Leiden. Figure 5, *Springer-Verlag*, Heidelberg. Figure 6, *Paul Parey*, Berlin. **Chapter 3:** Figure 3, *Paul Parey*, Berlin. Figure 4, *Alan R. Liss*, Inc., New York. **Chapter 4:** Figure 1, *Plenum Press*, New York. Figure 2, *Plenum Press*, New York. Figure 6, *Plenum Press*, New York. Figure 8, *Springer-Verlag*, Heidelberg. Figure 10, *Munksgaard*, Copenhagen. **Chapter 6:** Figure 3, *Elsevier Biomedical Press B.V.*, Amsterdam. **Chapter 7:** Figure 4, *Springer-Verlag*, Heidelberg. Figure 7, *Elsevier Biomedical Press B.V.*, Amsterdam. **Chapter 8:** Figure 2, *Bailliere Tindall*, London. Figure 3, *Journal of the Fisheries Research Board of Canada*, Fisheries and Oceans, Ottawa. Figure 4, *Academic Press*, Orlando, Florida. Figure 7, *Ecological Society of America*, Tempe, Arizona. Figure 8, *Ecological Society of America*, Tempe, Arizona. **Chapter 9:** Figure 3, *Ecological Society of America*, Tempe, Arizona. Figure 4, *Journal of the Fisheries Research Board of Canada*, Fisheries and Oceans, Ottawa. Figure 5, *Paul Parey*, Berlin. Figure 6, *Paul Parey*, Berlin. Figure 7, *Macmillan Journals Ltd.*, London. Figure 8, *Baillière Tindall*, London. **Chapter 12:** Figure 1, *Baillière Tindall*, London. Figure 6, *W. Junk Publishers*, Dordrecht. Figure 7b, *Paul Parey*, Berlin. Figure 8, *W. Junk Publishers*, Dordrecht. **Chapter 17:** Figure 1, *Memoires de Biospeliogie*, Saint Girons. Figure 2, *Memoires de Biospeliogie*, Saint Girons. Figure 5, *Elsevier Biomedical Press B.V.*, Amsterdam. Figure 8. *Paul Parey*, Berlin. Figure 9, *Paul Parey*, Berlin. Figure 10, *Paul Parey*, Berlin. Figure 11, *Paul Parey*, Berlin.

PART ONE: BASES OF BEHAVIOUR

INTRODUCTION

T.J. Pitcher

The opening section of the book aims to review the major determinants of the behaviours that are observed in mature teleost fish. The three chapters are about genes, motivation and development. These are fundamental issues in the science of animal behaviour, reviewed with reference to recent research on fishes: each of the authors addresses themselves to the question of the suitability of teleosts for such work. It is not intended to cover short-term causation via the central nervous system (see recent review by Guthrie 1983), neuroethology (Guthrie 1980), learning (Gleitman and Rozin 1971; Roper 1983), or hormones and fish behaviour (Munro and Pitcher 1983; Stacey 1983).

Chapter 1 covers the most fundamental area of the functional study of animal behaviour, its genetic basis. Behaviour is often far removed from the direct action of genes in the organism's coded DNA. David Noakes explores the nature of the links between genes and expressed behaviours, an issue of profound importance in the light of the distinction between causal and functional explanations, and uses examples from studies on teleost fishes. For biologists interested in functional explanations, he argues, it may not be worth while unravelling the full details of causation, provided we can ascribe differences in behaviour to genetic differences between individuals. Noakes describes how selection shapes fish behaviour, reviewing recent work on the inheritance of behaviour in fish, and discusses how we may make sense of the 'chocolate soup and jigsaw puzzle' which Nature has put before us.

Patrick Colgan, in Chapter 2, surveys ideas about how to deal with changes in responsiveness: the difficult field of the study of motivation. Why do animals respond differently to stimuli on different occasions? To some the answers are the pure detailed descriptions of physiology (the 'neurophysiologist's nirvana'), whereas to others physiological causation is no answer at all, the whole animal's behaviour emerging from its physiology in a holistic sense to meet the needs of survival. The functional goal is met by using whatever physiology happens to be available (see discussion in Huntingford 1984). As in Chapter 1, the split further reflects the distinction between causal and functional explanations. Colgan maps out the internal and external factors which influence the 'big four' motivational systems of hunger, fear, aggression and sex, and discusses the development, interaction and extensions of these systems using examples from research on teleost fishes. He takes the firm view that motivation is not 'physiology writ large', and reviews the range of models of motivation used by students

1

of animal behaviour, including the space-state approach, time-sharing, information theory, and stochastic and specific models. The author gives us strictures for the design of experiments on fish motivation, taking care to stimulate its natural environment and avoiding experiments where 'under precisely controlled conditions, the animal does as he damn pleases'. Patrick Colgan is critical of the current orthodoxy and leaves us with the thought that current analysis has 'scarcely gained a toehold on the terrain to be investigated'.

The development of behaviour in teleost fishes is reviewed by Felicity Huntingford in Chapter 3. She opens with a clear description of the development of behaviours in salmonids, one of the basal groups of teleosts, and cichlids, one of the most advanced groups. The chapter outlines the processes which shape the fishes' growing behavioural repertoire, from the first muscle twitches in the egg, to the range of feeding, antipredator, habitat selection, migration, agonistic, sexual and parental behaviours in the mature adult fish. Huntingford then concentrates on an analysis of the contemporary heuristic classification of factors which influence behavioural development, with detailed examples from experiments on teleost fishes. She rejects the distinction between learned and instinctive behaviours as unhelpful and inappropriate. The chapter examines sensitive periods during development, highly important for the reproductive homing of salmonids, discusses the concept of stability and plasticity during development in the face of environmental hazard, and concludes with a review of the current status of the term 'innate behaviour'. Huntingford rejects the term, but explains why some ethologists still prefer to retain it.

References

Gleitman, H. and Rozin, P. (1971) 'Learning and Memory', in W.S. Hoar and D.J. Randall (eds), *Fish Physiology* vol. 6, Academic Press, New York, Chapter 4, pp. 191-278.

Guthrie, D.M. (1980) *Neuroethology*, Blackwell, Oxford, 221, pp.

Guthrie, D.M. (1983) 'Integration and Control by the Central Nervous System', in J.C. Rankin, T.J. Pitcher and R.T. Duggan (eds), *Control Processes in Fish Physiology*, Croom Helm, London, Chapter 8, pp. 130-54.

Huntingford, F.A. (1984) *The Study of Animal Behaviour*, Chapman & Hall, London, 411 pp.

Munro, A.D. and Pitcher, T.J. (1983) 'Hormones and Agonistic Behaviour', in J.C. Rankin, T.J. Pitcher and R.T. Duggan (eds), *Control Processes in Fish Physiology*, Croom Helm, London, Chapter 9, pp. 155-75.

Roper, T. (1983) 'Learning as a Biological Process' in T.R. Halliday and P.J.B. Slater (eds), *Animal Behaviour 3: Genes, Development and Learning*, Blackwell, Oxford, Chapter 6, pp. 178-212.

Stacey, N. (1983) 'Hormones and Reproductive Behaviour in Teleosts', in J.C. Rankin, T.J. Pitcher and R.T. Duggan (eds), *Control Processes in Fish Physiology*, Croom Helm, London, Chapter 7, pp. 117-29.

1 THE GENETIC BASIS OF FISH BEHAVIOUR

D.L.G. Noakes

Introduction

Basic principles of inheritance (e.g. like begets like, children inherit grand-father's chin, the familial patterns of inheritance of haemophilia and other inborn conditions, and the selective breeding of domestic livestock and pet species) are widely accepted even though some of these are misunder-standings, or even misconceptions. A major problem which we have to face is that behaviour rarely bears the direct, one-to-one relationship to genes as do enzymes or perhaps even structural features of the phenotype (Dawkins 1982; Thomson 1984). The more complex and often more poorly defined features that we refer to as behaviour require elaboration.

A number of major review articles and textbooks more than adequately cover the general field of behaviour genetics. However, few give explicit consideration to any examples of fish species (an exception is Huntingford 1984). I will assume that readers will either be familiar with the basic prin-ciples of behaviour genetics, or be able to refer to other sources for that information. I will illustrate those principles with examples of fish behaviour, and aim to emphasise what this tells us about causation, func-tion, ontogeny and evolution. Behaviour is often recognised as the patterned output of muscles, glands or nerve cells, but I will use the term in a broader sense, to include all those things an animal does as an intact, functioning organism. Genetics of behaviour will include both heritable dif-ferences in behaviour and the mechanisms leading from genes to behaviour within individual animals (Gould 1974).

Methodology

As a hybrid field of science, behaviour genetics is characterised by vigour, independent assortment and random segregation in its activity and output. The vigour is exemplified by the number of books and major review articles (e.g. Manning 1975, 1976; Ehrman and Parsons 1976; Burghardt and Bekoff 1978: Barlow 1981; Immelman, Barlow, Petrinovich and Main 1981; Bonner 1982; Partridge 1983), and to a large extent there have been relatively independent and apparently random focal points of research activity and theoretical advance. We now have considerable understanding of genetic (causal) mechanisms at the molecular or cellular level (e.g. Dobzhansky, Ayala, Stebbins and Valentine, 1977), and a sub-stantial array of hypotheses concerning evolution and function (e.g. Dawkins 1982), but a relatively open chasm between. An understanding of the techniques of genetic studies of behaviour is necessary to appreciate this current situation.

3

The main techniques available for behaviour genetics studies include: studies of single gene differences and mutations, selection experiments, studies of hybrids and backcrosses, comparison of monozygous and dizygous twins, and studies of quantitative traits. I will discuss each of these techniques in subsequent sections of this chapter, with the exception of mono- and dizygous twins, and quantitative traits. Twin studies are most often applied to humans, and have rarely if ever been used in studies of fish. Studies of quantitative traits will be incorporated into other sections (notably selection and hybridisation).

There is something of a dichotomy in studies of behaviour genetics, between those interested primarily in the mechanisms and those interested more in functional questions. These different approaches are sometimes stated as 'how' and 'why', or 'proximate' and 'ultimate' questions, as related to causation and function, respectively (Tinbergen 1963). It is important to understand that these are fundamentally different types of question, require different arguments and evidence to answer them, and need not have any direct relationship linking them. For example, parental nest fanning behaviour by a male three-spined stickleback (*Gasterosteus aculeatus*) functions to enhance survival of embryos developing in his nest ('why'), but this fanning is caused ('how') by dissolved carbon dioxide in the water (among other factors) (Sevenster 1961). Individual behaviour genetics studies are not always explicitly directed towards either (or both) of these kinds of question. It is important to keep this distinction in mind, however, especially when interpreting or extrapolating from particular studies. To reach a conclusion regarding genetics, we must have the appropriate (i.e. genetic) evidence. Failure to appreciate this is the basis for one of the most persistent and misdirected controversies in biology, the so-called 'learned versus innate', or 'nature versus nurture' dichotomy. The term 'innate' has been equated with 'genetically determined' by some, but to others it has meant something that develops without the benefit of learning or experience. The former definition of innate requires genetic evidence, the latter does not. The potential for misunderstanding from these two different uses of the common term 'innate' is obvious.

On the Tracks of Turtles

Our viewpoint may dramatically change how we specify questions and judge answers. Consider the instructions for producing a circle on a computer video screen (Hayes 1984). We could describe one point moving so as to maintain a constant distance from another, fixed point. But an entirely different answer is possible, as the instructions to a point (the turtle in the computer language LOGO) to move so as to leave a circular trail. To the turtle, a circle would consist of a sequence of one step forward and then one turn to the right, repeated until it has turned through a total of 360°. The difference is between a global and a local perspective.

Ethologists typically have a global view of behaviour; animals behave in a territorial manner, or desert their mates, or migrate to a feeding area, for example. The causal links between genes and behaviour must be in local terms; genes specify enzymes, enzymes lead to proteins, and so on. It is the understanding of this causal connection between gene and behaviour that remains largely unknown.

Single Gene Differences

What Can Genes Be 'For'?

In principle, this approach is the right stuff of classical genetics, as the simplest and the most satisfying relationship between genes and behaviour. Phenotypic differences resulting from single gene differences are easy to trace experimentally, as in Mendel's original experiments, and led to the classic 'one gene, one enzyme' causal mechanism model. A comparison of individuals differing only in a single gene should allow us to map cause and effect directly and continuously, from nucleotide sequences to behavioural actions, but such has rarely been the case. Most behavioural traits appear to be associated with differences in a number of gene loci, and many genes appear to have pleiotropic effects on behaviour (Manning 1976).

Mendel's great insight was to study individual phenotypic characters, such as colour of blossoms, height of plant, colour and texture of seed coat, rather than the overall phenotype of the organism. As Barlow (1981) has suggested, it would be most appropriate to select modal (fixed) action patterns of behaviour for genetic analyses. Undoubtedly much of the complexity and uncertainty in studies of behaviour genetics has been a result of the failure to select independent units of behaviour. Of course, the question of what comprises basic units of behaviour is neither new nor trivial. Independent units of behaviour are seldom as conspicuous or unambiguous as size or colour. We rarely have a sufficiently detailed knowledge of the genotype of any fish species we study to make other than broad inferences as to genetic similarities or differences between individuals. Other possibilities do exist, however, along different lines, to allow us to ascribe observable behavioural differences to single gene (or at least a few genes) differences.

The common guppy (*Poecilia reticulata*), and some related species known to followers of the aquarium hobby as live-bearers (family Poeciliidae), and a few species in other families, including most notably goldfish (*Carassius auratus*) and carp (*Cyprinus carpio*) of the Cyprinidae, have been selectively bred in captivity to produce desired pigmentation patterns on the fins or body, or to elaborate or perpetuate monstrosities (e.g. so-called telescope eyes, lionheads, and veil-tails in goldfish). These are often produced in highly inbred lines, so we can reasonably assume that individuals differing in only one of these pigmentation or other features are probably very similar in genotype. Since these fish are typically raised under carefully regulated conditions in captivity, environmental differences between individual fish are also likely to be small. The social behaviour of guppies, especially male courtship, has been frequently studied (Keenleyside 1979; Farr and Peters 1984). There may be substantial differences in behaviour, apparently as a result of relatively small genetic differences among males (and/or females). However, we do not have much information as to how these genetic differences produce the behavioural differences, and since these are artificially selected strains of an inbred species, the differences may not have any significant ecological or evolutionary consequences.

Another possibility is a species kept in large numbers under intensive culture conditions so that individual gene mutants are more likely to be noticed (inbred strains of laboratory mice (*Mus musculus*) are a good example of this). The fish species kept in sufficiently large numbers to allow screening for behavioural mutants are often those kept for food production (e.g. rainbow trout (*Salmo gairdneri*)). These are usually held in such crowded conditions that behaviour can seldom be observed and ascribed to a recognisable individual, and in any event attention would probably not be given to any attributes of the fish other than survival and feeding. Interestingly, some recent studies of such cases not only provide evidence of single gene differences associated with behavioural differences but also give some clue as to the causal mechanisms.

Genetic differences, as measured by isozyme frequencies, have been widely surveyed and established for many different fishes (see, for example, Allendorf, Mitchell, Ryman and Stahl 1977; Allendorf and Phelps 1981). Geographically (and/or genotypically) different stocks of some fish species may also have significant behavioural differences. For example, stocks of rainbow trout differ in swimming performance and responses to water current (Figure 1.1), brook charr (*Salvelinus fontinalis*) differ in susceptibility to angling by humans, survival and growth, Pacific salmon (*Oncorhynchus* spp.) differ in homing ability, and Atlantic salmon (*Salmo salar*) may differ in growth, age at maturity and other characteristics (Raleigh 1971; Bams 1976; Thomas and Donahoo 1977; Kelso, Northcote and Wehrhahn 1981), to name a few examples. These differences are not likely to be one gene, one behaviour, cause-effect relationships, however. The behavioural measures are relatively complex, and the precise genetic constitution of the animals in question is often not known.

However, the adaptive significance of particular, single gene differences is known for some cases. Rainbow trout differing in lactate dehydrogenase (LDH) phenotype (several LDH isozyme variants are known) have significantly different swimming performances at varying ambient dissolved oxygen conditions (Klar, Stalnaker and Farley 1979). Since swimming performance is critical to fish, especially salmonids, such differences clearly could be of major ecological and evolutionary significance. The killifish (*Fundulus heteroclitus*) shows comparable differences in swimming

Figure 1.1: Directional Responses to Water Current of Young Rainbow (*Salmo gairdneri*) and Cutthroat (*Salmo clarki*) Trout, Hatched, Raised and Tested in Controlled Laboratory Conditions. Genetically determined differences in movement with respect to water current correspond to appropriate responses in natural habitat of each stock. In their native habitats, trout from inlet streams have to migrate downstream to reach the lake where they feed; trout from outlet streams have to move upstream to reach the lake

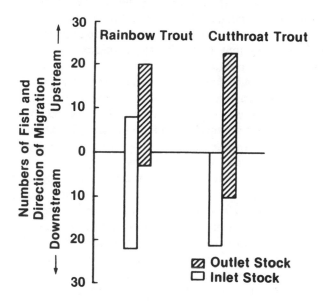

Source: 'Innate Control of Migration of Salmon and Trout Fry from Natal Gravels to Rearing Areas', R.F. Raleigh, *Ecology*, *52*, 291-7. Copyright © 1971 by Ecological Society of America. Reprinted by permission.

endurance at some temperatures, and in time of hatching from the egg membrane, relative to LDH isozyme differences between individuals (DiMichele and Powers 1982). Far-reaching effects of single gene differences in fishes have recently been reported by Allendorf, Knudsen and Leary (1983) in rainbow trout. Trout differing in phosphoglucomutase (PGM) liver isozyme differ significantly in time of hatching and developmental rate. Consequently they differ in size and time of sexual maturity. All these phenotypic differences can have significant adaptive consequences, and are often cited as central to theoretical arguments concerning life-history tactics (e.g. Noakes and Balon 1982). Another example of single genes affecting life-history traits of a fish species is among the poeciliid fishes. As a result of detailed genetic studies, considerable detail is now known about the genetics of sex determination and colour patterns in *Xiphophorus* species. Recently, Kallman (1983) has not only discovered a gene that controls the onset of sexual maturation, but has also demonstrated that the gonadotroph cells of the pituitary gland mediate this effect.

A note of caution should be raised here. Since single gene differences and single gene effects have such intrinsic appeal, it is tempting to look for examples either to illustrate the principle or as material for detailed study. Schemmel's (1980) study of feeding behaviour in blind and sighted populations of characids (*Astyanax mexicanus*) illustrates a potential limitation. Analysis of feeding behaviour in initial crossings between these two forms suggested a monofactorial inheritance, but studies of *F*2 and backcross generations revealed that these differences in feeding behaviour were controlled by polygenes (Figure 1.2). (See also Chapter 17 by Parzefall, this volume.) Unfortunately, because of limitations of crossings possible in many cases, we may be limited to *F*1 generation data, and so we should interpret results in these cases with some caution.

We also know that the action of genes may be affected by a variety of genetic and non-genetic factors. It is often considered that heritability, an estimate of the amount of the phenotypic variance among individuals that may be ascribed to additive genetic variance, can be taken as an absolute indicator of whether some difference is determined by genetic or environmental factors. This is not necessarily the case, for two quite different reasons. Estimates of heritability depend upon the environmental conditions under which the measurements were made. Heritability could be high in one set of conditions and low in another. Furthermore, the effects of genes (and environments) can depend upon the presence or absence of other genes (or environmental factors), and it may not always be obvious how these effects may be related. For example, there is a class of alleles in *Drosophila*, known as temperature-sensitive mutants, whose activity depends upon environmental temperature, as the name implies. There is also evidence that even within a given genotype, alleles may be selectively activated or deactivated at different times during ontogeny (Barlow 1981). In hybrid charrs (*Salvelinus namaycush* × *S. fontinalis*) for example, there is evidence that maternal and paternal alleles may be activated at different times during ontogeny (Berst, Ihssen, Spangler, Ayles and Martin, 1980).

The second major limitation to interpreting the genetic basis of behavioural differences from heritability estimates arises if a particular behavioural action pattern is totally determined by a genetic difference (i.e. one gene, one behaviour), and that particular allele is present in all individuals. For example, a particular trait has been selected for and has been driven to fixation. Since there would be no variability among individuals that could be ascribed to genetic differences, the heritability estimate would be zero. Thus heritability estimates may have some value, and they have a certain intuitive appeal, but their meaning and limitations must be appreciated.

The study of mutations has also been used as a technique, ideally as it might involve single gene differences. The incidence of mutations can be increased by exposing parents or gametes to mutagenic agents such as irradiation or chemicals. Most such mutations tend to be deleterious, or so

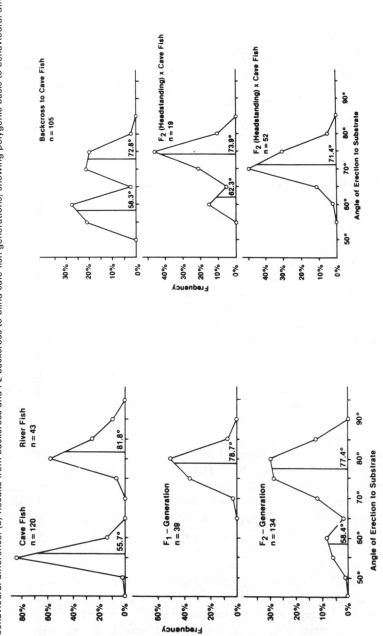

Figure 1.2: Frequency Distribution of Angles of the Longitudinal Body Axis to the Substrate during Feeding Behaviour in Total Darkness by Characids (*Astyanax mexicanus*). (a) Results for genetically blind cave fish, visually normal river fish, and their *F1* and *F2* hybrids, showing apparent monofactorial inheritance of this behavioural difference. (b) Results from backcross and *F2* backcross to blind cave-fish generations, showing polygenic basis to behavioural difference

· Source: 'Studies on the Genetics of Feeding Behaviour in the Cave Fish *Astyanax mexicanus f. anoptichthys*. An Example of Apparent Monofactorial Inheritance by Polygenes.' C. Schemmel, *Zeitschrift für Tierpsychologie, 53,* 9-22. Copyright © 1980 by Paul Parey. Reprinted by permission.

grossly aberrant as to be clearly maladaptive, and so are of interest primarily for study of causal mechanisms (Partridge 1983). Studies of cichlids (*Cichlasoma nigrofasciatum*) and guppies using mutagenic X-irradiation have shown quantitative behavioural differences (Schroder 1973; Werner and Schroder 1980). These results confirm the pattern known from other animals (Bentley 1976), including the polygenic model proposed to account for the observed effects.

That single gene differences should be associated with behavioural differences in fish is hardly unexpected. In fact, were it not the case, it would be remarkable indeed, since the correlational and causal links between single genes and behaviour have been widely established and generally accepted in virtually every other kind of animal studied (Ehrman and Parsons 1976). The remarkable feature of the examples from fishes is the basic and extensive effect single genes appear to have not only on behaviour in the narrow and immediate sense, but also on functional aspects in a more general sense. This latter point is of special interest as it relates to theoretical arguments suggested in terms of 'genes for strategies' in an evolutionary context (Dawkins 1976).

This hypothesising is necessary and reasonable, but is often justly criticised as not only unrealistic but also largely unfounded (Dawkins 1982). Although it is interesting to ask what we might expect if there were a 'gene for territoriality' or some such behaviour, critics have rightly pointed out that most such behavioural strategies can hardly be considered as single elements (or modal action patterns; Dawkins 1976, 1982; Barlow 1981). The connection between a single gene and behaviour must be an indirect and complex one at best, and some would suggest that it is unreasonable to expect anything other than pleiotropic effects of genes and polygenic influences on behaviour, not to mention the confounding effects of ontogeny, experience and environmental influences on behaviour. Can we, then, justifiably postulate genes for, say, 'territoriality' versus 'female mimicry' in the reproductive behaviour of bluegill sunfish (*Lepomis macrochirus*) (Dominey 1981)? Of course, such questions must be answered empirically. (See also Chapter 13 by Magurran, this volume.) Current evidence in fact suggests that the question is not as unreasonable as it might seen. Obviously, the difference between a male sunfish acting as a territorial resident and one entering a territory as a female mimic is enormous, in both qualitative and quantitative terms. But as the rainbow trout example shows, an entire suite of life-history features can differ as a result of a single gene difference. This could be the result of either a cascade of pleiotropic effects, or a switch-gene effect in development with one locus influencing several other loci.

It is almost a trivial exercise to suggest or even demonstrate that either genetic or environmental differences can produce behavioural differences in animals. The fundamental question remains as to how the potential for behaviour is genetically encoded and expressed during development

(Manning 1976). Ontogeny is typically a subtle and complex interplay of both genetic and environmental influences, and careful, controlled studies are required to unravel the causal threads (Bateson 1983; Slater 1983; and see Chapter 3 by Huntingford, this volume).

Behaviour but Not Genes

What Makes no Difference

It is interesting and informative to consider examples of differences in fish behaviour that quite clearly are not a consequence of genetic differences. Sex, or more particularly sex change, is a particularly convincing case. Many species of fish regularly undergo a functional sex change during ontogeny (Warner 1978); in fact some fish families are characterised by such sequential hermaphroditism, either protogyny or protandry (see Chapter 10 by Turner, this volume). In either case, they reproduce sexually, so that there are marked differences in gender-typical behaviour, including of course the elements of courtship behaviour. It is remarkable, therefore, that when such fish undergo a functional sex change, their behavioural phenotype also changes completely and correspondingly (e.g. Shapiro 1981), although the genetic constitution of the individual remains unchanged.

Some species are normally functional, simultaneous hermaphrodites: a few are self-fertilising, the others cross-fertilise. The latter are particularly interesting, as they undergo typical courtship behaviour between members of a pair (Barlow 1975), often with the two fish alternating male and female roles during a single courtship sequence. It is often assumed that sexual differences in behaviour, especially courtship behaviour, are likely under strong selection pressure and hence genetically determined, but this clearly need not be so.

There is abundant evidence from other fish species (including salmonids, poeciliids, cichlids and cyprinodontids, at least) that the functional sex (including gender-typical behaviour) of an individual can be determined by exogenous hormonal treatment during the embryonic or larval periods. This presents interesting questions concerning causal mechanisms, particularly involving ontogeny, and a comparison of gender-typical behaviour between individuals of the same genotype but different functional phenotype could well lead to useful insights. Similarly, species such as *Rivulus marmoratus*, or the various triploid *Poecilia* and *Poeciliopsis* species complexes that normally reproduce as a series of genetically identical clones, are material with considerable potential for future studies. It might be as interesting to know what range of behaviour could be produced by a given genotype as to know what differences can be produced by different genotypes.

Perhaps less dramatic, but more commonly encountered, is the plasticity frequently seen in the social behaviour of many fish species. For example, depending on factors such as the number of conspecifics, the amount and spatial distribution of food, and water flow, the behaviour of juvenile brook charr can range from a dominance hierarchy, through individual territories, to a crowded shoal (McNicol and Noakes 1984), often over a relatively short period of time (see also Chapter 12, by Pitcher, this volume).

Selection

What Comes Naturally ... or Artificially

Evidence relating genes to behaviour from selection experiments is frequently encountered in domestic species, or those whose propagation is controlled by humans. Darwin was well aware of this (1859, 1868), and Mendel used artificially selected strains for his breeding experiments with peas (Bateson 1913). The bewildering array of domesticated strains of plants and animals, including some fish species, is the most obvious everyday demonstration of the genetic basis for phenotypic variations. Of course no species can escape natural selection, and so similar evidence is potentially present in wild populations.

In selection experiments virtually any desired response can be produced (so long as the character subjected to selection has the necessary genetic variance). Detailed analyses of the actual mechanisms producing such responses to artificial selection almost always involve unpredictable twists, especially if behavioural characters are among those being selected. A classic example is Manning's (1976) study of mating speed in *Drosophila melanogaster*. Artificial selection was carried out to produce flies that mated faster than average (= unselected) for one strain, and slower than average for the other. Highly significant differences in mating speed were produced, but the basis for this effect was as much a difference in general locomotor activity and responsiveness as an intrinsic difference in actual sexual behaviour. The effect has been encountered so frequently that it can be viewed as fundamental. A particular behavioural phenotype can be achieved by more than one particular causal mechanism, and it may not always be obvious what that mechanism might be (Manning 1976).

Evidence of natural selection acting on behavioural phenotypes is typically more subtle, and requires either careful analysis of field data or insightful experiments. It could be argued that the adaptive nature of most characteristics of animals, including behaviour, is evidence that natural selection has moulded those characteristics to the particular features of the animal's niche. At worst, such arguments can easily become circular, *post hoc* rationalisations or tautologies, sometimes referred to as 'just-so stories' (e.g. Gould and Lewontin 1979). Combined with critical experiments to test specific predictions they can become rigorous and satisfactory analyses.

For example, evolutionary theory can lead to specific predictions as to

the patterns of genetic relatedness and behaviour we might expect in particular circumstances. Female guppies carry their developing embryos
inside their bodies, and give birth to them as juveniles (Noakes, 1978).
Under some conditions, females will cannibalise young, in both laboratory
and field conditions. Evolutionary theory would predict that females who
avoided cannibalising their own (genetic) offspring would be favoured.
Laboratory tests show that when presented with their own and unrelated
young, females preferentially cannibalise young from other females
(Loekle, Madison and Christian 1982). We do not know the mechanism
involved, but this difference in behaviour must be related to genetic differences among the fish, as an example of kin recognition (Colgan 1983).

The guppy has been a particularly rich and rewarding source of correlational and experimental data. The species lives naturally in freshwater
steams in Trinidad, with different, isolated populations exposed to varying
regimes of a variety of predators. Life-history characteristics (e.g. age and
size at maturity, size and number of offspring) vary predictably with the
predation regime to which each guppy population is subjected, and these
differences are genetically determined (Reznick 1982a, b). Genetically
determined behavioural differences between guppy populations are just as
striking. They respond differently to aerial predators, and show varying
degrees of development of schooling behaviour, depending upon the
predation regime to which they are exposed in nature (Table 1.1) (Seghers
1974a, b). However, the detailed experimental evidence from both the
laboratory and field on the effects of natural and sexual selection on colour
patterns of males is one of the most convincing arguments for the genetic
basis of adaptive differences in this species (Endler 1983). Not only can
differences in the details of male colour patterns be correlated with local
predation and courtship regimes in the field, but comparable differences
can be produced in a few generations in artificial selection experiments in
the laboratory simulating the range of conditions known from the field
(Figure 1.3).

Hybrids and Backcrosses

The study of hybrids can be productive both as a first step to demonstrate a
genetic basis for a behavioural difference, and as a more specific technique
for investigating the details of genetic mechanisms. The intermediacy of
hybrids and the patterns of simple Mendelian inheritance are typical
examples of the evidence from these studies. Hybrids may be produced
between inbred strains within the same species, or between genetically
distinct species. Crosses between inbred lines often depend on relatively
minor, quantitative differences that may be difficult to trace with certainty
beyond the $F1$ generation, even though the animals are fully interfertile.
On the other hand, crosses between distinct species are often limited by

Figure 1.3: Changes in the Number of Colour Spots on Male Guppies (*Poecilia reticulata*) in Laboratory Pond Experiments. K = no predators present, R = *Rivulus hartii* present (a weak predator); C = *Crenicichla alta* present (a dangerous predator); F = foundation population; S = start of experiments, with predators added to R and C ponds only; I and II are dates when censuses were taken of guppies in each pond. Increasing danger from predators produces a decrease in number of colour spots (= decreased conspicuous pattern) on males

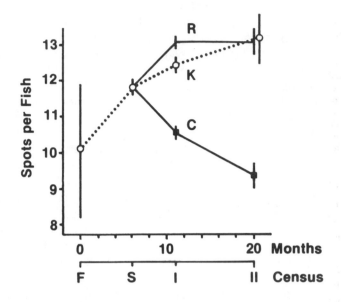

Source: 'Natural Selection on Color Patterns in *Poecilia reticulata*' by J.A. Endler, *Evolution, 34,* 76-91. Copyright © 1980 by the Society for the Study of Evolution. Reprinted by permission.

hybrid sterility that prevents *F*2 or backcross generations, even though in these cases the behavioural differences may be much greater and more clearly distinguished.

Schemmel (1980) has produced a detailed analysis of the genetic basis for a behavioural difference between forms of *Astyanax mexicanus* (a tropical characid), as mentioned earlier (and in Chapter 17 by Parzefall, this volume). Results of his *F*2 and backcross generations demonstrated that the apparent monofactorial pattern seen in the *F*1 generation was produced by a polygenic system. The caution to be drawn from these results is particularly important, since, as noted above, hybrid studies are seldom carried beyond the *F*1 generation.

Our studies of social behaviour in juvenile charrs (*Salvelinus* spp.) have the advantages of a striking contrast in behaviour between distinct species and yet *F*2 and backcross generations are easily available (Ferguson and Noakes 1982, 1983a, b; Ferguson, Noakes and Romani, 1983). The social behaviour of one of the parent species, the brook charr (*Salvelinus fontinalis*), consists of aggressive defence of fixed territories. The other

Table 1.1: Schooling Behaviour in Populations of Guppies (*Poecilia reticulata*) from Various Streams in Trinidad

Population (stream),	Major fish predators	Schooling behaviour
Guayammare	characids and cichlids*	Well developed
Lower Aripo	characids and cichlids*	well developed
Petite Curucaye	*Rivulus hartii* (high density)	intermediate
Upper Aripo	*Rivulus hartii* (medium density)	poorly developed
Paria	*Rivulus hartii* (low density)	absent

*Characids (mainly *Hoplias malabaricus*) and cichlids (mainly *Crenicichla alta*) exert heavy predation pressure on all life stages of the guppy, and so represent a serious, continuing threat. The major impact of *Rivulus hartii* is on immature guppies, and so this predator is a less serious threat. High density of *Rivulus hartii* was more than 50 individuals caught per hour; low density was less than 10 individuals caught per hour. (From 'Schooling Behavior in the Guppy (*Poecilia reticulata*): an Evolutionary Response to Predators', B.H. Seghers, *Evolution*, 28, 486-9. Copyright © 1974 by the Society for the Study of Evolution. Reprinted by permission.)

parent species, the lake charr (*Salvelinus namaycush*), is very much less aggressive, shows no indication of territoriality and tends to move about frequently. Our studies of hybrids and backcrosses showed a general tendency towards intermediacy of social behaviour, but with a significant maternal influence in all cases (Figure 1.4). The evidence supported a polygenic basis for these differences, which is not surprising given the complex measure of social behaviour. More interestingly, there was also a general pattern for measures of mobility, aggression and feeding behaviour to covary in these fish, suggesting that perhaps natural selection has favoured the inheritance of integrated suites of behavioural characteristics, perhaps through linkage of the genes involved.

A particularly clear case of a genetically determined difference in behaviour within a fish species in nature is that of Sevenster and t'Hart (1974) on the sexual behaviour of male three-spined sticklebacks. During the sexual (= courtship) phase of the reproductive cycle, males perform a number of activities directed towards their nests. The one action of interest here is 'creeping through', performed by the male wriggling his way through a tunnel he has formed in the nest in his territory. Sevenster and t'Hart (1974) found that most males typically performed this action at relatively infrequent intervals during courtship (about once every 10 min). However, some males, referred to as 'creeping-through maniacs', performed the action twice or more in rapid succession, at brief intervals. This abnormal behavioural phenotype occurred at low frequency among wild populations. They showed, through breeding experiments in the laboratory, that this difference was genetically determined, although environmental conditions also seemed to affect the threshold for the maniacs (Table 1.2) (see also Chapter 16 by Fitzgerald and Wootton, this volume).

The classic study of courtship behaviour in swordtails (*Xiphophorus helleri*) and platyfish (*X. maculatus*) by Clark, Aronson and Gordon

Figure 1.4: Differences in Social Behaviour of Parental Brook Charr (*Salvelinus fontinalis*), Lake Charr (*S. namaycush*) and Various Hybrid and Backcross Charrs. The female parent of each cross is named first in each case (B = brook charr, L = lake charr, B × L means the *F*1 hybrid from a brook charr female and a lake charr male, for example). The canonical variable 1 (from a discriminant analysis) was comprised of measures of lateral display, charge and forage by the fish. Brook charr are more aggressive (high levels of display and charge) and forage less than lake charr. Hybrids are intermediate, but closer to maternal parent, in all cases

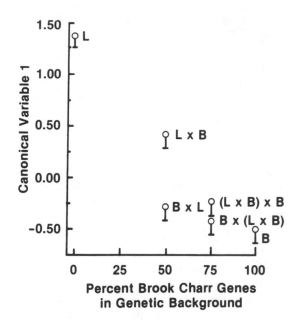

Source: 'Behaviour-genetics of Lake Charr (*Salvelinus namaycush*) and Brook Charr (*S. fontinalis*): Observation of Backcross and *F*2 Generations', M.M. Ferguson and D.L.G. Noakes, *Zeitschrift für Tierpsychologie, 62*, 72-86. Copyright © 1983 by Paul Parey. Reprinted by permission.

(1954) dealt largely with the study of hybrids (and is the only behaviour genetics study with fishes usually cited in textbooks). They found both quantitative and qualitative differences between male courtship behaviour in these species, with a general tendency towards intermediacy of behaviour in hybrids. The authors suggested a polygenic basis for the differences in male courtship behaviour between these species. This courtship behaviour functions as an effective reproductive isolating mechanism between the species in nature, and under all but the most extreme artificial laboratory conditions (no choice of mates offered, males of one species and females of the other confined together). Even in these extreme cases, less than 20 per cent of the infrequent copulation attempts resulted in insemination. Although few other cases have been studied in such detail (e.g. Liley 1966), there is often an assumption that these results typify the general case, i.e. species-typical differences in courtship behaviour, with a

Table 1.2: Comparison of Creeping Through by Male Three-spined
Sticklebacks (*Gasterosteus aculeatus*) in Standard Laboratory Tests

Strain	Total number of males tested	Number of creeping-through maniacs
T*	25	21
V*	40	0
*F*1 hybrids	8	0
*F*2 hybrids	43	10
Backcrosses	10	0

*T — Fish collected near Den Helder, V — fish collected from near Vaassen, The Netherlands.
(From 'A Behavioural Variant in the Three-spined Stickleback', P. Sevenster and P. t'Hart, in
J.H.F. van Abeelen (ed.), *Frontiers of Biology, vol. 38, The Genetics of Behaviour*, pp. 141-6.
Copyright © 1974 by Elsevier Biomedical Press, Amsterdam. Reprinted by permission.)

polygenic basis of inheritance, are an important reproductive isolating
mechanism (Keenleyside 1979).

Causal Mechanisms

Chocolate Soup and Jigsaw Puzzles

Selection (whether natural or artificial) can readily produce shifts in genetically determined behavioural differences, but there is no necessary or even
predictable relationship between causal mechanisms and behavioural
consequences. One strain of fish may grow efficiently because it has a particular set of enzymes regulating intermediary metabolism, whereas
another strain achieves the same result through a decrease in general
locomotory activity, sparing energy for metabolic processes. Even if we can
establish the causal thread from gene to behaviour, it will obviously include
many intermediate steps. Other genes and gene products are quite possibly
involved, and for many kinds of behaviour, non-genetic environmental
inputs are also likely to be involved. Given this complexity, it is perhaps
surprising that we can make any sense of the system at all!

The difference of a single enzyme in cases such as phenylketonuria in
humans has widespread and cumulative phenotypic effects, including some
behavioural differences. These cases are well understood as a consequence
of their nature, as single (typically recessive) mutant alleles, inherited in
simple Mendelian fashion. Behaviour is far removed from the direct action
of genes, so we can seldom expect to have such clear-cut examples, yet we
believe that the relationship between genes and behaviour can and does
exist. It is neither easy nor obvious to parse the continuous, ongoing stream
of behavioural output into basic constituent units. The confounding complexity of behavioural measures often degenerates into an indecipherable
and unpalatable blend of intermediate states and actions (a chocolate
soup).

The causal mechanism may be of no interest or particular importance to someone interested in functional questions. The causal mechanisms in some cases may not even be worth the effort required to decipher them. It may be sufficient to know that differences in courtship behaviour can be attributed to genetic differences between individuals. If the principle can be assumed to be common in different cases, the one more amenable to the study at hand will surely be chosen for study. I have cited several examples under the heading of single genes that establish the causal relationship between genes and behaviour. In a few of those cases we even know the single gene difference and at least the immediate gene product involved. Rainbow trout differing only in LDH or PGM phenotype differ significantly in swimming performance, for example. This could result from any of a number of steps in intermediary metabolism, oxygen uptake and utilisation. Though it might be of some ultimate interest to decipher each and every step in the causal nexus between production of a particular PGM variant, and the corresponding differences in life-history measures such as time of sexual maturation, it would hardly be a simple undertaking. This effect might fit into arguments relating evolutionary shifts in life histories to small, genetically determined changes in early ontogeny (e.g. Noakes 1981; Noakes and Balon 1982), but it remains a hypothetical construct at present.

To borrow an analogy from Manning (1976), we are trying to assemble a jigsaw puzzle composed of a large, but unknown, number of pieces. We do not have the completed picture to guide us, and we are presented with random pieces from the variety of experimental and analytical approaches. An extreme strategy might be to try each available piece in every possible combination with every other piece. The futility of this approach can be appreciated only through experience (unless one has the attention span, memory capacity and imagination of a large digital computer!). It soon becomes obvious that a much more fruitful strategy must incorporate some effort to formulate an image of what the final picture might be, and an attempt to recognise patterns among the pieces. Dwelling unduly upon the precise size or shape of a given individual piece might by chance provide just the right fit to one other piece and reveal a little more of the puzzle. More likely it will produce an expert knowledge of one isolated piece, and a fruitless search for a complimentary combination. It is much more likely that we can achieve steady progress if we try to judge what the basic pattern should be, and what kinds of pieces we should group together at different sides of the picture in anticipation of sketching the outlines at some broader scale in a greater perspective.

Some remarkable advances have been made in behaviour genetics and the field continues to prosper. Perhaps one of the most exciting areas that could yield future results would be that of combining the sophisticated techniques of genetic engineering and recombination with behaviour genetic analysis. We would no longer have to content ourselves with deciphering patterns given to us by nature, or searching among random mutations produced by experimental intervention. Single, specified genetic alterations could be employed in an experimental approach to test specific predictions for both causal and functional relationships. The other area of continuing excitement and productivity is that of evolutionary aspects. To some this is unfortunately synonymous with the term sociobiology, yet perhaps it is more productive to consider this as a way of formulating the overall picture for our jigsaw puzzle. To biologists the question of function must always be a central one, whatever aspects of behaviour we are considering.

As a group, teleost fish perhaps do not lend themselves particularly to studies of causal mechanisms; nematodes, flies and a few other species of choice have that field solidly to themselves. The ultimate questions of function seem less likely to emerge so readily from those subjects, so perhaps fish offer some promise in this regard. They are (mostly) not too big to be unmanageable nor too small to be microscopic, not too long-lived for our patience nor too short-lived for ontogenetic studies, and with an ecological and evolutionary diversity sufficient to provide material for almost anyone's interests. Considering the opportunistic nature of most behaviour genetics studies on fish thus far, the results have been quite rewarding.

Acknowledgements

I thank Tony Pitcher for providing the opportunity to write this chapter and for his constructive comments on an earlier draft, a referee for a perceptive and helpful review, and Eugene Balon, Jeff Baylis, Moira Ferguson, Jim Grant, Chris Nunan, Willem Pot, Skuli Skulason and Tina Soto for their comments on the manuscript, but especially Pat and Jeff for sharing access to Eddy (our Apple) to make the writing possible. My research has been supported by the Canadian National Sportsmen's Fund, the Natural Sciences and Engineering Research Council of Canada, the Ontario Ministry of Natural Resources, and a Faculty Leave Award from the University of Guelph. In conclusion, in all the publicity concerning 1984, there seems little awareness that 100 years ago, on 6 January 1884, Gregor Johann Mendel, Abbot of the convent at Brno in Moravia, died. Obviously we have progressed much beyond his understanding in almost every aspect of genetics, but the fundamental principles of the approach he pioneered are remarkably unchanged.

References

Allendorf, F.W. and Phelps, S.R. (1981) 'Use of Allelic Frequencies to Describe Population Structure', *Canadian Journal of Fisheries and Aquatic Sciences*, 38, 1507-14.

Allendorf, F.W., Mitchell, N., Ryman, N. and Stahl, K.G. (1977) 'Isozyme Loci in Brown Trout (*Salmo trutta*): Detection and Interpretation from Population Data', *Hereditas, 86*, 179-90.

Allendorf, F.W., Knudsen, K.L. and Leary, R.F. (1983) 'Adaptive Significance of Differences in the Tissue-specific Expression of a Phosphoglucomutase Gene in Rainbow Trout', *Procedings of the National Academy of Sciences, USA, 80*, 1397-1400.

Bams, R.A. (1976) 'Survival and Propensity for Homing as Affected by Presence or Absence of Locally Adapted Paternal Genes in Two Transplanted Populations of Pink Salmon (*Oncorhynchus gorbuscha*)', *Journal of the Fisheries Research Board of Canada, 33,* 2716-25.

Barlow, G.W. (1975) 'On the Sociobiology of Some Hermaphroditic Serranid Fishes, the Hamlets, in Puerto Rico', *Marine Biology, 33*, 295-300.

Barlow, G.W. (1981) 'Genetics and Development of Behavior, with Special Reference to Patterned Motor Output', in K. Immelmann, G.W. Barlow, L. Petrinovitch and M. Main (eds), *Behavioral Development*, Cambridge University Press, Cambridge,, pp. 191-251.

Bateson, P.P.G. (1983) 'Genes Environment and the Development of Behaviour', in T.R. Halliday and P.J.B. Slater (eds), *Animal Behaviour*, vol. 3, *Genes Development and Learning*, W.H. Freeman, New York, pp. 52-81.

Bateson, W. (1913) *Mendel's Principles of Heredity*, Cambridge University Press, Cambridge.

Bentley, D. (1976) 'Genetic Analysis of the Nervous System', in J.C. Fentress (ed.), *Simpler Networks and Behavior*, Sinauer Associates, Sunderland, Mass., pp. 126-39.

Berst, A., Ihssen, P.E., Spangler, G.R., Ayles, G.B. and Martin, G.W. (1980) 'The Splake, a Hybrid Charr *Salvelinus namaycush* × *S. fontinalis*', in E.K. Balon (ed.), *Charrs. Salmonid Fishes of the Genus Salvelinus*, W. Junk, the Hague, pp. 841-88.

Bonner, J.T. (ed.) (1982) *Evolution and Development*, Springer-Verlag, New York.

Burghardt, G. and Bekoff, M. (1978) *Development of Behavior*, Garland, New York

Clark, E., Aronson, L.R. and Gordon, M. (1954) 'Mating Behavior Patterns in Two Sympatric Species of Xiphophorin Fishes: their Inheritance and Significance in Sexual Isolation', *Bulletin of the American Museum of Natural History, 103*, 141-225

Colgan, P.W. (1983) *Comparative Social Recognition*, Wiley-Interscience, New York

Darwin, C.R. (1859) *The Origin of Species*, John Murray, London

Darwin, C.R. (1868) *The Variation of Animals and Plants under Domestication*, John Murray, London

Dawkins, R. (1976) *The Selfish Gene*, Oxford University Press, Oxford

Dawkins, R. (1982) *The Extended Phenotype. The Gene as the Unit of Selection*, W.H. Freeman, San Francisco

Dobzhansky, T., Ayala, F.J., Stebbins, G.L. and Valentine, J.W. (1977) *Evolution*, W.H. Freeman, San Francisco

Dominey, W. (1981) 'Maintenance of Female Mimicry as a Reproductive Strategy in Bluegill Sunfish (*Lepomis macrochirus*)', *Environmental Biology of Fishes, 6*, 59-64

DiMichele, L. and Powers, D.A. (1982) 'Physiological Basis for Swimming Endurance Differences between LDH-B Genotypes of *Fundulus heteroclitus*', *Science, 216*, 1014-16

Ehrman, L. and Parsons, P.A. (1976) *The Genetics of Behavior*, Sinauer Associates, Sunderland, Mass.

Endler, J.A. (1983) 'Natural and Sexual Selection on Color Patterns in Poeciliid Fishes', *Environmental Biology of Fishes, 9*, 173-90

Farr, J.A. and Peters, K. (1984) 'The Inheritance of Quantitative Fitness Traits in Guppies, *Poecilia reticulata* (Pisces: Poeciliidae). II Tests for Inbreeding Effects', *Heredity, 52*, 285-96

Ferguson, M.M. and Noakes, D.L.G. (1982) 'Genetics of Social Behaviour in Charrs (*Salvelinus* species)', *Animal Behaviour, 30*, 128-34

Ferguson, M.M. and Noakes, D.L.G. (1983a) 'Behaviour-genetics of Lake Charr (*Salvelinus namaycush*) and Brook Charr (*S. fontinalis*): Observation of Backcross and *F2* Generations', *Zeitschrift für Tierpsychologie, 62*, 72-86

Ferguson, M.M. and Noakes, D.L.G. (1983b) 'Movers and Stayers: Genetic Analysis of Mobility and Positioning in Hybrids of Lake Charr *Salvelinus namaycush* and Brook Charr *Salvelinus fontinalis*', *Behavior Genetics, 13*, 213-22

Ferguson, M.M., Noakes, D.L.G. and Romani, D. (1983) 'Restricted Behavioural Plasticity of Juvenile Lake Charr, *Salvelinus namaycush*', *Environmental Biology of Fishes, 8*, 151-6

Gould, J.L. (1974) 'Genetics and Molecular Ethology', *Zeitschrift für Tierpsychologie, 36*, 267-92

Gould, S.J. and Lewontin, R.C. (1979) 'The Spandrels of San Marco and the Panglossian Paradigm: a Critique of the Adaptationist Programme', *Proceedings of the Royal Society of London B, 205*, 581-98

Hayes, B. (1984) 'Computer Recreations', *Scientific American, 250*, 14-20

Huntingford, F. (1984) *The Study of Animal Behaviour*, Chapman & Hall, London, 411 pp.

Immelmann, K., Barlow, G.W., Petrinovitch, L. and Main, M. (eds) (1981) *Behavioural Development*, Cambridge University Press, Cambridge

Kallman, K.D. (1983) 'The Sex Determining Mechanism of the Poeciliid Fish, *Xiphophorus montezumae*, and the Genetic Control of the Sexual Maturation Process and Adult Size', *Copeia, 1983*, 755-69

Keenleyside, M.H.A. (1979) *Diversity and Adaptation in Fish Behaviour*, Springer-Verlag, Berlin

Kelso, B.W., Northcote, T.G. and Wehrhahn, C.F. (1981) 'Genetic and Environmental Aspects of the Response to Water Current by Rainbow Trout *Salmo gairdneri* Originating from Inlet and Outlet Streams of Two Lakes', *Canadian Journal of Zoology, 59*, 2177-85

Klar, G.T., Stalnaker, C.B. and Farley, T.M. (1979) 'Comparative Physical and Physiological Performance of Rainbow Trout, *Salmo gairdneri*, of Distinct Lactate Dehydrogenase B2 Phenotypes', *Comparative Biochemisry and Physiology, 63A*, 229-35

Liley, N.R. (1966) 'Ethological Isolating Mechanisms in Four Sympatric Species of Poeciliid Fishes', *Behaviour Supplement, 13*, 1-197

Loekle, D.M., Madison, D.M. and Christian, J.J. (1982) 'Time Dependency and Kin Recognition of Cannibalistic Behavior among Poeciliid Fishes', *Behavioral and Neural Biology, 35*, 315-18

McNicol, R.E. and Noakes, D.L.G. (1984) 'Environmental Influences on Territoriality of Juvenile Brook Charr, *Salvelinus fontinalis*, in a stream environment,' *Environmental Biology of Fishes, 10*, 29-42

Manning, A.W.G. (1975) 'The Place of Genetics in the Study of Behaviour', in P.P.G. Bateson and R.A. Hinde (eds), *Growing Points in Ethology*, Cambridge University Press, Cambridge, pp. 327-43

Manning, A.W.G. (1976) 'Behaviour Genetics and the Study of Behavioural Evolution', in G.P. Baerends, C.M. Beer and A.W.G. Manning (eds), *Function and Evolution in Behaviour*, Oxford University Press, Oxford, pp. 71-91

Noakes, D.L.G. (1978) 'Ontogeny of Behavior in Fishes: a Survey and Suggestions', in G.M. Burghardt and M. Bekoff (eds), *Comparative and Evolutionary Aspects of Behavioral Development*, Garland, New York, pp. 103-25

Noakes, D.L.G. (1981) 'Comparative Aspects of Behavioral Development: a Philosophy from Fishes', in K. Immelmann, G.W. Barlow, L. Petrinovitch and M. Main (eds), *Behavioural Development*, Cambridge University Press, Cambridge, pp. 491-508

Noakes, D.L.G. and Balon, E.K. (1982) 'Life Histories of Tilapias: an Evolutionary Perspective', in R.S.V. Pullin and R.H. Lowe-McConnel (eds), *The Biology and Culture of Tilapias*, ICLARM, Manila, Philippines, pp. 61-82

Partridge, L. (1983) 'Genetics and Behaviour' in T.R. Halliday and P.J.B. Slater (eds), *Animal Behaviour*, vol. 3, *Genes, Development and Learning*, W.H. Freeman, San Francisco, pp. 11-51

Raleigh, R.F. (1971) 'Innate Control of Migration of Salmon and Trout Fry from Natal Gravels to Rearing Areas,' *Ecology, 52*, 291-7

Reznick, D. (1982a) 'Genetic Determination of Offspring Size in the Guppy (*Poecilia reticulata*)', *American Naturalist, 120*, 181-8

Reznick, D. (1982b) 'the Impact of Predation on Life History Evolution in Trinidadian Guppies: Genetic Basis of Observed Life History Patterns', *Evolution, 36*, 1236-50

Schemmel, C. (1980) 'Studies on the Genetics of Feeding Behavior in the Cave Fish *Astyanax mexicanus* f. *anoptichthys*: an Example of Apparent Monofactorial Inheritance by Polygenes', *Zeitschrift für Tierpsychologie, 53*, 9-22

Schroder, J.J. (1973) (ed.) *Genetics and Mutagenesis of Fish*, Springer-Verlag, New York

Seghers, B.H. (1974a) 'Schooling Behavior in the Guppy (*Poecilia reticulata*): an Evolutionary Response to Predation', *Evolution, 28*, 486-9

Seghers, B.H. (1974b) 'Geographic Variation in the Responses of Guppies (*Poecilia reticulata*) to Aerial Predators', *Oecologia, 14*, 93-8

Sevenster, P. (1961) 'A Causal Analysis of a Displacement Activity (Fanning in *Gasterosteus aculeatus* L.)', *Behaviour Supplement, 9*, 1-170

Sevenster, P. and t'Hart, P. (1974) 'A Behavioural Variant in the Three-spined Stickleback', in J.H.F. van Abeelen (ed.), *Frontiers of Biology*, vol. 38, *The Genetics of Behaviour*, Elsevier, New York, pp. 141-6

Shapiro, D.Y. (1981) 'Behavioural Changes of Protogynous Sex Reversal in a Coral Reef Fish in the Laboratory', *Animal Behaviour, 29*, 1185-98

Slater, P.J.B. (1983) 'The Development of Individual Behaviour', in W.H. Freeman, New York, *Animal Behaviour*, vol. 3, *Genes, Development and Learning*, T.R. Halliday and P.J.B. Slater (eds),pp. 82-113

Thomas, A.E. and Donahoo, M.J. (1977) 'Differences in Swimming Performance among Strains of Rainbow Trout *Salmo gairdneri*', *Journal of the Fisheries Research Board of Canada, 34*, 304-7

Thomson, K.S. (1984) 'Reductionism and Other Isms in Biology', *American Scientist, 72*, 388-90

Tinbergen, N. (1963) 'On Aims and Methods of Ethology', *Zeitschrift für Tierpsychologie, 20*, 410-29

Warner, R.R. (1978) 'The Evolution of Hermaphroditism and Unisexuality in Aquatic and Terrestrial Vertebrates', in E.S. Reese and F.J. Lighter (eds), *Contrasts in Behavior*, Wiley, New York, pp. 77-101

Werner, M. and Schroder, J.H. (1980) 'Mutational Changes in the Courtship Activity of Male Guppies *Poecilia reticulata* after X-Irradiation', *Behavior Genetics, 10*, 427-30

2 THE MOTIVATIONAL BASIS OF FISH BEHAVIOUR

Patrick Colgan

Introduction

For Aristotle, in *Historia Animalium*, the motivation of fish ranged from enjoyment of tasting and eating to madness from pain in pregnancy, and for Francis Day, summarising for the 1878 Proceedings of the London Zoological Society in the wake of Darwin's epochal *The Expression of the Emotions in Man and Animals*, fish were variously moved by disgrace, terror, affection, anger and grief. The study of motivation thus has long been central in the analysis of fish behaviour. In its daily activities of finding food, avoiding predators, fighting and reproducing, a fish is a motivationally diverse animal. This study focuses on the internal proximate causes of behaviour, variously labelled as drives, instincts, or causal systems (see McFarland 1974). Researchers such as Skinnerians, wary of the ontological status of such notions, have emphasised the dynamics of performance and the concomitant controlling variables in the environment. Though the existence and nature of causal systems are empirical matters, conventional wisdom about the economy of nature suggests that they are likely solutions to the common problems encountered by fish, and they are heuristically very convenient. Motivational systems are more than physiology writ large, and require investigation in their own right.

Of Tinbergen's (1963) four aims of ethology (causation, survival value, ontogeny, and evolution), the study of motivation occupies much of the first. Its centrality is seen in the number of allied behavioural topics: physiological substrates (including sensorimotor (see Chapters 4 to 7, this volume), neural and hormonal mechanisms), ontogeny (see Chapter 3, this volume), rhythms, communication, and behavioural ecology (see Chapters 8 to 17, this volume). A common experimental paradigm in motivational research is to measure responses to a fixed external stimulus on several occasions, and then attribute changes in responsiveness to changes in motivation. Thus overt behaviour is viewed as the outcome of internal and external cues. (The composite effects of external cues such as temperature and substrate, referred to as heterogeneous summation (Tinbergen 1951), have been shown, for instance, in salmonid spawning behaviour (Fabricius 1950).) However, on any particular occasion, the situation may become complicated since the presentation of a stimulus or the performance of a response may itself change motivation (e.g. see Toates and Birke 1982). In classical terminology, stimuli can both release and motivate responses. In such cases one may often scrutinise the correlations among various features of responses, such as frequencies, latencies, durations or intensities, in the search for underlying causal mechanisms.

Thus motivational analysis must overcome several hurdles. Although behaviour may be fundamentally deterministic, stochastic models often seem to be the most appropriate analytical tools due to uncontrollable variation in responsiveness (e.g. see Metz in McFarland 1974). Substantiating such models generally requires data too extensive to be easily obtainable since motivational processes are usually non-stationary (i.e. the values of the underlying parameters change with time). Significant differences may exist between individuals (see Chapter 13 by Magurran, this volume). Additionally, many different measures of behaviour are possible, and these often change independently of each other (e.g. in feeding by sticklebacks, *Gasterosteus aculeatus*: Tugendhat 1960a).

In the shadow of these difficulties, this chapter considers research on the motivation of teleost behaviour, with emphasis on reports appearing since Baerends' (1971) magistral review. Although the diversity of motivational research reflects the efforts of workers with often quite disparate interests, three key questions are relevant to the investigation of any behaviour: first, how many internal causal factors underlie the behaviour, and how do they interact? Secondly, how are these factors influenced by the appearance of external cues and by deprivation from such cues? Thirdly, how do they change over time, both during the performance of the behaviour and over longer periods? Attempts to answer each question must, of course, involve experimental treatments appropriate to the natural history of the study species. For convenience of presentation, feeding, fear, reproduction and aggression, the four major causal systems of behaviour which have been widely discussed, will be reviewed separately, and then their interactions examined. But first the relation of motivation and ontogeny is considered.

Motivation and Ontogeny

A fish may respond differently to the same stimulus on different occasions because of motivational (non-structural) changes or structural changes affecting its capacity to act. The distinction is thus between changes in mechanisms and changes in the activation of mechanisms in performance. Ontogenetic changes in motivation and structure result from both maturation, which involves intrinsic processes, and experience with the environment. Little attention has been paid to motivational ontogeny in fish. In one of the few studies made, the aggressiveness of normally raised *Haplochromis burtoni* (an African cichlid), is increased by a patterned stimulus dummy but not a neutral dummy (Wapler-Leong 1974): that of individuals reared as isolates is augmented by the two stimuli.

Two major structural changes germane to a consideration of motivation are learning and imprinting. The meaning of 'learning' unfortunately varies with the user. It is generally taken to refer to a long-term change in the likelihood of a particular response following a particular stimulus, over successive associations of the stimulus and response ('when followed by appropriate reinforcement' according to reinforcement theorists). Such changes are variously interpreted as the outcome of conditioning of one or more types (see Domjan 1983). 'Imprinting' refers to the development of social and habitat attachments as the result of experience during a brief critical period in early life. Although imprinting was originally contrasted with traditional learning as investigated in the laboratory, it shares many features with such learning (see Bateson 1979). Habituation, the waning of a response after repeated presentation of the same stimulus, lies intermediate between motivational dynamics and learning in its features and time scale. A valid taxonomy of behavioural processes will be possible only when the physiological bases (at present mostly unknown) underlying these phenomena are understood.

Learning research has largely been concerned with testing the validity of general theories about this central process. One key question in such testing revolves around the motivational basis for mechanisms of reinforcement. The pattern of responses in Siamese fighting fish (*Betta splendens*) when operantly reinforced to swim through a tunnel, depends on whether the reinforcer is food or the opportunity to display to a conspecific (Hogan, Kleist and Hutchings 1970; Figure 2.1). As the number of responses required for a food reinforcement increased, the total number of responses over 12 h increased, whereas the number of reinforcements did not vary significantly. For display, on the other hand, tunnel responses were constant whereas reinforcements fell. It is therefore likely that different mechanisms underlie the two behavioural systems. More recent work with this species (e.g. Bols and Hogan 1979) emphasises this conclusion with the finding that a display reinforcer induces both approach and avoidance tendencies.

Social imprinting can involve offspring and sexual attachments (for review see Colgan 1982). Although species recognition by parents of different life stages is well developed in several cichlid species examined, there is no indication of imprinting by parents on offspring, unlike what occurs in some species of birds. Sexual imprinting (the establishment of sexual preferences as a result of early social experience) has been demonstrated in a variety of tropical-aquarium species, with polymorphic forms being convenient subjects, and experimental designs including cross-fostering and isolation procedures.

Habituation has been studied for many responses. In goldfish (*Carassius auratus*) habituation to an aversive buzzer is slowed by the injection of an endorphin analogue (Olson *et al.* 1978). (Endorphins constitute an important class of analgesic compounds in vertebrate brains.) In guppies (*Poecilia reticulata*) changes in behaviour associated with repeated exposure to an open field result from the differential habituation of fear

Figure 2.1: Responses and Reinforcements for Siamese Fighting Fish under Various Fixed-ratio Schedules for Display and Food. (Open points represent all eight fish; closed points represent three fish completing the series.) (Hogan *et al.* 1970.)

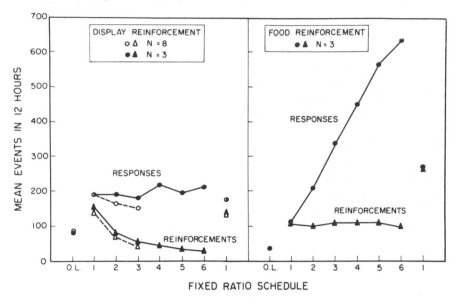

and exploratory responses (Warren and Callaghan 1976). Similarly, components of arousal and fright habituate independently in slippery dicks, *Halichoeres bivittatus* (Laming and Ebbesson 1984). The bulk of habituation research, however, has dealt with aggression (e.g. paradise fish (*Macropodus opercularis*): Brown and Noakes 1974; goldfish, and roach (*Rutilus rutilus*): Laming and Ennis 1982; convict cichlids (*Cichlasoma nigrofasciatum*): Gallagher, Herz and Peeke 1972). Siamese fighting fish; Klein, Figler and Peeke, 1976, Chantrey 1978) and sticklebacks (Peeke 1983) have been favourite species. Figure 2.2 presents some typical habituation curves and illustrates the common finding that habituation is often preceded by an initial increment in responding, a 'warm-up effect'. In colonies of bluegill sunfish (*Lepomis macrochirus*) the aggression between males on neighbouring nests habituates and can be reinstated or dishabituated by altering the appearance of the body covering (Colgan, Nowell, Gross and Grant 1979).

Thus a variety of ontogenetic processes influence the mechanisms producing behaviour.

Feeding

Non-motivational (i.e. extrinsic) aspects of feeding include the physical aspects of the environment; the densities, distributions, and availabilities of

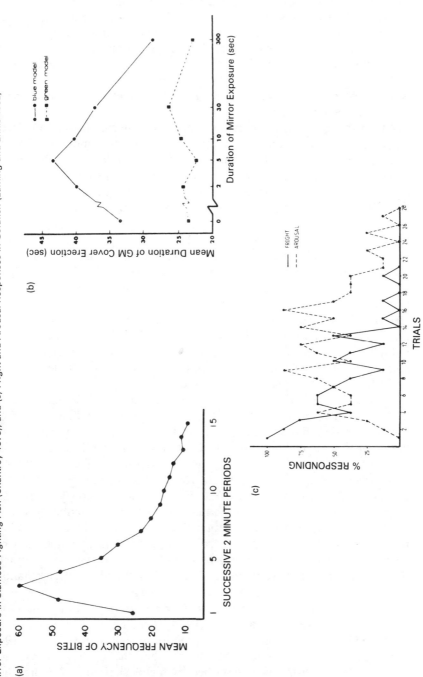

Figure 2.2: Habituation of (a) Biting in Sticklebacks (Peeke 1983), (b) Gill-cover Erection in Response to Two Stimulus Dummies Following Various Durations of Mirror Exposure in Siamese Fighting Fish (Chantry 1978), and (c) Fright and Arousal Responses in Goldfish (Laming and Ennis 1982)

prey types; feeding competitors; and learning. The last involves prey features, preferences, and switching (i.e. the disproportionate consumption of the common food type when several are available: see Visser 1981). Many of these aspects are dealt with in Chapter 8 by Hart, this volume. Compared to homeothermic animals, the energy requirements of fish are much lower and hunger rises more slowly. It is widely accepted that feeding motivation includes both a gastric factor based on gut fulness and a systemic factor reflecting metabolic balance (e.g. Holmgren *et al.* in Rankin, Pitcher and Duggan 1983). How these factors operate to produce observed feeding behaviour is the question at issue.

Much research has attempted to infer mechanisms underlying feeding performance from aggregate measures such as the amounts of various available food types consumed under different conditions by the end of a meal. Legions of field workers pass the time tallying stomach contents. In operant psychology, Herrnstein has developed a 'matching law' describing the relation of numbers of feeding acts to numbers of food reinforcements, but its relevance to natural behaviour is questionable since it applies only to variable-interval reinforcement schedules, which may be rare outside operant laboratories (see Kamil and Yoerg 1982). (Under variable-interval schedules of reinforcements, an animal is reinforced for the first response after a randomly varying interval of time.) Food preferences are often transitive (i.e. if food type A is preferred to B, and B to C, then A is pre-ferred to C) and comparison of consumption totals enables tests of quantitative aspects of such transitivity. In centrarchid fish, transitivity can be modelled as the outcome of preferences for individual food types (Colgan and Smith 1984). The amounts eaten, both when only one type is available and when two are available, can be predicted.

However, as in other areas of motivational analysis, the elucidation of hunger mechanisms is more effectively performed using fine-grained data on the duration and sequencing of individuals acts and the choices generated by this behavioural flow. Much attention has therefore come to focus on satiation curves (Figure 2.3) generated by fish feeding amid abundant food, where changes in performance reflect changes in internal state only. In general, fish feed more selectively (Figure 2.3a) and slowly (Figure 2.3b, c) as a meal proceeds, notwithstanding the contrary predictions of classical optimal-foraging theory (see Chapter 8 by Hart, this volume). Dill (1983) has summarised the effects of satiation on searching, handling, and ratio measures of behaviour such as attacks per approach. He points out that these effects are adaptive by increasing searching and decreasing handling time when low food availability leads to hunger. Hunger influences the feeding tactics of Australian salmon (*Arripis trutta*) preying on krill of various densities (Morgan and Ritz 1984). Feeding often occurs in bouts, separated by relatively long intervals (e.g. for bluegill sunfish, Lester 1976). Subsequently, our laboratory has found that the feeding behaviour of pumpkinseed sunfish is well described by highly skewed probability distri-

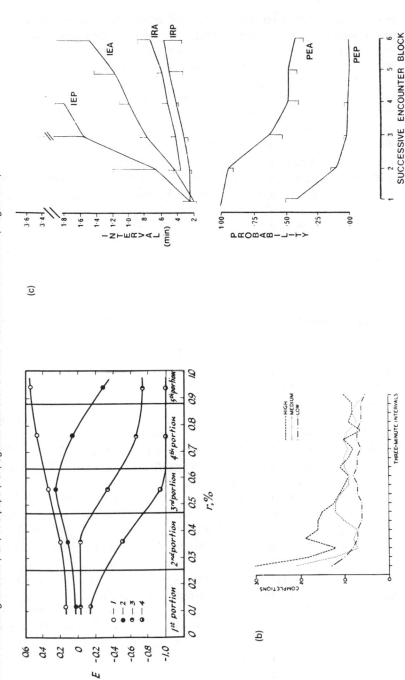

Figure 2.3: Satiation Curves in Fish Feeding. (a) Carp (*Cyprinus carpio*) feeding electivity, E, measuring preference for four prey types as a fraction of the total meal (*r*%) (Ivlev 1961). (b) Sticklebacks grouped according to initial feeding rates on *Tubifex* worms (Tugendhat 1960a). (c) Pumpkinseed sunfish (*Lepomis gibbosus*): length of the interval following an encounter in which a housefly adult was eaten (IEA) or refused (IRA) or a pupa was eaten (IEP) or refused (IRP), and probability of the fish eating an adult (PEA) or a pupa (PEP) against successive blocks of about 450 encounters (Colgan 1973)

butions, such as gamma, whose means are influenced by the individual fish, the food type, and satiation. Goldfish achieve nearly optimal depletion of different food patches by comparing the current feeding rate with that previously experienced elsewhere (Lester 1984).

An even more detailed investigation of feeding behaviour requires the examination of the behaviour involved in the treatment of each food item. Accepting and rejecting food items during a meal have marked and opposite influences on feeding behaviour in sticklebacks (Thomas 1974, 1977). After an acceptance, fish search more intensively in the immediate vicinity; such 'area restricted searching' has also been noticed in tetrapods. In contrast, after a rejection a stickleback is more likely to leave the area. It appears that, in addition to the effects of satiation extending over an entire meal, acceptances and rejections result in respective short-term positive and negative changes in feeding motivation. These changes are adaptive if prey are patchily distributed. Further careful scrutiny of such aspects of behaviour is needed for a comprehensive understanding of feeding motivation.

Fear and Avoidance

Fear and avoidance have been intensively examined in some animal groups. For instance, Hogan (1965) has shown that they are mutually inhibitory in young chicks. By way of contrast, these phenomena have been little studied in fish, perhaps because fish ethologists are themselves less motivated by fear than their non-ichthyological counterparts! Research comparing avoidance learning in different species has investigated, for example, the effect of varying the interval between a conditioned stimulus (generally a light) and a shock using goldfish (Bitterman 1965). Avoidance conditioning is better in goldfish for which the light has been associated with food or shock, and poorer in goldfish habituated to the light, compared with controls (Braud 1971). In convict cichlids, fear increases with social isolation (Gallagher *et al.* 1972), and plays a role in the prior-residence effect in dominance relations (Figler and Einhorn 1983). The lack of habituation of escape responses to test stimuli in the blue chromis, *Chromis cyaneus* (Hurley and Hartline 1974) is clearly adaptive. Compared with uninfested fish, sticklebacks infested with cestode larvae recover from a frightening stimulus more quickly (Giles 1983). This quicker recovery may increase the predation risk of the fish and hence be adaptive to the parasite in reaching its definitive avian host.

Reproduction

As reviewed in Chapters 10 and 11, there has been much investigation of functional aspects of mating and parental care in teleosts. In considering underlying motivation, Baerends, Brower and Waterbolk (1955) provided ethology with a classic illustration of the interaction of external stimuli and internal causation in the courtship behaviour of male guppies (see Figure 2.4). Males prefer large females over small ones, which is adaptive since brood size increases with female size. The state of arousal of a courting male is reflected in coloured skin patches, the colour intensity of which is under neural control. Together, female size (the external stimulus) and internal arousal combine to determine the courtship acts of the male, with more intense acts requiring higher combinations. Such coloration patterns are thus most useful in the study of courtship motivation.

Valuable physiological data are available indicating the mechanisms involved in the motivation of reproductive behaviour. In terms of chemical messages, a variety of hormones operate within individuals, and pheromones between individuals (see Liley and Stacey 1983). Refractory periods (pauses following mating) are known in many species of animals. In the lemon tetra (*Hyphessobrycon pulchripinnis*) males are physiologically limited in their ability to produce sperm and so to fertilise eggs during the daily spawning period, which takes up the first two hours of the morning (Nakatsuru and Kramer 1982).

In promiscuous species, the Coolidge effect refers to the reinstatement of copulatory behaviour in a previously satiated male by presenting a new mate. The effect has been most investigated in mammals, and involves dishabituation in the male and differential behaviour by mated and unmated females. In the live-bearer *Poecilia sphenops*, copulation attempts by males are more frequent after a new female is made available (Franck 1975). Possibly the effect is adaptive for a male by enhancing the number and variety of his offspring. On a longer time span, many species show reproductive rhythms, generally associated with such environmental cues as temperature, tides and rains (Schwassmann 1980).

In his classic study of the organisation of courtship behaviour in the bitterling (*Rhodeus amarus*), Wiepkema (1961) introduced the use of factor analysis for the detection of clusters of activities. Much of the variation in the data can be accounted for by distinguishing sexual responses occurring in courtship, agonistic responses involving aggression and flight, and non-reproductive responses such as feeding and comfort movements. Important external cues include a mating partner and a mussel in which eggs are laid. The number of underlying causal factors may vary among species, however. In the courtship of the blue chromis observed naturally above reefs, Boer (1980) detected only two causal factors, aggression and nesting, which determined both social responses and coloration.

Like a brilliant but brief-lived firecracker, Nelson (1964, 1965) illuminated temporal aspects of reproductive behaviour. In the glandulocaudine fish *Corynopoma riisei*, courtship sequences occur randomly in time, with different activities having different durations and taking place with prob-

Figure 2.4: Courtship in Male Guppies Is the Outcome of an External Stimulus, Female Size, and Internal Arousal, as Reflected in Coloration, Calibrated Using the Relative Frequencies of Copulation Attempts (CA) and Sigmoid Displays (S and Si) (Baerends *et al.* 1955). Si = sigmoid intention movement; P_f = posturing behaviour

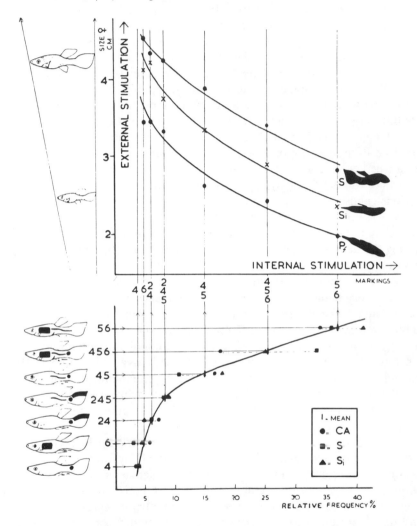

abilities dependent on the preceding activity. Near fertilisation, behavioural sequences become more determinate, as the females are influenced by the cumulative effect of male actions. In stickleback courtship (see Chapter 16 by Fitzgerald and Wootton, this volume), stimulation by female dummies affects the occurrence of zigzagging, fanning, and creeping through the nest. These changes can be adequately modelled using two variables, excitation and threshold, which rise during stimulation and creeping through, respectively, and fall otherwise. The Markovian nature of courtship sequences in *Barbus nigrafasciatus* led the investigators to conclude that 'barbs do it (almost) randomly' (Putters, Metz and Kooijman 1984).

Courtship behaviour is often important for the ethological islation of closely related species. Motivational analysis indicates how this isolation is achieved proximately. In the closely related *Cichlasoma citrinellum* and *C. zaliosum* of Central America, courtship requires at least four days before spawning takes place (Baylis 1976). Three phases can be distinguished: a phase of pair formation, an intervening phase, and a phase of preparation of the spawning site and spawning itself. Changes in the organisation of the behaviour result from differential frequencies of the occurrences of activities. There are also differences between the two species, especially early in courtship, due to differences in the thresholds for sexual and aggressive behaviour. Males and females of pumpkinseed and bluegill sunfish, and of their hybrids, generally discriminate the taxon of courting partners (Clarke, Colgan and Lester 1984). The courtships of the parental crosses (pumpkinseed × pumpkinseed and bluegill × bluegill) are least similar, and those of other crosses are intermediate.

In many species of fish, spawning is followed by care of the brood by one or both parents. This may involve fanning the eggs, which require oxygen during their development, consuming diseased eggs, and, especially, defending the eggs and often the brood against predation. In blue gouramis (*Trichogaster trichopterus*) broodiness can be induced by the repeated presentation of conspecific eggs (Kramer 1973). In convict cichlids, the duration of fanning declines over the period of incubation from a high level immediately after spawning (Mertz 1967). Fanning is the result of two randomly organised generators, one controlling the frequency of its initiation and the other its duration, which jointly enable this activity to intercalate flexibly with other activities such as foraging and defence.

Aggression

Beyond reproduction, fish are motivated to engage in other social behaviour, both affiliative and agonistic. Indeed, such behaviour may involve other species, as in interspecific shoaling (Ehrlich and Ehrlich 1973; and see Chapter 12 by Pitcher, this volume), and territoriality (Thresher 1978). Just as courtship behaviour is the result of internal and external factors, so is agonism. (For the role of hormones see Munro and Pitcher in Rankin *et al.* 1983.) In *Haplochromis burtoni*, presentation of a stimulus dummy has been reported to increase by a fixed amount the attack rate against smaller conspecifics (Heiligenberg, Kramer and Schulz 1972; Figure 2.5).

Along the same lines, aggression in male pumpkinseed sunfish defending nests, as reflected in various responses such as approaching stimulus dummies, is the outcome of both motivational state and external stimuli

Figure 2.5: Biting Rate in a Five-minute Interval in *Haplochromis burtoni* Is Raised a Fixed Amount Following Stimulus Dummy Presentation (Open Circles) Compared with Baseline Measured in the Preceding Interval (Closed Circles) (Heiligenberg *et al.* 1972)

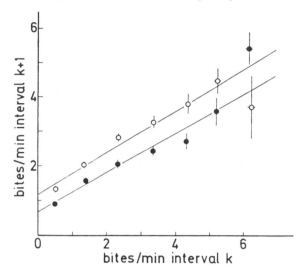

(Colgan and Gross, 1977; Figure 2.6). The motivational state of the defending fish changes greatly over the four phases of a typical ten-day nesting cycle (Figure 2.6a). Aggressiveness begins low in the initial two-day nesting phase during which nests are established. Then it rises for the three-day spawning phase (in which females lay eggs in the nest which the male fertilises) and the four-day brooding phase (in which the male cares for the hatching fry). Finally, it drops in the one-day vacating phase when nests are abandoned. External stimuli were manipulated by presenting painted plywood dummies of conspecifics in aggressive, normal and subordinate postures. A contingency-table analysis of the frequencies of approaches (Figure 2.6b) and other responses, with dummy posture and nesting phase as factors, indicated that both external and internal cues played a role in producing behaviour, and enabled a scaling of the relative effects of the dummies and nesting phases.

Several studies have investigated the motivational factors underlying agonistic behaviour. Using factor analysis, as similarly employed by Wiepkema (1961) and described earlier, Balthazart (1973, 1974) found that two independent factors, attack and flight, accounted for the variation in his data on the frequencies of different acts in *Oreochromis* (*Tilapia*) *macrochir*. The outcome of a fight appears to be decided early in the encounter, with subsequent behaviour being an artefact of aquarium confinement. Thresher (1978) has described two components of territorial aggression in the threespot damselfish, *Eupomacentrus planifrons*. One determines the size of area defended and the second the vigour of defence.

Figure 2.6: Aggression in Male Pumpkinseed Sunfish, as Measured by Such Responses as Approach, Is the outcome of an External Stimulus, a Conspecific Wooden Dummy, and Internal Arousal, as reflected in Phase of Nesting (Colgan and Gross 1977). (Dummy postures were AGG (△): aggressive NOR (●): normal, and SUB (▲): subordinate; nesting phases were Ne: nesting, Sp: spawning, Br: brooding, and Va: vacating. See text for details.)

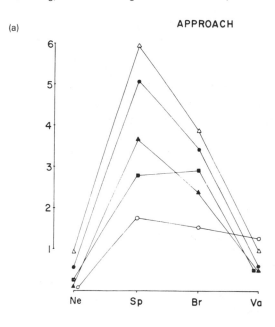

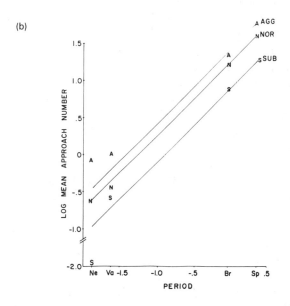

Congruent with Thresher's two-component theory are the findings by Colgan, Nowell and Stokes (1981) that pumpkinseed sunfish defending nests treat conspecific dummies differently in terms of the frequencies of their aggressive responses but similarly in terms of the spatial occurrence of those responses. The spatial aspects of nest defence are partially determined by the location of the nest rim, and are well modelled using catastrophe theory, with aggressive behaviour viewed as the joint outcome of the reproductive phase of the defending fish and the distance of an intruder from the nest centre.

Whereas the levels of feeding and sexual motivation clearly increase in the absence of the appropriate cues (food and partners), the extent to which aggression similarly changes has been a matter of controversy, both for fish and other species (especially for the human case, for obviously critical reasons). Like its freshwater cousin the Siamese fighting fish, the yellowtail damselfish (*Microspathodon chrysurus*) seeks opportunities for aggressive display, which can be used as a reward for conditioning (Rasa 1971). In isolation, individuals show increased comfort behaviour, with each activity following a separate time course: this has been interpreted as a means of achieving behavioural homeostasis in the face of an accumulation of endogenous aggressive energy deprived of its usual releasers. However, based on research on orange chromides (*Etroplus maculatus*), Reyer (1975) concluded that external stimuli, not internal drive, produce aggressive behaviour. Certainly social isolation has been found to decrease aggressiveness in a number of species (e.g. *Haplochromis burtoni*: Heiligenberg and Kramer 1972).

These conflicting results could be due to differences among species. However, data on the effects of social isolation in the paradise fish could offer a solution (Davis 1975). By increasing reactivity to both social and non-social stimuli, isolation has general effects on behaviour, and not necessarily specific effects on the readiness for social display. Thus a generalised hyperactivity due to isolation combined with a differential habituation to specific stimuli could account for these apparently contradictory reports.

It is interesting to note that the boldness of sticklebacks towards pike (*Esox lucius*) covaries with the intensity with which they attack conspecifics when measured in conditions ranging from non-reproduction through nesting to defending newly hatched young (Huntingford 1976). This covariance suggests that there is a common causal system producing timidity.

Causal Interactions

The stream of behaviour constantly reflects choices among options, and thus the conflict among causal systems for the 'behavioural final common path' (McFarland and Sibly 1975) demands close attention. The expression of any one causal system is likely to be interfered with by other systems. This has been noticed in many studies, such as Wiepkema's (1961) on bitterling reproduction. Dawkins (1976) has suggested random switches between hierarchically organised systems to generate behaviour, but the available data do not support this conclusion (Colgan and Slater 1980). McFarland and co-workers (see, for example, McFarland 1974) have elaborated possible mechanisms, including competition and time-sharing. In competition, changes in the causal factors of a second activity oust an ongoing activity, whereas in time-sharing, an ongoing activity terminates itself and disinhibits (releases) the second activity. The manner in which nesting male sticklebacks divide their time between the nest and females has been investigated by Cohen and McFarland (1979). (The behavioural ecology of sticklebacks is discussed in Chapter 16, this volume.) Males were kept in aquaria with three linearly adjoining compartments. They were allowed to nest in one end-compartment, and were presented with females in bottles in the other. Doors between compartments could be opened or closed to enable access or to interrupt ongoing behaviour. Dropping snails into the nests, which the fish removed, also interrupted behaviour. Principal components analysis of the behaviour revealed three groups, associated with nesting, sexual, and aggressive activities, which compete for dominance. Seasonal variations in motivational levels were also noted. The authors interpret the results as supporting a concept of time-sharing between dominant and subordinate activities. Recently, D.L.G. Noakes (personal communication) has conducted similar experiments with nesting sticklebacks, interrupting their activities and presenting male and female conspecifics. The results are also compatible with a time-sharing interpretation. Houston (1982) has challenged the underlying logic of this concept; McFarland (1983) has rebutted; and I consider that the details of what is really being asserted remain, as they have been, unclear.

Displacement acts, defined by their irrelevance in the flow of behaviour (Tinbergen 1952), are a major aspect of conflict behaviour. Their role in the evolution of communicative behaviour has often been invoked as one of the key elements in the phylogeny of motivation. Currently, the entire study of communication — in terms of causal mechanisms, the exchanged information about strength and intentions, and the adaptiveness of signals — is undergoing intense scrutiny (see Caryl (1982) and references therein). The present discussion will be limited to the motivation of displacement responses. Under the disinhibition hypothesis, displacement responses are viewed as the outcome of balanced conflict between two dominant behavioural tendencies, each cancelling the inhibition exerted by the other on the displacement activity. More recent work has suggested that other explanations may also apply. Wilz (1970a, b) conducted experiments on displacement responses in nesting male sticklebacks. In these males the tendencies to lead the female to the nest and to chase her from it are in conflict. The reactions of males to a female dummy were recorded, with the aggressive level of the male subjects being manipulated by presenting them with a stimulus male in a bottle. Particular attention was paid to 'dorsal pricking', in which the male nudges the female dummy towards the surface of the water with his dorsum. Wilz showed that dorsal pricking tends to occur in response to an approach by a female when the male is aggressive or when sexual stimulation is lacking. Dorsal pricking causes the female to wait while the male goes off to the nest, performs displacement activities (creeping through and fanning), and returns to court. Females failing to wait are generally attacked; attacks also result if the displacement activities are blocked. Thus dorsal pricking provides the male with an opportunity to perform displacement responses and thus to switch from primarily aggressive motivation to primarily sexual motivation. Wilz argues that such self-regulation of motivation is the function of these responses. Given these successive motivational changes, the time scale for measurement is critical. Wilz (1972) subsequently showed that aggression inhibits nest gluing, an activity associated with sexual behaviour, in the short term, but over the long term increases it.

The antagonism between sexual and aggressive behaviour has been examined in other species. When approached by a gravid female prepared to spawn, nesting male pumpkinseed and bluegill sunfish gradually shift from chiefly aggressive motivation (attacking the female) to chiefly sexual motivation (courting the female) (Ballantyne and Colgan 1977). This shift (and the underlying conflict) is indicated by sounds produced by the male rubbing his pharyngeal dental pads together, and hence these sounds play a key communicative role during courtship. (See also Chapter 5 by Howkins, this volume.) Similarly, by exposing test convict cichlids to stimulus individuals of either sex, and factor analysing the resulting data, Cole, Figler, Parente and Peeke (1980) showed a mutual inhibition of sex and aggression.

Students of both the motivation of feeding and its functional aspects, often in the context of optimal foraging theory, have been impressed by the manner in which feeding is compromised by other activities. Experimental sticklebacks subjected to electric shocks while feeding show differences in behaviour compared with unshocked control fish (Tugendhat 1960c). Feeding is interrupted in a manner dependent on the location of the shocks. Fish shocked at the entrance to the feeding areas return less frequently to the shock-free living area, whereas those shocked for taking food items return more frequently. The behavioural conflict is also reflected in changes in the frequency of dorsal spine raising and comfort movements. The total amount of feeding decreases as shock intensity increases. In a more recent study, experimental fish exposed to an avian predator began feeding more slowly, attacked less dense prey, and made fewer total bites compared with control fish not so exposed (Milinski and Heller 1978; see Chapter 9 by Milinski, this volume). These findings agree with a model, based on optimal control theory, proposed by the authors. Further investigations are required to discover the exact quantitative relations involved. Young, trained carp arrive at a feeding place more quickly when swimming in a shoal habituated to the experimental conditions than in a frightened one (Köhler 1976). This difference reflects the conflict of a conditioned reaction and mutual attraction among the members of the shoal. In fighting fish the interaction of feeding and aggressive display is affected by fear (Hogan 1974). Fish living in a T-maze offering food or a mirror image prefer food, but those transferred to the maze show a greater tendency to choose the mirror image. This differential effect of fear on feeding and aggression, in response to unfamiliar surroundings, illustrates the need to allow for environmental influences on motivational state in both field and laboratory studies.

The relation between feeding and aggression is of paramount importance in assessing the motivation of predatory species. Likewise, the motivation of cannibals and of feeding specialists such as cleaners removing ectoparasites from hosts, paedophages feeding on young, and lepidophages engulfing mouthfuls of integument are of special interest for the blend of trophic and social motivation (e.g. see Sazima 1983). In sunfish, observations on aggression and predation in socially dominant and subordinate individuals reveal that these two motivational systems are independent (Poulsen and Chiszar 1975). However, variation in hunger has complex effects on different components of aggressive behaviour, which thus cannot be regarded as a unitary system (Poulsen 1977). Siamese fighting fish direct more biting attacks toward small guppies and more threat displays towards large ones (Baenninger and Kraus 1981). Hunger level influences only biting attacks, and mirror exposure only threat displays. Lenke (1982) has argued for the existence of a motivational system for cleaning behaviour separate from feeding in the cleaner wrasse (*Labroides dimidiatus*), since this behaviour is both facultative and unaffected by food deprivation. Various injected amounts of different hormones are reported to have different effects on cleaning behaviour, thus demonstrating the endocrinal background of this motivation. However, more convincing data are needed to establish this system. For their part, hosts are motivated by a conflict between tactile reinforcement and aversive stimulation resulting from the removal of parasites (Losey 1979). In our laboratory we have studied how male guppies divide their time between feeding and courting females. As expected, males are less attracted to food as satiation proceeds, especially as the number of stimulus females is increased. On interactions between feeding and other activities, females of the cichlid *Pseudocrenilabrus multicolor*, which care for young in the mouth, feeding is inhibited by the broodiness of the female and by stimuli from the brood itself (Mrowka 1984).

The use of stimulus dummies for motivational analysis has been cultivated to a high art under Baerends' leadership at Groningen. The interactions between attack, escape and courtship can be manipulated in the jewel cichlid (*Hemichromis bimaculatus*) through the use of such dummies (Rowland 1975a, b). Different dummies differentially elicit responses associated with each causal system. For instance, red dummies produce strong sexual arousal, which inhibits aggressive activities such as biting. Recent experience also influences behavioural priorities in the fish. An unpaired male courted a dummy, whereas he attacked it after pairing with a female. In a similar vein, communication in *Pseudotropheus zebra*, a cichlid from Lake Malawi, involves the interaction of attack and escape tendencies, as assayed by dummies of various colours and sizes (Vodegel 1978). A quantitative model accounts for the influence on these tendencies, of dummy size and accessibility, the distance between the fish and dummy, habituation, and which activities have been performed. Subsequent research on three cichlids endemic to Lake George revealed that responsiveness to territorial intruders is the joint outcome of four components: interest in the intruder, avoidance, attraction to the shelter, and persistence (Carlstead 1983a, b). Attacks and displays are especially affected by avoidance and persistence. The levels of the components vary across species, and change as an encounter with a conspecific or environmental disturbance proceeds.

Conclusions

The research reviewed in this chapter indicates a variety of approaches for analysing the causal basis of behaviour into its external and internal components. It is clear that such analysis has scarcely gained a toehold on the terrain to be investigated. The characteristics of the different causal systems and their interactions in a diversity of teleost species must be investigated. The work of the past two decades has revealed many motivational subsystems within and beyond the 'big four', hunger, fear, aggression and sex. For instance, there appear to be four systems operating in a female Siamese fighting fish as she responds to a conspecific (Robertson 1979). Surely, as with finding new species and races in an unexplored region, the number of motivational systems discovered will approach an asymptote. The analysis of problems such as individual differences (often found in the research reviewed above (e.g. Vodegel 1978), and considered in Chapter 13 by Magurran, this volume) requires consideration of their motivational substrates. Beyond research into specific ontogenetic processes such as learning and imprinting, there is a need for examination in teleosts of the ontogeny of motivation, understood as the developmental patterning of causal systems, similar to that carried out for some tetrapod species. Similarly, the phylogeny of motivation merits attention: how does diversification evolve to enable complex interactions such as communication and interspecific cleaning behaviour? In the area of research between the levels of physiology and behaviour, further studies linking motivation with underlying physiological processes are required to sort through the current conceptual ragbag. At the interface of behaviour and ecology, the adaptiveness of causal mechanisms needs to be assessed. In the past, causation and function have often been confused. Such confusion is understandable: for instance, competition among causal systems and tradeoffs among objectives in behavioural ecology are obverse problems. But the separateness of causation and function must be realised: causal analysis does not indicate adaptiveness, and functional analysis does not reveal causation.

Methodologically, the above research reflects the breadth of useful analytical approaches to the problem of motivation: multivariate statistics, information theory, stochastic processes, and specific models. Conversely, motivational differences have direct significance for the methodology of behavioural research. For instance, it is generally hoped that observations from the field and the laboratory are comparable, but this is not always the case. Courtship sequences in sunfish are longer in aquaria than in natural colonies for several reasons: interruptions by other males are generally eliminated; the female cannot entirely escape the attention of the male; and, of particular relevance here, in the laboratory the female is placed with the male by the experimenter, hoping she is ready to spawn, whereas in the field she freely enters and leaves the colony (Clarke *et al.* 1984). Therefore field and small aquarium comparisons must be made with caution.

The available data on the motivation of fish behaviour and the analytical tools for its study provide a guiding framework for research: motivational organisation is viewed as a set of clusters of activities causally linked by transition probabilities dependent on both internal and external cues (see Itzkowitz 1979), and durations spent within a particular state are similarly influenced. A steadfast use of this framework will perhaps solve the recondite problems alluded to in Duboś (1971) dictum 'Under precisely controlled conditions an animal does as he damn pleases.'

Summary

The study of the internal proximate causes of responses is a central topic in fish behaviour. The motivation of fish, along with their learning capacities, develop over the lifetime of the individual. Feeding motivation is reflected in satiation over a meal and preferences between available food types. Fear and avoidance operate in learning and social interactions. Motivation for reproductive and agonistic behaviour is dependent on external cues and internal state, especially hormonal levels. Analyses of courtship, parental and aggressive activities indicate the operations of several causal factors. These major causal systems mutually interact in complex patterns to influence behaviour performance. The nature of these systems, the concepts to deal with them, and the relations with physiology and ecology all require much further investigation.

Acknowledgements

The writing of this chapter has been supported by Queen's University at Kingston and the Natural Sciences and Engineering Research Council of Canada, and was written during a leave-of-absence from the Biodegraders hockey team. Data in preparation was kindly made available by D.F. Frey and D.N.G. Noakes. For comments and criticisms I am grateful to J.A. Brown ('Yes, but what about ontogeny?!'), J.A. Hogan, P.H. Johansen ('It's probably all nonsense!'), M.H.A. Keenleyside, S. Muldal, and the Magi Systemae.

References

Baenninger, R. and Kraus, S. (1981) 'Some Determinants of Aggressive and Predatory Responses in *Betta splendens*', *Journal of Comparative and Physiological Psychology, 95,* 220-7

Baerends, G.P. (1971) 'The Ethological Analysis of Fish Behaviour', in W.S. Hoar and D.J. Randall (eds) *Fish Physiology,* vol. 6, Academic Press, New York, pp. 279-370

Baerends, G.P., Brower, R. and Waterbolk, H.T. (1955) 'Ethological Studies in *Lebistes reticulatus* (Peters). 1. An Analysis of the Male Courtship Pattern', *Behaviour, 8,* 249-334

Ballantyne, P.K. and Colgan, P.W. (1978) 'Sound Production during Agonistic and Reproductive Behaviour in the Pumpkinseed (*Lepomis gibbosus*), the Bluegill (*L. macrochirus*), and their Hybrid Sunfish, 1. Context', *Biology of Behaviour, 3,* 113-35

Balthazart, J. (1973) 'Analyse Factorielle du Comportement Agonistique chez *Tilapia macrochir* (Boulenger 1912)', *Behaviour, 46,* 37-72

Balthazart, J. (1974) 'Non-stationnarité et Aspects Fonctionnels du Comportement Agonistique chez *Tilapia macrochir* (Boulenger 1912) (Pisces: Chichlidae)', *Acta Zoologica Pathologica Antverpiensia, 58,* 29-40

Bateson, P.P.G. (1979) 'How Do Sensitive Periods Arise and What Are they For?' *Animal Behaviour, 27,* 470-86

Baylis, J.R. (1976) 'A Quantitative Study of Long-term courtship. 2. A Comparative Study of the Dynamics of Courtship in Two New World Cichlid Fishes', *Behaviour, 59,* 117-61

Bitterman, M.E. (1965) 'The CS-US Interval in Classical and Avoidance Conditioning', in W.F. Prokasy (ed.), *Classical Conditioning,* Appleton-Century-Crofts, New York, pp. 1-19

Boer, B.A.D. (1980) 'A Causal Analysis of the Territorial and Courtship Behaviour of *Chromis cyanea* (Pomacentridae, Pisces)', *Behaviour, 73,* 1-50

Bols, R.J. and Hogan, J.A. (1979) 'Runway Behavior of Siamese Fighting Fish, *Betta splendens,* for Aggressive Display and Food Reinforcement', *Animal Learning and Behaviour, 7,* 537-42

Braud, W.G. (1971) 'Effectiveness of "Neutral", Habituated, Shock-related, and Food-related Stimuli as CSs for Avoidance Learning in Goldfish', *Conditioned Reflex, 6,* 153-6

Brown, D.M.B. and Noakes, D.L.G. (1974) 'Habituation and Recovery of Aggressive Display in Paradise Fish (*Macropodus opercularis* (L.))', *Behavioural Biology, 10,* 519-25

Carlstead, K. (1983a) 'The behavioural organization of Responses to Territorial Intruders and Frightening Stimuli in Cichlid Fish (*Haplochromis* spp.)', *Behaviour, 83,* 18-68

Carlstead, K. (1983b) 'Influences of Motivation on Display Divergences in Three Cichlid Fish

Species (*Haplochromis*)', *Behaviour*, *83*, 205-28

Caryl, P.G. (1982) 'Telling the Truth about Intentions', *Journal of Theoretical Biology*, *97*, 679-89

Chantrey, D.F. (1978) 'Short-term Changes in Responsiveness to Models in *Betta splendens*', *Animal Learning and Behaviour*, *6*, 469-71

Clarke, S.E., Colgan, P.W. and Lester, N.P. (1984) 'Courtship Sequences and Ethological Isolation in Two Species of Sunfish (*Lepomis* spp.) and their Hybrids', *Behaviour*, *91*, 93-114.

Cohen, S. and McFarland, D. (1979) 'Time-sharing as a Mechanism for the Control of Behaviour Sequences during the Courtship of the Three-spined Stickleback (*Gasterosteus aculeatus*)', *Animal Behaviour*, *27*, 270-83

Cole, H.W., Figler, M.H., Parente, F.J. and Peeke, H.V.S. (1980) 'The Relationship between Sex and Aggression in Convict Cichlids (*Cichlasoma nigrofasciatum* Günther)', *Behaviour*, *75*, 1-21

Colgan, P.W. (1973) 'Motivational Analysis of Fish Feeding', *Behaviour*, *45*, 38-66

Colgan, P.W. (1983) *Comparative Social Recognition*, Wiley Interscience, New York

Colgan, P.W. and Gross, M.R. (1977) 'Dynamics of Aggression in Male Pumpkinseed Sunfish (*Lepomis gibbosus*) over the Reproductive Phase', *Zeitschrift für Tierpsychologie*, *43*, 139-51

Colgan, P.W. and Slater, P.J.B. (1980) 'Clustering Acts from Transition Matrices', *Animal Behaviour*, *28*, 965-6

Colgan, P.W. and Smith, J.T. (1984) 'Experimental Analysis of Food Preference Transitivity in Pumpkinseed Sunfish (*Lepomis gibbosus*)', *Biometrics*, *41*, 227-36.

Colgan, P.W., Nowell, W.A., Gross, M.R. and Grant, J.W.A. (1979) 'Aggressive Habituation and Rim Circling in the Social Organization of Bluegill Sunfish (*Lepomis macrochirus*)', *Environmental Biology of Fishes*, *4*, 29-36

Colgan, P.W., Nowell, W.A. and Stokes, N.W. (1981) 'Nest Defence by Male Pumpkinseed Sunfish (*Lepomis gibbosus*: Stimulus Features and an Application of Catastrophe Theory', *Animal Behaviour*, *29*, 433-42

Davis, R.E. (1975) 'Readiness to Display in the Paradise Fish *Macropodus opercularis*, L., Belontiidae: the Problem of General and Specific Effects of Social Isolation', *Behavioural Biology*, *15*, 419-33

Dawkins, R. (1976) 'Hierarchical Organisation: a Candidate Principle for Ethology', in P.P.G. Bateson and R.A. Hinde (eds), *Growing Points in Ethology*, Cambridge University Press, Cambridge, pp. 5-54

Dill, L.M. (1983) 'Adaptive Flexibility in the Foraging Behavior of Fishes', *Canadian Journal of Fisheries and Aquatic Sciences*, *40*, 398-408

Domjan, M. (1983) 'Biological Constraints on Instrumental and Classical Conditioning: Implications for General Process Theory', *Psychology of Learning and Motivation*, *17*, 215-77

Dubos, R. (1971) 'In defense of biological freedom', in E. Tobach *et al.* (eds), *The Biopsychology of Development*, Academic Press, New York, pp. 553-60

Ehrlich, P.R. and Ehrlich, A.M. (1973) 'Coevolution: Heterotypic Schooling in Caribbean Reef Fishes', *American Naturalist*, *107*, 157-60

Fabricius, E. (1950) 'Aquarium Observation on the Spawning of the Char (*Salvelinus alpinus*)'. *Report of the Institute of Freshwater Research, Drottningholm*, *34*, 14-48

Figler, M.H. and Einhorn, D.M. (1983) 'The Territorial Prior Residence Effect in Convict Cichlids (*Cichlasoma nigrofasciatum* Günther): Temporal Aspects of Establishment and Retention, and Proximate Mechanisms', *Behaviour*, *85*, 157-83

Franck, D. (1975) 'Der Anteil des "Coolidge-Effektes" an der isolationsbedingten Zunahme sexueller Verhaltensweisen von *Poecilia sphenops*', *Zeitschrift für Tierpsychologie*, *38*, 472-81

Gallagher, J.E., Herz, M.J. and Peeke, H.V.S. (1972) 'Habituation of Aggression: the Effects of Visual Social Stimuli on Behavior between Adjacently Territorial Convict Cichlids (*Cichlasoma nigrofasciatum*)', *Behavioural Biology*, *7*, 359-68

Giles, N. (1983) 'Behavioural Effects of the Parasite *Schistocephalus solidus* (Cestoda) on an Intermediate Host, the Three-spined Stickleback, *Gasterosteus aculeatus* L', *Animal Behaviour*, *31*, 1192-4

Heiligenberg, W. and Kramer, U. (1972) 'Aggressiveness as a Function of External Stimulation', *Journal of Comparative Physiology*, *77*, 332-40

Heiligenberg, W., Kramer, U. and Schulz, U. (1972) 'The Angular Orientation of the Black Eye-bar in *Haplochromis burtoni* (Cichlidae, Pisces) and its Relevance to Aggressivity', *Zeitschrift für vergleichende Physiologie, 76*, 168-76

Hogan, J.A. (1965) 'An Experimental Study of Conflict and Fear: an Analysis of Behavior of Young Chicks toward a Mealworm. Part 1. The Behavior of Chicks which Do Not Eat the Mealworm', *Behaviour, 25*, 45-97

Hogan, J.A. (1974) 'On the Choice between Eating and Aggressive Display in the Siamese Fighting Fish (*Betta splendens*)', *Learning and Motivation, 5*, 273-87

Hogan, J.A., Kleist, S. and Hutchings, C.S.L. (1970) 'Display and Food as Reinforcers in the Siamese Fighting Fish (*Betta splendens*)', *Journal of Comparative Physiological Psychology, 70*, 351-7

Houston, A.I. (1982) 'Transitions and Time-sharing', *Animal Behaviour, 30*, 615-25

Huntingford, F.A. (1976) 'A Comparison of the Reaction of Sticklebacks in Different Reproductive Conditions towards Conspecifics and Predators', *Animal Behaviour, 24*, 694-7

Hurley, A.C. and Hartline, P.H. (1974) 'Escape Response in the Damselfish *Chromis cyanea* (Pisces: Pomacentridae): a Quantitative Study', *Animal Behaviour, 22*, 430-7

Itzkowitz, M. (1979) 'On the Organization of Courtship Sequences in Fishes', *Journal of Theoretical Biology, 78*, 21-8

Ivlev, V.S. (1961) *Experimental Ecology of the Feeding of Fishes*, Yale University Press, New Haven

Kamil, A.C. and Yoerg, S.I. (1982) 'Learning and Foraging Behavior', in P.P.G. Bateson and P.H. Klopfer (eds), *Perspectives in Ethology*, vol. 5, Plenum Press, New York, pp. 325-64

Klein, R.M., Figler, M.H. and Peeke, H.V.S. (1976) 'Modification of Consummatory (Attack) Behaviour Resulting from Prior Habituation of Appetitive (Threat) Components of the Agonistic Sequence in male *Betta splendens* (Pisces, Belontiidae)', *Behaviour, 58*, 1-25

Köhler, D. (1976) 'The Interaction between Conditioned Fish and Naive Schools of Juvenile Carp (*Cyprinus carpio*, Pisces)', *Behavioural Processes, 1*, 267-75

Kramer, D.L. (1973) 'Parental Behaviour in the Blue Gourami *Trichogaster trichopterus* (Pisces, Belontiidae) and its Induction during Exposure to Varying Numbers of Conspecific Eggs', *Behaviour, 47*, 14-32

Laming, P.R. and Ennis, P. (1982) 'Habituation of Fright and Arousal Responses in the Teleosts *Carassius auratus* and *Rutilus rutilus*', *Journal of Comparative and Physiological Psychology, 96*, 460-6

Laming, P.R. and Ebbessan, S.O.E. (1984) 'Arousal and Fright Responses and their Habituation in the Slippery Dick, *Halichoeres bivittatus*', *Experientia, 40*, 767-9

Lenke, R. (1982) 'Hormonal Control of Cleaning Behaviour in *Labroides dimidiatus* (Labridae, Teleostei)', *Marine Ecology, 3*, 281-92

Lester, N.P. (1976) 'Motivational Implications of the Temporal Pattern of Feeding in Bluegill Sunfish (*Lepomis macrochirus*)', Unpublished MSc thesis, Queen's University at Kingston, Canada

Lester, N.P. (1984) 'The feed: Feed Decision: How Goldfish Solve the Patch Depletion Problem', *Behaviour, 89*, 175-90

Liley, N.R. and Stacey, N.E. (1983) 'Hormones, Pheromones and Reproductive Behavior in Fish', in W.S. Hoar and D.J. Randall (eds.), *Fish Physiology vol. 8*, Academic Press, New York, pp. 1-63.

Losey, G.S.J. (1979) 'Fish Cleaning Symbiosis: Proximate Causes of Host Behaviour', *Animal Behaviour, 27*, 669-85

McFarland, D.J. (1974) (ed.) *Motivational Control Systems Analysis*, Academic Press, London

McFarland, D.J. (1983) 'Time-Sharing: a Reply to Houston (1982)', *Animal Behaviour, 31*, 307-8

McFarland, D.J. and Sibly, R.M. (1975) 'The Behavioural Final Common Path', *London Royal Society Philosophical Transactions, 270B*, 265-93

Mertz, J.C. (1967) 'The Organization and Regulation of the Parental Behavior of *Cichlasoma nigrofasciatum* (Pisces: Cichlidae), with Special Reference to Parental Fanning', Unpublished PhD thesis, University of Illinois

Milinski, M. and Heller, R. (1978) 'Influence of a Predator on the Optimal Foraging Behaviour of Sticklebacks (*Gasterosteus aculeatus* L.)' *Nature*, London, *275*, 642-4

Morgan, W.L. and Ritz, D.A. (1984) 'Effects of Prey Density and Hunger State on Capture of Krill, *Nyctiphanes australis* Sars, by Australian salmon, *Arripis trutta* (Block & Schneider)', *Journal of Fish Biology, 24*, 51-8

Mrowka, W. (1984) 'Brood Care Motivation and Hunger in the Mouthbrooding Cichlid *Pseudocrenilabrus multicolor*', *Behaviour Processes, 9*, 181-90

Nakatsuru, K. and Kramer, D.L. (1982) 'Is Sperm Cheap? Limited Male Fertility and Female Choice in the Lemon Tetra (Pisces, Characidae)', *Science, 216*, 753-5

Nelson, K. (1964) 'The Temporal Patterning of Courtship Behaviour in the Glandulocaudine Fishes (Ostariophysi, Characidae)', *Behaviour, 124*, 90-144

Nelson, K. (1965) 'After-effects of Courtship in the Male Three-spined Stickleback', *Zeitschrift für vergleichende Physiologie, 50*, 569-97

Olson, R.D., Kastin, A.J., Mitchell, G.F., Olson, G.A., Coy, D.H. and Montalbane, D.M. (1978) 'Effects of Endorphin and Enkephalin analogs on fear habituation in goldfish', *Pharmacology, Biochemistry and Behavior, 9*, 111-14

Peeke, H.V.S. (1983) 'Habituation Sensitization, and Redirection of Aggression and Feeding Behaviour in the Three-spined Stickleback (*Gasterosteus aculeatus* L.)', *Journal of Comparative Psychology, 97*, 43-51

Poulsen, H.R. (1977) 'Predation, Aggression and Activity Levels in Food-deprived sunfish (*Lepomis macrochirus* and *L. gibbosus*): Motivational Interactions', *Journal of Comparative and Physiological Psychology, 91*, 611-28

Poulsen, H.R. and Chiszar, D. (1975) 'Interaction of Predation and Intraspecific Aggression in Bluegill Sunfish *Lepomis macrochirus*', *Behaviour, 55*, 268-86

Putters, F.A., Metz, J.A.J. and Kooijman, S.A.L.M. (1984) 'The Identification of a Simple Function of a Markov Chain in a Behavioural Context: Barbs Do it (Almost) Randomly', *Nieuw Arch. Wiskunde, 2*, 110-23

Rankin, J.C., Pitcher, T.J. and Duggan, R.T. (1983) (eds) *Control Processes in Fish Physiology*, Croom Helm, London

Rasa, O.A.E. (1971) 'Appetence for Aggression in Juvenile Damselfish', *Zeitschrift für Tierpsychologie Suppl., 7*

Reyer, H.-U. (1975) 'Ursachen und Konsequenzen von Aggressivität bei *Etroplus maculatus* (Cichlidae, Pisces)', *Zeitschrift für Tierpsychologie, 39*, 415-54

Robertson, C.M. (1979) 'Aspects of Sexual Discrimination by Female Siamese Fighting Fish (*Betta splendens* Regan)', *Behaviour, 70*, 323-36

Rowland, W.J. (1975a) 'The Effects of Dummy Size and Color on Behavioral Interaction in the Jewel Cichlid, *Hemichromis bimaculatus* Gill', *Behaviour, 53*, 109-25

Rowland, W.J. (1975b) 'System Interaction of Dummy-elicited Behavior in the Jewel Cichlid, *Hemichromis bimaculatus* Gill', *Behaviour, 53*, 171-82

Sazima, I. (1983) 'Scale-eating in Characoids and Other Fishes', *Environmental Biology of Fishes, 9*, 87-101

Schwassmann, H.O. (1980) 'Biological Rhythms: their Adaptive Significance', in M.A. Ali (ed.), *Environmental Physiology of Fishes*, Plenum Press, New York, pp. 613-30

Thomas, G. (1974) 'The Influences of Encountering a Food Object on Subsequent Searching Behaviour in *Gasterosteus aculeatus* L', *Animal Behaviour, 22*, 941-52

Thomas, G. (1977) 'The Influence of Eating and Rejecting Prey Items upon Feeding and Food Searching Behaviour in *Gasterosteus aculeatus* L', *Animal Behaviour, 25*, 52-66

Thresher, R.E. (1978) 'Territoriality and Aggression in the Threespot Damselfish (Pisces: Pomacentridae): an Experimental Study of Causation', *Zeitschrift für Tierpsychologie, 46*, 401-34

Tinbergen, N. (1951) *The Study of Instinct*, Clarendon Press, Oxford

Tinbergen, N. (1952) "Derived" Activities: their Causation, Biological Significance, Origin, and Emancipation during Evolution', *Quarterly Review of Biology, 27*, 1-32

Tinbergen, N. (1963) 'On Aims and Methods of Ethology', *Zeitschrift für Tierpsychologie, 20*, 410-33

Toates, F.M. and Birke, L.I.A. (1982) 'Motivation: a New Perspective on Some Old Ideas', in P.P.G. Bateson and P.H. Klopfer (eds), *Perspectives in Ethology*, vol. 5, Plenum Press, New York, pp. 191-241

Tugendhat, B. (1960a) 'The Normal Feeding Behavior of the Three-spined Stickleback (*Gasterosteus aculeatus* L.)', *Behaviour, 25*, 284-318

Tugendhat, B. (1960b) 'The Disturbed Feeding Behavior of the Three-spined stickleback: 1. Electric Shock Adminstered in the Food Area', *Behaviour, 16*, 159-87

Visser, M. (1981) 'Prediction of Switching and Counter-switching Based on Optimal Foraging', *Zeitschrift für Tierpsychologie, 55*, 129-38

Vodegel, N. (1978) 'A Causal Analysis of the Behaviour of *Pseudotropheus zebra*', Unpublished PhD thesis, University of Groningen, The Netherlands

Wapler-Leong, C.-Y. (1974) 'The Attack Readiness of Male *Haplochromis burtoni* (Cichlidae, Pisces) reared in isolation', *Journal of Comparative Physiology, 94*, 219-25

Warren, E.W. and Callaghan, S. (1976) 'The Response of Male Guppies (*Poecilia reticulata*, Peters) to Repeated Exposure to an Open Field', *Behavioural Biology, 18*, 499-513

Wiepkema, P.R. (1961) 'An Ethological Analysis of the Reproductive Behaviour of the Bitterling (*Rhodeus amarus* Bloch)', *Archives Neerlandaises de Zoologie, 14*, 103-99

Wilz, K.J. (1970a) 'Causal and Functional Analysis of Dorsal Pricking and Nest Activity in the Courtship of the Three-spined Stickleback *Gasterosteus aculeatus*', *Animal Behaviour, 18*, 115-24

Wilz, K.J. (1970b) 'The Disinhibition Interpretation of the "Displacement" Activities during Courtship in the Three-spined Stickleback, *Gasterosteus aculeatus*', *Animal Behaviour, 18*, 682-7

Wilz, K.J. (1972) 'Causal Relationships between Aggression and the Sexual and Nest Behaviours in the Three-spined Stickleback (*Gasterosteus aculeatus*)', *Animal Behaviour, 20*, 335-40

3 DEVELOPMENT OF BEHAVIOUR IN FISH

Felicity A. Huntingford

Questions about the Development of Behaviour

A newly fertilised egg does not behave; an adult fish responds to its environment with a repertoire of complex, adaptive behaviour patterns (Dawkins 1983). This chapter is about what happens in-between. When in the developmental process do co-ordinated behaviour patterns arise, and how does this come about? When and how are social relationships established, and what is their role in the life of a young fish? How do external stimuli control behaviour in fish of various ages? When and how do different behaviour patterns take on the motivational relationships that characterise adult behavioural systems? To answer these questions we need to know what young fish actually do at each stage in the development from egg to adult. Equally we might ask how it comes about that different species, sexes and individuals show distinct behavioural responses. This second and very difficult question requires a knowledge of the factors that influence the sequence of developmental events, and how they do so.

Fish as Subjects for Developmental Research

Teleost fish have a number of characteristics which make them suitable subjects for developmental studies. In the first place, although their behaviour is often spectacular, their repertoire of action patterns is not unmanageably complex. It is therefore possible to produce an accurate description of the complete behavioural repertoire of a particular species (an 'ethogram') so that in a number of cases we know where behavioural development is heading. Secondly, the huge number of teleost species (at least 22000) means that the comparative approach (Huntingford 1984) can be put to good use; we can, for example, compare the development of behaviour in species with and without parental care (see Chapter 10 by Turner, and Chapter 11 by Sargent and Gross, this volume) to find out how interactions with parents influence the way behaviour develops. A further advantage of fish as subjects for developmental studies is that they generally produce many offspring whose larval and juvenile stages are free living. This means that fish are potentially accessible for observation and experimental manipulation of the environmental conditions in which they develop. On the other hand, young fish are not easy to rear and their small size makes experimental manipulation of the developing nervous system difficult. Overall, however, the development of behaviour in fish has proved a fruitful area of research, and the aim of this chapter is to describe some of these studies and see if they provide any insights into the process of development. Like other reviews of the subject (Noakes 1978), this one concentrates on the behavioural ontogeny of the two most intensively studied groups, the salmonids and the cichlids.

The Early Development of Behaviour in Salmonids

Because of their commercial importance, salmonids have received a lot of attention from scientists. However, because they are long lived and often have complex life histories, most detailed behavioural studies have been carried out on salmonids in the first few months of life. Figure 3.1 summarises the main behavioural changes that occur in the Atlantic salmon (*Salmo salar*) from fertilisation to an age of about 40 days. The first stages of development, up to a few weeks after hatching (which occurs after about 110 days), normally take place under several centimetres of gravel, where the eggs are buried after fertilisation. The first observed movements occur about one-third of the way into embryonic life in the form of weak contractions of the developing heart. These gradually increase in strength and frequency until co-ordinated contractions of the fully formed heart occur. Shortly after heart movement starts, the anterior muscles above the yolk sac being to twitch, causing very slow and irregular bending movements of the trunk, initially with no orderly relationship between different parts of the body. These bending movements become gradually stronger, extending to muscle blocks along the whole of the body, with the two sides co-ordinated to produce S-shaped swimming movements. Trunk movements initially occur spontaneously, but later are elicited by tactile stimulation. Towards the end of embryonic life the jaws, operculae and the pectoral fins begin to move spasmodically, when the axial muscles contract. Shortly before hatching, movements of the jaw and operculae and of the pectoral fins are fully co-ordinated and independent of contractions of the axial musculature. Hatching is the result of swimming movements which are sufficiently frequent and violent to split the egg membrane and allow the young fish to emerge (Abu-Gideiri 1966).

At first the newly hatched larvae lie mostly on their side, but later they rest upright with the head pointing downwards. As the yolk sac becomes smaller, the body axis gradually becomes horizontal; the young fish initially lie close to the ground but later support the front of their body with the pectoral fins. Lateral movements of the body and the tail occur in all these positions, producing forward movement along and finally off the substrate. At this stage swimming may be elicited by mechanical and visual stimuli, and the young fish orient away from light, towards the ground and into the current. On emergence from the gravel, young salmonids periodically make sudden vertical movements by flexing their tails and pushing against the

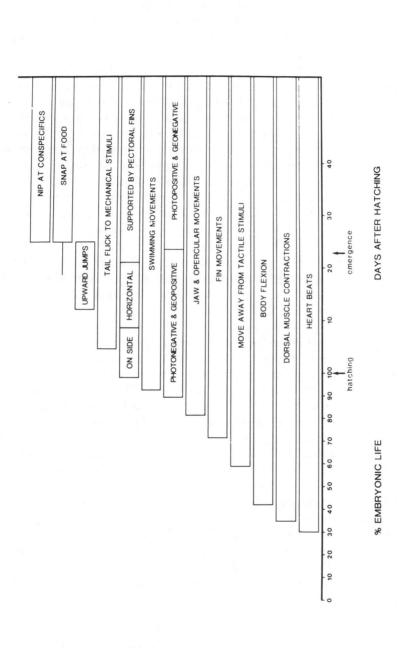

Figure 3.1: A Simplified Representation of Early Morphological and Behavioural Development in the Atlantic Salmon. (Modified from Abu-Gideiri 1966 and Dill 1977.)

substrate. These movements, which bring the young fish out of the gravel and into the stream, coincide with a change to positively phototactic behaviour. Emergence takes place just before the yolk supply is exhausted and when development of a functional gut is complete. Atlantic salmon parr tend to remain in contact with the substrate and to swim into the current. Even before emergence, the young fish will dart and snap at a solid object moved in front of its head; from a few days after leaving the gravel, these biting movements are common. Initially bites are poorly co-ordinated, not very successful and directed indiscriminately at any small solid object; later, they become more effective and directed only at likely food items. The young fish begin to chase and nip at their companions at the same time as they start to feed and with similar movements; the fish that are being chased give escape responses (Dill 1977).

The subsequent behaviour of young salmonids varies from species to species and has been less minutely documented. In the presence of a predator, even very young sockeye (*Oncorhynchus nerka*) and Atlantic salmon parr maintain a minimum distance from the larger fish. Should the predator approach, the parr form schools (see Chapter 12 by Pitcher, this volume) and, if it attacks, they escape rapidly and remain motionless either at the water surface or on the bottom (Ginetz and Larkin 1976; Jacobssen and Jarvik 1976).

In Atlantic salmon and rainbow trout (*Salmo gairdneri*) agonistic inter-actions between young fish become more frequent with age but their form changes; chasing and nipping become less common and are replaced by stereotyped fin displays and head-down postures; these elicit avoidance without overt physical contact (Dill 1977; Cole and Noakes 1980). Agonistic behaviour gradually becomes restricted to specific areas as the fish establish and defend feeding territories (Keenleyside and Yamamoto 1962). As Atlantic salmon parr get older, the size of food they take and the area they need to find it in increases. They leave their territories, move into deeper, more open water, and feed together in a dominance-structured group (Wankowski and Thorpe 1979a).

This may represent the start of the complex physiological and behavioural changes of smolt transformation which some salmonids (Atlantic and coho but not pink salmon (*O. gorbuscha*)) undergo after 1 to 4 years of living and feeding in streams. During smolt transformation, the fish become silvery, change shape and finally abandon their territories. Most smolts then cease to swim into the current and are carried down-stream in schools. This takes them to the sea where they feed and grow for a variable period of time before returning to their home stream (which they recognise by its smell) as mature adults ready to breed (Hasler and Scholz 1983). Some of the larger, faster growing male Atlantic and coho salmon may mature precociously without migrating (Dalley, Andrews and Green 1983; Thorpe, Morgan, Talbot and Miles 1983). Precocious males are usually in a minority, but in some populations 70 to 100 per cent of male parr may be mature. In contrast to normally developing parr, these males retain their tendency to swim into the current and remain on the spawning grounds. They produce viable sperm and surreptitiously fertilise the eggs produced by mature females, but pay the price of reduced body condition and higher mortality rates. (See Chapter 13 by Magurran, this volume.)

The Development of Behaviour in Cichlids

Because they are easy to keep in the laboratory and have a complex and variable repertoire of social responses, the behaviour of cichlid fish has been intensively studied. Much of this work has focused on their reproductive and parental behaviour (see Chapter 10 by Turner, this volume), and it was a natural step to extend these investigations beyond the stage at which the young are deserted by their parents into the relatively short period from independence to adulthood.

The early embryonic stages of cichlids are similar to those of salmonids. The pattern of development of behaviour after hatching in the substrate-brooding orange chromide (*Etroplus maculatus*) is summarised in Figure 3.2. In this species, the young remain in a shoal for their first few weeks of life, close to their parents. The young respond to their parents from the moment of hatching and swim towards them, either nipping at their sides (micronip) or curving their body so that their side contacts the parent briefly (glance). When micronipping, the fry are feeding on mucus secreted by their parents' scales; glancing movements probably stimulate mucus production. In another substrate-brooding cichlid, *Cichlasoma citrinellum*, these responses are initially directed towards both parents, but after a few days the young fish makes contact preferentially with their father (Noakes and Barlow 1973). The frequency of interactions between parents and young decreases up to the time the parents desert; the young then remain in the shoal where social encounters occur, initially during competition for food items. To return to the orange chromide, the distance between the young in the shoal gradually increases, and individuals concentrate their activities in a specific area. By 55 days, larger individuals defend small territories within which they feed. Glance and micronip gradually disappear, to be replaced by a variety of other behaviour patterns.

At about 25 days, young orange chromides begin to charge at each other. A charge is like a glance, but involves harder contact with the other fish. Initially the recipient shows no response, but by day 35 frontal display (see below) and the beginning of a charge are enough to elicit retreat. From about 35 days, three new acts are seen. Ramming, a fierce slap with the side of the body, is first shown in encounters in the shoal, but eventually occurs only in a territorial context where it serves to terminate a fight. Lateral display, a static glance, can lead to carouseling, mutual glance-like movements oriented to the opponent's tail resulting in a circular chase. As

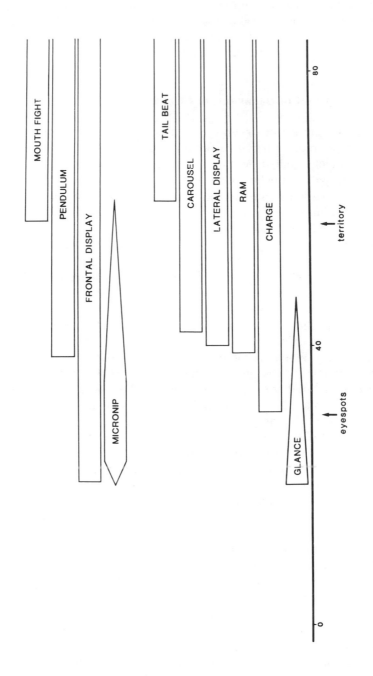

Figure 3.2: Simplified Summary of Behavioural Development in the Orange Chromide (*Etroplus maculatus*); the Occurrence of Ten Behaviours Plotted against Age. (Modified from Wyman and Ward 1973.)

the fish get older, this complicated movement increases in diameter, becomes more fixed in form, and ceases to involve any physical contact. At 64 days, lateral display in territorial disputes may turn into tail beating; initially the body posture is not maintained as the tail moves, but later, vigorous tail movements are performed on the spot.

Other new acts in the repertoire involve a frontal position with the mouth open and median fins elevated; in other words they look like variations on the micronip theme. Frontal display is the earliest to appear, initially being no more than a brief hesitation with fins raised at the end of a micronip. Later, frontal display is used in sibling interactions as a preliminary to charging, and eventually it elicits retreat even when it is performed alone. If the recipient of a frontal display responds with the same behaviour, the two opponents may perform a back-and-forth pendulum movement. Once territories are established frontal display becomes an important component of territorial defence, during which mutual frontal display can give rise to mouth fighting.

Thus, in a period of about 70 days, young orange chromides change from living in a closely packed shoal, interacting with their parents by means of just two behaviour patterns, to maintaining feeding territories by an extensive repertoire of discrete and efficient behaviour patterns, finely tuned to the behaviour of other fish. Some 100 days after the agonistic repertoire is complete, fish that have territories begin to breed. During the complex courtship sequence, movements that are very similar to those already described are used in quite different functional contexts. For example, potential mates glance against and nip at each other, skim and dig with similar movements at the spawning site, and may show long bouts of mutual tail beating (Wyman and Ward 1973).

What Have These Two Examples Shown?

These two studies describe the development of the various functional behavioural systems that we observe in adult fish, namely habitat selection, feeding and anti-predator responses, together with agonistic, sexual and parental behaviour. Together, they illustrate a number of general points:

(1)   Strong, co-ordinated movements emerge gradually from the weak, spasmodic muscle twitches of very young embryos. These early co-ordinated responses themselves change in form and may be replaced by recognisably different acts. In the orange chromide, the component of the glance which is directed towards another fish becomes more marked to produce a charge; holding the movement at the outermost extent of the circle results in lateral display (Figure 3.3). Thus, as the young fish get older, their repertoire is enlarged by emphasising or modifying different components of existing actions

Figure 3.3: Proposed Sequence for the Development of Adult from Juvenile Behaviour Patterns in the Orange Chromide. (Modified from Wyman and Ward 1973 in Huntingford 1984.)

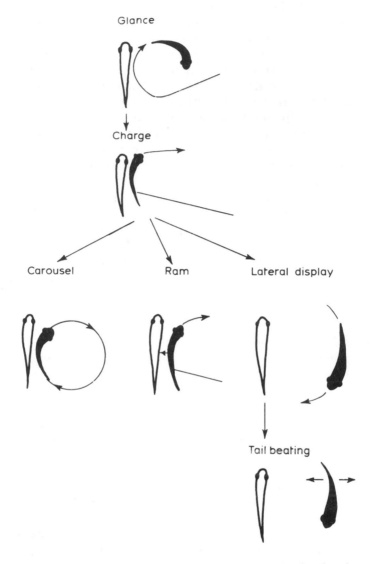

to form a number of new behaviour patterns. In the development of agonistic behaviour, this often results in overtly aggressive acts with physical contact being replaced by stereotyped movements that act at a distance.

(2)   In addition to these changes in form, the systems controlling the various behaviour patterns alter as development progresses. A change in internal control mechanisms might show up, for example, as a shift in the sequences in which they are performed (see Colgan,

Chapter 2, this volume). The frontal display of the young orange chromide initially occurs at the end of a micronip, later reliably precedes a charge, and finally occurs on its own. Some differences in the way these acts are controlled must underlie such changes. The effect of external stimuli may also change with age, as when the initially photonegative response of Atlantic salmon fry becomes photopositive at emergence. In a social context, charge and frontal display elicit no response in a very young cichlid, but both acts can cause an older opponent to retreat.

(3)   In both cichlids and salmonids, periods of rapid behavioural reorganisation are interspersed with periods when change is more gradual. These times of rapid change often correspond to important events such as emergence from the redd, migration to a new habitat, or becoming independent of parents.

(4)   The end result of all these changes is an adult fish with a complete repertoire, but there is more to the process of development than simply constructing functional adults. Much of the behaviour of a young fish serves an immediate and vital function in survival. The complex sequence of responses to light and gravity keeps salmon fry safe in the gravel until they need to come out to feed; micronip and glance collect and maintain a food supply for young cichlids as well as being the precursors of movements used by older fish.

(5)   The course and timing of developmental events are not identical for all members of a species, and the differences between individuals may be the result of the way they behaved at an earlier stage. For example, young salmonids that initially grow faster, perhaps because they collect or compete for food more efficiently, often have a different developmental history from their slower growing companions. They mature precociously, do not migrate, remain on the spawning grounds to breed, and die young (Dalley 1983; Thorpe *et al.* 1983).

What Causes Behaviour to Change during Development?

The previous sections described a number of behavioural changes that occur during development, and have given some answers to the first set of questions spelled out in the introduction. The next stage is to identify and characterise the causes of these changes, to elucidate the factors and processes that control the course of development, and so determine how adult animals behave. From the examples given here, a number of possible explanations can be suggested for the behavioural changes that occur during development. The simplest would be that they come about because the eliciting stimuli in the external environment have altered. On the other hand, behaviour patterns may appear in the repertoire or alter in form because the nervous system or some other structure necessary for their performance reaches a certain stage of maturation, or because physiological conditions within the fish change. Behaviour may change during development because what an animal experiences at one stage alters what happens subsequently. These various possibilities are now discussed in turn. They are not mutually exclusive; indeed, changes in the nervous system or some other aspect of structure or physiology may be the mechanism by which experience modifies the course of development.

Behavioural Changes May Result from Alteration in External Stimuli

The gradual reduction in the frequency of parent-contacting movements that occurs as young cichlids become independent may come about because their parents are producing less mucus. This may actually represent an accelerating process, since parents receiving fewer contacts are stimulated to produce less mucus, hence the rapid nature of the behavioural changes that occur at this time. The appearance of charging movements in cichlids corresponds to the time when their companions start to develop the colour patterns typical of this species. Here we have behavioural shifts that probably occur because of changes in the external stimuli that the environment provides, although of course the reduction in mucus levels and appearance of colour patterns are themselves developmental changes which require explanation.

Behavioural Changes Accompanying Development of the Nervous System

Most studies of development of the nervous system in fish are confined to the embryonic stage, but here clear correlations exist between the state of the nervous system and the behaviour of the developing fish. The very earliest movements of the heart and some of the first twitches in the dorsal musculature occur before any differentiation of the nervous system takes place, and are clearly myogenic in origin. As embryonic life approaches the half-way stage, the major motor systems appear in the spinal cord. These become gradually more extensive, with motor neurones contacting the motor end-plates of the anterior muscles by 55 per cent of embryonic life. It is at this time that trunk flexions first occur, becoming more extensive as more somites gain functional motor connections. By 66 per cent of embryonic life, interneurones appear: the earliest make synaptic contact with motor neurones at different points and on the two sides of the body, allowing the co-ordinated movements which produce undulating swimming. Initially trunk movements are spontaneous, but as the sensory system, which appears about half-way through embryonic life, makes connection with the skin, they can be elicited by tactile stimulation. A conspicuous feature of the developing nervous system at this time is the Mauthner cells, a pair of large interneurones which run from the hindbrain to the motor neurones of the spinal cord. When these are functional, the developing fish is able to produce rapid directed movements in response to and away from mechanical stimulation. When functional connections with the optic midbrain are complete, an event that coincides with the development of rods and cones in the eye, visual control of swimming movements (necessary for effective feeding) is possible. At a late stage in embryonic life, large interneurones can be seen near the base of the pectoral fins, and the trigeminal nerve is formed; these developments may explain why co-ordinated and independent fin and jaw movements appear at this time. Here we have examples of behavioural capacities emerging in parallel with growth of the sensory and neural structures necessary for their performance (Abu-Gideiri 1966; Armstrong and Higgins 1971).

Behavioural Changes Resulting from Non-neural Morphological Changes

The changes in body posture that occur in the period immediately after hatching in Atlantic salmon can be explained very simply in terms of a morphological development, namely shrinkage of the yolk sac. More interestingly, the gradual appearance of co-ordinated movements of the mouth during feeding and respiration can be related to structural developments in the head. The lower jaw of salmonids can be depressed by a number of separate mechanisms, two of which are shown in Figure 3.4. When the *levator arcus palatini* contracts, the roof of the skull swings up and outwards, pulling up the hyoid bone as it does so. The hyoid is attached to the back of the lower jaw (mandible) by a long ligament (the mandibulo-hyoid ligament) so that when the hyoid is pulled up, the lower jaw opens. When the *levator operculum* muscle contracts, this pulls on a series of bones: the opercular, the sub-opercular and the interopercular.

Figure 3.4: Side View of the Skull of a Rainbow Trout, Showing Two Mechanisms for Depressing the Lower Jaw. (Modified from Verraes 1977.) Broken lines represent ligaments; arrows represent bone movements

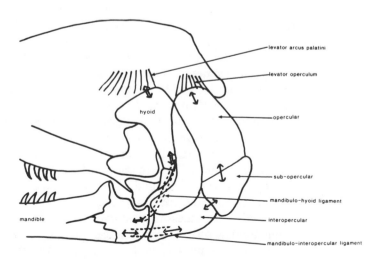

This last bone is also attached to the mandible by a ligament (the mandibulo-interopercular ligament), so that a pull on the interopercular depresses the lower jaw.

The mandibulo-hyoid ligament is present at hatching, when it makes contact at both ends with cartilage. Weak contraction of the developing *levator arcus palatini* occurs long before overt jaw movements are observed, and this is thought to stimulate ossification of the cartilage near the points of attachment of the ligament. Once the bones are fully formed, effective lowering of the jaw, as used to generate a respiratory current, is possible. The mandibulo-interopercular ligament and the opercular bones do not develop until much later, and therefore the second mechanism for lowering the jaw, which plays a major role in ingesting food, does not become active until later, at the time the young fish have depleted their yolk sacs. Thus, although unco-ordinated contractions occur in the developing muscles at an early stage, the two patterns of co-ordinated and effective jaw movement only appear in the repertoire when the relevant system of bone and ligaments is connected up (Verraes 1977). The subsequent increase in the size of particle that young salmon choose to ingest is paralleled by an increase in the size and suction pressure of their gape (Wankowski and Thorpe 1979b).

Behavioural Changes Resulting from Alterations in Non-Neural Physiology

About a month before smolt transformation, coho salmon experience a rapid increase in the blood concentration of the hormone thyroxin. Injection of thyroid-stimulating hormone or thyroid extract into Atlantic and coho parr produces the morphological changes associated with smolting. It also reduces the level of aggressiveness and induces downstream migration. In contrast, levels of the hormone prolactin decrease before smolting, and injection of this substance prevents the physiological changes associated with development of salinity tolerance in smolts (Hasler 1983; Hasler and Scholz 1983). It has been suggested that the resulting changes in osmotic balance reduce swimming activity and thus contribute to the fish's downstream displacement (Thorpe 1982). The appearance of agonistic behaviour in the paradise fish (*Macropodus opercularis*) parallels histological changes in the gonads (Davis and Kessel 1975). These authors suggest that the increase in levels of agonistic behaviour and appearance of reproductive movements in the orange chromide may also correlate with increased gonadal activity, and in particular with increased production of gonadal hormones. In these examples, the changes in behaviour that are observed during development relate to and may well be caused by alterations in the hormonal state of the young animal.

Although young salmonids have access to food particles and are capable of ingesting and digesting food as soon as they emerge from the redd, feeding movements occur infrequently until several days later. Up to this time the yolk is not completely used up and so feeding may not occur because the fish have no need for food. One might therefore ascribe this behavioural shift to a deprivation-induced change in feeding motivation.

Behavioural Changes Caused by Experience during Development

Many of these morphological and physiological changes which cause developmental shifts in behaviour are themselves the result of the circumstances impinging on the young animal. These influences may come from within the animal itself; growth of the bones which allows jaw movements in young salmon occurs under the influence of the early twitches of the relevant muscles; movement and growth deplete the yolk sac, which alters the levels of food in the blood and may make the fish start feeding. Alternatively, developmental changes may be caused by influences external to the animal. Exposure to light is known to influence the way the visual system and visually guided behaviour develop in other groups of animals (see Lund 1978 for mammals; Copp and McKenzie 1984 for amphibians). In eyeless cave fish (*Astyanax spp.*), the visual pathways develop in the absence of any stimulation and contain a severely reduced number of nerve cells. The optic tectum, which in normal fish processes visual information, is effectively deprived of sensory input. This region of the brain undergoes compensatory innervation by neurones associated with the lateral line system; as a result, the blind fish have particularly well developed responses to water movements (Kutz, Lasek and Kaiserman-Abramof 1981; and see Chapter 17 by Parzefall, this volume).

There is a rather special kind of influence impinging on the developing animal from its external environment. This is the experience of particular beneficial or harmful contingencies following performance of a given action or, in other words, learning. The increasingly discriminating response of young salmonids to the various small objects they encounter may come about as a result of a learning process in which getting food rather than a mouthful of inedible rubbish provides reinforcement for snaps aimed at appropriate objects. Equally, since male *Cichlasoma citrinellum* produce more mucus than females, the development in the fry of a preference for the father may come about because they learn to associate the visual cues provided by the male with a superior food supply. The gradual tightening up in the performance of a number of agonistic acts in young orange chromides may be the result of a general improvement in strength and co-ordination, but it could also come about as a result of learning that certain acts are more effective in driving off a rival. After a rainbow trout has experienced frontal display and charging in quick succession on a few occasions, frontal display may become a conditioned stimulus which can elicit retreat on its own (Chiszar and Drake 1975). Here are a number of behavioural changes that may come about through some sort of learning process in the developing animal.

Experimental Studies

Thus we have a number of candidate explanations, which may not be mutually exclusive, for the gradual changes in behaviour that occur as an egg develops into a mature fish. However, since these suggestions are mostly based on correlations, it is not easy to distinguish cause and effect. For example, although young *Cichlasoma citrinellum* may develop a preference for their father because they learn that males provide more mucus, the bias might equally well develop for some other reason and may itself stimulate more copious mucus production in the preferred parent. These two possibilities were distinguished by an experiment; young cichlids raised without the opportunity to contact their parents' bodies fail to develop a preference for their father, which supports the learning hypothesis (Noakes and Barlow 1973). This study illustrates the form that most experimental studies of development take. To find out whether a particular factor is involved in determining the course of behavioural development, we alter the state of that factor in some way; if this changes what the young animals and/or adults do, then we can conclude that the factor is important and start to investigate the nature of its influence on development. If the course of development proceeds unaltered in spite of the experimental manipulation, we can conclude that the particular factor we manipulated was not essential for the normal process of behavioural development.

This conclusion has to be worded carefully, since negative results of deprivation experiments are notoriously hard to interpret (Bateson 1981; Huntingford 1984). Thus it is not easy to be sure that we really have removed a particular environmental influence. Sticklebacks (*Gasterosteus aculeatus*) that have had no experience of a predator show co-ordinated escape responses when they first encounter one; however, during early development, male sticklebacks snap at their young, so unless the sticklebacks are reared as orphans, even predator-naive fish have the experience of being chased. To complicate matters further, the appearance of normal behaviour in subjects which really have been deprived of a particular influence does not prove that it has no effect in normal development, nor does it show that development is independent of *any* environmental influences. With these provisos in mind, however, such experiments can help us to identify and characterise factors that influence the course of behavioural development.

The developing system can be manipulated in two general ways, by altering the genetic constitution of the zygote or by tampering with the environment in which it develops. It is quite clear that differences in genetic make-up can result in behavioural changes in fish (see Chapter 1 by Noakes, this volume). However, there is very little information available about exactly how genes exert their effects on the development of the nervous system and behaviour in fish, or indeed for most other groups of animals, except possibly fruit flies, nematodes and Siamese cats (Partridge

1983). Much more is known about the effects of environmental manipulations, and although fewer studies of this sort have been carried out on fish than on birds and mammals, there is still a substantial literature on this subject.

Effects of Early Experience on the Development of Social Behaviour

Sticklebacks, and cichlids (*Haplochromis burtoni*), reared in total isolation with no experience of the normal adult colour patterns of their species, attack models with a red breast and black eye-bar, respectively, when tested as adults (Cullen 1961; Fernald 1980). In these cases, where the natural stimuli which elicit overt attack are simple and conspicuous, specific experience of these stimuli does not seem to be necessary for the normal development of correctly directed attack. However, experience can modify the strength of the response elicited. For example, male *H. burtoni* reared to maturity with either their own or a closely related species spend more time attacking the one with which they were reared (Crapon de Caprona 1982).

Several studies have shown that most of the major components of the agonistic repertoire appear in isolated fish at the normal age. However, these are often poorly co-ordinated, wrongly oriented and not strung together into normal sequences (Tooker and Miller 1980). Another striking and common result is that fish reared in isolation tend to use overtly aggressive acts, rather than non-contact displays. These findings suggest that whereas interaction with other fish is not necessary for the differentiation of agonistic movements during development, it is critically important if the fish are to perform them effectively and in the correct context, and to respond appropriately to them. Notice that this does not mean that *no* practice is necessary for the normal development of the movements. Young fish in isolation often direct their movements towards inanimate objects in the environment, as when orange chromides glance against stones and nip at objects on the walls of the tank, which may provide the experience necessary for development of other behaviour patterns.

Male *Astatotilapia strigigena* that have been reared in isolation will court models and females of other species, something which normally reared males rarely do (Setz 1940, quoted in Baerends 1971). Similarly, male platyfish (*Xiphophorus maculatum*) reared in isolation from soon after birth perform normal sexual movements but fail to associate preferentially with females, unlike males reared with conspecifics. Isolated males, and males reared with both red and black companions, show no preference when offered a choice between red and black females. In contrast, males housed just after sexual maturity with fish of just one colour prefer females of that colour. The females themselves show no preference for black or red males, regardless of their rearing conditions (Ferno and Sjolander 1973). Here again, whereas experience of conspecifics is not necessary for development of the movements used in sexual interactions, in males it does influence the context in which these movements are used.

Young of some species of cichlid (e.g. *Aspidogramma reitzigi*, Kuenzer 1962, quoted in Baerends 1971) reared in isolation from the egg stage prefer models with the colour pattern typical for their species from the moment of hatching, although this preference becomes stronger if they are allowed contact with their parents. Isolated young of other cichlid species hatch without any preference for the normal parental colour; here, a preference develops over a few weeks' association with the parents. For example, adult *Cichlasoma meeki* have grey bodies and red throats. Young reared in isolation show no preference for red or grey, but over the first 3 weeks of their life young kept with their parents gradually become less responsive to all but red and grey models (Baerends and Bearends-van Roon 1950).

There is some contradictory evidence that these early filial preferences may influence later mate selection; in platyfish, as described earlier, experience at a later stage (closer to maturity) influences choice of mating partners. Young *H. burtoni* were reared to maturity and subsequently offered a choice between adults of the two species, with either visual or chemical cues available. Given visual cues, males prefer females of their own species, regardless of previous experience. Males given a choice based on chemical cues, and females given visual cues, prefer potential mates of the species with which they have been reared (Crapon de Caprona 1982). In that early experience results in filial preferences and later experience may influence mate choice, these examples clearly have something in common with the filial and sexual imprinting for which young precocial birds are famous (Bateson 1983, and see Chapter 106 by Turner, this volume). However, in the case of the response of cichlids to their parents, there is no clear evidence that learning fails to occur after a certain age.

Studies of the Development of Behaviour in Salmonids

Although hatchery rearing is not a controlled isolation experiment, it does deprive the fish of particular aspects of the normal developmental environment, including the presence of predators. Comparing the behaviour of hatchery-reared and wild salmon may therefore throw some light on the role of these factors in the normal development of their behaviour. For example, hatchery-reared Atlantic and sockeye salmon show the full repertoire of responses to a fast-moving predator, keeping their distance, forming shoals, and jumping away when attacked. In hatchery sockeyes, these responses are less well developed than in wild-caught fish, and in hatchery Atlantic salmon no escape response is given to a slow-moving predator. Nevertheless, a brief encounter with a predator is sufficient to strengthen the response (Ginetz and Larkin 1976; Jacobssen and Jarvik 1976).

Hatchery-reared fish released indiscriminately make poor returns following migration, partly because just before smolting the young fish rapidly learn the smell of the stream in which they live: after smolting, such learning does not occur. The memory is extremely long lived, being used several years later when the fish return to the stream to breed. Here again, a memory of a particular constellation of stimuli is formed at a specific time in the animal's life, and this influences preferences at a later date: this phenomenon is called olfactory imprinting. The period in which this normally occurs corresponds to the presmolting surge in thyroid hormone levels, and the capacity for olfactory imprinting can be artificially advanced by injections of TSH (thyroid-stimulating hormone: Hasler and Scholz 1983). Thyroid hormones have been implicated in the process of imprinting in chicks (Chandrasekhar, Moskovin and Mitskevich 1979), are taken up by the brain of newly smolted steelhead trout (migratory *Salmo gairdneri*), (Scholz, White, Muzi and Smith 1985), and facilitate learning in rainbow trout at smolting (Hasler and Scholz, 1983). Thus, thyroid hormones are clearly implicated in the process of olfactory imprinting in salmonids, although their precise role remains to be elucidated. (See also Chapter 6 by Hara, this volume.)

What Have these Experimental Studies Shown?

It is clear from the examples discussed so far that a variety of different factors, both genetic and environmental in origin, determine the course of development of any particular item of behaviour. Their effects may range from the extremely specific (olfactory imprinting causes a certain chemical to trigger a particular pattern of orientation in homing salmon) to the more general (experience of light alters a range of visually guided behaviour). In addition, these influences, genetic or environmental, with general or specific consequences, can work in all sorts of different ways.

It may be helpful to distinguish between inducing, facilitating, maintaining and predisposing factors in development (Table 3.1 and see Bateson 1983). Inducing factors are drastic influences which switch development from one path to another (e.g. exposure to different chemicals produces salmon with very different migration patterns); facilitating factors speed up processes which will occur in their own time anyway (e.g. naive salmon shoal, but experience of a predator strengthens this response), and maintaining factors preserve the status quo (e.g. once a platyfish develops a preference for females of a particular colour, periodic exposure to such females prevents this preference from waning). Finally, predisposing (or enabling) factors do not themselves bring about a particular change in behaviour but allow another, later, influence to do so (e.g. high levels of thyroxin do not cause salmon to remember a smell but they are a necessary prerequisite for this learning to occur). These different types of influence grade into one another, and at the end of the day each case has to be analysed in its own right. However, Bateson's taxonomy stresses the diver-

Table 3.1: A Three-way Classification of the Kinds of Process that Can Influence the Way Behaviour Develops. (After Bateson 1978, 1983.)

ORIGIN OF INFLUENCE	NATURE OF INFLUENCE	NATURE OF CONSEQUENCES
	Inducing	Specific ⟷ General
	Facilitating	Specific ⟷ General
Genome	Maintaining	Specific ⟷ General
	Enabling	Specific ⟷ General
	Inducing	Specific ⟷ General
	Facilitating	Specific ⟷ General
Environment	Maintaining	Specific ⟷ General
	Enabling	Specific ⟷ General

sity of factors controlling behavioural development and may help us to identify features of development shared by different behavioural systems and by different species.

Not all things are learned equally well by all individuals. For example, male platyfish learn from the appearance of their companions the characteristics of their future sexual partners; this experience does not influence which fish they attack, nor does similar experience influence what females do. Similarly, male *Haplochromis burtoni* learn the smell of potential mates, whereas females learn what they look like (Crapon de Caprona 1982). Rainbow trout may be predisposed to associate the visual cues provided by an approaching conspecific with the painful physical consequences of receiving the nip that normally follows such an approach (Chiszar and Drake 1975). This would be a constraint on the cues that can be associated with reinforcement, similar to that described for rats by Garcia, Ervin and Koelling (1966).

Nor are things learned equally well at all times. There are periods when the developing animal is particularly susceptible to the influence of environmental factors: for example, just after birth when all stimuli are being experienced for the first time in some young cichlids, and just before smoltification when the natal environment has proved itself a good one in salmon. Sensitive periods are clearly characteristic of behavioural development in fish just as they are in birds and mammals; in salmon, at least, the time course of the sensitive period has been characterised. In precocial birds, development of a preference for one object biases the animal against making contact with any others, thus bringing the sensitive period to an end (Bateson 1983). A similar process may well be at work in young cichlids but it is not at all clear what ends the sensitive period for olfactory imprinting in salmon. The developmental processes which underlie such sensitive periods are probably not different from those that act at other times. However, their effects (the formation of preferences which cut the developing animal off from further experience of the same kind) provide particularly dramatic examples of early experiental influences on subsequent development. In addition, just because developmental events occur so rapidly at these times, they may provide valuable insights into the control of behavioural development.

These various kinds of constraint on learning (Hinde and Stevenson-Hinde 1973) are part of a broader phenomenon, the stability of many developing systems. Whereas some behaviour patterns are profoundly influenced by specific experiences, others develop to more or less the same end point in a broad range of environmental conditions. Where and on what a salmon chooses to feed is influenced throughout its life by continually shifting patterns of food availability; in contrast, what a male stickleback will attack is not altered even in the profoundly abnormal rearing condition of complete social isolation. In extreme cases, there almost seems to be an element of self-regulation in the system, with animals seeking out or providing themselves with stimuli when the normal source of these is denied them. Thus isolated orange chromides perform glancing movements against inanimate objects, and do so at a very high frequency (Wyman and Ward 1973). This may provide enough experience for some degree of differentiation of the agonistic repertoire to occur.

Instincts and Learning

It is this last kind of phenomenon, the impressive resistance of many behaviour patterns to developmental perturbation, that gave rise to the concept of an instinct (or 'innate' behaviour). Thus, in the early ethological literature, a dichotomous classification of behaviour patterns was made into learned responses as opposed to instincts. An instinct develops into its complete, species-specific form under the influence of the genome and without the need for specific environmental contingencies. This dichotomous classification has been criticised on many occasions (Hinde 1970; Lehrman 1970; Bateson 1981) and so far in this chapter the term 'instinct' has been avoided deliberately. When analysis of behavioural development takes the form of careful longitudinal studies of exactly what happens at each stage of development (and the literature on fish behaviour is particularly rich in such accounts), attention is correctly focused on the processes of development as well as its outcomes. It then becomes quite obvious that a particular item of behaviour develops under the continually interacting influence of maturational events initiated by the genome and various aspects of the external environment. The concept of an instinctive behaviour pattern which develops under the influence of the genes alone does not seem very helpful.

One can trace the movements of ramming in cichlids back to the glancing movements of the newly hatched young, through the body movements of the embryo to the earliest twitches of the trunk muscles. Practising against another fish or an inanimate object may cause the movements of a glance to be differentiated into a ram, and social experience allows ramming to be integrated into an effective agonistic repertoire. Once we know all this, it is clearly meaningless and unnecessary to classify ramming as either learned or instinctive. To use Hebb's analogy (1953), does the area of a rectangle depend on its length or its breadth?

Two useful things remain from the original dichotomy. In the first place, we can assign *differences* in behaviour to genetic or environmental effects. Notwithstanding the complex developmental history of ramming, one cichlid may show more of it than another, either because it inherited from its parents certain alleles that enhance aggressive motivation, or because by chance it won its first fight; of course, both of these may be the case. Secondly, as we have pointed out, the development of certain behaviour patterns does occur in a wonderfully stable way and it is this that has led some ethologists to retain the concept of an innate behaviour pattern (Alcock 1984). Such stability may come about for a number of different reasons. For example, development of a particular behaviour may depend on a given environmental influence, but circumstances are such that this influence is reliably available; in real life, most young cichlids grow up in a world containing companions. Even if the environment is unpredictable, development may be stable if the animal is predisposed to respond to or learn about a limited range of stimuli. In addition, development may be resistant to disruption of a critical aspect of its developmental environment because other environmental features can produce the same effect. At its most extreme, this may involve the developing animal compensating in some way for deficiencies in its environment. Whichever is the case, the resulting developmental stability is remarkable and requires explanation. I feel that labelling such behaviour patterns 'instinctive' does not help, and can hinder, progress towards an understanding of how they develop.

Summary

As a fertilised egg develops into an adult fish, weak, unco-ordinated muscle twitches are replaced by efficient, co-ordinated behaviour patterns. These may themselves subsequently be modified in form, in causation, and/or in function, until they give rise to the full behavioural repertoire of the adult animal. Observational and experimental studies show that these developmental changes are the result of a continuous interaction between maturational processes within the developing animal and various aspects of its external environment. Although complex, this interaction is amenable to experimental investigation. Such studies have shown that many behaviour patterns in fish develop with remarkable stability in spite of environmental perturbation. However, the concept of instinctive behaviour is of limited value in explaining such stability.

Acknowledgements

I would like to thank Neil Metcalfe, John Thorpe and a referee for helpful comments on an earlier draft of this chapter.

References

Abu-Gideiri Y.B. (1966) 'The Behaviour and Neuroanatomy of Some Developing Teleost Fishes', *Journal of Zoology*, *149*, 215-41

Alcock, J. (1984) *Animal Behaviour*, 3rd edition, Sinauer, Sunderland, Mass.

Armstrong, P.B. and Higgins, D.C. (1971) 'Behavioural Encephalisation in the Bullhead Embryo and its Neuroanatomy', *Journal of Comparative Neurology*, *143*, 371-84

Baerends, G.P. (1971) 'The Ethological Analysis of Fish Behaviour', in W.J. Hoar and D.J. Randall (eds), *Fish Physiology*, vol. 6, Academic Press, New York, pp. 279-370

Baerends, G.P. and Baerends-van Roon, J. (1950) 'An Introduction to the Study of the Ethology of Cichlid Fishes', *Behaviour Supplement*, *1*, 1-242

Bateson, P.P.G. (1978) 'How Does Behaviour Develop?', in P.P.G. Bateson and P.H. Klopfer (eds), *Perspectives in Ethology*, vol. 3, Plenum Press. New York, pp. 55-66

Bateson, P.P.G. (1981) 'Ontogeny of Behaviour', *British Medical Bulletin*, *37*, 159-64

Bateson, P.P.G. (1983) 'Genes, Environment and the Development of Behaviour', in T.R. Halliday and P.J.B. Slater (eds), *Animal Behaviour*, vol 3, Blackwell, Oxford, pp. 52-81

Chandrasekhar, K., Moskovin, G.N. and Mitskerich, M.S. (1979) 'Effect of Methylthiouracil and Triiodothyronine on Development of the Central Nervous System in Chick Embryos', *General and Comparative Endocrinology*, *37*, 6-14

Chiszar, D. and Drake, R.W. (1975) 'Aggressive Behaviour in Rainbow Trout (*Salmo gairdneri* Richardson) of Two Ages', *Behavioural Biology*, *13*, 425-31

Cole, K.S. and Noakes, D.L. (1980) 'Development of Early Social Behaviour of Rainbow Trout, *Salmo gairdneri*', *Behavioural Processes*, *6*, 97-112

Copp, S. and McKenzie, R.L. (1984) 'Effects of Light Deprivation on Development of Photo-positive Behaviour' in *Xenopus laevis* tadpodes', *Journal of Experimental Zoology*, *23*, 219-227

Crapon de Caprona, M.D. (1982) 'The Influence of Early Experience on Preferences for Optical and Chemical Cues Produced by Both Sexes in the Cichlid Fish *Haplochromis burtoni*', *Zeitschrift für Tierpsychologie*, *58*, 329-61

Cullen, E. (1961) 'The Effect of Isolation from the Father on the Behaviour of Male Three-spined Sticklebacks to Models', *USAFRDC, Final Report, Qutr. Af.*, 61(052)-29, 1-23

Dalley, E.L., Andrews, C.W. and Green, J.M. (1983) 'Precocious Male Atlantic Salmon Parr (*Salmo salar*) in Insular Newfoundland', *Canadian Journal of Fisheries and Aquatic Science*, *40*, 647-52

Davis, R.E. and Kessel, J. (1975) 'The Ontogeny of Agonistic Behaviour and the Onset of Sexual Maturation in the Paradise Fish, *Macropodus opercularis* (Linnaeus)', *Behavioural Biology*, *14*, 31-9

Dawkins, M.S. (1983) 'The Organisation of Motor Patterns', in T.R. Halliday and P.J.B. (Slater (eds) *Animal Behaviour*, vol. 1, Blackwell, Oxford, pp. 175-99

Dill, P.A. (1977) 'Development of Behaviour in Alevins of Atlantic Salmon, *Salmo salar*, and Rainbow Trout, *S. gairdneri*', *Animal Behaviour*, *25*, 116-21

Fernald, R.D. (1980) 'Response of Male Cichlid Fish, *Haplochromis burtoni*. Reared in Isolation towards Models of Conspecifics', *Zeitschrift für Tierpsychologie*, *54*, 85-93

Ferno, A. and Sjolander, S. (1973) 'Some Imprinting Experiments on Sexual Preferences for Colour Variants in the Platyfish (*Xiphophorus maculatus*)', *Zeitschrift für Tierpsychologie*, *33*, 418-23

Garcia, J., Ervin, F.R. and Koelling, R.A. (1966) 'Learning with Prolonged Delay of Reinforcement', *Psychonomic Science*, *51*, 121-2

Ginetz, R.M. and Larkin, P.A. (1976) 'Factors Affecting Rainbow Trout (*Salmo gairdneri*)

Predation of Migrant Fry of Sockeye Salmon (*Oncorhynchus nerka*)', *Journal of the Fisheries Research Board of Canada, 33,* 19-24

Hasler, A.D. (1983) 'Synthetic Chemicals and the Homing of Salmon', in J.C. Rankin, T.J. Pitcher and R.T. Duggan (eds), *Control Processes in Fish Behaviour,* Croom Helm, London, pp. 103-13

Hasler, A.D. and Scholz, A.T. (1983) *Olfactory Imprinting and Homing in Salmon,* Springer-Verlag, Berlin

Hebb, D.O. (1953) 'Heredity and Environment in Animal Behaviour', *British Journal of Animal Behaviour, 11,* 43-7

Hinde, R.A. (1970) *Animal Behaviour,* McGraw-Hill, New York

Hinde, R.A. and Stevenson-Hinde, J. (1973) *Constraints on Learning,* Academic Press, London

Huntingford, F.A. (1984) *The Study of Animal Behaviour,* Chapman & Hall, London

Jacobssen, S. and Jarvik, T. (1976) 'Anti-predator Behaviour of Two-year-old Hatchery-reared Atlantic Salmon (*Salmo salar*) and a Description of the Predatory Behaviour of Burbot (*Lota lota*)', *Zool. Revy, 38*(3), 57-70

Keenleyside, M.H. and Yamamoto, F.T. (1962) 'Territorial Behaviour of Juvenile Atlantic Salmon (*Salmo salar*)', *Behaviour, 19,* 139-69

Kutz, M.I., Lasek, R.I. and Kaiserman-Abramof, I.R. (1981) 'Ontophyletics of the Nervous System: Eyeless Mutants Illustrate how Ontogenetic Buffer Mechanisms Channel Evolution', *Proceedings of the National Academy of Science (USA), 78,* 397-401

Lehrman, D.S. (1970) 'Semantic and Conceptual Issues in the Nature-Nurture Problem', in L.R. Aronson, E. Tobach, D.S. Lehrman and J.S. Rosenblatt (eds), *Development and Evolution of Behaviour,* W.H. Freeman, San Francisco, pp. 17-52

Lund, R.D. (1978) *Development and Plasticity of the Brain,* Oxford University Press, New York

Noakes, D.L.G. (1978) 'Ontogeny of Behaviour in Fishes: a Survey and Suggestions', in G.M. Burghardt and M. Bekoff (eds), *The Development of Behaviour,* Garland, New York, pp. 103-25

Noakes, D.L.G. and Barlow, G.W. (1973) 'Ontogeny of Parent-contacting in Young *Cichlasoma citrinellum*', *Behaviour, 46,* 221-57

Partridge, L. (1983) 'Genetics and Behaviour', in T.R. Halliday and P.J.B. Slater, *Animal Behaviour,* vol. 3, Blackwells, Oxford, pp. 11-57

Scholz, A.T., White, R.J., Muzi, M. and Smith, T. (1985) 'Radioactive T3 in Developing Trout Brains', *Aquaculture* (in press)

Thorpe, J.E. (1982) 'Migration in Salmonids, with Special Reference to Juvenile Movements in Freshwater', in E.L. Brannon and E.O. Salo (eds), *Proceedings of the Salmon and Trout Migratory Behaviour Symposium,* School of Fisheries, Unversity of Washington, pp. 86-97

Thorpe, J.E., Morgan, R.I.G. Tablot, C. and Miles, M.S. (1983) 'Inheritance of Developmental Rates in Atlantic Salmon', *Salmo salar* L.', *Aquaculture, 33,* 119-28

Tooker, C.P. and Miller R.J. (1980) 'The Ontogeny of Agonistic Behaviour in the Blue Gourami (*Trichogaster trichogaster*)', *Animal Behaviour, 28,* 973-88

Verraes, W. (1977) 'Postembryonic Ontogeny and Functional Anatomy of the Ligamentum Mandibulo-hyodeum and the Ligamentum Interoperculo-mandibular with Notes on the Opercular Bones and Some other Cranial Elements in *Salmo gairdneri*', *Journal of Morphology, 151,* 111-20

Wankowski, J.W.J. and Thorpe, J.E. (1979a) 'Spatial Distribution and Feeding in Atlantic Salmon, *Salmo salar,* Juveniles', *Journal of Fish Biology, 14,* 239-47

Wankowski, J.W.J. and Thorpe, J.E. (1979b) 'The Role of Food Particle Size in the Growth of Atlantic Salmon', *Journal of Fish Biology, 14,* 351-70

Wyman, R.L. and Ward, J.A. (1973) 'The Development of Behaviour in the Cichlid Fish *Etroplus maculatus*', *Zeitschrift für Tierpsvchologie, 33,* 461-91

PART TWO: SENSORY MODALITIES

INTRODUCTION

T.J. Pitcher

Accurate and up-to-date information about the environment is crucial to an animal's success in taking optimal decisions about food, predators or mates. The early ethologists recognised the importance of such sensory inputs to behaviour and carried out the extensive investigations to show precisely which stimuli triggered appropriate behaviours. We now know a lot more about how sense organs and sensory analysis works to detect signals from other fish and the fish's habitat. In teleost fish, information from the environment about sound, light, water movements and water-borne molecules is transduced into nervous messages by various derivatives of the marvellously versatile neuromast cells, evolutionary modifications of cilia. To some extent information is filtered and analysed in the sense organs, but it is then passed to the CNS for more sophisticated signal processing. Behavioural decisions, moderated by motivational state, can then be rapidly made. The four chapters in this section of the book cover the main sensory modalities with which teleost fishes operate, describe how they work to extract significant cues from a fish's environment, and discuss the roles which information provided by each sense plays in the behaviour and survival of the fish.

Chapter 4 reviews the role of vision, perhaps the major sense of teleost fish. Simon Guthrie begins with an analysis of how the characteristics of natural waters affect vision and the kinds of objects which fish can detect both above and in the water. He moves on to discuss how the trade-off between conspicuousness and camouflage is affected by spectral shift. A description of the teleost visual hardware follows, from the eye, lens and accommodation, retinal rods and cones, to the optic tectum of the midbrain. Unlike mammals, which have an optic lobe in the cerebral cortex of the forebrain, in fish the tectum is the major visual analysis area in the brain. It seems to work in a closely analogous way, however, and Guthrie describes some of his own work on feature-extracting cells. The link between ecology and the pigments in retinal cones is briefly reviewed, along with work on colour and movement detection by the visual system. Visual discrimination abilities are dealt with next, the upper contours of objects turning out to be the most important for shape discrimination. The author then discusses recent experiments on visually mediated behaviours in shoaling, mating and territorial behaviours, including the 'poster' colours of coral-reef fishes, colour and movement signals in courtship, predation and predator recognition. Guthrie concludes that greater understanding of the higher levels of image processing in deeper tectal layers, along with an elucidation of the precise roles of the elements of visual signals used by fish, are two important goals for the future.

Chapter 5 begins by noting Isaac Walton's advice to anglers to be silent 'lest they be heard' by their quarry. In this chapter, Tony Hawkins covers the role of sound in fish behaviour, a field which until recently suffered from neglect in relation to other sensory modalities. After setting out the basic physics of sound under water, Hawkins distinguishes between water turbulence and sound, and between effects near and distant from the source, and describes the instruments required to measure sound pressure (volume), frequency (pitch), particle displacement and temporal structure. Sound can travel vast distances under water, faster than in air, and many underwater habitats are inherently noisy. A salutary warning about the acoustic design of tanks used for experiments in fish behaviour is given. Hawkins describes, with the aid of some beautiful drawings, the range of mechanisms by which teleosts make sounds. He then covers, with examples, the functions of acoustic communication in alarm, mating, shoaling behaviour, species and individual recognition. Shoaling fish stay silent to avoid the attention of predators: the interception of damselfish and haddock mating calls obeys the same rules as the familiar examples in toads. Hearing ability in teleosts is well developed and, as in birds, the temporal structure of sound is resolved better than in man, although the range of frequencies is narrower. Cod, goldfish and salmon can tune an auditory filter to any frequency of interest to help distinguish a signal from background noise. The anatomy and physiology of the sound-reception hardware is described, including details of the hair-cell transducers (modified cilia), the otolith organs (there is no mammalian cochlea-like organ), and acoustic amplifiers in the form of swimbladder and bone modifications. Hawkins closes with a discussion of the recent finding that teleosts have accurate directional hearing.

Toshiaki Hara reviews the role of olfaction in fish behaviour in Chapter 6, distinguishing at the outset between senses of taste and smell. Fish are remarkably sensitive to water-borne molecules, but solubility is more significant in aquatic habitats than the volatility important for air-borne olfaction. Hara describes the teleost olfactory system. The paired nasal pits contain folded lamellae bearing olfactory receptor cells on which amino acids bind at specific sites. Likewise derived from cilia, but with fewer associated structures than in auditory, visual or lateral line receptors, the olfactory neuromasts have direct connections with the forebrain. Hara follows the 'orderly sequence' of molecular, membrane and neural events: coded spike-trains of nervous impulses convey olfactory information to the decision-making CNS. Amino acids are detected in tiny concentrations, as low as 0.1 nmol litre^{-1}. The chapter continues with a critical account of olfaction and the classic experiments on homing in salmonids. Hara argues that the 'homestream odour' and 'juvenile pheromone' hypotheses are not incompatible, and that there are still some questions to be answered about reproductive homing and odour imprinting. Olfaction in feeding behaviour is described next, along with the fascinating topic of fish feeding stimulants and attractants. Anglers may note that 'fingerprint' mixtures of straight-chain amino acids are the most effective. In reproductive behaviour, steroid glucuronides released at ovulation act as pheromones keying courtship and mating, but not a single fish pheromone has been fully identified chemically. Hara concludes with an account of the fright substances in cyprinid fishes, leaving us with an intriguing problem in the evolution of this behaviour.

In Chapter 7, Horst Bleckmann describes a teleost sense of which humans do not have direct experience: the lateral line, which is sensitive to minute water movements and acts as a close-range object detector. Some lateral lines have become specialised as detectors of electric fields, perfected in electric communication in some teleosts. Bleckmann opens by reviewing the historical controversy about the lateral line, which was variously thought to be for touch, pressure, hearing, balance, taste, mucus secretion and swimbladder gas production. The action of the neuromast transducer system is described, including specialised versions found in several groups of fishes. Sensitivity is reduced by efferent nerves during the fish's own power strokes. The system responds to near-field water displacements produced by a sound source and to tiny water currents set up by the fish's own motion which are reflected from stationary objects. Bleckmann reviews a contemporary controversy over near- and far-field effects. The author's own investigations of prey-spotting by fish feeding at the 'deadly clinging trap' of the water surface are covered next: using their lateral lines, fish such as topminnows can accurately compute targets from the curvature of surface-wave fronts and other signal characteristics. The role of the lateral line in reproductive, agonistic and shoaling behaviour is discussed, and Bleckmann moves on to describe the lateral line organs specialised for electroreception in at least five groups of teleost fishes. He reviews passive electroreception and geomagnetic navigation, and concludes with a section on the fascinating subject of communication of electric teleosts.

There is, unfortunately, not room to discuss 'minor' senses such as touch or temperature detection (see Murray 1971), or the recently discovered and still controversial magnetic sense in fishes. For an entry into the magnetic-sense literature, see Walker (1984) and Quinn and Groot (1983).

References

Murray, R.W. (1971) 'Temperature Receptors', in W.S. Hoar and D.J. Randall (eds), *Fish Physiology, vol. 5*, Academic Press, New York, Chapter 5

Quinn, T.P. and Groot C. (1983) 'Orientation of Coho Salmon as Affected by Internal and External Magnetic Fields', *Canadian Journal of Fisheries and Aquatic Sciences, 40*, 1598-1606

Walker, M.W. (1984) 'A Candidate Magnetic Sense Organ in the Yellowfin Tuna', *Science, 224*, 751-2

4 ROLE OF VISION IN FISH BEHAVIOUR

D.M. Guthrie

Despite the generally poor quality of underwater images, fish depend a great deal on vision as a source of sensory information. All but a few (mainly cave-dwelling species) have well-developed eyes, and in those forms that inhabit clear-water environments, the variety of colour patterns and specific movements that they display invites comparison between them and the most visually oriented species among birds and mammals. Because of the physical nature of light and its complex interactions with the environment, a variety of different properties of visible objects can be recognised, differing either in type (i.e. brightness, hue, texture, contour, etc.), or degree (such as patch size or pattern grain). Comparative properties such as colour contrast or brightness contrast can also be identified. The extent to which particular visual properties are important depends on (a) the type of visually mediated behaviour, and (b) the restrictions to visual signalling imposed by the aquatic medium.

Visually mediated behaviour ranges in complexity from the simple alerting or attentive state evoked by any novel but non-specific visual event, to the triggering of an elaborate fixed action pattern by means of a highly specific visual signal which consists of a precise grouping of visual properties. The structure of the visual signal enables it to remain effective even when a high concentration of suspended matter in the water alters the perceived properties of hue and texture. At the same time, the more complex signals must require more specialised processing by the nervous system and longer reaction times. It is perhaps slightly ironic that most of the waters of very high optical purity inhabited by fishes are so lacking in nutrients that they support only a meagre fauna. Wastwater in Cumbria is perhaps an example of such a water in the UK.

The Optical Properties of Natural Waters and their Effects on Underwater Visibility

(1) Generalised Habitats

In a few oceanic waters and a number of dystrophic lakes there is very little suspended matter. Solar radiation can penetrate to considerable depths (below 100 m) but ultraviolet and long wavelengths are attenuated by the water molecules, resulting in maximum transmission being between 400 and 500 nm. In consequence these waters have a bluish appearance.

Most waters contain appreciable quantities of mineral and organic matter (especially chlorophyll), which absorbs most of the light at depths of

75

25 m or so and narrows and shifts the spectral composition to a band between 500 and 600 nm, giving a greenish or yellowish cast to the water. A majority of coastal waters, lowland (oligotrophic and eutrophic) ponds, and rivers fall into this category. Less common are those waters, mainly fresh, carrying high concentrations of silt, or plant breakdown products which give them a brownish or reddish appearance. Here little light penetrates below 3 m, and this is mostly at wavelengths above 600 nm. The so-called 'black' (Muntz 1973) or 'infra-red' (Levine, Lobel and MacNichol 1980) waters are typified by some of the tropical fresh waters in South America. These are heavily stained but transparent rivers, usually of high acidity, and are contrasted with uncoloured 'white' waters, which may form part of the same river system, as in the case of the Rio Negro. Heavily peat-stained tarns in Northern Britain (e.g. Wise Een Tarn, Cumbria) provide a somewhat similar example. Opaque reddish waters are found where red clays form part of the watershed (Pahang and Kalang Rivers, Malaysia). Clearly, fish living in waters that approximate to these types operate under special visual conditions.

(2) Lines of Sight (see Figure 4.1)

As Levine *et al.* (1980) have pointed out, the amount and spectral quality of light entering a fish's eye will differ according to the line of sight involved. The overhead view will involve the shortest optical path length and thus luminance attenuation and spectral shift will be least. Further, objects overhead above the water surface, within a solid angle of 97°, can be viewed (Snell's window). Horizontal lines of sight will be subject to lower light levels and greater spectral shifts due to the longer optical path involved. Objects viewed in this direction will have their apparent contrast reduced by veiling light scattered from particles between eye and object. Further, as the object recedes from the eye, the effect of the veiling light becomes greater, the spectral reflectance of the object is increasingly shifted towards that of the dominant spacelight, and object brightness approximates to background. The downward line of sight from the eye involves the longest optical pathway, with the greatest effects on both the brightness and spectral properties of objects viewed. In shallow water, most of the visual background is provided by the bottom of the pond or river.

(3) Temporal and Spatial Changes in Underwater Visibility in Natural Waters

Stratification in optical properties is perhaps most noticeable in fresh waters where a seasonal cycle occurs. The most well known of these changes involves the warm-water layer that forms in lakes and ponds as sunlight hours increase in spring and early summer. The separation of a lighter surface layer (the epilimnion) due to the action of the sun produces a boundary layer across which the temperature changes abruptly. In large lakes the zone of greatest change (thermocline) occurs between 5 m and

Figure 4.1: Pathways of Light Entering a Fish's Eye. Arrows indicate from the left: a — direct overhead path (high intensity, daylight spectrum); b — long path scattered from bottom (low intensity, modified spectrum); c — light scattered from suspended particles (veiling light); d — light scattered from submerged object; e — path as in d, but light largely absorbed by suspended matter and therefore not reaching the observer; f — as c but longer light path (spectral shift greater, brightness less); g — as b but shorter path reduces the spectral shift produced by differential absorption and reflectance; h — as g and b but total reflectance from water surface has occurred, and spectral shift greater due to long path through turbid epilimnion. Stippled areas indicate epilimnion rich in organic particles at surface, and zone of humic particles and 'Gelbstoff' near bottom. Dotted line indicates limits of total reflectance (Snell's window)

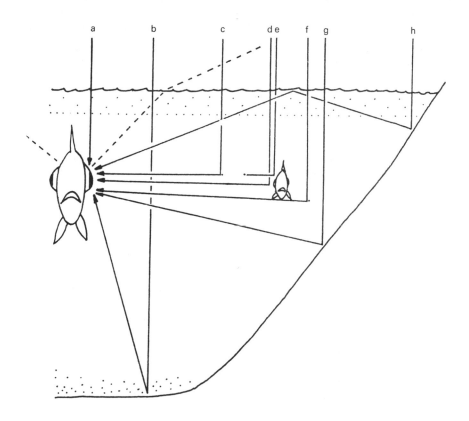

20 m, but in small, shallow ponds the transition zone may be less than 1 m from the surface. The change in temperature between surface waters and the hypolimnion is of the order of 10°C. In the four examples of Austrian lakes given by Ruttner (1952), the values are 12-14°C. In eutrophic lakes the hypolimnion becomes depleted of oxygen, relative to the surface layers (< 50 per cent). These differences are due to a rapid growth of planktonic organisms in the surface layers, with a consequent increase in turbidity. As long as wind and wave action are limited, stratification is maintained.

Thermal stratification may also be associated with the segregation of particles that accumulate from other sources. In Green Lake, New York (Brunskill 1970) significant concentrations of particulate calcite $(1.5\,\mathrm{mg\,litre^{-1}})$ are found near the surface (5 m) in June, which later disperse. At its peak there is six times as much calcite at 5 m as at 45 m.

Although the thermal stratification in marine habitats is less precise than it is in fresh waters during settled midsummer weather, the formation of a layer of less dense water containing higher concentrations of particles is very general. Over the deep bottom formations of the Nittinat Fan off Vancouver Island, high turbidity layers (scattering index over 7 units, range 0-9 units) were found from the surface down to 100-200 m by Baker, Sternberg and MacManus (1974), but the light-scattering depth profile was complex (see Figure 4.2). At one site, for instance, there was an isolated band of light-scattering values at about 650 m. The horizontal light-scattering patterns are also complex because of water movements and underwater contours. At the edge of the Oregon coast, Zaneveld (1974) found that the surface zone of relatively turbid water turned downwards over the shore to a depth of 50-100 m. The clearest water was some way off shore. A turbid shoreline zone caused by water movements can also be

Figure 4.2: Spatial Variations in Water Turbidity Affecting Underwater Visibility. Turbidity measurements of the deep ocean layers off Vancouver Island, Canada. The light-scattering profile was obtained by means of a nephelometer with an integral light source. Note strongly scattering layers at surface and near the bottom, and distinctive mid-depth layers. Light scattering values are arbitrary units, but 1 unit equals approximately $9\,\mu\mathrm{g\,litre^{-1}}$ solids

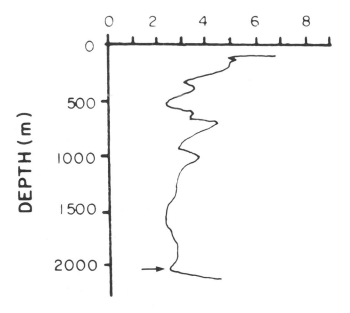

Source: Baker *et al.* (1974).

observed in freshwater lakes because of wave action.

Striking anomalies of water density are sometimes observed by divers. Limbaugh and Rechnitzer (1951) found 'mirror pools' of cold, high-density water off the Californian coast capable of reflecting surrounding objects, as well as moving tongues of cold water 60f (1.83m) offshore. They point out that these thermal discontinuities affect underwater visibility as a result of (a) differences in turbidity, (b) differences in refractive index, and (c) the formation of 'streamlines' by water movements at the refractive boundaries. 'Streamlines' can often be seen where a clear river water empties into a bay.

In fresh water, floating vegetation both shadows out phytoplankton and reduces the effect of wave action on soft bottom sediments, resulting in greater water clarity. The water column near the bottom of lakes or oceanic depths often contains an accumulation of particulate matter derived from the surface or stirred up from bottom sediments. Two- to four-fold increases have been recorded by Kullenberg (1974) for North-west African waters, and for North Pacific waters by Baker *et al.* (1974). The concentration of humic substances in Esthwaite water, UK (Tipping and Woof 1983) is generally highest near the bottom at 14m (bottom 15m).

Towards the end of the year in freshwaters much of the marginal and floating vegetation dies back and rots away, and phytoplankton breaks down. If a sample of water is taken in October and November (in the Northern Hemisphere) after the equinoctial gales have caused mixing to take place, a layer of bright yellow material will settle out of a sample. This probably consists of chlorophyll breakdown products and humic substances from the degradation of lignin and cellulose. The early German limnologists referred to this accurately as 'Gelbstoff' (yellow substance). Since then the term has often been applied indiscriminately to phytoplankton and other suspended matter containing chlorophyll. Suspended humic matter is particularly associated with bog-lakes and moorland lakes with runoff from sphagnum meadows. Tipping and Woof (1983) describe the autumn (October/November) building up of humic substances in Esthwaite Water, a shallow eutrophic lake in the English Lake District, at depths of 5m and 14m. Concentrations are slightly higher at the bottom than near the surface at most times of year, other than the period between November and January. Changes in concentration at the bottom of the water column seem to lead those at the top. These substances appear to absorb strongly in the ultraviolet, and absorbance at 320nm is maximal in the autumn. Ultraviolet receptors are discussed below. In bog-lakes, Heath and Franks (1982) found that UV-sensitive aggregations containing humic compounds, phosphorus and iron were actively broken down by sunlight, so that accumulations that built up at night were dispersed by noon.

One of the major sources of increased turbidity in waters receiving suspended matter from rivers is the effect of precipitation. In fresh waters, run off from streams on clays and sedimentary rocks produces high concentrations of coarsely particulate matter with broad effects on visibility, whereas water from old igneous rock formations tends to contain small amounts of mineral particles but may be deeply stained by the decayed sphagnum (peat) characteristic of such areas.

An extreme example of the effects of runoff is provided by the changes in suspended solids for the water in Conowingo Bay near the exit of the Susquehanna River (Maryland) following hurricane Agnes (Schubel 1974). The normal level of solids is about 10 mg litre^{-1} (clear oceanic water provides a range of 1.0-0.05 mg litre^{-1} so these waters are usually fairly turbid). On 26 June 1972, the solids concentration was a maximum of 10000 mg litre^{-1} and remained between 700 and 1400 mg litre^{-1} for some days. Although this is an extreme example, rapid changes in turbidity due to precipitation affect fresh waters and estuaries very generally (see Figure 4.3).

Turbidity changes also accompany freezing and thawing in many lakes and rivers. Stewart and Martin (1982) followed changes in light transmission in a narrow glacial lake with winter ice cover. Hemlock Lake, North America, is frozen over from mid-January to mid-March. In June the clearest water in this shallow lake (12 km long, mean depth 78.5 m) is near the bottom (over 50 per cent transmission), but the autumn overturn results in the upper part of the water column becoming most transparent. The transmissivity pattern for December was one of slight turbidity throughout the lake, with high turbidity near the inflow stream as in June. Between February and March the upper part of the water column clears, and a more turbid layer collects near the bottom and near the inflow. Transmissivity ranged from less than 10 per cent to over 50 per cent.

These examples indicate that in many aquatic habitats comparatively large changes in underwater visibility occur, and will significantly modify visual communication by fishes. Some of the ways in which the visual system of fishes takes account of these effects are considered in following sections.

(4) Exposure of Fish to Surface Predators

For many species of teleosts the water film provides a convenient trap for insect prey. At the same time, feeding at the surface is potentially hazardous, mainly because of the attentions of diving birds. Some protection from surface predators is provided by wave action. At reasonable wind speeds, capillary and gravity waves result in the distortion, blurring and loss of apparent brightness of objects viewed directly from above. In sunshine the moving pattern of reflected bright sky and dark sky provides a confusing patterned surface, which may be coupled with a complex shadow pattern on the bottom. Convex parts of the water surface will focus light into bright patches, and concave regions will produce darker areas, resulting in the familiar dappled pattern observed from above. In addition, the low levels of subaquatic reflectance under these conditions are enhanced for the overhead observer in air by the stronger illumination provided by

Figure 4.3: Temporal Turbidity Changes Affecting Underwater Visibility in the Waters of Conowingo Bay at the Mouth of the Susquehanna River in 1972. Note that the normal monthly variations in the concentration of dissolved solids are considerable, of the order of 2 to 10 times prior to the arrival of hurricane Agnes in June.

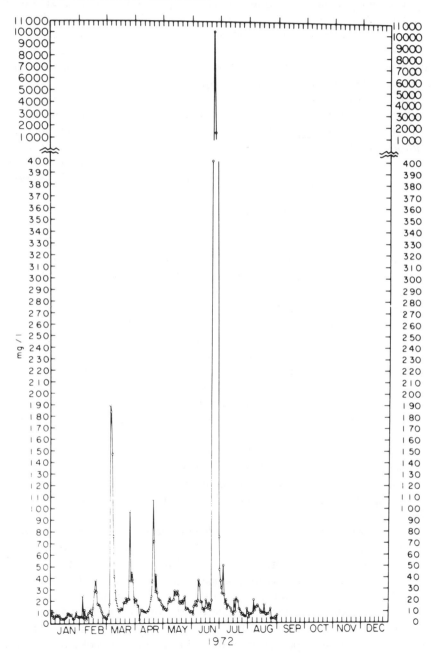

Source: Schubel (1974).

the superaquatic scene. Birds coming over still water will remain hidden from fish until they appear overhead and are seen through Snell's window (see Chapter 12 by Pitcher, this volume), but where the surface is broken by wavelets, the image of the bird will appear intermittently at angles greater than 45° to the vertical. The thermoclinal surface layer of turbid water and the colouring produced by water runoff also protect fish from surface predators, yet most fish species approach the surface with caution and react with a startle response to the movement of any objects overhead (C-type: Eaton and Bombardieri 1978).

Helfman (1981) has pointed out that many species of temperate lake fish aggregate beneath floating surface objects. He suggests that this allows them to escape detection by an aquatic observer in open water due to a high-increment threshold response for the observer who will be exposed to high ambient light levels. The poor illumination of the target must also be involved. At the same time the sheltering fish have good viewing conditions with a low-increment threshold and reduced effects of glare, while remaining protected from surface predation.

(5) Conspicuousness vs. Camouflage

For most animals there is a balance of selective forces between signalling effectively to conspecifics, and having these signals intercepted by predators. Endler (1977) has illustrated this very clearly in his study on wild populations of the guppy (*Poecilia reticulata*). Red or orange spots are much more abundant where the only predator is the freshwater crayfish, which is insensitive to longer wavelengths. Stickleback males with brighter scarlet patches are more successful in obtaining females, but are more heavily predated (McPhail 1969). For many fish species the turbid conditions often found in natural waters do offer the opportunity to signal to conspecifics at close range, whereas at a distance they become inconspicuous through loss of contrast and shift of spectral properties described in (1). Endler (1977) has also pointed out that patch-size contrast alters with distance. That is to say that the dimensions of the contrasting areas or patches presented by the surface of the fish and patterns on it may match those of the background at one distance, but as the distance between fish and background changes, the match is lost due to a difference in apparent patchsize.

The nature of fish communities is also of importance. In the clear warm water of a coral reef, several hundred fish species may be present, supported by a large invertebrate fauna (see Chapter 14 by Heffmann, this volume). Visibility is relatively good and territories are often physically identifiable. We find diverse and striking patterns in many species. As Levine *et al.* point out, there may be trends towards a particular type of colour or pattern within a family (Levine *et al.* 1980). Some groups are relatively sombre in appearance. In the cold, turbid waters of a pond in the UK, only five or six species may be present, and most of these are relatively dull coloured, with few species possessing striking surface patterns. Distinguishing features among these species appear to the human observer to be the dorsal silhouette (overhead line of sight), and ventral fin colours (downward line of sight). This signalling repertoire would perhaps be too limited for the competitive conditions of a coral-reef community.

As described subsequently, the theory of offset colours suggests that red would be most conspicuous as a contrast colour in greenish fresh water, and yellow (Lythgoe and Northmore 1973) would provide the best long-range conspicuousness in blue oceanic waters. Levine *et al.* point out that black and white (achromatic) will also remain conspicuous under most conditions of lighting. A survey of body colours in four groups of fish from different habitats has been attempted by Levine and his colleagues. The habitats were shallow and deep reef waters, and clear and turbid fresh waters. They assessed the significance of a colour by combining (a) relative area, (b) frequency of occurrence, and (c) saturation or density, to give a 'colour importance index' (maximum score 70 units). Highest values in all habitats were found for achromatic patterns. In reef water, there were few greens, and colours that matched the blue spacelight were common, but as predicted from data on the spectral shift in blue waters, yellow and yellow-orange were quite common (about 25 per cent of maximum). In shallow reef habitats, however, reds were found, presumably as the effect of the spectral shift was slight. In clear fresh waters, again blues and yellows were notable (yellow 80 per cent maximum), but in turbid fresh water, as would be expected from spectral-shift data, green and reds predominated. There was therefore a reasonable match between the hypothetically expected distribution and that observed, although the very high value for achromatic colours (silver, grey, black and white), and the high value for yellow in the clear fresh water was a little surprising.

What might be termed camouflage features are found in many species that live in well-lighted environments or swim close to the water surface. This consists of reverse counter shading so that the back is dark (melano-phores) and the belly lighter (guanin). Furthermore, in many species (like the herring), the scales reflect the prevailing spacelight, thus matching the fish to its background (Denton 1970).

Predatory fish species like pikes and muskellunge (*Esox* spp.) and bass (*Micropterus* spp.) appear reasonably well camouflaged to the human observer when seen at intermediate ranges, by virtue of low-contrast flank patterns of spots or patches, and rely for their success on high acceleration and the ability to strike from cover. The most interesting examples of camouflage are represented by the angler fishes (*Lophius*) and frog-fishes (*Histrio*), which have to submit to close-range scrutiny by other fishes. In extreme examples the skin texture resembles the surface of a sponge, the outline of the head is broken up by protuberances, and the pupil of the eye (a higher-contrast feature) is reduced to a very small aperture.

The Visual System

(1) The Eye

Paired image-forming eyes are found in most species. In addition the median pineal body is an effective light sensor in many forms. Benthic species living at depths down to 1000 m tend to have large eyes and luminescent organs. Abyssal forms do have small eyes, but those without eyes are mostly confined to caves (see Chapter 17 by Parzefall, this volume). The teleost eye conforms to the vertebrate type in that it consists of a subspherical chamber containing an inverted retina (with outwardly facing receptors) and a focusable lens. Because the cornea/water interface does not contribute significantly to bringing images of objects to a focus, the fish lens is a highly refractive sphere lying far forward in the eyeball. It follows from this that (a) accommodation is by the movement of the lens (rather than by changing the shape of the lens, as in mammals) under the action of external muscles, and (b), because of the position of the lens, the iris cannot be 'stopped down' very far, and adaptation to different light intensities is by movement of the outer segments of retinal receptors relative to the pigment layer (retinomotor response). In many teleosts, changes in pupillary diameter under natural conditions are very small and rather sluggish. Charman and Tucker (1973) point out that pupil size in the goldfish shows little change. Ali (1959) states that the iris of Pacific salmon is not capable of photomechanical changes under light conditions varied experimentally by a factor of $\times$ 4000. We observed that several hours of light or dark adaptation effected no measurable change in the diameter of the pupil of the perch eye (D.M. Guthrie and J.R. Banks, unpublished observations); measurement accuracy $\pm$ 0.3 per cent). However, under the action of drugs a reduction of pupil diameter of 16 per cent occurred after 2 h (= 30 per cent reduction of pupil area). This lack of pupillary function, even in species like the perch, which are active under diverse light conditions, may be correlated on the one hand with the forward position of the spherical lens within the eye, and on the other with the retinomotor mechanism. There are, however, a few specialised teleosts (eels, flatfish, stargazers) in which the iris is much more mobile. In the eel (Young 1950) the pupil diameter can be reduced by 45 per cent by the combined effects of drugs and light. Even so, these changes are much smaller than those of diurnal mammals, such as cats, dogs and primates. In humans (de Groot and Gebhard 1952) the pupil diameter is reduced by 75 per cent under strong illumination (equivalent to a 91 per cent reduction in area).

In lower teleosts like the goldfish, pike and trout, the movement is produced by a single muscle pulling the lens backwards and inwards, but in percoid fishes (bass, perch) a complex set of four or more muscles is present (Schwassmann and Meyer 1973; Guthrie 1981). Sivak (1973) measured accommodatory movements in a number of species and found that the transverse excursions were similar in most of them (equivalent to 10-12 dioptres) including the trout, but that rostro-caudal excursions were much greater in the percoid fishes (perch: 40 dioptres). Sivak and Howland (1973) were able to demonstrate by cinematrography that quite large accommodatory movements of the lens could occur within 30 milliseconds in the rock bass (*Ambloplites*) in response to natural objects. We have observed similar rapid accommodatory movements in the perch (D.M. Guthrie and J.R. Banks, unpublished observations). Sivak and Kreutzer (1981) aiso looked at the spherical aberration of the lenses of various vertebrates with the aid of a helium-neon laser. It had been suggested that fish lenses might have to be intrinsically better corrected, due to the lack of corneal correction. In fact there was wide variation from the slightly corrected lens of the rock base (*Ambloplites*) to the poorer optical quality of the goldfish lens. Sroczynski (1977; see also his earlier papers) has calculated the spherical aberration of the lenses of pike (*Esox lucius*), rainbow trout (*Salmo gairdneri*) and roach (*Rutilus rutilus*), from focal-length measurements made on freshly dissected lenses. Spherical aberration was highest in the roach (spherical aberration area, SLA $= 0.0092 \pm 0.0012$), and lowest in the pike (SLA $= 0.0040 \pm 0.0008$). Sroczynski suggests that spherical aberration is not the only determinant of image quality, but also points to the standard deviation of focal-length samples as an indicator of the level of variation in the refractive properties of the different layers of the lens. Here again the roach produced the highest values; the pike had the lowest. This suggests that the image quality may be better in a visual predator like the pike than in species less dependent on vision, such as the omnivorous roach. Chromatic aberration has been studied by Sivak and Bobier (1978) in *Lutjanus, Chilomycterus* and *Astronotus.* Their paper resumés some information on other species. Chromatic aberration was significant (4-5 per cent, equivalent to 7-8 dioptres), and would be considerably reduced by the yellow filters found in the eyes of some species. Muntz (1973, 1976) showed that many reef species have a yellow cornea (trigger fish and puffer fish). Wrasse have yellow lenses in addition. Some freshwater species (e.g. cichlids) may possess yellow pigment in the retina as well as the features listed above. As in air, shorter wavelengths are likely to be subject to most scattering, so that yellow filters may filter out shorter wavelengths and probably contribute to image sharpness. Whereas the benefit of these filters for reef fishes is easy to see, the very heavy pigmentation found in cichlids living in a long-wavelength-dominated environment is less clear. There is a fairly extensive older literature on the refractive properties of the fish lens, and the reader should refer to Charman and Tucker's (1973) paper on the goldfish eye for an introduction to it.

Attempts have been made to estimate the visual acuity of the fish eye using minimum cone separation and lens diameter. Tamura (1957) applied this to 27 marine species (mostly belonging to percoid families) and obtained values ranging from 4.2′ (minutes of arc) for the grouper

Epinephelus which are reef-dwelling predators, to 15.4′ in *Chloro-phthalamus*, a bottom-dwelling benthic form. Schwassman (1974), using minimal stimuli and electrophysiological methods, obtained a value of 4′ for *Lepomis* (a percoid sunfish) and 15′ for the goldfish. These differences between percoids and a cypriniform like the goldfish may be correlated with the different requirements for mid-depth pursuit predators and planktivores on the one hand, and a partly herbivorous bottom feeder like the goldfish.

A further specialisation related to underwater vision is worth mentioning. Especially in rock pools and other shallow littoral habitats, the cornea of some species (gobies, blennies) may contain oblique layers that reflect away direct rays from the overhead sun while remaining transparent to reflected rays from objects in front of the eye (Lythgoe 1974).

(2) The General Structure of the Retina

The generalised vertebrate pattern of three tiers of cells, receptors, bipolars and ganglion cells connected perpendicularly by horizontal cells and amacrine cells, is clearly seen in fishes. However, except for one or two species, a pit-like fovea is absent, and instead there are regional variations in the density of retinal cells. Often there is a high density 'area', or *area centralis*. In the perch (Ahlbert 1968) there are about four to five times as many cones per unit area in the upper temporal part of the retina as in anterior regions, and a similar difference is found with regard to ganglion cells (D.M. Guthrie, unpublished). This region appears to correspond to the projection of the major binocular fixation point.

The structural complexity of the fish retina as compared with that of higher vertebrates appears to be quite elevated. Three or more cone types and rods, two types of bipolar cell, three to five kinds of horizontal cell, and six or more types of ganglion cell have been observed in different teleost species (Cajal 1893; Wagner 1975). For detailed comparisons between species the reader should refer to Ali and Anctil (1976).

(3) Receptor Adaptations

In the typical duplex fish retina, the arrangement of the cones may be irregular, or, as in many of the advanced predatory species, it forms a regular mosaic with rosettes of double cones aligned in rows. In some deep-ocean species, clusters of receptors are arranged in pigment-separated groups. The larger cones often form opposed pairs, and according to whether they contain the same pigment or a different one are termed twins or doubles, respectively. The single cones tend either to lie at the centre of a cone rosette (long central singles), or irregularly at the edge of a cone group (short additional or accessory singles). The rods are visible in a tangential section of the retina of a fish like the perch as 10 to 20 small profiles accompanying each cone pair.

The absorption curves of receptor pigments are clearly of considerable interest in relation to colour vision and the spectral quality of available light. Following the pioneering work of Liebmann and Entine in 1964, a number of studies employing microspectrophotometric methods to characterise single receptors have been published (see Levine *et al.* 1980 for a review). It has been suggested that different cone pigments first evolved in order to assist discrimination of brightness contrast. The spacelight-matching pigment responds best to dark objects or those viewed in silhouette, and a pigment with peak absorbance offset from this point is most sensitive to the other spectral elements reflected by a bright object (Figure 4.4). For hue discrimination, as Levine *et al.* point out, pigments are likely to be most effective if their absorbance curves overlap and are steep. This is indeed so in many instances. As for the distribution of pigments among morphological types of cone, the following generalisations have emerged. Where cone pairs are present (most species), they contain middle- or long-wave sensitive pigments. Normally either a single type of twin cone (same pigment) or a single kind of double cone (different pigment) is present, but this is not always the case, as in *Cichlasoma* sp. and *Oreochromis* (Levine *et al.* 1980) where twin cones are present containing different pigments. The long central cones usually contain pigment that absorb in the medium wavelengths, whereas the small single accessory cones are almost always most sensitive to short wavelengths. Rod pigments are rather conservative, with peak absorbances in different species within the range 480-530 nm. At mesopic light levels (in the transition zone between rod and cone vision) many species have three cone pigments and a rod pigment functionally available to them. Occasionally, four cone pigments are found. In some of the freshwater benthic species, only a single long-wave cone is present; a further reduction is to be found in marine benthic species where only rods occur.

The wide variation in cone pigments revealed by microspectrophotometry suggested a possible correlation with the ecological characteristics of different habitats (Loew and Lythgoe 1978). As explained earlier, a model system for vision in a spectrally filtered medium might consist of a receptor with peak absorbance close to the dominant wavelength or spacelight and one or more receptors with pigment absorbances offset from this dominant wavelength, and capable of responding to spectral reflection contrasting with the spacelight. The collection of pigments possessed by a species would be expected to be clustered about a mode determined by the dominant wavelength characteristic of that environment. A number of studies have demonstrated a broad correspondence of this kind, in particular a recent study by Levine *et al.* (1980) involving 55 species. Levine and his co-workers found that species living in coastal marine habitats with ambient light dominated by wavelengths in the 525-540 nm band have rods and cone pairs with pigments with peak absorbance within the range. Single cones provide an offset pigment with peak absorbances around

Figure 4.4: Munz and McFarland Suggested that Provision of More than One Cone Type in Teleosts Originated with Contrast Detection. (a) This diagram illustrates the fact that a light object, especially one near the surface, will relfect a broad daylight spectrum (upper insert) at high intensity and thus will be visible to short-wave-absorbing cones against the spectrally more restricted spacelight. Cones absorbing at medium wavelengths will respond best to the difference in brightness between dark object and spacelight background. For objects in deeper water the position is altered, as in clearer waters the transmission curve will be shifted towards the shorter wavelengths (lower insert). Under these conditions the short-wave cones will be most effective at detecting a dark object against the background. (b) This diagram illustrates the way in which, for objects near the surface, the offset channel (short-wave channel, SW) is able to respond to the shoulder of the normal daylight distribution curve. L.o., light object; S.l., spacelight; D.o., dark object; SW, shortwave cone (absorption curve in (b)); MW, medium wavelength absorbing cone (absorption curve in (b)); λ, wavelength spectrum; i, light intensity; solid lines: most effective transmission.

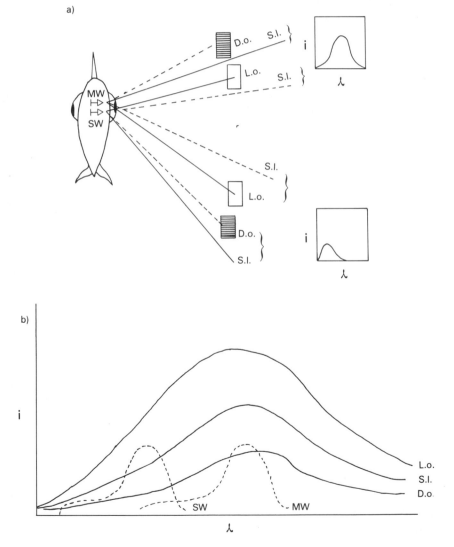

460 nm. Three tropical marine species and two clear freshwater species (cichlids) were also examined and found to have a rather similar range of pigments, mostly falling between 450-550 nm, although the double cones absorbed at rather shorter wavelengths than those of the temperate marine species. A much larger group of 43 tropical freshwater species were also examined (see Figure 4.5). Perhaps the most striking feature here was that in 26 species, receptors occurred with absorbance peaks between 600 and 630 nm (most double cones). These are associated with the effect of suspended organic and inorganic matter found in most fresh waters, which produces a dominant spacelight at the longer wavelengths. The double cones' absorbances tend to straddle, or to coincide with the dominant wavelengths, and the offset pigments in these species fall within the range 450-550 nm.

Within this group of freshwater species there was considerable variation, and Levine and his colleagues divided it into four ecological groups. Group I consisted of diurnal surface-living species. At the surface the filtering effect of particulate matter is relatively slight, and the double cones absorb in the range 460-575 nm, with offset single cones absorbing at wavelengths near 415 nm. In group II the pigment range is shifted upwards by about 40 nm at either end of the range. These are mid water species, and the effect of the longer light path is to shift the dominant wavelengths upwards. The crepuscular and predaceous species of group III tend not to have the short-wave cone found in groups I and II, and their single cones often contain the same pigment as the rods. It is likely that at low light levels there is insufficient short-wave light present for a separate short-wave channel, and that at mesopic levels rods and spacelight-matching long-wave cones provide an adequate means of contrast detection. A fourth group of benthic species were also examined. These live in the muddy haze near the bottom, and are often nocturnal forms relying on gustatory and mechanoreceptive sense more than on sight. The retina contains only single cones and rods, with pigment absorbance peaks arranged as two clusters between 525-550 nm and 600-620 nm. The visual pigment of the rod-only retina of the fish of the deep oceans matches closely the prevailing wavelength of downwelling light at 485 nm, allowing objects to be discerned in silhouette. Until recently, the range of peak absorbances for teleost cones was believed to be 415-630 nm. This range accorded with the major component of available light in many natural waters. Further, the trichromatic system found in many species seemed to provide economical and effective colour vision. However, following the initial observations of Avery, Bowmaker, Djamgoz and Downing (1983) on roach (*Rutilus rutilus*), it has been established that many cyprinids have a tetrachromatic system including a cone absorbing in the near ultraviolet (Harosi and Hashimoto 1983; Fukurotani and Hashimoto 1984; Harosi 1984). The ultraviolet cone absorbs maximally at 355-360 nm. Observations on other teleosts (Fukurotani and Hashimoto investigated 29 species), suggest a range of 350-385 nm. The cones in

Figure 4.5: Distribution of Cone Types in Four Groups of Tropical Freshwater Fishes Differing in General Ecology. Group I: diurnal surface dwellers. Group II: midwater species. Group III: crepuscular and predaceous species. Group IV: bottom dwellers.

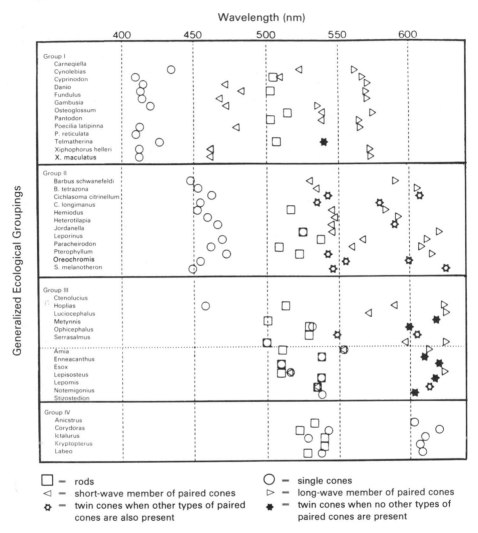

Source: Levine *et al.* (1980).

question are miniature single cones. One of the surprising features of this discovery is that the freshwater cyprinids in which this ultraviolet channel occurs are in many cases mid-depth or bottom feeders (roach, goldfish) living in rather turbid waters. Lythgoe (1984) quotes figures for *winter* reservoir water which suggest transmission of less than 5 per cent surface light at a depth of 1 m in the 362 nm band.

(4) The Central Projection of the Visual Pathway

The optic nerve contains at least four different kinds of fibre (including efferents) as detected by conduction velocity measurements (Vanegas, Essayag-Millan and Laufer 1971; Schmidt 1979). Most of these fibres pass across the midline to the opposite side of the brain, and terminate in the upper layers of the optic tectum. A few fibres are now known to pass ipsilaterally, and some of those that project contralaterally end in anterior nuclei rather than in the tectum. There are some six nuclei lying in the anterior subtectal area, and four of these receive direct retinal input. Pathways from the tectum pass to some of these nuclei, to a median nucleus involved with eye movments (the nucleus isthmi), and to subtectal areas like the torus semicircularis (see Figure 4.6). Several transverse commissures link the two optic tecta, and are believed to be involved in binocular function.

In the goldfish each optic tectum contains about 2 million intrinsic cells (Meek and Schellart 1978), and these are deployed as 15 or so recognisable and distinct morphological types arranged in successive layers and coincident with the fibrous layers (see Meek and Schellart for details of

Figure 4.6: The Central Projection of the Visual Pathway in Teleosts, based on *Perca* and *Eugerres*. Most retinal afferents project to the contralateral tectum, where information from the retina acquires more complex spatial properties and becomes associated with cerebellar and other brain inputs. Some of the subtectal nuclei are believed to be involved in a number of functions to do with image stabilisation. The torus semicircularis is a sensory-associated area for visual and auditory mixing. Torus semicirc — torus semicircularis, Torus long. = torus longitudinalis.

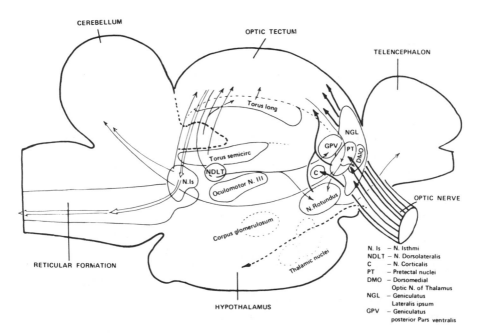

CEREBELLUM

OPTIC TECTUM

TELENCEPHALON

Torus long

NGL

GPV PT

Torus semicirc

DMO

NDLT

C

N.Is

Oculomotor N. III

OPTIC NERVE

N.Rotundus

Corpus glomerulosum

RETICULAR FORMATION

Thalamic nuclei

HYPOTHALAMUS

N. Is — N. Isthmi
NDLT — N. Dorsolateralis
C — N. Corticalis
PT — Pretectal nuclei
DMO — Dorsomedial
Optic N. of Thalamus
NGL — Geniculatus
Lateralis ipsum
GPV — Geniculatus
posterior Pars ventralis

goldfish and other species). Other species examined possess similar cell types so that the tectum can be regarded as a very conservative structure. The commonest neurone types have their cell bodies within a close-packed layer, the periventricular layer (PVC). Broadly speaking, a point-to-point 'through' system can be seen as separate from a system running parallel to the fibre layers and interconnecting the through pathways, as in the retina. Degeneration and retrograde-staining experiments indicate that the outer fibre layers contain most of the retinal afferents, and projections from other brain regions enter either in deep layers, or in the most superficial layers (stratum marginale).

(5) The Neural Physiology of Fish Vision

(a) Chromatic Processing.   Some of the most important early studies on vertebrate retinal physiology were made on the large receptor cells and horizontal cells of fishes. Svaetichin (1953) demonstrated opponent processes recording from what were probably horizontal cells, and Tomita, Kaneko, Murakami and Pautler (1967) provided evidence of three separate wavelength-sensitive channels from cones. The most significant generalisations, however, have come from the analysis of recordings from retinal afferents (Daw 1968; Spekreijse, Wagner and Wolbarsht 1972). Daw found that in the goldfish 68 per cent of cells are double opponent, that is to say their receptive fields have a roughly circular centre and surround structure, with chromatic opponency at the centre (often red 'on', green 'off'), but also opponency between the centre and the surrounding annulus. Three main types of colour opponent cell were observed by Daw, but Spekreijse *et al.* identified 12 different combinations of red-, blue- and green-sensitive channels, also in the goldfish. Long-wavelength 'on' centres were the commonest, and medium- and short-wave-sensitive channels were always complementary. Short-wave-sensitive (blue) centres were also observed. In the perch tectum (*Perca fluviatilis*), Guthrie (1981) found that only 18 per cent of cells had double opponent chromatic properties. The majority were red 'on' centre cells, but blue 'on' centre cells were also seen. A few cells responded best either to wavelengths below 400 nm (380 nm), or to light in the spectral range 650-700 nm. Specific responses to near ultraviolet were occasionally seen, and are interesting in the light of more recent observations on UV-senstive cones (see above). The response to very long wavelengths corresponds quite well to the characteristic reflectance curve of perch ventral fins (Guthrie 1983).

The significance of double opponency is twofold. First, it preserves narrow spectral tuning under different conditions of brightness, and is therefore crucial in hue resolution; and secondly it is effective in mediating responses to colour contrast borders moved across the cell's receptive field. (This latter may also be important in colour constancy under different spacelights: goldfish have recently been shown to have colour constancy.)

(b) Non-chromatic Processing. For predation and for obstacle avoidance, the accurate registration of objects in space is clearly necessary, and we find that object position is preserved in local tectal unit responses (the retinotopic map). It is preserved most accurately by the small simple receptive fields of retinal afferents (2.5-12.5°, goldfish: Wartzok and Marks 1973; Sajovic and Levinthal 1982). Intrinsic tectal cells tend to have larger receptive fields (30-160°: Sutterlin and Prosser 1970; O'Benar 1976; Guthrie and Banks 1978) and field structure is irregular, or multicentre (Schellart and Spekreijse 1976). Even regular annulate receptive fields may have some shape-selective properties (see Guthrie 1981), but preferences for oriented edges have not been generally observed.

Some degree of directional preference was observed among movement-sensitive afferents in the goldfish by Wartzok and Marks (1973). In the upper nasal eye field there was a preference for anteriorward movement, and Zenkin and Pigarev (1969) noted a similar preference in movement detectors in the pike (*Esox*). This coincides with the stimulus most likely to be involved in predation, which also demands accurate coding of object velocity. Information about the velocity of an object can be used in two ways: to drive pursuit movements, or to provide data for object identification. In the first case, direct coding of velocity as impulse frequency may be adequate, but for recognition purposes tuned units would seem to be required. There is some evidence for tuned units from work on goldfish afferents with reasonably sharp peaks within the range 5-20°s^{-1} (Wartzok and Marks 1973). Pike afferents seemed to fall into three response categories corresponding to object velocities of 0.5-5°s^{-1}, 3-40°s^{-1} and 50°s^{-1} (Zenkin and Pigaret 1969).

Certain tectal cells respond, but habituate rapidly, to small movements almost anywhere in the eye field (trout: Galand and Liege 1975; goldfish: Ormond 1974; O'Benar 1976; perch: Guthrie and Banks 1978). These 'novelty' units are probably associated with the attentive phase of visual behaviour (in the perch these lie in the deep tectal layers).

Localised electrical stimulation of the tectum causes eye movements and body movements directed towards appropriate parts of visual space. This suggests that heightened local activity in the tectum is sufficient to drive a specific motor programme (Meyer, Schott and Schaeffer 1970; El-Akell 1982).

We know very little about the function of the subtectal areas. Although it is now clearly demonstrated (Page and Sutterlin 1970; Schellart 1983) that mixed modality units, responding to both visual and auditory stimuli, are found in the torus semicircularis, their significance is unclear. In some species (e.g. cichlids) courtship movements are reinforced by specific sounds (Nelissen 1978).

(c) The Pineal. This structure is a light-sensitive organ that is often well developed in fishes. In most forms it is a sac-like median lobe of the

diencephalon projecting obliquely forwards towards the cranial roof. Cells within the pineal secrete the hormone melatonin, an amino acid derivative that causes darkening of the skin by expansion of the melanophores, most usually during the hours of darkness. Pineal fibres are also believed to project to the hypothalamic-pituitary region and have more general neuro-endocrine effects, reducing activity and affecting territorial aggression (Rasa 1969).

In trout and tuna, the head of the pineal is large and lies below a transparent cartilaginous window in the frontal bone (Rivers 1953; Morita 1966), but in other species (most percoid species, for instance), there is no obvious frontal window and the pineal lobe is small and delicate. The electrical activity of the pineal cells, receptors and ganglion cells has been successfully recorded by a variety of methods (Dodt 1963; Morita 1966; Hanyu 1978). There is general agreement that photic stimuli inhibit on-going spontaneous activity, and that the degree of inhibition is related to light intensity. Conversely, dimming produces an increase in discharge frequency. In some species (e.g. of *Plecoglossus, Pterophyllum*) there is a phasic response followed by a steady tonic discharge which is related to the maintained light level, although Hanyu found that the quantitative relationship varied for different cells in *Plecoglossus*. The spectral sensitivity of the pineal cells peaks at about 500 nm in the trout (Morita 1966), and this and the slow response are rod-like characteristics. Hanyu sees the pineal as a dimming or dusk detector in its tonic mode, responding to maintained light levels, and as a detector of shadows by means of the dynamic or phasic response.

Studies on Visual Behaviour

(1) Classical Conditioning Methods for Determining Aspects of Visual Discrimination

(a) Spatial Proportions.   A considerable body of work, mainly provided by the German school of Gestalt behaviourists, has come into existence over the last 50 years (see Northmore, Volkmann and Yager 1978; Guthrie 1981). Most of these studies involve rewarded single-choice conditioning using a pair of geometrical figures. The trained subject (i.e. one which has reached a certain score of correct responses) is then tested with another pair of figures which (or one of which) are slightly altered versions of the two shapes used initially. The ease of transfer, or degree of maintenance of correct choices, measures the extent to which the fish recognises the alterations that have been made. The problem with many of these transfer experiments is that the precise level of recognition is difficult to assess since two elements of recognition are involved simultaneously in the transfer, A/B differences and A/A' differences, supposing the transfer is A/B to A'/B. To some extent the ability to generalise opposes the ability

to notice that the A′ is difference from A. Both processes may be valuable in nature.

The results suggest that teleosts can discriminate between a variety of simple solid geometrical forms. This can be disrupted by altering the direction of figure/ground contrast in some species like the perch, although such contrast reversal did not affect transfer in the trout and the minnow. Several studies indicate again that the upper contour of a figure is more important than the lower in shape recognition. Points, acute angles or knobs seem to make for easier recognition than small differences in the form of curves or obtuse angles. Interestingly in relation to spatial frequency, Schulte's carp (Schulte 1957) found great difficulty in discriminating between four stripes and seven stripes occupying the same area. Most of these tests do not involve fine detail and depend heavily on one group of cyprinoid fish. Herter's (1929) work with the perch (*Perca fluviatilis*) suggests that this species can respond to relatively small differences in test figures (see Guthrie 1981).

(b) Hue Discrimination. The question of whether fishes could distinguish similar objects of the same brightness differing only in hue was largely resolved by the work of von Frisch in 1912-25, working with minnows, and Hurst (1954) using the bluegill (*Lepomis*). Later studies showed that all teleost species tested were capable of making discriminative choices on the basis of hue alone. It has to be emphasised that most behavioural tests have been conducted using visually active freshwater species now known to possess two or more cone types.

The minimum spectral interval that can be discriminated on the basis of behavioural tests has been determined for the goldfish and lies between 10 and 20 nm at the point (about 600 nm) of greatest sensitivity for fish and man; this can be compared with intervals of 1-2 nm for man (Northmore *et al.* 1978). The spectral range of teleosts is mentioned since behavioural testing has usually been limited (especially at the shortwave end) to the mammalian range of 400-650 nm, but seems often to match this. Spectral sensitivity curves based on the lowest intensities to which teleosts respond at different wavelengths not only provide information about spectral range, but also may indicate the presence of separate wavelength-dependent channels. Trimodal curves have been obtained even in those species like the shanny (*Blennius pholis*) and the perch believed to possess only two classes of cone. In his study on the perch, Cameron (1974) suggests that a 'differencing' channel, comparing outputs from red- and green-sensitive cones, is responsible for the relatively high sensitivity found at 400 nm (third peak). Of especial interest is the evidence provided by Burkamp (1923) for colour constancy, in that ambient spectral conditions may vary regionally as described above. Colour constancy is an important property of the primate visual system (Zeki 1980) which allows hue recognition to remain effective despite widely varying spectral conditions.

(2) Field-based Studies

(a) Aggregation Responses.   A large proportion of fish species form conspecific shoals, pairs of territorial groups (see Chapter 12 by Pitcher, this volume), and there is general agreement (Breder 1951; Hemmings 1966) that visual signals are important in shoal formation and group cohesion. The elements in the visual appearance of a species that determine aggregation behaviour can be discovered by experimentation using models, or less certainly by observation of the responses of the subject species when confronted by a variety of different species differing in certain details of appearance from its own. The starting point has inevitably to be the appearance of the fish as seen by the human observer, and to this extent it is impossible to avoid a subjective element in many of these studies. Nevertheless, a number of studies support the idea that specific visual features operate to trigger behaviour of this kind and broadly conform to Tinbergen's ideas on releasers and sign stimuli (Tinbergen 1951).

Keenleyside (1955) found that if the black-and-white tip of the dorsal fin was removed in *Pristella riddlei*, a small characid, the deprived fish was no longer accepted into the shoal. It should be pointed out that several other allied species have a dorsal fin spot and that it tends to be flicked in a manner that makes it conspicuous (see Figure 4.7).

Moller, Servier, Squire and Boudinot (1982) suggested that the vertical bars of some weakly electric species (*Gnathonemus petersi, Brienomyrus niger*) were used as a shoal aggregation device. These fish are relatively poor sighted, but Tayssedre Moller (1983) found that they would exhibit a following response to a striped drum. Patfield (1983), using live fish in adjacent tanks, demonstrated the ability of perch to aggregate with conspecifics, rather than with other local common species (*Rutilus* sp., *Leuciscus* sp.), and studied the effect of the vertical dark bars, dorsal fin spot and red ventral fins on the aggregation responses. The dorsal fin spot appeared to provide an important stimulus for aggregation. In seven out of eight trial series, the 'buttout' aggregation response was much stronger towards normal perch than towards perch from which the spot had been snipped out. Roach with a spot added did not exert a significant effect. Results from tests involving ventral fins were non-significant. However, weak aggregation (above control) responses resulted from exposing perch to roach (painted intradermally) with dark bars. More recent studies with models in our laboratory suggest that the dark vertical bars do play a significant part in recognition. These results are in line with the additive effect of such features shown by the results of Stacey and Chiszar, mentioned later.

Katzir (1981) found that the black-and-white striped humbug damselfish (*Dascyllus aruanus*) would aggregate with black-and-white still photographs of conspecifics, rather than with pictures of allied species (*D. marginatus* and *D. trimaculatus*). Experiments with photomontages in which stripes were deleted showed that the long second stripe was the most

Figure 4.7: Achromatic Features of High Contrast from a Number of Teleost Species, but Mainly from characids. In *Hyphessobrycon* and *Rasbora*, characteristic flicking movements of the fin draw attention to the feature

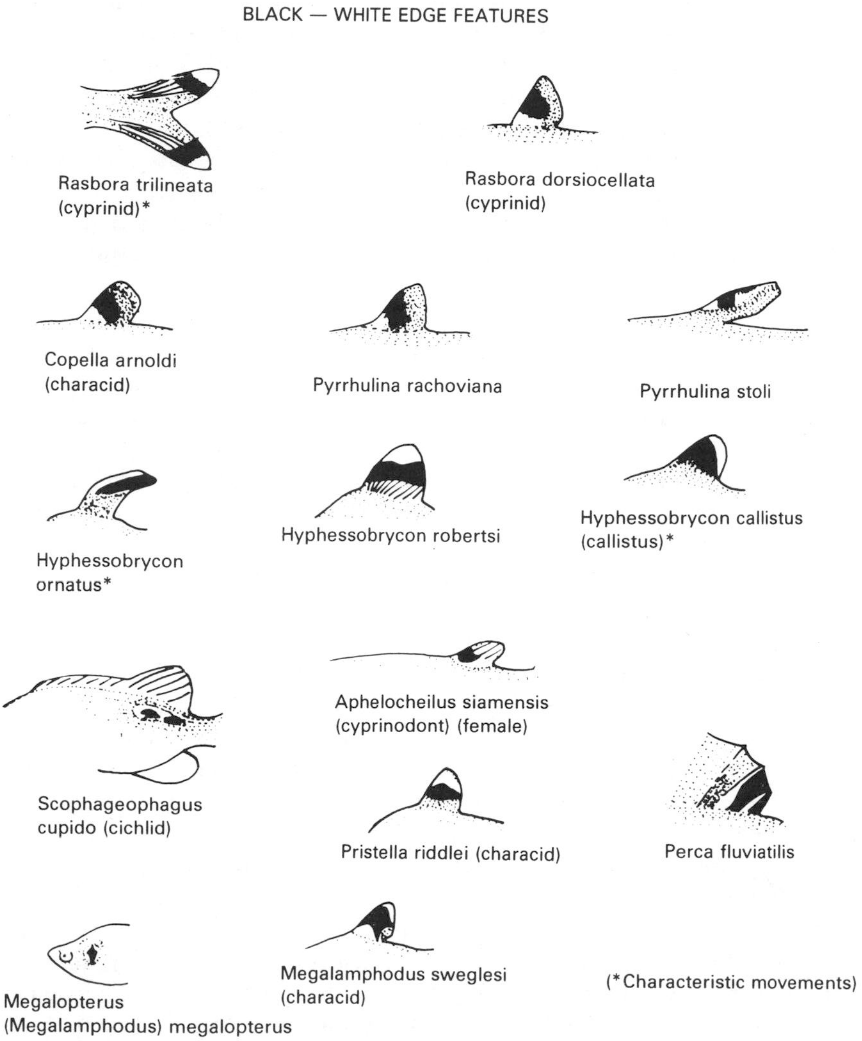

BLACK — WHITE EDGE FEATURES

Rasbora trilineata
(cyprinid)*

Rasbora dorsiocellata
(cyprinid)

Copella arnoldi
(characid)

Pyrrhulina rachoviana

Pyrrhulina stoli

Hyphessobrycon
ornatus*

Hyphessobrycon robertsi

Hyphessobrycon callistus
(callistus)*

Scophageophagus
cupido (cichlid)

Aphelocheilus siamensis
(cyprinodont) (female)

Pristella riddlei (characid)

Perca fluviatilis

Megalopterus
(Megalamphodus) megalopterus

Megalamphodus sweglesi
(characid)

(*Characteristic movements)

powerful feature in inducing aggregation.

There is a good deal of anecdotal evidence that species which form stable pairs during the breeding season are able to recognise their mates by visual means. The cichlid *Hemichromis bimaculatus* was shown by Noble and Curtis (1939) to form monogamous pairs capable of individual recognition, and Barlow (1982) was able to demonstrate this in the natural habitat by identifying face patterns of the reef species

(*Oxymonacanthus* sp.). Fricke (1973) used a two- or four-choice experiment to demonstrate individual recognition in the laboratory with *A. bicinctus*. Attack rates of one partner against the other partner, or foreign individuals, in adjacent tanks, were monitored. For the two-choice experiment, discrimination was at the level of 390:4, and where two other foreign individuals were visible in separate compartments (four choices) it was at the level of 615:0. If the partner was dyed with bromocresol green or enclosed in a green plastic jacket, it was unrecognised, and in the event of a new partner being provided, effective recognition was acquired within 24 h.

The responses to visual signals of adult fishes can be regarded as the function of mature nervous and endocrine systems. The responses of fry have to be viewed a little differently. Work in this field has the interest that it may reveal innate elements of recognition-processing dependent on the function of the inexperienced nervous system. Furthermore, the retina is enlarged as the fish grows by marginal accretions of new cells and the image formed on the retina at this stage is presumably less detailed than that formed by the adult eye.

Baerends and Baerends-van Roon (1950), working with mouth-brooding cichlids (*Tilapia*), found that a variety of simple rules described fry behaviour. They preferred dark objects to light; the lower side of objects, and concave surfaces. No detailed form discrimination was needed for them to locate the mother's mouth. Kuenzer (1968) came to roughly the same conclusion working with the fry of *Nannacara anomala*. These would aggregate with a long oval model of a particular shade of grey more readily than they would with other simple shapes, but a very simplified representation of the mother with its characteristic flank patches was more effective. Precisely detailed models produced no increase in response. When exposed to a variety of geometrical figures, a shape preference diagram could be constructed on the basis of two choice tests. The rules governing some of these choices were obscure, but Kuenzer (1975) expanded his testing scheme with an exhaustive analysis of the visual preferences of the fry of *Hemihaplochromis multicolor*. These tests involved discrimination of colour, grey shades, size, form, dimensionality, contrast with background, spot and annulus features. Responses were categorised as orientation, approach/contact or flight. Curiously perhaps, in view of the bright colours of the parents, grey shades (pale) were preferred to any colour when presented as discs, but blue or green (not red) spot features were preferred over grey. The largest test disc sizes offered were preferred, but preferences for any particular geometrical figure were weak. Face-like configurations were not preferred to the standard stimulus, a disc with a central spot. A noticeable feature of most responses was a decided choice of what might be termed constant contrast. That is to say, if the major feature was made darker, a darker ground was chosen by the fry. Where figure/ground contrast was maintained over a range, medium to high values were chosen. Kuenzer then introduced variations in background brightness, and again values were chosen that suggested that the fish preferred a constant level of background/figure plus ground contrast. There seemed also to be a preferred proportionality between the area of the spot and of the surrounding annulus, in relation to spot contrast. As might be expected, paler spots resulted in larger preferred spot sizes, darker spots smaller preferred spot sizes. Sometimes flight reactions were produced by the largest targets (20 mm): black spot and grey annulus.

The results of these tests suggest that, as with the experiments of Baerends and Baerends-van Roon (1950), very simple models can trigger the aggregation responses, but that some preferences for levels of contrast, colour, etc. do exist, which probably correspond to those most likely to be encountered under natural conditions. In contradiction to these results are those of Hay (1978) using convict cichlid *Cichlasoma nigrofasciatum* fry. He employed both simplified models and precise replicas of fish species. The aggregation preferences envinced by the fry clearly demonstrated an ability to respond positively to both generalised and detailed features possessed by the parent models when compared with models of 'foreign' species. Discrimination was best developed in the very young fry (1-3 days), but as the tendency to flee from larger fish developed, positive responses to detailed models of the parents disappeared. Though these reactions are probably mainly innate, there was a specific enhancement of response with exposure to the appropriate model, suggesting some kind of rather weak imprinting mechanism. Indeed, Hay uses the term 'filial imprinting' to describe it.

(b) Territorial Responses. Aggregation responses can be used for discrimination experiments, but much more powerful agents are available to the experimental behaviourists in the territorial behaviour of fishes. In sunfish and sticklebacks a nest site provides a localised resource, and intruding male conspecifics are driven off as part of the programme of breeding behaviour. One of the most important studies in this field is that of Stacey and Chiszar (1978) on the breeding male pumpkinseed sunfish (*Lepomis gibbosus*), a centrarchid. They used a variety of cut-out models representing intruding males, with key features deleted. Their general conclusions were that all the seven key features had an effect in determining aggression. The most powerful features are the red iris of the eye, and the red patch on the earflap. The blue-speckled body pattern and the yellow chest patches are subsidiary in their ability to promote attacks. Conversely, the black eye and tab features slightly reduce aggression, and the vertical bars displayed by the female strongly reduce aggression. A black dorsal fin spot (which is like the one found in perch, see above) characteristic of an allied species, the bluegill (*L. macrochirus*), greatly increases aggression.

From this example, with its 'features hierarchy', we pass to a variety of aggressive manoeuvres involving larger and less well defined territories.

Most of the studies that are available to us assume a more or less constant level of motivation in the defender.

In the striped cleaner fish *Labroides dimidiatius*, Potts (1973) analysed the visual content of conspecifics by means of six different models supported on wires placed at the cleaning site. The ratio between the number of encounters followed by attacks and those resulting in no response varied from 0.1 for a white cut-out to 10 for a striped model. Neither of the two striped models was quite similar to the *Labroides* pattern, but the most successful model mimicked the central feature of the flank pattern — a white bar bordered by black bars. A vertical-bar model produced a marginally better score than some of the other models, and it could be argued that this model possessed the correctly contrasting type of pattern but with incorrect orientation.

The importance of the orientation of features has seldom been analysed in detail, but Heiligenberg, Kramer and Schultz (1972) did examine this, using models to determine the most effective orientation of the black eyebar in male *Haplochromis burtoni*. In territorial males of this species a bar extends downwards from the eye at an angle of 45° to the snout profile and this triggers attacks in other males. The problem of orientation is complicated by the head-down posture adopted by the fish at close range (the conflict position). Their findings were as follows:

(i)   Bars at 180° or 0° were more effective than those at 45°.

(ii)   The shape of the attack rate/bar angle curve was similar for vertical and horizontal positions of the fish.

(iii)   Vertical positions of the dummy were more effective than horizontal for four out of five bar angles. They noted that even a dummy without an eyebar would provoke attacks if held vertically. Eyebar and body angle behave largely as independent additive features. The finding that there is a more effective signal than the natural one has been noted in herring gulls and fireflies, and must be the result of other constraints. It is clear that within rather wide limits, the eyebar angle is important and attack rate doubles across the bar angle range A smaller increment is due to body position. The results indicate that eyebar angle is assessed as a feature related to the geometry of the fish rather than to horizontal and vertical planes. When the male has established a territory, orange shoulder patches appear, and these inhibit attack readiness in other males. Heiligenberg *et al.* (1972) developed a theoretical scheme to explain these interactions (Figure 4.8). It was suggested earlier that Hay's results with fry support the idea of an acquired element in the recognition process. Fernald (1980) came to the conclusion from the study of *Haplochromis burtoni* reared in isolation that eyebar recognition was an innate process, but, as discussed in Chapter 3 by Huntingford, this volume, Crapon de Caprona (1982), using similar material, found that (at least for the males) early experience was required for the correct interpretation of the complex colour patterns.

What might be termed the generalised territorial behaviour of reef fishes has been the subject of enthusiastic study by several authors. The brilliance and variety of the colour patterns of these species, combined with the high clarity of the reef water, suggest that visual signals are particularly important. Lorenz (1962) put forward the idea that brilliant 'poster' colours found in certain reef fishes were related to their degree of territoriality. For example, the relatively non-competitive trigger fish (*Odonus niger*) is relatively inconspicuous, whereas the bizarrely patterned Picasso fish *Rhinecanthus aculeatus* is very competitive and quarrelsome. It also came to be recognised that competition for a resource was likely to be most fierce between related forms with similar requirements. Zumpe (1965) studying three species of butterflyfish (*Chaetodon*) found 192 attacks on conspecifics as compared with 20 attacks on foreign species. The pattern found in these species conforms to a generic type, but is otherwise recognisably different to the human observer. Ehrlich, Talbot, Russell and Anderson (1977), also working with *Chaetodon* (*C. trifascialis*) but using models, came to a similar conclusion, that is to say, the level of aggression displayed towards models of other species was directly proportional to their similarity to conspecifics.

Myrberg and Thresher (1974), using the damselfish (*Eupomacentrus planifrons*), also found that conspecifics were driven off at the greatest range (2.75-4.0 m) whereas an allied species like *E. patilius* could approach to within 1 m; other genera could approach still closer. The authors point out that *E. planifrons* is a drab species, yet very aggressive, and this conflicts with the advertisement-aggression idea of Lorenz. Other studies confirm the lack of correlation between bright colours and territorial aggression (see Fine 1977).

In contrast, the results of Thresher's later study (1976) on 22 species of intruders into the territory of *E. planifrons* found that only 5 per cent of attacks were on conspecifics. The species attacked at close range belonged to different genera and seemed to be competitive for shelter, nest holes, etc. Another class, attacked at longer range, fed on similar benthic algae, and this resource was presumably more significant.

The visual consequences of the need to recognise conspecifics is that comparatively small differences of pattern and colour have to be recognised rather than large differences of outline, head shape, and the like.

(c) Visual Signals and Courtship Behaviour

(i) *General*: the more striking aspects of the physical appearance of fishes seem often to be related to communication between conspecifics in the context of breeding behaviour. The stereotyped nature of this kind of behaviour implies highly specific, stable and effective recognition processes

Figure 4.8: Diagrammatic Representation of the Quantitative Relationships Involved in the Territorial Defence of Male *Haplochromis burtoni*. The black eyebars tend to produce a fast (E1) and a slow (E2) effect on attack response, and the orange shoulder patches inhibit (E3) responses. The inputs are seen as acting through feedback-stabilised amplifiers (boxes). The profiles indicate the kind of time-dependent process involved. Slow excitation (E2) has its main effect via an internal motivational system, V1-V4

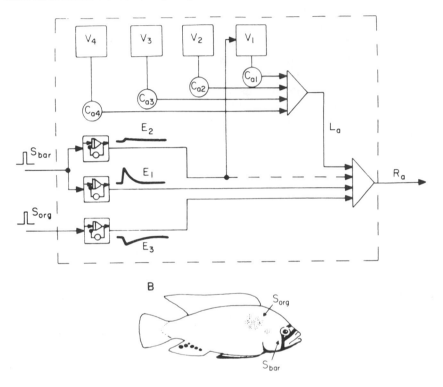

Source: Heiligenberg *et al.* (1972).

which are genetically determined and subject to little modification as a result of experience. The powerful motivational systems involved in this kind of behaviour probably aid experimentation.

(ii) *Sexual dimorphism*: two kinds of visual signal are involved in courtship, signals usually consisting of movements that initiate courtship responses, and recognition of the sexual status of the partner. Sexual status may be conveyed by morphological differences such as temporary breeding colours, or sexual dimorphism which is a constant feature of the species. Sexual dimorphism of the kind that is obvious to the human observer is rare in most teleost species and the lack of it may prove a source of difficulty to fish farmers.

In *Megalamphodus swegelsii* the dorsal fin is lance-shaped in the male and triangular in the female; the anal fin is rounded in males of *Nannosto-*

mus marginalis and angular in females. Sexual dimorphism is striking in the dragonet (*Callionymus lyra*). The male has sail-like fins and turquoise stripes; the female short fins and drab colours. Males of *Puntius arulius* have exposed rays at the margin of the dorsal fin, whereas in the female the membrane extends to the edge. In the spined loach (*Cobitis taenia*) the male fish have much larger pectoral fins than the females, and these fins bear an enlarged scale, the Canestrini scale. Similar morphometrical differences can be observed in certain other species. Studies made in our laboratory on the freshwater perch using measurements made from projected images of fresh specimens suggested that small differences in the morphometry of the head are sexually related. In males the eye is relatively larger, the maxilla longer, etc., but it is difficult to believe that these can be used in recognition by the fish themselves.

During the reproductive season the males of many species exhibit breeding colours. The scarlet belly patch of the male stickleback is the most well known, and other European freshwater and littoral marine species like the minnow, the goby and the charrs (*Salvelinus* spp.) exhibit red patches or coloration. As we have seen, these reds and oranges show up well against a greenish spacelight. In both the stickleback and the minnow, the red colour seems to be a signal used to drive off competing males, and it is the bluish-green iridescence of the flanks in both species (Pitcher, Kennedy and Wirjoatmodjo 1979) that is effective in encouraging the females. In some species the males are much darker than the females: the black-faced blenny (*Tripterygion tripteronotus*) has a dark head (in the male only) that blackens at spawning, and in several species of goby (*Gobius fluviatilus*) the males go entirely dark at spawning and male surgeon fish (acanthurids) darken in the breeding season. The degree of darkening in *Tilapia* expresses the degree of dominance of that male. As we have seen earlier, achromatic hues are believed to show up well. Changes in silhouette are rarer. In the strongly territorial *Blennius pavo*, males develop a pad of fat on the head (as well as blue spots), and in the poeciliid *Neoheterandria*, males develop a flank spot as an appeasement signal (MacPhail 1969).

Alterations in female coloration are perhaps less common. Females of the pumpkinseed sunfish and those of the blue panchax (*Aplocheilus lineatus*) develop dark vertical bars during the breeding season. In *Lepomis* it is possible using models to show that this feature inhibits male aggression (Stacey and Chiszar 1978), and in *Lepomis cyanellus* and *L. macrochirus* such vertical banding can be produced by focal electrical stimulation of midbrain areas (Bauer and Demski 1980). Does the position of these distinctive dimorphic features matter?

Picciolo (1964) showed that the blue anterior-ventral patch of male dwarf gourami (Colisa lalia) is critically located for sexual recognition, by displacing it to other sites on models, and Keenleyside (1971) also demonstrated the importance of position with regard to the eye and opercular patch of the long-ear sunfish (*Lepomis megalotis*).

(iii) Courtship movements: courtship displays by males have been described in many different species. Usually it can be seen that there is a repetitive or rhythmic component in the movement. One of the most easily observed among tropical freshwater species is the courtship dance of a common aquarium species, zebrafish (*Brachydanio rerio*). The male moves through a series of abrupt turns and in an elliptical pattern around the female. The movement is striking and clearly distinguishable from the normal movements. To the human observer it has the effect of emphasising the flank and fin markings. Minnows perform a low-level waggle dance, hiding the red belly that signals aggressive intent to other males, but exposing the turquoise flank patches to the female. In the labrid (*Halichoeres melanochir*), there is a mating system similar to a lek, sexual dimorphism is slight, but there is a complex routine of rhythmic movements that leads to spawning (Moyer and Yogo 1982).

These dance-like movements are often followed up by the male fish poking or butting the female in the flanks (*Poecilia, Gasterosteus* and *Phoxinus*), and in some species the male eventually curls himself around the body of the female (*Macropodus* sp., *Barbus* sp.).

A specialised sequence occurs in the mouth-brooding cichlid *Haplochromis burtoni*. The male drags his anal fin, which bears egg-like markings, in front of the female; when she comes forward to pick up the eggs, she takes in the sperm which is expelled near the male's anal fin, and fertilisation occurs in the mouth. The representation of the egg row on the male's fin appears an exact one to the human eye.

To sum up, courtship behaviour often relies chronologically on (i) visual recognition of the opposite sex and of its breeding condition; (ii) the state of readiness conveyed by the male by means of specific movements; (iii) the participatory behaviour of the female. The latter is often less striking than that of the male, but maintains continuity (by the female staying near spawning or nesting site). These visual elements may be strongly supported by pheromones or other cues. Sound emission may complement colour changes during courtship or territorial behaviour as, for example, in cichlids. Nelissen (1978) found that among a group of six cichlid species that he studied there was an inverse relationship between the number of colour patterns and the number of sound patterns. Needless to say, visual signals are relatively unimportant in some species, such as the goldfish, where pheromones play the dominant role.

(d) Visual Differentiation of Social Role and Development Status.   In many species there is little visible sign of developmental status apart from size. This is true of most European freshwater species. At the other extreme are cichlids, and a number of reef genera (such as *Pomacanthus*) in which juveniles are conspicuously different from adults. There is insufficient space here for detailed descriptions of the range of colour patterns that occur within a species and provide signals to conspecifics, but the

reader should refer for an example to the tables of Voss (1977). These depict the appearance of some 20 cichlid species under 20 headings corresponding to differences of sex, age, breeding role and emotional state. It is worth noting that the transitions from one form to another may be as significant as the patterns themselves (Figure 4.9).

Although there is some evidence for dominance in shoaling species, colour differences related to this are less well known. However, McKaye and Kocher (1983) found that the dominant individual in a foraging shoal of *Cyrtocara moorei* is a vivid blue colour distinguishable from the others, which have only the basic flank pattern (three dark spots on a pale blue ground).

(e) Visually Mediated Aspects of Predation.   A good deal of interest has been focused on prey selection in relation to optimal foraging strategy (see Chapter 8 by Hart, this volume). Hairston, Li and Easter (1983) showed that as bluegill sunfish (*Lepomis macrochirus*) grow, they appear to be able to detect smaller and smaller prey. Cone separation in the retina does not alter, but the angle between adjacent cones becomes smaller. The improvement in acuity increases the range at which a given prey size can be detected. Gibson and Easter (1982) point out that this wider range of available prey should allow larger fish to be more selective than smaller ones, with perhaps a bias towards the more valuable, i.e. larger, prey. O'Brien, Slade and Vineyard (1976) found that large bluegills *did* select large prey when there was abundant prey. More recently it has been suggested that other factors such as turbidity and effective water volume (shallows) may have a decisive effect on strategy.

Coates (1980a) demonstrated how quickly the approach strategy of predators was altered according to the expected outcome. The five species tested all learned to avoid the non-reward species (*Dascyllus aruanus*). Confer *et al.* (1978) provide some valuable comparative data concerning the distances at which four species of fish (bluegills, lake trout, brook trout, pumpkinseed) appear to respond to prey of different sizes (Figure 4.10). As might be expected, reaction distances (RD) increased as light levels were raised, up to a critical level, but for the pumpkinseed RD was much larger (+75 to 115 per cent) than for the brook trout at the two lowest light levels and similar prey size. For lake trout the relationship between turbidity (estimated as extinction coefficient) and RD was a linear one.

(f) Recognition of Predators by Fish.   Many species show a visually mediated startle response to large moving objects. This is probably controlled by the action of the giant Mauthner cells (Eaton and Bombardieri 1978). More specific responses to predator species have been studied by Coates (1980b) in the humbug (*Dascyllus aruanus*). Selected predatory species presented in transparent containers had the effect of halving the number of *Dascyllus* swimming above their 'home' coral head, as compared

Figure 4.9: Main Colour Patterns of two Species of Cichlids, *Thysia ansorgii* and *Pelvicachromis pulcher*. (i) A₁ — neutral livery. (ii) A₂ — livery of territorial males (two versions in *P. pulcher*). (iii) Females, early stages of sexual parade, *T. ansorgii* — A₃; *P. pulcher* — B₃. (iv) C₁, *T. ansorgii* alternative to A₂ — egg-laying livery (males). (v) Egg-guarding livery, *T. ansorgii* — D, males, C₂, females; *P. pulcher* — B2 (females). (vi) Fry-guarding livery, *T. ansorgii* — D males, E females. (vii) Fear, inferiority, *T. ansorgii* — A₃, B₂; *P. pulcher* — A₁, B₁ fo;llowed by C. (viii) Young — special form (*T. ansorgii*); *P. pulcher* — D. arg, Silver spot in female; rg, red; bl, blue; nr, black; grenat, garnet

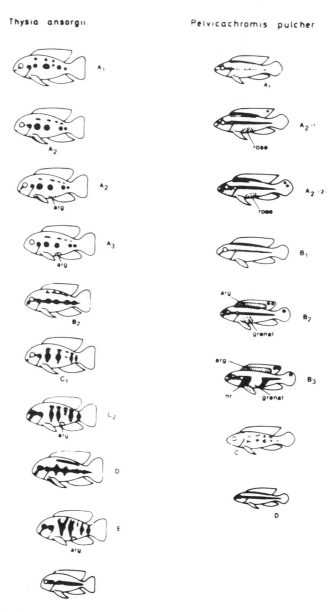

Figure 4.10: Effective Visual Acuity as Determined by the Range of the Initial Predatory Response to Prey of Different Sizes in Three Species of Trout and Two Species of Sunfish

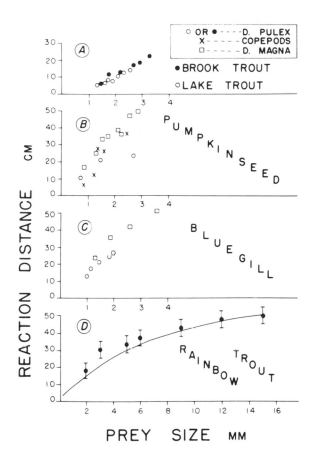

Source: Confer *et al.* (1978).

with either an empty container or one containing a non-predatory species presented at the same ranges. Despite the tendencies of the humbugs to retreat, they would also make attempts to drive off the predator, but it was notable that this aggressive behaviour was rarer when the predators were presented at close range (one-fifth the occurrences at long range for five predator species). The very striking looking *Pterois volitans* was seldom attacked at any range.

One of the most interesting papers in this field resulted from a study by Karplus, Goren and Algom (1982) on the recognition by the reef fish *Chromis caeruleus* of predator's faces where recognition must depend on frontal views of the approaching fish. A large-eyed model with a large upturned mouth (piscivore) was contrasted with a small-eyed model with a small downturned mouth (non-piscivore). Subject position was significantly affected by the piscivore model. If the piscivore model was contrasted with models with small upturned mouths, or large downturned mouths, then the subject moved closer to the model with non-piscivore features. A disruptive pattern of patches applied to the piscivore model significantly reduced the aversive response of the subject. In this context, considerable interest has been aroused among primate ethologists by the discovery by Perrett and Rolls (1983) of neurones in the monkey temporal lobe specific for face recognition.

The camouflage adopted by would-be predators has two forms: (i) aggressive (Peckhammian) mimicry, or (ii) resemblance to background. In the scale-eater *Proboluchus heterostomus* the visual appearance of the predator closely resembles that of the harmless species *Astyanaxe fasciatus*, and this allows it to approach its victims (Sazima 1977). A more versatile example is provided by the paedophagous cichlid *Cichlasoma orthognathus*, which adopts a stripe when stalking *C. pleurotaenia*, and remains all silver when stalking *C. encinostomous* (McKaye and Kocher 1983).

Summary

The physical properties of natural waters produce varying limitations on the visual signalling capabilities of fishes. In fish communities rich in species, especially where the water is reasonably transparent (coral reefs; Lake Malawi) the complex colour patterns, contours and textures distinguish individual species, and it can be clearly demonstrated that species respond differentially to these signals.

Studies on the fish eye show that its ability to resolve detail roughly matches the degradation of edges caused by suspended particles. Contrast direction and hue may therefore have become more important as a means of object recognition. One of the most successful areas of study has been the correlation of receptor pigments of a species with the spectral characteristics of its environment. The bulk of physiological work has so far been centred on the retina, where opponent processes rather similar to those of primates appear to mediate colour contrast and hue discrimination. Our understanding of central visual processing, i.e. beyond the afferent tectal layers, remains fragmentary and offers an interesting though daunting prospect for future work.

Conditioning experiments have provided some useful basic information about shape significance and hue discrimination, but have seldom been related to natural forms or colours. Field and tank trials using pre-existing responses (territorial aggression, shoaling aggregation) and natural stimuli have been among the most exciting studies made in the last 15 years. From the point of view of discriminatory mechanisms the order of effectiveness under given environmental conditions of hue, texture, contour, etc. is not clearly elucidated for any one example, and remains a goal for the future.

References

Ahlbert, I.-B. (1968) 'The Organization of the Cone Cells in the Retinas of Four Teleosts with Different Feeding Habits', *Arkiv für Zoologie, 22*(11), 445-81

Ali, M.A. (1959) 'The Ocular Structure, Retinomotor and Photo-behavioural Responses of Juvenile Pacific Salmon', *Canadian Journal of Zoology, 37,* 965-96

Ali, M.A. and Anctil, M. (1976) *Retinas of Fishes: an Atlas,* Springer-Verlag, Heidelberg

Avery, J.A., Bowmaker, J.K., Djamgoz, M.B.A. and Downing, J.E.G. (1983) 'Ultra-violet Sensitive Receptors in a Freshwater Fish', *Journal of Physiology, London, 334,* 23

Baerends, G.P. and Baerends-van Roon, J.M. (1950) 'An Introduction to the Study of Cichlid Fishes', *Behaviour (Supplement), 1,* 1-242

Baker, E.T., Sternberg, R.W. and MacManus, D.A. (1974) 'Continuous Light-scattering Profiles and Suspended Matter over the Nittinat Deep Sea Fan', in R.J. Gibbs, (ed.), *Suspended Solids in Water,* Plenum, London, pp. 155-72

Barlow, G.W. (1982) 'Monogamy amongst Fishes'. Abstract from *4th Congress of European Ichthyologists, Hamburg*

Bauer, D.H. and Demski, L.S. (1980) 'Vertical Banding Evoked by Electrical Stimulation of the Brain in the Anaesthetized Green Sunfish *Lepomis organellus* and Bluegills *Lepomis machochir*', *Journal of Experimental Biology, 84,* 149-60

Breder, C.M. Jr. (1951) 'Studies on the Structure of Fish Shoals', *Bulletin of the American Museum of Natural History, 198,* 1-27

Brunskill, G.J. (1970) 'Fayetteville Green Lake, New York', *Limnology and Oceanography, 14,* 133-70

Burkamp, W. (1923) 'Versuche über Farbenwiedererkennen der Fische', *Zeitung für Z. Sinnesphysiologie, 55,* 133-70

Cajal, S.R. (1893) 'La Retiné des Vertebres', *La Cellule, 9,* 17-257

Cameron, N.D. (1974) 'Chromatic Vision in a Teleost Fish: *Perca fluviatilis*', Unpublished PhD thesis, Sussex University

Charman, W.N. and Tucker, J. (1973) 'The Optical System of the Goldfish Eye', *Vision Research, 13,* 1-8

Coates, D. (1980a) 'Potential Predators and *Dascyllus aruanus*', *Zeitschrift für Tierpsychologie, 52,* 285-90

Coates, D. (1980b) 'The Discrimination of and Reactions towards Predatory and Non-predatory Species of Fish in the Humbug Damselfish', *Zeitschrift für Tierpsychologie, 52,* 347-54

Confer, J.L., Howick, G.L., Corzette, M.H., Kramer, S.L., Fitzgibbon, S. and Landesberg, R. (1978) 'Visual Predation by Planktivores', *Oikos, 31,* 27-37

Crapon de Caprona, M.D. (1982) 'The Influence of Early Experience on Preferences for Optical and Chemical Cues Produced by Both Sexes of a Cichlid Fish *Haplochromis burtoni*', *Zeitschrift für Tierpsychologie, 58,* 329-61

Daw, N.W. (1968) 'Colour Coded Ganglion Cells in the Goldfish Retina'. *Journal of Physiology, 197,* 567-92

de Groot, S.G. and Gebhard, J.W. (1952) 'Pupil Size as Determined by Adapting Luminance', *Journal of the Optical Society of America, 42,* 492-5

Denton, E.J. (1970) 'On the Organization of Reflecting Surfaces in Some Marine Animals', *Philosophical Transactions of the Royal Society of London B, 182,* 154-58

Dodt, E. (1963) 'Photosensitivity of the Pineal Organ in the Teleost Salmo irideus', *Experientia, 19,* 642-3

Eaton, R.C. and Bombardieri, R.A. (1978) 'Behavioural Functions of the Mauthner Cell', in D. Faber and H. Korn (eds.), *Neurobiology of the Mauthner Cell,* Raven Press, New York, pp. 221-44

Ehrlich, P.R., Talbot, F.H., Russell, B.C. and Anderson, G.R.V. (1977) 'The Behaviour of Chaetodontid Fishes with Special Reference to Lorenz's Poster Coloration Hypothesis', *Journal of Zoology, 183,* 213-22

El-Akell, A. (1982) The Optic Tectum of the Perch', Unpublished PhD thesis, University of Manchester

Endler J.A. (1977) 'A Predator's View of Animal Colour Patterns', *Evolutionary Biology*, *11*, 319-64

Fernald, R.D. (1980) 'Response of Male *Haplochromis burtoni* Reared in Isolation to Models of Conspecifics', *Zeitschrift für Tierpsychologie*, *54*, 85-93

Fine, M.L. (1977) 'Communication in Selected Groups: Communication in Fishes' in T.A. Sebeok (ed.), *How Animals Communicate*, Indiana University Press, Bloomington, pp. 472-518

Fricke, H.W. (1973) 'Individual Partner recognition in Fish: Field Studies on *Amphiprion bicinctus*', *Naturwissenschaften*, *60*, 204-6

Fukurotani, K. and Hashimoto, Y. (1984) 'A New Type of S-Potential in the Retina of Cyprinid: the Tetraphasic Spectral Response', *Investigations into Ophthalmic and Visual Science*, *25*, ARVO abstract 1

Galand, G. and Liége, G. (1975) 'Responses Visuelles Unitaires Chez la Truite', in M.A. Ali (ed.), *Vision in Fishes*, Plenum, New York, pp. 1-14

Gibson, R.M., Li, K.T. and Easter, S.S. (1982) 'Visual Abilities in Foraging Behaviour of Predatory Fish', *Trends in Neurosciences*, June, 1-3

Guthrie, D.M. (1981) 'The Properties of the Visual Pathway of a Common Freshwater Fish (*Perca fluviatilis*) in Relation to its Visual Behaviour', in P.R. Laming (ed.), *Brain Mechanisms of Behaviour in Lower Vertebrates*, Cambridge University Press, Cambridge, pp. 79-112

Guthrie, D.M. (1983) 'Visual Central Processes in Fish Behaviour', in J.P., Ewert, R.R. Capranica, and D.I. Ingle (eds.), *Recent Advances in Vertebrate Neuroethology*, Plenum, New York, pp. 381-412

Guthrie, D.M. and Banks, J.R. (1978) 'The Receptive Field Structure of Visual Cells from the Optic Tectum of the Freshwater Perch', *Brain Research*, *141*, 211-25

Hairston N.G., Li, K.T., and Easter, S.S. (1983) 'Fish Vision and the Detection of Planktonic Prey', *Science*, *218*, 1240-2

Hanyu, I. (1978) 'Salient Features in Photosensory Function of Teleostean Pineal Organ', *Journal of Comparative Biochemistry and Physiology*, *61A*, 49-54

Harosi, F.I. (1984) 'Ultra-violet and Visible Pigments in Two Cyprinids', *Investigations in Ophthalmic and Visual Science*, *25*, ARVO abstract 28

Harosi, F.I. and Hashimoto, Y. (1983) 'Ultra-violet Visual Pigment in a Vertebrate: a Tetrachromatic Cone System in a Dace', *Science*, *222*, 1021-3

Hay, T.F. (1978) 'Filial Imprinting in the Convict Cichlid Fish *Cichlasoma nigrofascinatum*', *Behaviour*, *65*, 13860

Heath, R.T. and Franks, D.A. (1982) 'U.V. Sensitive Phosphorous Complexes in Association with Dissolved Humic Material and Iron in a Bog Lake', *Limnology and Oceanography*, *27*(3), 564-9

Heiligenberg, W., Kramer, U. and Schultz, V. (1972) 'The Angular Orientation of the Black Eyebar in *Haplochromis burtoni* and its Relevance to Aggressivity', *Zeitschrift für vergleichende Physiologie*, *76*, 168-76

Helfman, G.S. (1981) 'The Advantage to Fishes of Hovering in the Shade', *Copeia*, *2*, 392-400

Hemmings, C.C. (1966) 'Factors Influencing the Visibility of Underwater Objects' in C.C. Evans, R. Bainbridge and O. Rackham (eds.), *Light as an Ecological Factor, Symposium of the British Ecological Society 6*, Blackwell, Oxford, pp. 359-74

Herter, K. (1929) 'Dressurversuche an Fischen', *Zeitschrift für vergleichende Physiologie*, *10*, 688-711

Hurst, P.M. (1954) 'Colour Discrimination in the Bluegill Sunfish', *Journal of Comparative and Physiological Psychology*, *46*, 442-5

Karplus, I., Goren, M. and Algom D. (1982) 'A Preliminary Experimental Analysis of Predator Face Recognition by *Chromis caerulaeus*', *Zeitschrift für Tierpsychologie*, *61*, 149-56

Katzir, G. (1981) 'Visual Aspects of Species Recognition in the damselfish *Dascyllus aruanus*', *Animal Behaviour*, *29*(3), 842-9

Keenleyside, M.H.A. (1955) 'Some Aspects of Schooling in Fish', *Behaviour*, *8*, 183-249

Keenleyside, M.H.A. (1971) 'Aggressive Behaviour of the Male Long Ear Sunfish (*Lepomis megalotis*)', *Zeitschrift für Tierpsychologie*, *28*, 227-40

Kuenzer, P. (1968) 'Die Auslösung der Nachfolgereaktion bei Erfahrungslosen Jungfischen von *Nannacara anomala*', *Zeitschrift für Tierpsychologie*, *25*, 257-314

Kuenzer, P. (1975) 'Analyse der Auslosende Reizsituationen für die Anschwimm-, Eindring- und Fluchtreaktion Junger *Hemihaplochromis multicolor*', *Zeitschrift für Tierpsychologie*, *47*, 505-44

Kullenberg, G. (1974) 'The Distribution of Particulate Matter in a North-west African Coastal Upwelling Area', in R.J. Gibbs (ed.), *Suspended Solids in Water*, Plenum, London, pp. 95-202

Levine, J.S., Lobel, P.S. and MacNichol, E.f. (1980) 'Visual Communication in Fishes', in M.A. Ali (ed.), *Environmental Physiology of Fishes*, Plenum, New York, pp. 447-75

Liebmann, P.A. and Entine, G. (1964) 'Sensitive Low Light-level Microspectrophotometric Detection of Photosensitive Pigments of Retinal Cones', *Journal of the Optical Society of America*, *54*, 1451-9

Limbaugh, C. and Rechnitzer, A.B. (1951) 'Visual Detection of Temperature-density Discontinuation in Water by Diving', *Science*, *121*, 11

Loew, E.W. and Lythgoe, J.N. (1978) 'The Ecology of Cone Pigments in Teleost Fishes', *Vision Research*, *18*, 715-22

Lorenz, K. (1962) 'The Function of Colour in Coral Reef Fishes', *Proceedings of the Royal Institute of Great Britain*, *39*, 282-96

Lythgoe, J.N. (1974) 'The Structure and Function of Iridescent Corneas in Fishes' in M.A. Ali (ed.), *Vision in Fishes*, Plenum, New York, pp. 253-62

Lythgoe, J.N. (1984) 'Visual Pigments and Environmental Light', *Vision Research*, *24*, 1-26

Lythgoe, J.N. and Northmore, D.P.M. (1973) 'Problems of Seeing Colours Underwater' in *Colour '73*, Adam Hilger, London, pp. 77-98

McKaye, K.R. and Kocher, T. (1983) 'Head-ramming Behaviour by Three Species of Paedophagous Cichlids', *Animal Behaviour*, *31*, 206-10

McPhail, J.D. (1969) 'Predation and the Evolution of a Stickleback (*Gasterosteus*), *Journal of the Fisheries Research Board Canada*, *26*, 3183-208

Meek, H. and Schellart, N.A.M. (1978) 'A Golgi Study of the Goldfish Optic Tectum', *Journal of Comparative Neurology*, *182*, 89-22

Meyer, D.L., Schott, D. and Schaeffer, K.D. (1970) 'Brain Stimulation in the Optic Tectum of Freely Swimming Codfish *Gadus morrhua*', *Pflügers Archiv gesamtliche Physiologie*, *314*, 240-52

Moller, P., Servier, J., Squire A. and Boudinot, M. (1982) 'Role of Vision in Schooling', *Animal Behaviour*, *30*, 641-50

Morita, Y. (1966) 'Enladungsmuster Pinealer Neurone des Zwischenhirns', *Pflügers Archiv gesamtliche Physiologie*, *289*, 135-67, 1

Moyer, J.T. and Yogo, Y. (1982) 'The Lek-like Mating System of *Halichoeres melanochir* at Miyake-jina, Japan', *Zeitschrift für Tierpsychologie*, *60*(3), 209-26

Muntz, W.R.A. (1973) 'Yellow Filters and the Absorption of Light by the Visual Pigment of Some Amazonian Fishes', *Vision Research*, *13*, 2235-54

Muntz, W.R.A. (1976) 'On Yellow Lenses in Mesopelagic Animals', *Journal of the Marine Biological Association, U.K*, *56*, 963-76

Myrberg, A.A. and Thresher, R.E. (1974) 'Interspecific Aggression and its Relevance to the Concept of Territoriality in Reef Fishes', *American Zoologist*, *14*, 81-96

Nelissen, M.H.J. (1978) 'Sound Production by some Tanganyikan Cichlid Fishes and a Hypothesis for the Evolution of the r Communication Hypothesis', *Behaviour*, *64*, 137-47

Noble, G.K. and Curtis, B. (1939) 'The Social Behaviour of the Jewel Fish *Hemichromis bimaculatus*', *Bulletin of the American Museum of Natural History*, *76*, 1-46

Northmore, D., Volkmann, F.C. and Yager, D. (1978) 'Vision in Fishes: Colour and Pattern' in D.I. Mostofsky (ed.), *Behaviour in Fish and Other Aquatic Animals*, Academic Press, New York, pp. 79-136

O'Benar, J.D. (1976) 'Electrophysiology of Neural Units in the Goldfish Optic Tectum'. *Brain Research Bulletin*, *1*, 529-41

O'Brien, W.J., Slade, N.A. and Vineyard, G.L. (1976) 'Optimal Foraging Strategy in Fishes', *Ecology*, *57*, 1304-30

Ormond, R.F.G. (1974) 'Visual Responses in Teleost Fish', unpublished PhD thesis, University of Cambridge

Page, C.H. and Sutterlin, A.M. (1970) 'Visual Auditory Unit Responses in the Goldfish Tegmentum'. *Journal of Neurophysiology*, *33*, 129-36

Patfield, I. (1983) 'Conspecific Recognition in the Perch', Unpublished PhD thesis, University of Manchester

Perrett, D.I. and Rolls, E.T. (1983) 'Neural Mechanisms Underlying the Analysis of Faces', in J.P. Ewert, R.R. Capranica and D.I. Ingle, (eds.), *Advances in Vertebrate Neuroethology*, Plenum, New York, pp. 543-68

Picciolo, A.K. (1964) 'Sexual and Nest Discrimination in Anabantid Fishes of the Genera *Colisa* and *Trichogaster*', *Ecological Monographs*, *34*, 53-77

Pitcher, T.J., Kennedy, G.J.A. and Wirjoatmodjo, S. (1979) 'Links between the Behaviour and Ecology of Freshwater Fishes', *Proc. 1st British Freshwater Fisheries Conference*, 162-75

Potts, G.W. (1973) 'The Ethology of *Labroides dimidiatus* on Aldabra', *Animal Behaviour*, *21*, 250-91

Rasa, O.A.E. (1969) 'Territoriality and the Establishment of Dominance by Means of Visual Cues in *Pomacentrus jenkinsii*', *Zeitschrift für Tierpsychologie*, *26*, 825-45

Rivers, R. (1953) 'The Pineal Apparatus of Tunas in Relation to Phototactic Movement', *Bulletin of Marine Science of the Gulf of the Caribbean*, *3*, 168-80

Ruttner, F. (1952) *Fundaments of Limnology*, University of Toronto Press, Toronto

Sajovic, P. and Levinthal, C. (1982) 'Visual Cells of the Zebra Fish Optic Tectum. Mapping with Small Spots', *Neuroscience*, *7*(10), 2407-26

Sazima, I. (1977) 'Possible Case of Aggressive Mimicry in a Neotropical Scale-eating Fish', *Nature, London*, *270*, 510-12

Schellart, N.A.M. (1983) 'Acousticolateral and Visual Processing and their Interaction in the Torus Semicircularis of the Trout *Salmo gairdneri*', *Neuroscience Letters*, *42*, 39-44

Schellart, N.A.M. and Spekreijse, H. (1976) 'Shapes of Receptive Field Centers in the Optic Tectum of the Goldfish', *Vision Research*, *16*, 1018-20

Schmidt, J.T. (1979) 'The Laminar Organization of Optic Nerve Fishes in the Tectum of Goldfish', *Proceedings of the Royal Society of London B*, *205*, 287-306

Schubel, J.R. (1974) 'Effect of Tropical Storm Agnes on the Suspended Solids of the Northern Chesapeake Bay', in J.R. Gibbs (ed.), *Suspended Solids in Water*, Plenum, London, pp. 113-32

Schulte, A. (1957) 'Transfer- und Transpositions-Versuche mit Monokular Dressierten Fischen', *Zeitschrift vergleichende Physiologie*, *393*, 432-76

Schwassman, H.O. (1974) 'Refractive State, Accommodation and Resolving Power of the Fish Eye' in M.A. Ali (ed.), *Vision in Fishes*, Plenum, New York, pp. 279-88

Schwassman, H.O. and Meyer, D.L. (1973) 'Refractive State and Accommodation in Three Species of *Paralabrax*', *Videnskabelige Meddelser fra Dansk Naturhistorisik forening i København*, *134*, 103-8

Sivak, J.G. (1973) 'Accommodation in Some Species of North American Fishes', *Journal of the Fisheries Research Board of Canada*, *30*, 1141-6

Sivak, J.G. and Bobier, W.R. (1978) 'Chromatic Aberration of the Fish Eye and its Effect on Refractive State', *Vision Research*, *18*, 453-5

Sivak, J.G. and Howland, H.C. 'Accommodation in the Northern Rock Bass (*Ambloplites rupestris*) in Response to Natural Stimuli', *Vision Research*, *13*, 2059-64

Sivak, J.G. and Kreutzer, R.O. (1981) 'Spherical Aberration of Crystalline Lenses Studied by Means of a Helium-neon Laser', *Vision Research*, *23*, 59-70

Spekreijse, J., Wagner, H.G. and Wolbarsht, M.T. (1972) 'Spectral and Spatial Coding of Ganglion Cell Responses in Goldfish Retina', *Journal of Neurophysiology*, *35*, 73-86

Sroczynski, S. (1977) 'Spherical Aberration of the Crystalline Lens in the Roach', *Journal of Comparative Physiology*, *121*, 135-44

Stacey, P.B. and Chiszar, D. (1978) 'Body Colour Pattern and Aggressive Behaviour of Male Pumpkinseed Sunfish (*Lepomis gibbosus*)', *Behaviour*, *64*, 271-304

Stewart, K.M. and Martin, P.J.H. (1982) 'Turbidity and its Causes in a Narrow Glacial Lake with Winter Ice Cover', *Limnology and Oceanography*, *27*(3), 510-17

Sutterlin, A.M. and Prosser, C.L. (1970) 'Electrical Properties of the Goldfish Optic Tectum', *Journal of Neurophysiology*, *33*, 36-45

Svaetichin, G. (1953) 'The Cone Action Potential', *Acta physiologica scanda Supplement* *106*, 565-600

Tamura, T. (1957) 'A Study of Visual Perception in Fish, Especially on Resolving Power and Accommodation', *Bulletin of the Japanese Society of Scientific Fisheries*, *22*, 536-57

Tayssedre, C. and Moller, P. (1983) 'The Optomotor Response in Weakly Electrical Fish',

Zeitschrift für Tierpsychologie, 60(4), 265-352

Tinbergen, N. (1951) *The Study of Instinct,* Clarendon Press, Oxford

Tipping, E. and Woof, C. (1983) 'Humic Substances in Esthwaite Water', *Limnology and Oceanography, 28*(1), 145-53

Thresher, R.E. (1976) 'Field Analysis of the Territoriality of the 3-spot damselfish *Eupomacentrus planifrons'. Copeia, 2,* 266-76

Tomita, T., Kaneko, A., Murakami, M. and Pautler, E.L. (1967) 'Spatial Response Curves of Single Cones in the Carp', *Vision Research, 7,* 519-31

Vanegas, H., Essayag-Millan, E. and Laufer, M. (1971)'Response of the Optic Tectum to Electrical Stimulation of the Optic Nerve in *Eugerres plumieri', Brain Research, 31,* 107-18

Von Frisch, K. (1925) 'Farbensinn der Fische und Duplizitat-theorie', *Zeitschrift für vergleichende Physiologie, 2,* 393-452

Voss, J. (1977) 'Les Livreés on Patrons de Coloration chez les Poissons Cichlids Africains', *Revue francaise d'aquarologie herpetologie, 4*(2), 43-80

Wagner, H.J. (1975) 'Patterns of Golgi Impregnated Neurones in a Predator-type Fish Retina', in F. Zetler, and R. Weiler, (eds.), *Neural Principles of Vision,* Springer, Berlin, pp. 7-25

Wartzok, D. and Marks, W.B. (1973) 'Directionally Selective Visual Units Recorded in Optic Tectum of the Goldfish', *Journal of Neurophysiology, 36,* 588-603

Young, J.Z. (1950) *The Life of Vertebrates* Oxford University Press, Oxford

Zaneveld (1974) 'Spatial Distribution of the Index of Refraction of Suspended Matter in the Ocean', in R.J. Gibbs (ed.), *Suspended Solids in Water,* Plenum, London, pp. 87-100

Zeki, S. (1980) 'Colour Vision and Colour Constancy in Primates', *Nature, London, 284,* 5775

Zenkin, G.M. and Pigarev, I.N. (1969) 'Detector Properties of the Ganglion Cells of the Pike Retina', *Biophysika, 14,* 763-72

Zumpe, D. (1965) 'Laboratory Observations on the Aggressive Behaviour of Some Butterfly Fishes (Chaetodontidae)', *Zeitschrift für Tierpsychologie, 22,* 226-36

5 UNDERWATER SOUND AND FISH BEHAVIOUR

A. D. Hawkins

It has long been known that sounds are important to fish. Isaac Walton advised anglers 'to be patient and forbear swearing, lest they be heard'. A wide range of species, including many that are commercially valuable, emit sounds as part of their social behaviour (Tavolga 1976), and several species have been shown to be acutely sensitive to underwater sounds. However, before we consider the acoustical behaviour of fish in more detail, we need to understand what sound is, how sounds are created, and how they are transmitted through water.

The Nature of Underwater Sound

Sound is essentially a local mechanical disturbance generated in any material medium such as air or water, and is a remarkably pervasive and ubiquitous form of energy which is difficult to shut out. Sounds are generated by the movement or vibration of any immersed object, and result from the inherent elasticity of the surrounding medium. As the source moves, kinetic energy is imparted to the medium and in turn is passed on, travelling as a propagated elastic wave within which the component particles of the medium are alternately forced together and then apart (Figure 5.1). This disturbance travels at a high speed, which depends on the density and elasticity of the material, and can travel great distances. Close to a source it is not easy to draw a distinction between sound and bulk movements of the medium itself. Fundamentally, sound propagation involves a transfer of energy without any net transport of the medium. Local turbulent and hydrodynamic effects do involve net motion of the medium, and neither depend upon the elasticity of the medium nor propagate at the velocity of sound (see Chapter 7 by Bleckmann, this volume). To a particular sense organ, the latter effects may be indistinguishable from sounds, however.

The to-and-fro displacements that constitute the sound are extremely small, of the order of nanometres. They are accompanied by an oscillatory change in pressure above and below the hydrostatic pressure, the sound pressure. In a free sound field, where there are no physical obstructions to passage of the sound, and where the advancing wavefront is an almost plane surface, the velocity of vibration of the particles (v, the first time derivative of the particle displacement) and the sound pressure (p) are directly proportional to one another, i.e. $v = p/\varrho c$, where c is the propagation velocity ($\mathrm{m\,s^{-1}}$), and ϱ is the density of the medium ($\mathrm{kg\,m^{-3}}$).

Figure 5.1: Sound Propagation through Water. The source sets up a wave which travels through the medium at a constant velocity. The wave can be characterised at a particular point by a to-and-fro motion of the component particles of the medium (the particle velocity, v) or by a variation in pressure above and below the ambient level (the sound pressure, p). If the source moves in a simple harmonic fashion, as here, both parameters vary sinusoidally with time. Distant from the source, v is in phase with p (solid line). Close to the source, v may lag p by up to 90° (dashed line)

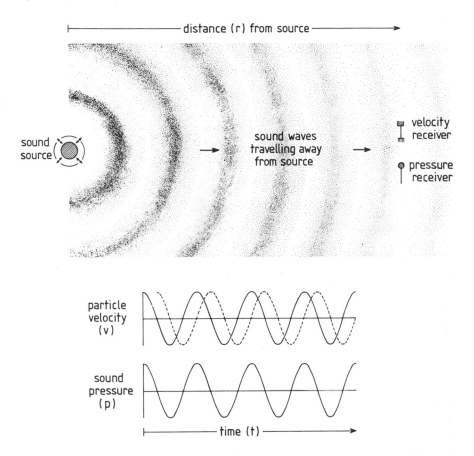

The product (ϱc) is termed the acoustic impedance, and is a measure of the acoustic properties of the medium (analogous to the resistance of an electrical conductor). The particle velocity is measured in metres per second and the sound pressure in pascals ($1\,\text{Pa} = 1\,\text{N}\,\text{m}^{-2} = 10\,\mu\text{bars} = 10\,\text{dyn}\,\text{cm}^{-2}$), but because a great range of amplitudes of both quantities are encountered in nature, it has become conventional to express sound levels in terms of a logarithmic measure — the decibel — relative to a reference quantity. Thus:

sound pressure level (SPL) $= 20\log_{10} p/p_{\text{ref}}$ dB where p is the measured sound pressure and p_{ref} is a reference pressure (normally taken as 1 Pa for water, the SPL being expressed as dB re 1 Pa). A ten-fold increase in sound pressure is equal to 20 dB, a hundred-fold increase is 40 dB, and a reduction of one-thousandth is −60 dB.

Many simple sound sources generate regular waves of motion and pressure, where the amplitudes of both pressure and motion vary with time in a sinusoidal manner. Examples are the tuning fork and bell, which generate sounds of a single frequency and wavelength (as in Figure 5.1), perceived as having a particular pitch. Other sources may generate complex sounds, with much more irregular waveforms, composed of a wide range of sine waves of differing frequency, amplitude and phase.

Sounds inevitably diminish in level as they propagate away from a source. Distant from the source, in a free acoustic field, both pressure and velocity decline with the inverse of the distance (i.e. by a factor of 2, or 6 dB, for a doubling of distance), and both parameters are in phase with one another. Close to a source, however, where the radiating wave fronts are no longer plane but spherical, the simple plane-wave equation no longer applies. The particle velocity is much higher for a given sound pressure, the so-called near-field effect (Figure 5.2). Within the near field, velocity declines with the inverse square of the distance (Harris 1964), the phase of velocity lagging that of pressure (by 90° close to the source). The extent of the near field depends on the nature of the source. For a simple monopole source (a pulsating sphere) the limit is reached when $r = \lambda/2\pi$, where r is the distance and λ is the wavelength of the sound ($\lambda = c/f$, where f is the frequency. (See also Chapter 7 by Bleckmann, this volume.)

Sounds also depart from the plane-wave equation close to a reflecting boundary. At a boundary with a 'soft' material, of low acoustic impedance like air, the local amplitude of particle motion is much higher. Conversely, close to a 'hard' sea bed, the amplitude is much lower. In a small tank in the laboratory, any sound source is completely surrounded by reflectors, and the acoustic conditions become very complex. [This problem is rarely considered in behavioural experiments on fish — Ed.] It is no longer possible to predict the particle velocities which accompany a measured sound pressure by simple application of the wave equations. For this reason, many experiments on the acoustical behaviour of fish must be performed in large bodies of water, where more predictable acoustic conditions prevail.

The to-and-fro motion of the particles, whether it is expressed as particle displacement, velocity or acceleration, differs from the sound pressure in that it is inherently directional, usually taking place along the axis of transmission. Thus, the particle velocity, displacement and acceleration are all vector quantities. A single particle motion detector can, if suitably constructed to resolve a signal into its components, detect the axis of propagation. Sound pressure, on the other hand, is a scalar quantity, acting in all directions.

Figure 5.2: The Relationship between Particle Velocity and Pressure Varies with Distance from a Sound Source. The velocity/pressure ratio increases closer to the source (a), and shows a phase lag behind pressure (b). Distant from the source (in the far field), velocity always bears the same proportion to pressure, and is in phase with pressure. The near-field effect is illustrated along the major axes of two types of source: a monopole, or pulsating sphere, and a dipole, or oscillating sphere. The spherical wave equation for a monopole is given. (After Siler 1969.)

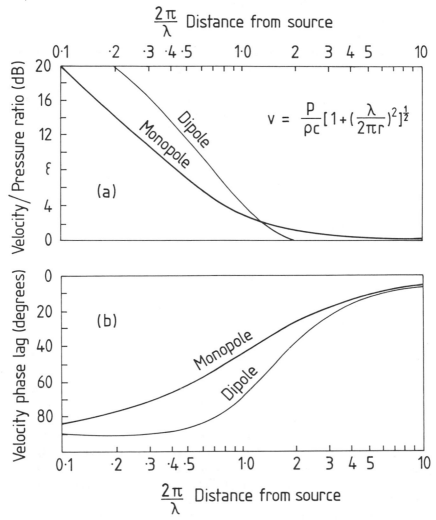

$$v = \frac{p}{\rho c}\left[1+\left(\frac{\lambda}{2\pi r}\right)^2\right]^{\frac{1}{2}}$$

Water has a greater density, lower elasticity and higher sound propagation velocity than air. It has a high acoustic impedance, which essentially means that, for a given sound pressure, the particle motion is smaller (approximately 3500 times less than in air). For a sound of given frequency, the wavelengths are much longer (by a factor of approximately 4.5), so that the near-field effect extends some distance from the source. In the absence of discontinuities or reflectors, however, water is a very good medium for propagating sound. Underwater explosions may be detected half-way around the world, and the sounds of motor vessels may travel many miles. Any animal moving in water almost inevitably generates sounds which are potentially detectable by other organisms. Indeed, one of the problems in listening to the sounds emitted by particular aquatic organisms is that they are often masked by sounds generated by others, some close and others distant.

Any animal capable of detecting sounds gains a number of advantages. First, because sounds propagate rapidly and can travel great distances, the detector is provided with an early notification of the presence of the source, even where there is no direct line of sight from the source to the receiver. Low-frequency sounds, in particular, may propagate around solid objects without being absorbed. Sounds can penetrate dense algal cover, or propagate around corners, providing an almost instantaneous warning of something which would otherwise be concealed. Moreover, because sounds can vary in their characteristics, depending on the nature of the source, fish can potentially gain important information about the object emitting them. They can determine whether the source is predator or prey, inert or alive. Indeed, by analysing the time structure of the sound or by resolving a complex sound into its component frequencies, the fish may be able to identify precisely the particular source. Perhaps more important, however, a sound receiver is potentially able to determine the direction and even the distance of any source. This ability may be especially significant in the sea, where light levels are low and long-distance vision is often impaired.

The Detection of Underwater Sounds

Man must detect underwater sounds by means of a hydrophone, an underwater microphone which converts the water-borne sound into an electrical signal which can subsequently be amplified, analysed or broadcast into air. Most hydrophones are sensitive to sound pressure, the pressure-sensitive element being composed of a waterproofed electrostrictive transducer of barium titanate or lead zirconate connected to a high-input impedance preamplifier (Urick 1975). The detected signals are usually filtered to eliminate noise at extremely low or high frequencies, and then recorded on magnetic tape and monitored on headphones. The temporal structure of the sound can be examined by displaying the sound-pressure waveform on an oscilloscope (Figure 5.3). Alternatively, a frequency analysis can be prepared, either mathematically (Figure 5.3) or by passing the signal through a number of parallel narrow-band filters and displaying the spectrum, showing the relative energy at different frequencies. A more elaborate analysis can be performed by means of a sound spectrograph (Pye 1982), which shows changes in the frequency structure with time (Figure 5.4).

Figure 5.3: The Call of the Tadpole Fish, *Raniceps raninus* (family Gadidae), Shown as the Variation in Sound Pressure with Time (the Oscillogram), and as the Spectrum, where the Sound is Broken Down into its Component Frequencies. Here the spectral analysis was performed by mathematical analysis of the oscillogram (by fast Fourier transform). The call is made up of a series of repeated pulses. (a) shows that the spectrum for a single pulse is simple, and typical of that for a heavily damped resonator; (b) shows that repetition of a series of similar pulses gives a spectrum consisting of several more-or-less regularly spaced frequency bands (the spacing of the bands, in hertz, being the reciprocal of the time interval between the pulses, in seconds)

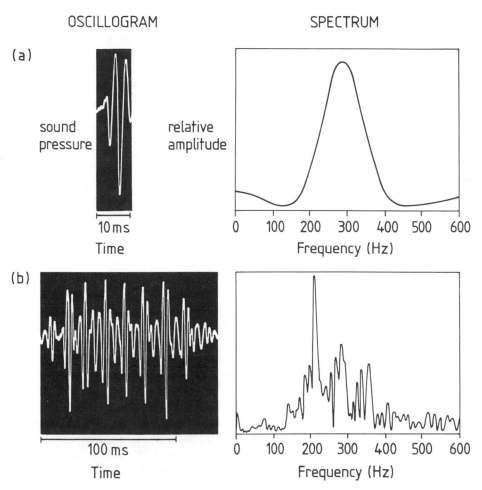

Sound-pressure hydrophones are calibrated in terms of the voltage produced at the output terminals by a sound of a given sound pressure (for example − 60 dB re 1 V (i.e. 1 mV) for a sound pressure of 1 Pa). If the particle velocity of the sound wave is required, then it can be calculated using the appropriate wave equation — though only if the sound is measured in a free sound field. Direct measurement of particle velocity or particle acceleration is possible by means of a suitable mounted seismic accelerometer,

Figure 5.4: Sound Spectrogram of a Long Spawning Call from a Male Haddock, *Melanogrammus aeglefinus*. The filter bandwidth for analysis was 45 Hz. The spectrogram demonstrates changes in the frequency spectrum of the call with time. The spectrum, like that for the call of the tadpole fish (Figure 5.3), consists of a series of equally spaced frequency bands and is typical of a call made up from a series of repeated pulses. During spawning, the pulses are produced at varying rates (shown later in Figure 5.7), resulting in changes in the spacing of frequency bands

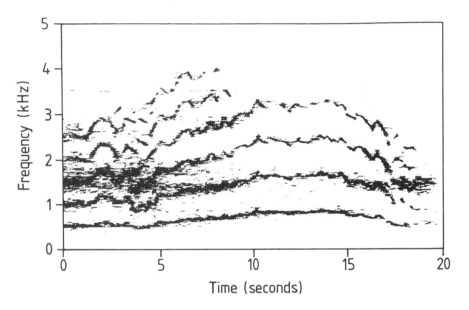

though such a sensor is relatively insensitive. Because particle motion is a vector quantity (see above), it is usually necessary to employ three such sensors, directed at mutual right angles.

The first impression gained when a hydrophone is lowered into a body of water is that the medium is very noisy. In a small pond or rock pool, any footsteps or vibrations imparted to the adjacent ground are readily audible. To obtain quiet conditions in an aquarium tank it is necessary to acoustically isolate the tank from the floor, and to switch off any pumps or machinery in the vicinity. Airborne noise is less of a problem because most of it is reflected at the air-water interface. In rivers and lakes, and in the sea, there is a continual and pervasive background noise resulting from turbulence, water flow, and breaking white-caps and spray, together with the sounds of distant storms and precipitation at the surface. Ship traffic is clearly audible, often at great distances, and superimposed on this background noise are the sounds made by aquatic organisms. Among the latter are the high-frequency calls and sonar signals of aquatic mammals, and the low-frequency sounds of aquatic crustaceans and fish. Many of the sounds of biological origin, like the rasp of a marine snail's radula or the grinding of a sea urchin's teeth, may be purely incidental, but collectively they all contribute to the ambient noise. There have been several detailed studies of sea noise, because of its importance for the detection of submarines, sur-face craft and underwater weapons (Urick 1975).

Sounds Produced by Fish

Sound-producing Mechanisms

In a recent survey, Myrberg (1981) listed over 50 families of fish contain-ing sound producers. Some of these vocal fish may have emitted their sounds accidentally or in response to stimuli such as electric shocks, but many fish produce calls as part of a particular behaviour pattern, and the sounds are believed to elicit a change in the behaviour of other individuals of the same or different species. The sounds vary in structure, depending on the mechanism used to produce them (Schneider 1967). Generally, however, they are predominantly composed of low frequencies, with most of their energy lying below 3 kHz. So far, no ultrasonic sounds (above the range of human hearing) have been recorded from fish, though such sounds are produced by marine mammals.

Stridulatory sounds are made by fish rubbing hard parts of the body together. Characteristically they are rasps and creaks, often made up from a series of very rapidly produced and irregular transient pulses, containing a wide range of frequencies. Members of the grunt family, Pomadasyidae, produce a sharp, vibrant call by grating a dorsal patch of pharyngeal denticles against smaller ventral patches, and in the trigger fish (family Balistidae) the fused anterior spines of the dorsal fin produce a grating sound when moved against their socket. Some catfish of the family Siluridae produce a squeak when the enlarged pectoral spines are moved. Other fish clap or thump different parts of the body together, like the grouper (*Mycteroperca bonaci*), which bangs the opercula or gill covers against the body to produce a low-pitched thump.

Hydrodynamic sounds are produced by fish which are actively swim-ming, or rapidly turning. Though much of the disturbance recorded on a hydrophone in the vicinity is generated by water turbulence, a booming or rushing sound is often detectable at several metres, and may include ele-ments derived from the internal stresses set up within the body of the animal.

With all these mechanisms the gas-filled swimbladder may play a sub-sidiary part, and may impart a hollow resonant quality to the sound. If the swimbladder of a sound-producing white grunt, *Haemulon plumieri*, is deflated, the sound generated by stridulation of the pharyngeal teeth is less loud, and loses its grunt-like quality. The swimbladder is more directly involved in the production of sounds by some physostomatous fish (with open swimbladders), such as the eel (*Anguilla anguilla*), where gas is released from the swimbladder into the oesophagus, giving a sharp pop or squeak.

Figure 5.5: The Sound-producing Apparatus in Gadoid Fish. (a) The drumming muscles in the haddock (*Melanogrammus aeglefinus*) are dorsally attached to the wall of the swimbladder overlying strong lateral wings (parapophyses) extending from the anterior vertebrae. Ventrally, the muscle fibres insert upon the tough outer tunic of the gas-filled swimbladder. (b) In the cod (*Gadus morhua*), the muscles are attached dorsally directly to the vertebral parapophyses and ventrally to the swimbladder. (c) In the tadpole fish (*Raniceps raninus*), the muscles are attached dorsally to the sessile ribs of the 2nd vertebra, the fibres running rostro-ventrally to insert on a sheet of connective tissue, continuous with the peritoneum and enclosing the two anterior cornua of the swimbladder, on either side of the oesophagus. (d) In the burbot (*Lota lota*), the only freshwater gadoid, a pair of thin muscle sheets are attached entirely to the swimbladder, as is also the case in the lythe, *Pollachius pollachius* (e)

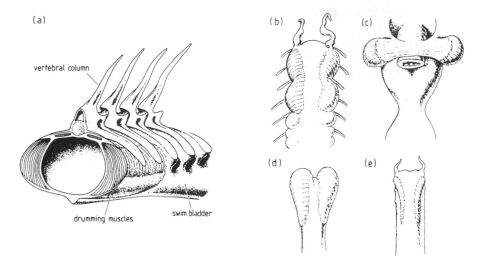

In perhaps the most specialised and characteristic sound-producing mechanism, paired striated muscles from the body wall compress the swimbladder. These muscles are derived from the trunk musculature, and may simply overlie the swimbladder, as in the drums (Sciaenidae), or they may be attached partly or wholly to the organ. Within a family, such as the gurnards (Triglidae) or codfishes (Gadidae), there can be substantial differences between the conformation of the muscles in individual species (Figure 5.5) and even between fish of different sexes, the muscles often being more highly developed in the male. The muscle fibres themselves are specialised. They are often red or yellow in colour, with a high myoglobin content and a rich blood supply. Their diameter is thin, they contain a well-developed sarcoplasmic reticulum, and they may show innervation by a large number of nerve fibres, with many nerve terminals along their length. They contract very rapidly, with a high degree of synchrony, and can be stimulated to contract repeatedly at a very fast rate without going into a sustained state of contraction or tetany. In the haddock each synchronous contraction of the paired muscles results in a brief thump or pulse of sound, the repeated contraction giving rise to longer calls in which the individual

Figure 5.6: Calls Produced by Different Members of the Family Gadidae from the Northern North Sea. The timebase in each case is 100 ms. All these calls were produced during aggressive behaviour outside the spawning season. (a) A short series of closely repeated knocks from the haddock; (b) a grunt from the cod; (c) a grunt from the lythe; (d) a grunt from the tadpole fish. Note that the various grunts, like the longer call of the haddock, are made up of rapidly repeated pulses, the rate of production of pulses rising to over 100 per second in the case of the lythe

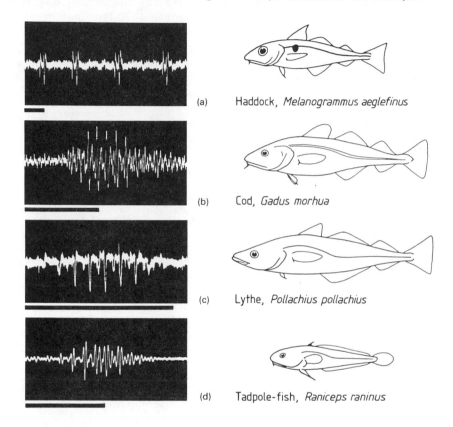

(a)   Haddock, *Melanogrammus aeglefinus*

(b)   Cod, *Gadus morhua*

(c)   Lythe, *Pollachius pollachius*

(d)   Tadpole-fish, *Raniceps raninus*

pulses may still be detectable or may run together to give a grunt. In different species from the same family, the differing patterns of contraction of the muscles give rise to quite different calls (Figure 5.6). The presence of this muscular mechanism in a range of unrelated families, and its absence from closely related species, suggests that it has evolved independently in different fish. In the gurnards, the muscles that develop as part of the trunk musculature are innervated by branches of the anterior spinal nerves. This origin suggests that the mechanism evolved from the incidental production of sound through the contraction of originally unspecialised muscles, perhaps during swimming, or accompanying rapid flexure of the trunk. In more specialised forms, the muscles can be contracted independently, to produce sound without any accompanying movement of the fish.

The involvement of the swimbladder in sound production by so many fish suggests that it plays a key role in the generation of sounds. To generate low-frequency sounds efficiently, it is necessary to move a large body of water. This can most readily be achieved by causing a volume change in the medium (Harris 1964). Compression of the gas in the swim-bladder by contraction of muscles is an effective way of achieving this. We shall see later that the swimbladder is resonant and tends to pulsate at a particular frequency, but the organ is heavily damped and the pulsations rapidly die out. The resultant sound pulse is short, and contains a range of frequencies centred about the resonant frequency, but a longer call can be generated by repeated contraction of the muscles.

Figure 5.7: Courtship Behaviour and Associated Calls in the haddock (*Melanogrammus aeglefinus*). The timebase for the calls is 50 ms. (a) The male approaches a maturing female, with its fins erect, uttering a short series of repeated knocks. (b) A sexually active male swims along the bottom, in tight circles, with exaggerated body movements and fins erect. A heavy pattern of pigmentation is shown, and long calls of rapidly repeated knocks are uttered. (c) The male leads a ripe female up through the water column, its tail moving from side to side, with all fins extended. A continuous rasping call is produced. (d) The male mounts the female from below, the continuous call reaching a hum. Sound production ceases as eggs and sperm are released into the water. Note that the calls of the male consist of series of repeated pulses, the repetition rate varying at different contexts. The female remains silent throughout courtship

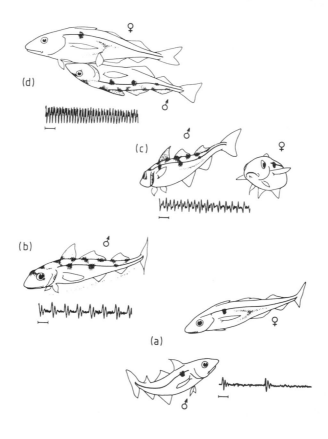

Examples of sounds emitted by fish using this mechanism are illustrated in Figures 5.3, 5.4, 5.6 and 5.7. In an individual species, like the haddock, different calls may be produced in different contexts, as shown in Figure 5.7, the calls essentially varying in their patterns of pulse modulation. Such variations will be apparent to a receiver capable of resolving the individual pulses in time. However, the variations also result in changes in the frequency structure, as shown by the spectrogram in Figure 5.4. This analysis, performed with a narrow-tuned filter, is incapable of resolving rapid variations with time but shows a number of spaced frequency bands, the spacing varying with the temporal spacing of the individual pulses. There is evidence (reviewed by Hawkins and Myrberg 1983) that fish can discriminate the calls of their own species and that this is done by recognising the particular pattern of pulses. For example, members from each of four species of damselfish of the genus *Eupomacentrus* can distinguish their own courtship chirps from those of other species by attending solely to the duration of the interpulse intervals.

Communication by Sounds

It is evident that the great majority of sounds emitted by fish are produced in a social context, and involve interaction between individuals. There may be some exceptions. There is a bathypelagic fish which is believed to produce a low-frequency echo-location call, perhaps serving to locate the sea bed, though the evidence for this is unconvincing (Griffin 1955). In general, however, the calls from fish have been recorded during an encounter between the vocalist and another fish of the same or a different species, and it is usually tacitly assumed that the call involves communication between the animals. Myrberg (1981), in seeking rigorous definition of this term 'communication', decided that it described the transfer of information between individuals, with the functional intent of gaining an adaptive advantage for the sender. This definition accepts that communication may involve individuals of differing species, and does not rule out the possibility that both sender and receiver may obtain mutual benefit.

Sound production commonly occurs in fish when an individual is disturbed by a predator or subjected to a noxious stimulus, as shown by the gurnard (*Trigla lucerna*). When disturbed, this species utters a short grunt while at the same time erecting prominent dorsal-fin spines and unfolding the large brightly marked and coloured pectorals. Generally, calls produced in this context are sharp, with a sudden onset, and they are often accompanied by a strong visual display, perhaps analogous to the 'flash' display shown by some insects. Though there is no documented evidence for fish that such calls drive away predators, or reduce the likelihood of a successful attack, if we argue that these calls are analogous to the well-studied startle displays and calls from insects and mammals (see Edmunds 1974), it is likely that they do serve this function. The same calls may also alert other fish, and help them to escape predators. The advantages to the sender in these circumstances are not clear, unless the fish alerted are kin-related. Myrberg termed this phenomenon 'interception', an attentive listener reacting to the call of another to its own advantage.

One of the most common contexts of sound production is during reproductive activity, where the calls may directly influence the behaviour of prospective mates. Sometimes the sounds accompany complex visual displays (Figure 5.7), and it is possible that they form only part of a more complex signalling system. Often the calls are produced by the male fish, which in many cases shows territorial behaviour. For example, male toadfish (*Opsanus tau*) occupy well-defined sites on the sea bed and utter long and characteristic 'boatwhistle' calls, even in the apparent absence of prospective mates. It has been established that female fish may approach the sender of the sounds, giving rise to an increased rate of call production by the male (Gray and Winn 1961). Similarly, male haddock will occupy the floor of an aquarium tank for long periods, swimming in an exaggerated manner and developing a characteristic pattern of pigmentation, while emitting a continual train of sound pulses. A male will often increase the rate of pulse production as another fish approaches, and may subsequently rise from the floor of the tank and lead the fish upwards, flickering the vertical fins while swinging from side to side, and increasing the rate of pulse production still further (Figure 5.7). If the other fish is a female, this behaviour may lead to a spawning embrace. It is notable that in this species the drumming muscles are more highly developed in the males, especially in the spawning season. The female remains silent throughout courtship, though any males that approach the active male may engage it in an aggressive bout in which both participants produce sounds. From the contexts in which these reproductive calls occur it seems likely that they serve to advertise the presence and reproductive readiness of the sender to the females, and may even arouse reproductive activity in the latter. The calls may provide the basis for mate selection by the female (see Chapter 10 by Turner, this volume). Myrberg has remarked that the preponderance of territorial species producing the calls suggests that this kind of behaviour is characteristic of species whose sexes may be separated by considerable distances, or living in habitats where visual and chemical signals are inadequate. Several studies, including that of Tavolga (1958) on the frillfish goby (*Bathygobius soporator*), have confirmed by the playback of recorded calls that the sounds may elicit a response even in the absence of the visual and other signals which normally accompany them. However, other workers have emphasised the limited response obtained in playback alone, and have stressed that sound is only part of a more complex assemblage of signals emanating from the fish. Though there is evidence from birds that sound production by the males may assist in stimulating reproductive maturation in the female, there is only a single observation on a mouth-breeding cichlid, *Oreochromis mossambicus*, that points to this possibility in fishes (Marshall 1972).

There are several recorded instances of sound production by one male stimulating others to be vocal. Thus, in the haddock, another male may approach a sound-producing male and may engage it in an aggressive display, both fish emitting sounds. Among colonial males of the various damselfish (*Eupomacentrus* spp.) courtship chirping by an individual male may initiate chirping among neighbouring males on the reef, and some may move towards the territory of the initiator (Myrberg 1972). These are almost certainly examples of the interception of a call. Since an imitator may lead an approaching female away from the initiator, they point to the adaptive significance of intercepting the calls of others (Krebs and Davies 1978; and see Chapter 10 by Turner, this volume).

Aggressive fish are often vocal. Male croaking gouramis (*Trichopsis vittatus*) may participate in prolonged bouts, with butting, chasing and lateral displays by one male to another. However, there are numerous examples of fish producing sounds while competing for other resources, for example food or space. Female haddock, though silent during courtship, will readily produce sounds during competitive feeding outside the breeding season. Valinsky and Rigley (1981) have confirmed that sound production can provide significant benefits in such circumstances. Experimentally muted territorial residents of the loach, *Botia horae*, were unable to deter intruders from entering their shelter sites, despite appropriate visual displays. Intact and sham-operated fish were successful in repelling intruders.

It has been suggested that the social aggregation of fish (shoaling: see Chapter 12 by Pitcher, this volume) may be facilitated by sounds (Moulton 1960). Where fish swim in co-ordinated groups or schools, the motions set up in the water by the swimming fish may well be important in maintaining the cohesion of the school under poor visual conditions. Saithe (*Pollachius virens*) can school while temporarily blindfolded (Pitcher, Partridge and Wardle 1976), and cutting the posterior lateral line nerve indicates that the lateral line system of sense organs may play a significant role during normal schooling. The response in this case is probably to bulk motion of the water, or turbulence, rather than sound *per se* (see Chapter 7 by Bleckmann, this volume), but sounds may be important in maintaining social cohesion in other species, or over greater distances. So far, however, few sounds have been recorded from schooling species. Indeed, within a family like the Gadidae, sound production is often absent from the more actively schooling forms like saithe. One disadvantage in using sound to promote cohesion of the school once assembled is that predators may intercept the sounds, perhaps eliminating any anti-predator advantages provided by the shoaling habit. Various studies have shown that predatory fish, and especially sharks, may home in on the incidental sounds produced by struggling or injured prey.

Hearing in Fish

The Hearing Abilities of Fish

The realisation that many fish were sound producers, and that sound provided an effective channel for communication under water, prompted an early interest in the hearing abilities of fish. By the end of the 19th century, the morphology of the fish ear had been well described, but critical experimental studies of the hearing characteristics of fish awaited two developments. The first was the application of conditioning techniques, pioneered by von Frisch and his associates, which enabled fish to be trained to respond unambiguously to sound stimuli. The second was the development of controlled electronic means for delivering and measuring underwater sounds. Since the early days, the hearing of a wide range of teleosts and elasmobranchs has been examined, both by means of conditioning experiments under controlled conditions, and by direct observation of the responses of free-ranging and captive fish to sounds. Of particular value have been experiments to determine threshold values for particular sound stimuli, where fish were conditioned to respond to high sound levels and the stimulus was then progressively reduced to determine the limiting level for the response.

These experiments have not been performed without difficulty. A special problem is the need to perform acoustical studies under suitable conditions, where sounds can be presented without distortion and subsequently measured with precision. Most aquarium tanks are deficient in this respect for the reasons outlined earlier. A range of special tanks have been constructed by different workers, and hydrophones sensitive to particle velocity have been developed to adequately monitor the sound stimuli. Some workers have performed their experimental studies in midwater in large bodies of water, well away from reflecting boundaries, where measurements of the sound pressure can be used to calculate particle velocities. The various techniques have been reviewed by Hawkins (1981).

The sensitivity of a fish to sound is conveniently expressed as an audiogram, a curve showing the thresholds or minimum sound levels to which the fish will respond over a range of frequencies. Examples for several species are given in Figure 5.8. In general, most thresholds for fish have been determined in terms of the measured sound pressures. In some experiments, however, the ratio of sound pressure to particle velocity has been varied. In these circumstances the thresholds for some species follow the sound pressure; but in others the thresholds follow the particle velocity. Thus the units used to express the auditory thresholds in any audiogram differ, depending on the key stimulus. In Figure 5.8, the audiograms for the cod (*Gadus morhua*) and catfish (*Ictalurus nebulosus*) are given in terms of sound pressure, and the thresholds for the Atlantic salmon (*Salmo salar*) and dab (*Limanda limanda*) are given in terms of particle velocity. In an open body of water, distant from the sound source, the two parameters are

Figure 5.8: Audiograms for Four Species of Teleost Fish, Showing Auditory Thresholds or Minimum Sound Levels Detectable by the Fish, at a Range of Different Frequencies. The thresholds for the cod *Gadus morhua*) (Chapman and Hawkins 1973) and freshwater catfish (*Ictalurus nebulosus*) (Poggendorf 1952) are given in terms of sound pressure. Those for the salmon (*Salmo salar*) (Hawkins and Johnstone 1978) and dab (*Limanda limanda*) (Chapman and Sand 1974) are expressed in terms of the particle velocity. Though the thresholds for all four species are directly comparable at a substantial distance from the sound source, in the near field, where the particle velocity increases steeply in relation to sound pressure, the sensitivity of the dab and salmon will increase relative to the other species. The sound-pressure thresholds are given in decibels relative to 1 Pa (see text for details), and particle-velocity thresholds are given in decibels relative to $6.4935 \times 10^{-5} ms^{-1}$ (the velocity corresponding to a pressure of 1 Pa in the far field)

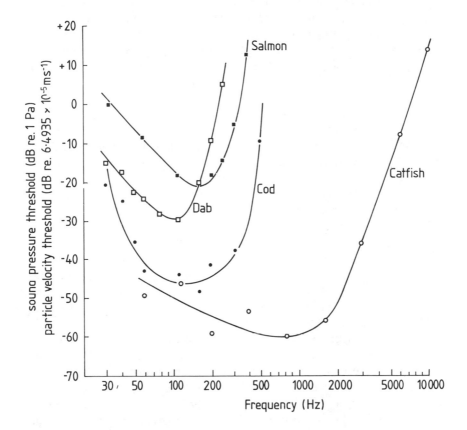

of course proportional to one another and the audiograms are directly comparable, but within the near field, where particle velocity increases steeply against sound pressure, the sensitivities differ.

Several conclusions can be drawn from the audiograms published for a wide range of fish. In general, fish are sensitive to a rather restricted range of frequencies compared with terrestrial vertebrates, and especially mammals and birds. Even the best fish are relatively insensitive to sound at frequencies above 2 or 3 kHz whereas man retains a sensitivity above

15 kHz, and some mammals — including aquatic forms — can detect frequencies of over 100 kHz. Within their resticted frequency range, however, many fish are acutely sensitive to sounds, especially those such as the cod and catfish, which respond to sound pressure. Indeed, it has been established that in the sea the cod is not limited by its absolute sensitivity, but by its inability to detect sounds against the background of ambient noise, even under relatively quiet sea conditions (Chapman and Hawkins 1973). Any increase in the level of ambient noise, either naturally as a result of a storm, or imposed artificially by replaying broad-band white noise, results in an increase in the auditory threshold (a decline in sensitivity), as shown in Figure 5.9. Thus, many of the differences in sensitivity seen in the audiograms of different species are probably the result of different noise levels prevailing during the experiments. These differences are most apparent at low frequencies, where noise levels are much more variable, and also for fish studied in quiet laboratory tanks rather than in the field.

Some fish are inherently less sensitive to sounds and are only rarely limited by naturally occurring noise levels. These insensitive species, including the salmon (Figures 5.8 and 5.9) and flatfish such as the dab and plaice (*Pleuronectes platessa*), have been shown to be sensitive to particle velocity (Figure 5.10). We have already seen that particle-velocity amplitudes in water are very low. Even the best man-made velocity hydrophones are very insensitive compared with their pressure counterparts.

Whether a fish is sensitive to sound pressure or particle velocity depends upon the presence of a gas-filled swimbladder, this organ playing a key role in hearing as well as in sound production. A linkage between the swimbladder and the ear is characteristic of those fish that are sensitive to sound pressure. Indeed, both the absolute sensitivity of fish and their frequency range appear to depend upon the degree of association between the swimbladder and the ear. Thus, the cypriniform or ostariophysan fish, which have a close connection between the two, show a very acute sensitivity to sounds and an extended frequency range, as in the catfish (*Ictalurus nebulosus*) (Figure 5.8). In the cod, the anterior portion of the swimbladder is simply placed close to the ear and this species is both less sensitive and has a more restricted frequency range than the catfish. Within a particular family like the Holocentridae the audiograms may vary, depending on the degree of association between the swimbladder and the ear (Figure 5.11).

We have seen that the hearing of a fish like the cod may be affected by the prevailing level of ambient noise. Where thresholds are masked in this way it can be shown that not all frequency components of the background noises are equally effective at impairing detection (Hawkins and Chapman 1975). If high-level noise is transmitted in a relatively narrow frequency band, successively tuned to different frequencies, the degree of masking for a particular tone is strongly dependent upon the frequency of the noise.

Figure 5.9: Changes in Auditory Threshold for the Cod and Salmon at 160Hz, Accompanying Variations in the Level of Background Noise in the Sea. The solid symbols represent measurements made at different natural ambient noise levels, and the open symbols represent measurements made at higher levels, broadcast from a loudspeaker. The higher the noise level, the higher the threshold through auditory masking. Note that the hearing of the cod is often impaired at quite moderate ambient levels, whereas the hearing of the salmon is only masked at much higher levels (corresponding to sea states greater than 4). The data for the cod (*Gadus morhua*) are taken from Chapman and Hawkins (1973) and those for the salmon (*Salmo salar*) from Hawkins and Johnstone (1978)

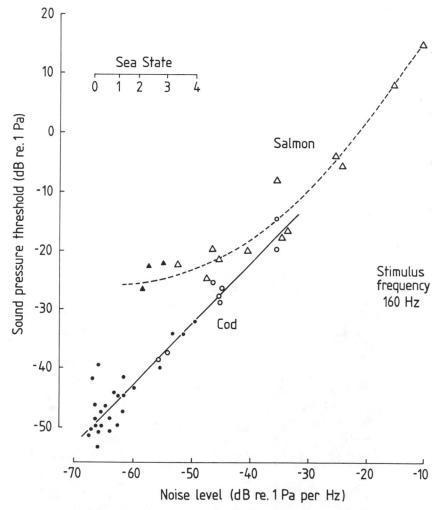

Detection of the tone is masked most effectively by noise at the same or immediately adjacent frequencies. This observation provides evidence that interaction between the stimulus and the noise is confined to a narrow range of frequencies on either side of the stimulus. The cod, the goldfish (*Carassius auratus*) and even the salmon appear to possess an auditory filter, capable of being tuned to any frequency of interest, thereby

Figure 5.10: An Experiment on the Sacculus of the Plaice (*Pleuronectes platessa*) Has Confirmed that the Otolith Organ Essentially Responds to Particle Motion Rather than Sound Pressure (Hawkins and MacLennan 1976). The fish was placed at the centre of a standing wave tank (a), where the amplitudes of particle velocity and sound pressure could be varied independently by adjusting the amplitudes and phases of two opposing sound projectors (b). (c) Changes in particle velocity resulted in pronounced changes in the receptor potentials recorded at the saccular macula (the saccular microphonics). Changes in the sound pressure had no effect upon the potentials. In these oscillograms the microphonics are on the upper beam and sound pressure is on the lower beam

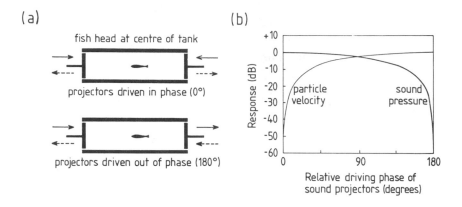

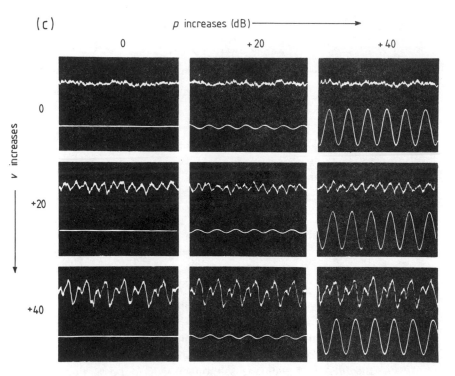

Figure 5.11: Audiograms of Four Species of Squirrelfish (Family Holocentridae). *Myripristis kuntee* yields lower sound-pressure thresholds and responds over a wider frequency range than other holocentrids. The greater sensitivity of this species is associated with a linkage between the gas-filled swimbladder and the ear. Anteriorly, the swimbladder is divided into two lobes, the medial faces of which make strong contact with a thin membrane in the wall of the auditory bulla, lateral to the sacculus of the ear. In other holocentrids, the anterior of the swimbladder lies close to the skull or in contact with it (as in *Holocentrus ascensionis*). The audiograms for *Adioryx xantherythrus* and *Myripristis kuntee* are taken from Coombs and Popper (1979), and those for *Adioryx vexillarius* and *Holocentrus ascensionis* from Tavolga and Wodinsky (1963). The morphology of the swimbladder and rear part of the skull is redrawn from Nelson (1955)

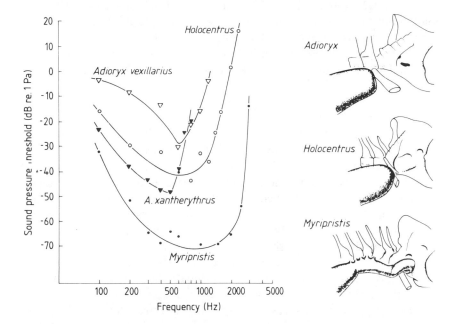

improving the ability of the fish to detect signals in the presence of high levels of noise (Tavolga 1974; Hawkins and Chapman 1975; Hawkins and Johnstone 1978). The filter is narrower in the cod than it is in the salmon, the latter (with its reduced sensitivity) being less likely to have its hearing impaired by ambient noise.

The presence of a frequency-selective auditory filter suggests that fish may also be able to distinguish between different frequencies. Behavioural discrimination experiments have confirmed that this is so (Dijkgraaf 1952; Fay and Popper 1980). The goldfish can separate tones whhch differ in frequency by as little as 3-5 per cent, an ability which is poorer than that of man but broadly comparable to that of many mammals and birds. However, there may be significant differences between fish. Cypriniformes like the goldfish and the minnow (*Phoxinus phoxinus*) appear to be better than others like the marine goby (*Gobius niger*) and the freshwater bullhead (*Cottus gobio*) at discriminating frequency.

Fish are also able to discriminate signals that differ in amplitude. The goldfish can discriminate 300 Hz tone pulses which differ by 4 dB, and the cod and haddock discriminate 50 Hz tone pulses differing by as little as 1.3 dB. We have seen that many fish calls differ in their patterns of pulse modulation. It is therefore especially interesting to know whether fish can resolve the individual pulses, distinguishing calls by their temporal patterning, or whether the pulses are run together by the auditory system, so that the calls must be distinguished in some other way (perhaps through differences in their frequency spectrum). In fact, the goldfish appears to be able to resolve short-duration pulses much better than man. Fay (1982) has pointed out that the teleost ear may be well adapted for preserving the fine temporal structure of sounds. The goldfish can also discriminate between sounds that differ in phase; that is, between sounds that begin with a compression and those that begin with a rarefaction, an ability not possessed by man.

In a medium like water where light levels are low and vision is often impaired, an ability to locate the position of a sound source in space is likely to be especially important. Despite initial doubts, it is now firmly established that some fish have this ability (the evidence is reviewed by Schuijf and Buwalda 1980). The only species for which extensive experimental data are available is the cod. Field experiments have shown that cod are able to discriminate between spatially separated loudspeakers in both the horizontal and vertical planes, and also that they can orientate towards particular sources. Field observations on predatory fish, and especially sharks, have shown that fish may locate and track down their prey by means of sound, often over large distances (Myrberg, Gordon and Klimley 1976). In several respects the auditory localisation abilities of fish may exceed those of terrestrial vertebrates. Cod can discriminate between sound sources at different distances (Schuijf and Hawkins 1983) and can discriminate between diametrically opposed loudspeakers in both the horizontal and vertical planes under circumstances which are ambiguous or confusing for man (Buwalda, Schuijf and Hawkins 1983). Living in an environment where vision at a distance is often precluded, fish may depend heavily upon the ear for information about their distant surroundings.

Anatomy and Organisation of the Auditory System

The main sound receptors of fish are the otolith organs of the inner ears. The ears are paired structures embedded in the cranium on either side of the head (Figure 5.12) close to the midbrain. There are no obvious external structures to indicate their presence. In the clupeoid fishes there is indirect connection to the exterior via the lateral line system (see below) and in some other fish there is a small opening by way of a narrow endolymphatic duct.

Each ear is a complicated structure of canals, sacs and ducts filled with endolymph, a fluid with a particular ionic composition and special viscous properties. In elasmobranchs and teleosts the ear has three semicircular

Figure 5.12: The Paired Ears of Fish are Membranous Structures Embedded in Fluid-filled Spaces within the Cranium, on Either Side of the Midbrain. The position of the two ears is shown for the cod in lateral (A) and dorsal (B) views. The position of the swimbladder in relation to the ears is also shown

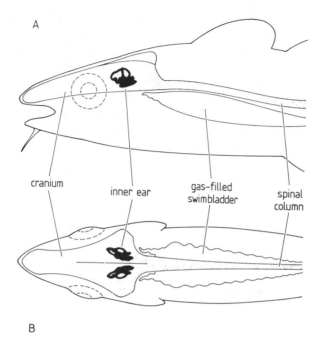

canals, each incorporating a bulbous expansion, the ampulla, occluded by a jelly-like flap or diaphragm, the cupula. The canals are arranged orthogonally (at mutual right angles). Angular accelerations of the head cause the endolymph to lag behind the movements of the canal, deflecting the cupula and stimulating a population of sensory hair cells mounted on the crest of a saddle-shaped wall extending across the ampulla, the crista.

Three expanded sacs within the ear are linked with one another, and with the semicircular canals (Figures 5.12 and 5.13). The most superior sac, the utriculus, communicates directly with the lumen of the semicircular canals, and with them forms the pars superior. The sacculus communicates with the utriculus by a very small aperture, and also with a posterior diverticulum, the lagena. The sacculus and lagena together constitute the pars inferior. In the teleosts, each of these sacs contains a dense mass or stone of calcium carbonate and other inorganic salts within a protein matrix, sitting upon a sensory membrane or macula containing many mechanoreceptive hair cells. The body of the otolith is separated from the delicate hair cells by a thin otolithic membrane, which may extend over parts of the macula not covered by the otolith. In elasmobranchs (and in higher vertebrates), the otolith is replaced by a jelly-like cupula containing many small spherules of calcium carbonate, the otoconia. There is an additional macula with an unloaded cupula in elasmobranchs and some teleosts, the macula neglecta, which is often adjacent to the endolymphatic duct.

The various sacs and their enclosed cupulae and otoliths vary in size, orientation and shape from one species to another. The sacculus is generally the largest, but in some catfish, and in the Clupeiformes, the utriculus may exceed the sacculus in size. In the holostean *Amia calva*, and in many Cypriniformes, the lagena is particularly large. The utricular macula and the otolith that surmounts it usually lie predominantly in the horizontal plane, and the sacculus and lagena lie in different vertical planes (Figures 5.12 and 5.13); the maculae are often twisted. The attachment of the otoliths to the macula is only poorly understood. Though the otolithic membrane appears to be thin and fragile, it may serve in some species to suspend or restrain the movements of the otolith. Most otoliths have a complex, sculptured shape, and some appear to have flanges or keels which may be important in influencing their freedom of movement. The otoliths may rotate about their own axes, or move along a curvilinear path, rather than show simple linear translation.

Figure 5.13: The Main Parts of the Cod Ear, Shown Schematically. The sensory membranes or maculae of the three otolith organs are shown, with the orientation of the sensory hair cells indicated by arrows. (A) Lateral view of left ear. (B) Dorsal view of left ear

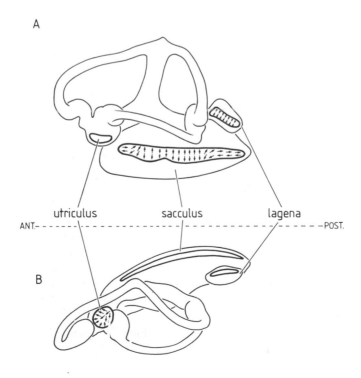

The ears are innervated by the eighth cranial nerve, which sends rami to each of the ampullary organs, and to the various maculae. The conformation of the different nerve branches is often complicated, with the finer ampullary rami combining with the larger rami to the otolith organs. The various branches may also be overlain by other nerve trunks, especially those of the lateral line, which issue from several cranial nerve roots.

The Hair Cell.   The epithelial, mechanoreceptive hair cells of the ear and lateral line neuromasts are found in all vertebrate classes, and show strong structural similarities throughout the group. The physiology of hair cells has been reviewed by Ashmore and Russell (1983). In fish, the hair cells are typically elongated and cylindrical, surrounded by supporting cells on a firm connective-tissue base. Bipolar afferent nerve fibres synapse with the base of the cells and pass through the basement membrane of connective tissue into the auditory nerve (Figure 5.14). Finer efferent fibres also terminate on the hair cells. The latter are strengthened at their apical ends by a cuticular plate surmounted by a sensory process made up of cilia embedded within the otolithic membrane. Many stereocilia, packed with microfilaments, are grouped together with steadily increasing length towards a longer, eccentrically placed kinocilium, containing nine double microtubules and two central single tubules. The positioning of the kinocilium towards one side of the cell gives the apical end of the cell a pronounced structural asymmetry.

The hair cell essentially responds to mechanical deflection of its sensory process. Electrical potentials exist at the apical and basolateral ends of the cell, due to differences in ionic content between the extra- and intracellular fluids. Deformation of the cell at a sensitive locus at the apical end results in ionic flow across the cell membrane, producing a progressive depolarisation which acts at the base to modulate the release of a chemical transmitter at the afferent synapse. The changing receptor potentials are carried primarily by potassium ions, with calcium as a necessary cofactor.

The transduction properties of the hair cells are influenced by their external mechanical connections, and also by the arrangement of the stereocilia. The stereocilia themselves contain the protein actin, which may influence their mechanical properties. Mechanical constraints and electrical resonances within the cell may tune the cells to particular frequencies. In fish the tuning would appear to be rather broad (see below), but in reptiles, birds and mammals it can be very sharp, with some of the hair cells tuned to high frequencies and others to low. There is evidence that the efferent nerve fibres to the cell may regulate the tuning.

An important feature of the hair cell is that it is directional in its response to mechanical stimulation. Depolarisation of the cell and excitation of the afferent fibres are most pronounced when the stereocilia are deflected in the direction of the kinocilium, and hyperpolarisation and inhibition of the afferent fibres result when the stereocilia are deflected by

Figure 5.14: Fine Structure of the Otolith Organ. (A) Schematic cross-section through the utriculus, showing the otolith mounted on the hair cells but separated from the cilia of the cells by an otolithic membrane. Bipolar afferent fibres and finer efferent fibres synapse with the hair cells. (B) The hair cells are directional in their physiological response to stimulation, having a distinct axis defined by the position of the kinocilium. Adjacent hair cells often have a common axis, and many hair cells often synapse with a simple afferent fibre

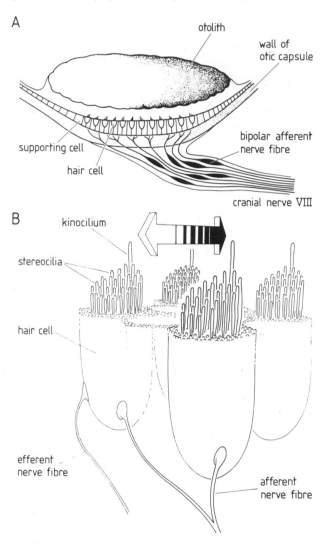

shearing forces acting along the same axis in the opposite direction (Figure 5.14). The response is non-linear and asymmetrical, the positive response resulting from movement towards the kinocilium gradually rising to a higher level than the negative response resulting from movement in the reverse direction. The physiological basis for this directionality is not yet established, but it may be related to the arrangement of the graded stereocilia and filaments that connect them, and perhaps with the mounting of the cuticular plate.

Scanning electron microscopy has revealed particular patterns in the arrangement of the polarised hair cells in the various sensory epithelia of the octavo-lateralis system. Hair cells in the cristae of the ampullae of the semicircular canals share the same axis and the same polarisation. Thus, all the afferent fibres are excited by fluid motion in one direction, and inhibited by motion in the reverse direction. A wider range of hair-cell polarisation patterns are found in the maculae of the otolith organs. Generally, each macula may be subdivided into a number of regions, within which all the hair cells are morphologically polarised in the same direction, but the pattern may vary greatly between the different maculae (Figure 5.13). Moreover, between families, and even species, the pattern can vary for any particular macula, reflecting major differences in the shape, size and orientation of the otolith itself. At least five different hair-cell patterns have been identified for the saccular macula (Platt and Popper 1981). Within any macula it is common to find a bidirectional arrangement of hair cells, whereby the presence of a group of hair cells polarised in one direction is usually paralleled by a group of cells polarised in the opposite direction (Figure 5.13). The pattern may be much more complicated, however. The utriculus, in particular, often has a wide range of hair-cell orientations. It must not be forgotten that most diagrams illustrating patterns of hair-cell polarisation are drawn in two dimensions whereas the maculae themselves are often curved or twisted.

Within each macula there are different 'types' of hair cell, with varying heights of ciliary bundles and differing relative lengths of the kinocilium and stereocilia. It appears that most afferent nerve fibres synapse only with hair cells of a particular polarisation, and respond only to stimulation of the cell in one direction. However, it is clear from the relatively large numbers of hair cells in relation to the few innervating fibres that there is a great convergence upon each fibre, and each fibre may innervate widely separated hair cells. There is much scope for differing patterns of response, depending upon the numbers, types and orientation of the hair cells synapsing with each afferent fibre.

Accessory Structures. In many fish the inner ears stand alone, with no ancillary structures or attachments. In others, however, there are well-defined structural linkages with gas-filled cavities. Best known are fish of the order Cypriniformes (the ostariophysan fish), where the anterior end of the swimbladder is coupled to the ear by a chain of movable bones, the Weberian ossicles. Expansion or contraction of the anterior chamber of the bilobed swimbladder results in motion of the ossicles. This motion subsequently causes fluid motion in a small sinus, filled with perilymph, which is then communicated to an endolymphatic transverse canal connecting with the lumen of both saccular chambers. Thus, motion of the anterior wall of the swimbladder results in a deflection of the saccular otolith and stimulation of the hair cells. There is evidence that the wall of the anterior chamber of the swimbladder is kept taut by maintenance of the gas at a slight excess pressure. Small changes in depth by the fish might be expected to move the ossicles to the limit of their range and restrict the functioning of this mechanism, but Alexander has argued that the high compliance of the swimbladder wall and its high viscosity cause it to act as a high-pass filter, accommodating changes in hydrostatic pressure while still enabling the system of ossicles to respond to rapid variations in pressure.

In the Clupeiformes there is a very different coupling with the ear, the swimbladder entering the cranial cavity. The system in the herring, (*Clupea harengus*) has been described by Allen, Blaxter and Denton (1976). The central feature is a pair of pro-otic bullae, each divided into gas-filled and liquid-filled parts by a membrane under tension. The upper part contains perilymph, connected with that of the labyrinth by a fenestra in the upper wall of the bulla. Lateral to the fenestra is a compliant membrane (the lateral recess membrane), positioned in the skull wall at the back of the lateral recess from which all the lateral line canals radiate. The gas-filled part of the bulla is connected to the swimbladder by a long gas-filled duct. When the fish changes depth, the main part of the swimbladder, which has a more compliant wall than the bulla membrane, accommodates the change in hydrostatic pressure by changing volume. Gas may then pass along the duct to equalise pressure between the swimbladder and bulla. By this means, the volume of the gas-filled part of the bulla remains constant as the fish changes depth. The swimbladder essentially act as a reservoir of gas for the pro-otic bullae.

Rapid motion of the membrane in the bulla generates motion in the perilymph which is transmitted to the sense organs of the inner ear and lateral line. Thus, pressure changes in the bulla lead to displacements of the perilymph which are transmitted to the macula of the utriculus, sacculus, and perhaps also the lagena, displacing the sensory processes of the hair cells.

Many other teleosts possess a modification of the anterior end of the swimbladder which may influence the functioning of the ear. In the Mormyridae the swimbladder enters the intracranial space, whereas in others, for example some of the Holocentridae and Sparidae (Perciformes), and the Moridae (Gadiformes), the swimbladder is attached extracranially to the skull, adjacent to the sacculus. In other fishes, alternative gas-filled spaces are utilised, including air-filled branchial cavities in the Anabantidae, and subpharyngeal cavities in the Channidae. Moreover, there is experimental evidence that, even in a species like the cod without direct connections between the swimbladder and the ear, the gas contained within the organ may play a part in hearing.

Where a connection exists between the swimbladder and the ear, there may be a further specialisation at the macular level. Thus, a highly specialised tripartite utricular macula is found in the Clupeiformes. On the other hand, in the Cypriniformes and Mormyridae, a very simple pattern of hair-cell orientation is encountered in the saccular macula, with two groups of vertically orientated hair cells in opposition to one another.

Hearing Mechanisms in Fish

The Role of the Otolith Organs

The simple otolith organ serves several functions. First, it serves as a gravity receptor, enabling the fish to determine its orientation with respect to the Earth's gravitational field. The heavy otolith tends to shift as the head of the fish tilts, deflecting the sensory processes of the hair cells (Figure 5.15). Such a system is also sensitive to linear acceleration, the otolith tending to lag behind the accelerating fish, or overshooting when the body rapidly comes to rest. In birds and mammals, the detection of these accelerational forces appears to be the main function of the otolith organs. In fish, however, surgical elimination experiments have confirmed that the otolith organs also play a role in sound reception.

If a fish without a swimbladder (e.g. the plaice) is placed in a standing wave tank, where the particle motion and sound pressure can be varied independently, the summed extracellular receptor potentials (or microphonics) from the hair cells of the sacculus respond only to changes in particle motion. Variations in the sound pressure have no effect (Figure 5.10). This evidence that the otolith organ is essentially driven by particle motion is supported by field experiments on the dab (*Limanda limanda*) and the salmon (*Salmo salar*). If auditory thresholds are measured for these fish at different distances from the sound source, the sound pressure thresholds measured within the near field are lower, confirming that the fish respond to the greater amplitudes of particle motion close to the source. Pumphrey suggested that the wave of particle motion passing through the body of these fish moves the tissues of the head, which have a similar acoustic impedance to water, but the dense otoliths or otoconia lag behind, creating a shearing force to stimulate the hair cells (Figure 5.15).

Particle motion amplitudes in water are very low, especially distant from a sound source, and though the hair cells are believed to be extremely sensitive to mechanical displacement, it is unlikely that such a system can provide for the detection of very weak sounds or operate over a wide frequency range. Certainly, fish like the dab and salmon are relatively insensitive to sounds and have a narrow frequency range. A simple mathematical model of the otolith and its suspension, suggested by de Vries (1957), predicts that the otolith is heavily damped, with a rather low natural frequency of vibration. Its amplitude of motion will progressively decline at frequencies above the natural frequency, which de Vries suggests occurs at

Figure 5.15: The Otolith Organs Appear to Serve Several Functions. In (a), the fish is stationary and the otolith is in its resting position. In (b) the fish is tilted with respect to the Earth's gravitational field and the otolith also tilts, deflecting the cilia of the hair cells. In (c) the fish is accelerating forwards, the heavy mass of the otolith lagging behind and deflecting the cilia. In (d) the fish is stimulated by a sound wave. The fish flesh is acoustically transparent and moves back and forth with the sound wave. The dense otolith lags behind, creating an oscillatory deflection of the cilia

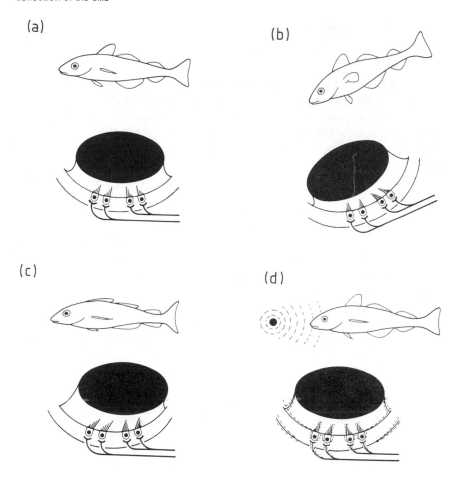

a few hundred hertz. In fact, the audiograms of the dab and salmon decline steeply above about 150 Hz.

The Role of Gas-filled Cavities

Those fish having a close association between the swimbladder and the ear are sensitive to sound pressure. If a dwarf catfish is placed in a tank with a strong gradient of particle motion, the same sound pressure audiogram is obtained at all positions. Moreover, in field experiments, the cod — unlike the dab and salmon — shows similar sound pressure thresholds at higher frequencies even within the near field of the source. Surgical experiments on the dwarf catfish have shown that interference with the Weberian ossicles or deflation of the swimbladder results in a decline in sensitivity to sounds. In the cod, which simply has the swimbladder close to the ear, deflation of the swimbladder results in a pronounced drop in the amplitude of microphonic potentials recorded at the saccular macula (Figure 5.16). Even more remarkable is the observation that placing a small inflated balloon close to the head of a fish lacking a swimbladder (the dab) gives an increased sensitivity, and a more extended frequency range (Chapman and Sand 1974).

It would seem that the gas-filled cavity acts as an acoustic transformer. Incident sound pressures cause the compressible body of gas within the organ to pulsate, generating a much higher amplitude of particle motion than would otherwise have existed. These locally high particle motions may be coupled directly to the otolith organs of the inner ear, or may simply propagate through the surrounding tissues to stimulate the otolith organs. Various authors have assumed that there is a close correspondence between the behaviour of the swimbladder and that of a free gas bubble in water. The latter can be regarded as a simple mass/spring system, where the spring factor is provided by the low elastic modulus of the contained

Figure 5.16: The Role of a Gas-filled Space in Hearing. (A) Sound pressure thresholds for two dabs, obtained with (•) and without (○) an air-filled balloon beneath the head of the fish. The arrow indicates the resonance frequency of the balloon. Redrawn from Chapman and Sand (1974). (B) Audiograms of two cod with empty (•) and full (○) swimbladders, respectively. Redrawn from Sand and Enger (1973). The audiograms were obtained by measuring the sound pressures necessary to evoke a given amplitude of saccular microphonic potential. They indicate changes in relative sensitivity rather than in the absolute sound pressure thresholds

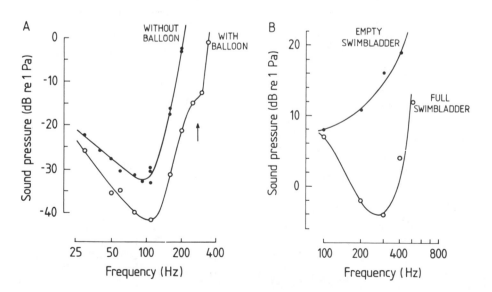

gas, and the mass results from the high inertia of the surrounding water. If such a bubble is exposed to sound pressures of equivalent amplitude but varying frequency, its mechanical response reaches a maximum at a particular frequency, the resonant frequency (Figure 5.17). The sharpness of the resonance depends upon the degree of damping of the bubble, whereas the resonant frequency depends upon the hydrostatic pressure, the volume of gas, and several other factors. If the gas-containing organ behaved acoustically as a free gas bubble, the resonance would provide the animal with a highly sensitive receiver, but there would be other effects. The resonant frequency would depend upon the size of the fish and its depth, and the tuning of the auditory system would vary accordingly. There would be a time delay in detecting sounds, since a lightly damped resonant system takes time to build up to its maximum response. This would have the effect of reducing the time resolution of the system, since it would be unable to preserve rapid amplitude modulations. Moreover, the auditory system would no longer preserve the phase of sounds which differed in frequency, since the phase of response of the gas bubble varies on either side of the resonance.

The actual properties of a gas-filled cavity can be measured directly, either optically or by means of appropriately placed hydrophones. Measurements on the swimbladders of intact, living cod show that the swimbladder is highly damped, and that the resonance frequency is generally well above the hearing range of the fish (Sand and Hawkins 1973). It provides moderate amplification over a wide frequency range, with little

Figure 5.17: Calculated Curves Showing the Pulsation Amplitude at Different Frequencies of a 1.5 cm Radius Gas Bubble at 20 m Depth in a Sound Field. The sound pressure is kept constant at −20 dB re/Pa. The amplitudes are shown for different degrees of damping of the motion (Q values are shown). The water displacements accompanying the same sound pressure in the absence of a gas bubble are shown as a dashed line for comparison. Note that the bubble amplifies the local particle motion over a wide range of frequencies

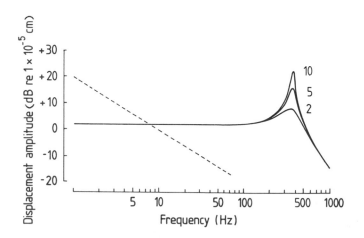

phase distortion, and preserves the ability of the auditory system to respond rapidly to amplitude modulations. If the fish is forced to change depth, the damping at first becomes lighter, but is rapidly restored by the fish. In the herring, with its auditory bullae, the response of the mechanical system extends up to 1 kHz, but falls off rapidly above this frequency and shows an increasing phase lag. Thus, there is evidence that the gas-filled chambers in fish are not highly tuned, with all the attendant disadvantages. Rather, they amplify the particle motion over a wide frequency range with minimal distortion.

The Analysis of Sound Quality

Because the human ear is able to separate sounds into their component frequencies, it is often assumed that other animals distinguish between sounds in a similar fashion. Fish are certainly capable of distinguishing between tones of differing frequency, and masking experiments show that some frequency filtering is performed within the auditory system. However, in fish there is no obvious morphological frequency analyser analogous to the mechanical filter provided by the cochlea of mammals. Behavioural studies of sound communication have indicated that fish discriminate between calls on the basis of differences in repetition rate and duration, rather than frequency or bandwidth (Fine 1978; Myrberg 1981). There has therefore been some controversy about whether fish distinguish beeween sounds through differences in their frequency spectra or their fluctuations in amplitude with time. Though these characteristics are not completely independent, a device for spectral analysis will be quite different from one for temporal analysis. The former will contain narrow frequency filters and show poor time resolution, whereas the latter will have a wide bandwidth but show good temporal resolution.

There have been several attempts to examine the analysis of sounds through study of the activity of afferent neurones from the otolith organs to the brain in fish. Many, though not all, of the neurones from all three otolith organs are spontaneously active even under quiet conditions. Some of these increase their discharge rates when stimulated with sound, and most show a high degree of synchrony of firing with the waveform of the sound stimulus, giving one or more spike discharge for every cycle of the stimulus, the spikes being locked to a particular phase of the stimulus cycle. Some neurones respond with a high synchrony to the stimulus, but without showing any change of discharge rate. The frequency response of the individual neurones can be measured by comparing the firing rate, or the degree of synchrony, at different frequencies and amplitudes. The response curves obtained are rather broad, but in several species different neurones are tuned to different frequencies or at least their upper frequency limits vary. However, there is little evidence that the tuning is sufficiently sharp to explain the degree of frequency discrimination obtained from fish in behavioural experiments, and certainly there is no evidence for an array of fibres tuned to different frequencies and innervating different parts of the sensory macula as exists in the cochlea of mammals and to a lesser degree the maculae of amphibians. There is some peripheral filtering of the received signals, but essentially the waveform of the received signal is coded by the discharge rate of the neurones, their firing retaining a high degree of synchrony with the stimulus waveform often down to very low signal levels.

Fay (1982) has recently drawn attention to the ability of the goldfish ear to detect rapid amplitude fluctuations in both tonal and noise signals. He suggests that the auditory system of the goldfish is particularly well adapted for temporal resolution, both amplitude modulation of a tonal signal, and the frequency of the signal itself being coded by the discharge patterns of the auditory neurones. In this respect, goldfish may differ significantly from man. Certainly the goldfish is able to discriminate much more rapid amplitude modulation, probably basing the discrimination on temporal variations in the signal rather than spectral cues.

Mechanisms of Directional Hearing

Van Bergeijk (1964) originally suggested that directional hearing in fish was entirely dependent upon the lateral line system (see Chapter 7 by Bleckmann, this volume). However, surgical elimination experiments on the cod have shown that both the ears are essential for directional detection at a distance from the source, as in terrestrial vertebrates. It is highly unlikely, however, that fish utilise the same direction-finding mechanisms as the latter. In man the directional cues result from the physical separation of the two ears. Differences in the length of the stimulus path between the source and each of the ears give rise to differences in the time of arrival or phase of the sounds. In addition, at higher frequencies, the head interferes with sound propagation, and has a shadowing effect, causing differences in stimulus intensity at the two ears which are dependent upon direction. For fish, the high velocity of sound in water and the close proximity of the two ears means that differences in stimulus timing are minimal. Moreover, most fish are small relative to the sound wavelength of interest to them, reducing sound shadowing effects, and fish flesh is similar in acoustic properties to the surrounding water, rendering the head effectively transparent. Together, these factors minimise interaural intensity differences. These difficulties, together with the linking of the two ears to the swimbladder, forming a single receptor, led van Bergeijk to assert that directional hearing by means of the ear was not possible.

There is evidence, however, that the otolith organs can provide a basis for the detection of direction. The hair cells of the inner ear have a definite axis of sensitivity, indicated by the position of the eccentrically placed kinocilium. There are orderly patterns of hair-cell orientation within each macula, and there is evidence that this segregation is preserved at the level of the primary afferent neurones (Hawkins and Horner 1981). The movements of the otolith are essentially driven by particle motion, which takes place along a radial axis from the source. It might, therefore, be expected that these movements will stimulate hair cells of differing orientation to a differing degree. By this means the fish should be able to determine the axis of propagation by a process of vector weighing, comparing the outputs of differently orientated groups of hair cells. It has been confirmed by electrophysiological experiments (Figure 5.18) that particular neurones respond best to stimulation along a particular axis, and that this axis differs from one neurone to another and between otolith organs (Hawkins and Horner 1981). Given that the fish has two ears, each containing three otolith organs that are potentially sensitive to sound, and each with its own distinctive pattern of hair-cell orientation, then a system is available which is potentially capable of determining the axis of propagation in three-dimensional space.

There are two flaws in this model of directional detection. First, detection of the axis of propagation does not in itself indicate the bearing of the source. Particle motion alternately takes place towards and away from the source, and the hair cells are inherently bidirectional so that a simple vector weighing of the kind proposed yields a 180° ambiguity in the detection of the source. Secondly, it is difficult to understand the role of the swimbladder in this process. It would appear that in fish such as the cod and goldfish, in addition to the direct vectorial input from the source, the otolith organs must also receive indirect stimulation from the swimbladder, which carries no directional information *per se* and which might be expected to dominate the direct signal.

In practice, experiments in midwater in the sea have shown that the cod can discriminate between opposing sound sources (180° apart) in both the horizontal and vertical planes. Significantly, however, the phase relationship between sound pressure and particle motion is crucial for the fish to perform this discrimination (Buwalda *et al.* 1983). If the sound is switched to a second opposing source, but with the expected inversion of the phase of sound pressure with respect to particle motion locally abolished (by the addition of a standing wave), the fish is no longer able to discriminate between the two sources. Alternatively, if the sound emanating from a single source is simulated to come from an opposing source by locally inverting the phase relationship, the fish responds as if the sound were coming from the opposing source. Thus, phase comparison between sound pressure and particle motion appears to provide the basis for eliminating the ambiguity in directional detection. This suggests that, far from interfering with the detection of the signals reaching the ear directly from the source, the signal re-radiated from the swimbladder is essential for the elmination of ambiguity.

The precise details of the auditory mechanisms used by the cod to discriminate direction need further examination. It is not yet clear whether all the otolith organs of the ear are implicated, nor is it clear whether certain

Figure 5.18: Directionality of a Single Unit Recorded from the Left Anterior Sacculus of the cod *Gadus morhua*. The polar diagrams represent the relative response recorded from a single afferent nerve fibre (in terms of spikes per hertz), at different angles in three mutually perpendicular planes. The response is highly directional in both the median vertical and horizontal planes

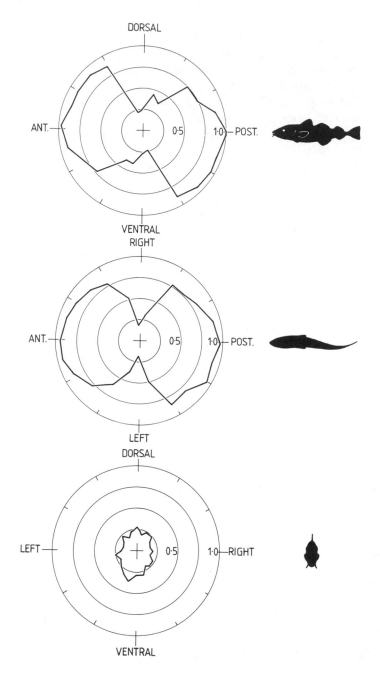

parts of the ear are isolated from stimulation via the swimbladder. It is particularly difficult to determine how far the model developed for the cod can be applied to other species of fish, including those lacking a swimbladder. A recent study on the shark *Chiloscyllium griseum* has confirmed that this species is sensitive to both particle motion and sound pressure, though the actual pressure to displacement transformer is unknown. What we do know is that the cod and probably many other species of fish are well able to locate sound sources in three dimensions and have a real acoustical sense of space. Their sense of hearing is undoubtedly of great importance in enabling them to find mates, to seek out prey and to avoid predators, often under conditions where other senses cannot operate.

Summary

Sounds are local mechanical disturbances that propagate rapidly and very effectively through water. Communication by means of sound appears to be widespread in fish, low-frequency calls being produced in a variety of social contexts including competitive and aggressive behaviour and courtship. Fish are acutely sensitive to sounds, though their hearing abilities are confined to low frequencies. They are able to discriminate between sounds of different amplitude and frequency, and between calls that differ in their pulse patterning — an ability that seems to be particularly important in enabling them to distinguish their own calls from those of other species. Fish are also able to determine the direction and even the distance of a sound source. Sounds are important to fish, and may enable them to seek out both prey, predators and their own kind, sometimes at great distances, under conditions where other senses may be impaired.

References

Allen, J.M., Blaxter, J.H.S. and Denton, E.J. (1976) 'The Functional Anatomy and Development of the Swimbladder-Inner Ear-Lateral Line System in Herring and Sprat', *Journal of the Marine Biological Association, UK, 56*, 471-86

Ashmore, J.F. and Russell, I.J. (1983) 'The Physiology of Hair Cells', B. Lewis, (*ed.*), *Bioacoustics: a Comparative Approach*, Academic Press, New York, pp. 149-80

Buwalda, R.J.A., Schuijf, A. and Hawkins, A.D. (1983) 'Discrimination by the Cod of Sounds from Opposing Directions', *Journal of Comparative Physiology, 150*, 175-84

Chapman, C.J. and Hawkins, A.D. (1973) 'A Field Study of Hearing in the Cod, *Gadus morhua*', *Journal of Comparative Physiology, 85*, 147-67

Chapman, C.J. and Sand, O. (1974) 'Field Studies of Hearing in Two Species of Flatfish, *Pleuronectes platessa* and *Limanda limanda*', *Comparative Biochemistry and Physiology, 47A*, 371-85

Coombes, S. and Popper, A.N. (1979) 'Hearing Differences among Hawaiian Squirrelfish (family Holocentridae) Related to Differences in the Peripheral Auditory System', *Journal of Comparative Physiology, 132A*, 203-7

de Vries, H. (1956) 'Physical Aspects of the Sense Organs', *Progress in Biophysics and Biophysical Chemistry, 6*, 207-64

Dijkgraaf, S. (1952) 'Uber die Schallwahrnehmung bei Meeresfischerei', *Zeitschrift vergleichende Physiologie, 34,* 104-22

Edmunds, M. (1974) *Defence in Animals,* Longmans, London, 257 pp.

Fay, R.R. (1982) 'Neutral Mechanisms of Auditory Temporal Discrimination by the Goldfish', *Journal of Comparative Physiology, 147,* 201-16

Fay, R.R. and Popper, A.N. (1980) 'Structure and Function in Teleost Auditory Systems', A.N. Popper and R.R. Fay (eds), *Comparative Studies of Hearing in Vertebrates,* Springer-Verlag, New York, pp. 3-42

Fine, M.L. (1978) 'Seasonal and Geographical Variation of the Mating Call of the Oyster Toad-fish *Opsanus tau* L.', *Oecologia, 36,* 45-7

Gray, G.A. and Winn, H.E. (1961) 'Reproductive Ecology and Sound Production of the Toadfish, *Opsanus tau, Ecology, 42,* 274-82

Griffin, D.R. (1955) 'Hearing and Acoustic Orientation in Marine Animals. Papers on Marine Biology and Oceanography', *Deep Sea Research Supplement,* 406-17

Harris, G.G. (1964) 'Considerations on the Physics of Sound Production by Fishes', W.N. Tavolga (ed.), *Marine Bio-acoustics,* Pergamon Press, New York, pp. 233-47

Hawkins, A.D. (1981) 'The Hearing Abilities of Fish', W.N. Tavolga, A.N. Popper and R.R. Fay (eds), *Hearing and Sound Communication in Fishes,* Springer-Verlag, New York, pp. 109-33

Hawkins, A.D. and Chapman, C.J. (1975) 'Masked Auditory Thresholds in the Cod, *Gadus morhua*', *Journal of Comparative Physiology, 103A,* 209-26

Hawkins, A.D. and Horner, K. (1981) 'Directional Characteristics of Primary Auditory Neurons from the Cod Ear', W.N. Tavolga, A.N. Popper and R.R. Fay (eds), *Hearing and Sound Communication in Fishes,* Springer-Verlag, New York, pp. 311-28

Hawkins, A.D. and Johnstone, A.D.F. (1978) 'The Hearing of the Atlantic Salmon, *Salmo salar*', *Journal of Fish Biology, 13,* 655-73

Hawkins, A.D. and MacLennan, D.N. (1976) 'An Acoustic Tank for Hearing Studies on Fish', A. Schuijf and A.D. Hawkins (eds), *Sound Reception in Fish,* Elsevier, Amsterdam, pp. 149-69

Hawkins, A.D. and Myrberg, A.A. (1983) 'Hearing and Sound Communication under Water', B. Lewis (ed.), *Bioacoustics: a Comparative Approach,* Academic Press, New York, pp. 347-405

Krebs, J.R. and Davies, N.B. (1978) *Behavioural Ecology,* Blackwell, Oxford, 494 pp.

Marshall, J.A. (1972) 'Influence of Male Sound Production on Oviposition in Female *Tilapia mossambica* (Pisces: Cichlidae)', *American Zoologist, 12,* 633-64

Moulton, J.M. (1960) 'Swimming Sounds and the Schooling of Fishes', *Biological Bulletin, 119,* 210-23

Myrberg, A.A. (1972) 'Using Sound to Influence the Behaviour of Free-ranging Marine Animals', J.E. Winn and B.L. Olla (eds), *Behaviour of Marine Animals,* vol. 2, Plenum, New York, pp. 435-68

Myrberg, A.A. (1981) 'Sound Communication and Interception in Fishes', W. Tavolga, A.N. Popper and R.R. Fay (eds), *Hearing and Sound Communication in Fishes,* Springer-Verlag, New York, pp. 395-452

Myrrberg, A.A., Gordon, C.R. and Klimley, P. (1976) 'Attraction of Free-ranging Sharks by Low Frequency Sound, with Comments on the Biological Significance', A. Schuijf and A.D. Hawkins (eds), *Sound Reception in Fish,* Elsevier, Amsterdam, pp. 205-28

Nelson, E.M. (1955) 'The Morphology of the Swimbladder and Auditory Bulla in the Holocentridae', *Fieldiana: Zoology, 37,* 121-30

Pitcher, T.J., Partridge, B.L. and Wardle, C.S. (1976) 'A Blind Fish Can School', *Science, 194,* 963-5

Platt, C. and Popper, A.N. (1981) 'Fine Structure and Function of the Ear', W.N. Tavolga, A.N. Popper and R.R. Fay (eds), *Hearing and Sound Communication in Fishes,* Springer-Verlag, New York, pp. 4-38

Poggendorf, D. (1952) 'Die Absoluten Hörschwellen des Zwergwelses (*Ameiurus nebulosus*) und Beiträge zur Physik des Weberschen Apparate der Ostariophysen', *Zeitschrift vergleichende Physiologie, 34,* 222-57

Pye, J.D. (1982) 'Techniques for Studying Ultrasound', B. Lewis (ed.), *Bioacoustics: a Comparative Approach,* Academic Press, New York, pp. 39-68

Sand, O. and Enger, P.S. (1973) 'Evidence for an Auditory Function of the Swimbladder in the Cod', *Journal of Experimental Biology, 59,* 405-14

Sand, O. and Hawkins, A.D. (1973) 'Acoustic Properties of the Cod Swimbladder', *Journal of Experimental Biology, 58*, 797-820

Schneider, H. (1967) 'Morphology and Physiology of Sound-producing Mechanisms in Teleost Fishes', W.N. Tavolga (ed.), *Marine Bio-acoustics*, vol. 2, Pergamon Press, Oxford, pp. 135-58

Schuijf, A. and Buwalda, R.J.A. (1980) 'Under-water Localisation — a Major Problem in Fish Acoustics', A.N. Popper and R.R. Fay (eds), *Comparative Studies of Hearing in Vertebrates*, Springer-Verlag, New York, pp. 43-77

Schuijf, A. and Hawkins, A.D. (1983) 'Acoustic Distance Discrimination by the Cod', *Nature, London, 302*, 143-4

Siler, W. (1969) 'Near- and Far-fields in a Marine Environment', *Journal of the Acoustics Society of America, 46*, 483-4

Tavolga, W.N. (1958) 'Underwater Sounds Produced by Two Species of Toadfish, *Opsanus tau* and *Opsanus beta*', *Bulletin of Marine Science, Gulf and Caribbean, 8*, 278-84

Tavolga, W.N. (1974) 'Signal/Noise Ratio and the Critical Band in Fishes', *Journal of the Acoustics Society of America, 55*, 1323-33

Tavolga, W.N. (1976) (ed.) *Sound Reception in Fishes*, Dowden, Hutchinson and Ross, Pennsylvania, 317 pp.

Tavolga, W.N. and Wodinsky, J. (1963) 'Auditory Capacities in Fish. Pure Tone Thresholds in Nine Species of Marine Teleosts', *Bulletin of the American Museum of Natural History, 126*, 177-239

Urick, R.J. (1975) *Principles of Underwater Sound for Engineers*, 2nd edn, McGraw-Hill, New York

Valinsky, W. and Rigley, L. (1981) 'Function of Sound Production by the Skunk Loach *Botia horae* (Pisces: Cobitidae)', *Zeitschrift für Tierpsychologie, 55*, 161-72

Van Bergeijk, W.A. (1964) 'Directional and Nondirectional Hearing in Fish', W.N. Tavolga (ed.), *Marine Bio-acoustics*, vol. 1, Pergamon Press, New York, pp. 281-99

6 ROLE OF OLFACTION IN FISH BEHAVIOUR

Toshiaki J. Hara

The chemical senses of teleosts play a major role in mediating physiological and behavioural responses to the fishes' environment. Chemical stimuli include biochemical products released by conspecifics and other organisms some of which may reveal the presence and location of food, mates, predators, or spawning sites. Although available information indicates that these chemical signals, including pheromones, are more widespread in the social interactions of fish than might have been suspected, investigations are still largely at a descriptive stage. Pheromones are strictly defined as substances that are secreted by an individual and received by other members of the same species, releasing a specific behavioural reaction, or leading to a developmental process. In fish, very few investigations have seriously considered whether the chemical interactions concerned are mediated by a pheromone in its strictest sense. Not a single teleost pheromone has been fully identified chemically (for detailed survey see Liley 1982).

Fish detect chemical stimuli through at least two different channels of chemoreception, olfaction (smell) and gustation (taste). The distinction between these two sensory modalities in fish is not always as clear as in terrestrial, air-breathing vertebrates, mainly because in fish both olfaction and taste are mediated by molecules dissolved in water. In terrestrial air-breathers, olfaction is defined as the detection via the nose of air-borne molecules emanating from a distance, while taste is a water-soluble chemical already in the mouth. Solubility rather than volatility is more relevant in fish chemoreception. Historically, whether taste and smell in fish are two distinct functions has long been controversial. It was Strieck (1924) who first provided the most convincing experimental evidence indicating that the sense of smell exists in fish. He trained minnows (*Phoxinus phoxinus*) to discriminate between odorous and taste substances. Trained fish were unable to discriminate odorous substances after the forebrain was removed, but they could still perceive taste substances.

Many findings point to olfaction as a general mediator of chemical signals involved in various teleost behaviours. Nevertheless, very little is known about their underlying physiological mechanisms. The main aim of this chapter is to summarise the basic principles of the olfactory system in teleosts and show how it is used to extract information about the chemical environment in order to control behaviour. For a more detailed description of ultrastructural, electrophysiological, biochemical and behavioural studies on fish chemoreception, the reader is referred to Hara (1982).

Anatomy of the Olfactory System

Peripheral Olfactory Organ

The olfactory organs of fishes show a considerable diversity, reflecting the degree of development and ecological habits. In the teleost fishes the paired olfactory pits (nasal cavity or olfactory chamber) are usually located on the dorsal side of the head (Figure 6.1A). Each nasal cavity generally has two openings, anterior inlet and posterior outlet. Unlike terrestrial vertebrates, there is no contact between the olfactory and respiratory systems in any teleost species. A current of water enters the anterior and leaves through the posterior naris as the fish swims through water. The floor of the nasal cavity is lined with the olfactory epithelium or mucosa, which is raised from the floor into a complicated series of folds or lamellae to form a rosette (Figure 6.1B). The arrangement, shape, and degree of development of the lamellae vary considerably among species. In the majority of the teleosts, the lamellae radiate from a central ridge (raphe) arising rostrocaudally from the floor of the cavity. The number of olfactory lamellae also varies greatly among species: from a few in sticklebacks to as many as 120 in eels and morays. The number of lamellae increases to some extent with growth of an individual, but remains relatively constant after the fish reaches a certain stage in development. Additionally, secondary folding of the lamella occurs in some species. Most notable are salmonids (Figure 6.1C); there are betweeen five and ten secondary foldings per lamella in adults but none exists in the parr.

No simple correlation has been established between the number of olfactory lamellae and the acuity of the sense of smell. Nevertheless, a marked sexual dimorphism has been evolved in deep-sea fishes, ceratoid angler fishes and *Cyclothone* spp. (Marshall 1967). In these groups the olfactory organ is large in the males but reduced in the females. Marshall speculates that these bathypelagic fishes generally use senses other than olfactory for procurement of food, and that the well-developed olfactory organ in males may be useful in searching for a mate.

Ultrastructure of the Olfactory Epithelium

The olfactory lamella is composed of two layers of epithelium enclosing a thin stromal sheet. The epithelium is separated into two regions, sensory and indifferent. The sensory epithelium shows various distribution patterns in different fish groups. The olfactory epithelium consists of three main cell types: (1) sensory receptor cells, (2) supporting or sustentacular cells, and (3) basal cells.

Sensory Receptor Cells.   Although there has been some argument as to the identification of the receptor cells, in teleosts at least two morphologically distinct receptor cell types generally exist: ciliated and microvillous (Figure 6.1D). Approximately 5-10 million olfactory receptor cells

Figure 6.1: Position of the Nose (A) in the Rainbow Trout (*Salmo gairdneri*), and Scanning Electron Micrographs of an Olfactory Rosette (B), Lamella (C), and Surface View of the Sensory Epithelium (D). CR, ciliated receptor cell; MR, microvillous receptor cell. (Scanning electron micrography courtesy of Dorthy Klaprat, Freshwater Institute, and Dr B. Dronzek and B. Luit, Department of Plant Science, University of Manitoba.)

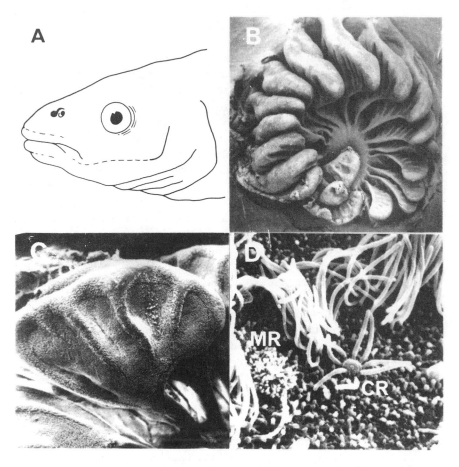

comprise the sensory epithelium on each side of the nasal cavity in an average teleost. The receptor cell is a bipolar neurone with a cylindrical dendrite (1.5-2.5 μm in diameter) which terminates at the free surface of the epithelium. This anatomical position contrasts strikingly with visual and auditory receptor cells, which are guarded by membranes, fluid baths, bones and other structures serving to transduce and process the signals before they are perceived (see Chapters 4 and 5 by Guthrie and by Hawkins, this volume). The distal end of the dendrite forms a swelling (olfactory knob or vesicle) which protrudes slightly above the epithelial surface. The proximal part of the perikaryon tapers to form an axon. The axons pass through the basement membrane, become grouped in the sub-mucosa, and form the olfactory nerve fascicles, which run posteriorly to end in the olfactory bulb. The ciliated receptor cell has 4 to 8 cilia radiating from an olfactory knob (Figure 1D). Each cilium measures 2-7 µm in length and 0.2-0.3 µm in diameter, much shorter than those found in air-breathing vertebrates. Usually the cilia show the 9 + 2 arrangement of microtubules, which is identical with that of common kinocilia (see Chapters 5 and 7 by Hawkins and by Bleckmann, this volume). The micro-villi number from 30 to 80 depending upon the species, and are 2-5 µm long and about 0.1 µm wide. Since the free ending of the receptor cells, whether ciliated or microvillous, is the only bare portion of the dendrite and is exposed directly to stimulant molecules, it probably plays a key role in the stimulus-transduction process.

An interesting feature of the olfactory system is the regenerative capacity of the receptor neurones. The receptor cells are continuously renewed in normal adults, and experimental severance of the olfactory nerve or treatment of the olfactory mucosa with toxicants (e.g. heavy metals) causes degeneration of the sensory neurones followed by the reconstitution of a new population of functional neurones (Evans, Zielinski and Hara 1982). This renewal process of the olfactory receptor cells may be considered to be an adaptation of the system to injury from the environment during the normal life of the animal.

Non-sensory Cells. The supporting cells are columnar epithelial cells extending vertically from the epithelial surface to the basal lamina, forming a mosaic interspersed with receptor cells. The free surface is flat, with relatively few irregular microvilli. Supporting cells adjoin receptor cells, ciliated non-sensory cells, and other supporting cells at the free surface by means of a junctional complex consisting mainly of an apical tight junction. Ciliated non-sensory cells are typical columnar epithelial cells with a wide flat surface (4-7 µm), from which a number of long kinocilia (20-30 µm in salmon) extend. The beating of these cilia creates weak currents over the lamellae, presumably assisting in water renewal and the transport of stimulant molecules in the olfactory organ. The basal cells are small and undifferentiated cells lying adjacent to the basal lamina and having no cyto-plasmic processes reaching the free surface. The basal cell in the sensory epithelium is assumed to be the progenitor of the receptor or supporting cells. In addition to the cell types described, mucous (goblet) cells are abundant in the indifferent epithelium.

In summary, the olfactory epithelium consists of three principal cell types: receptor, supporting and basal cells. The main function of the receptor cells is to detect, encode, and transmit information about the chemical environment to the olfactory bulb and higher brain centres. The basal cells appear to be stem cells and become active during cell turnover and reconstitution of the olfactory epithelium. The function of the supporting cells in olfactory perception is not defined, but they are likely to have a significance beyond mere mechanical support. Some olfactants cause the release of secretory products from supporting cells, changing the nature of the mucus bathing the epithelial surface.

Olfactory Bulb and Tract

The olfactory nerve fibres, unmyelinated axons of the receptor neurones, course to the olfactory bulb where they make a synaptic contact with the second-order bulbar neurones in the form of glomeruli (Figure 6.2). The axons within the olfactory nerve are arranged into several groups of bundles. In the carp (*Cyprinus carpio*), for example, the olfactory nerve consists of two main bundles, medial and lateral. The former is derived from the more rostral lamellae, and the latter from the more caudal. The fibres do not branch until they terminate.

The olfactory bulb in fishes is poorly differentiated and the lamination is not so distinct as in higher vertebrates. The mitral cell, the most characteristic cell found in the bulb, has a relatively large cell body and, unlike in higher vertebrates, usually more than one dendrite ending in different glomeruli (Figure 6.2). It is significant that the axon of a receptor cell does not terminate in more than one glomerulus, and that each glomerulus receives neural inputs only from a limited group of several olfactory receptor cells (Figure 6.2). This is in marked contrast to mammalian mitral cells, in which only a single main dendrite ends in each glomerulus. The glomerular synapses are always directed from the olfactory nerve to the mitral-cell dendrites.

Figure 6.2: A Schematic Representation of the Cellular Anatomy of the Peripheral Olfactory System and the Neural Organisation in the Olfactory Bulb of Teleosts. Olfactory Receptor cells (OR) lie in the olfactory epithelium. Bundles of receptor-cell axons form the fila olfactoria, which coalesce as the olfactory nerve (ON). Axon terminals branch at the olfactory bulb in the form of glomeruli (GLOM), where they synapse with processes of second-order neurones, mitral cells (M). Mitral-cell axons project centrally as the olfactory tract (OT). G, granule cells

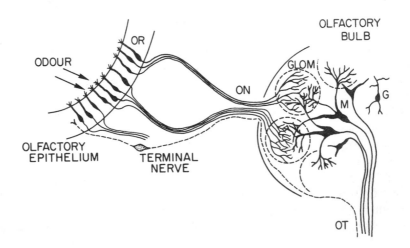

Figure 6.3: Diagram of The Central Olfactory Pathway in Teleosts (Horizontal View). MOT, medial olfactory tract; LOT, lateral olfactory tract; AC, anterior commissure; mtf, medial terminal field; ltf, lateral terminal field; ptf, posterior terminal field, ctf, contralateral terminal field. A dotted line in MOT represents a centrifugal fibre. Modified from Oka *et al.* (1982)

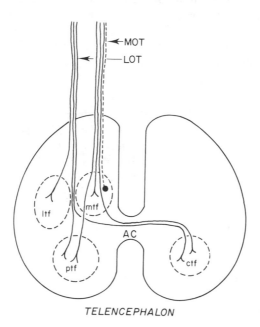

TELENCEPHALON

The axons of the mitral cells form the majority of the centripetal fibres of the olfactory tract through which information from the olfactory bulb is conveyed to the telencephalic hemispheres (Figure 6.3). The olfactory tract consists of two main bundles, lateral and medial. Both bundles are further subdivided into several small bundles. Some fibres run directly to the hypothalamus, and some cross in the anterior commissure. The centrifugal fibres originating in the telencephalon run backwards through the medial portion of the tract, terminate in the granule-cell layer, and form synapses with granule-cell dendrites of the bulb (Oka, Ichikawa and Ueda 1982). The mitral cells are thus under inhibitory control of higher centres through the centrifugal fibres. The olfactory tract fibres number about 10^4. The convergence of the primary olfactory nerve fibres (*a.* 5-10 million) upon secondary neurones would therefore be about 1000:1. This ratio is approximately the order that is estimated for the mammalian olfactory system. Such a high convergence ratio may be responsible for the high sensitivities found in the olfactory system.

The olfactory tract fibres appear to terminate bilaterally in the following telencephalic hemispheres (Figure 6.3): (1) medial terminal field in the area ventralis telencephali, (2) lateral terminal field in the ventrolateral part of the area dorsalis telencephali, and (3) posterior terminal field in the central part of the area dorsalis telencephali (Oka *et al.* 1982). Satou *et al.* (1983) propose that the olfactory system is composed of two separate systems: (1) the lateral olfactory system in which the lateral part of the olfactory bulb receives inputs mainly from the lateral bundle of the olfactory nerve and sends outputs to the lateral olfactory tract, and (2) the medial olfactory system in which the medial part of the olfactory bulb receives inputs mainly from the medial bundle of the olfactory nerve and sends outputs to the medial olfactory tract.

Terminal Nerve

The axons of another cranial nerve, the terminal nerve, course centrally into the telencephalon in association with the medial olfactory tract in a wide range of vertebrates including teleosts (cf. Figure 6.2). In goldfish, large neuronal cell bodies occur within the medial half of the olfactory nerves and the rostral part of the olfactory bulb (Demski and Northcutt 1983). The peripheral process of some ganglion cells of the terminal nerve end in the olfactory epithelium, and the central processes terminate in the ventral telencephalon.

Function of the Olfactory System

Interactions of Odorants with Receptors

Sensory information about the chemical environment is transmitted to the brain by olfactory receptor neurones through an orderly sequence of molecular, membranous and neural events. The processes are initiated by the impingement of odorous molecules upon the surface of the olfactory mucosa. Biochemical studies using a preparation enriched in plasma membranes from the olfactory mucosa have shown that binding of odorant amino acids is a physiologically relevant measurement of an early event in olfactory transduction (Brown and Hara 1982; Fesenko *et al.* 1983). In these studies a sedimentable membrane-mitochondria fraction derived from olfactory rosettes is incubated with radiolabelled amino acids, and specific binding activities are assayed by either a rapid filtration or centrifugation method. Binding meets many of the basic receptor criteria such as saturability, reversibility, affinity, stereospecificity, and quantity of sites. The extent of binding of a series of amino acids parallels their relative electrophysiological effectiveness. Recent studies using a cilia-enriched preparation have provided evidence indicating that at least two separate populations, one for neutral and the other for basic amino acids, of odorant binding sites exist in the rainbow trout olfactory system (Rhein and Cagan 1983). This supports the earlier hypothesis that odorant recognition sites are proteins and integral parts of the cilia. Isolation and characterization of odorant receptor macromolecules from these preparations may be possible in the near future by utilising affinity chromatography and affinity-labelling techniques.

Neurophysiology of Olfactory Reception

The molecular interaction, coupled with specific transmembrane ion conductance mechanisms, initiates generator current flow into the dendritic region of the receptor cell. The current then spreads electronically through the generator potential. This eventually leads to the generation of spike potentials by which the sensory information is transmitted to the higher-order olfactory brian. These electrical signals can be tapped, amplified and displayed at various levels of the olfactory system.

Receptor Activity.   When the nares of a fish are infused with water containing an odorous chemical, three types of response may be recorded from the olfactory epithelium using appropriate electrophysiological techniques. The first type is the electro-olfactogram (EOG), a slow negative voltage change recorded from the epithelial surface (Ottoson 1971). This is a summated receptor potential and represents an excitatory response generated by many receptor cells in response to odour stimulation. The second type is a unitary action potential recorded extracellularly from an olfactory neurone intraepithelially, or from olfactory nerve twigs at some distance from the nares. Many receptor neurones exhibit spontaneous activity, i.e. impulse discharges without odour stimuli, and respond to odour stimulation by, in the majority of cases, increasing impulse discharges. Generally, receptor cells are broadly tuned, and each cell may respond to many odorants differing in quality. The third type is a transmembrane voltage change recorded using an intracellular microelectrode. This provides insight into transmembrane mechanisms associated with cellular activation by odour; however, it is extrmely difficult to obtain such recordings mainly because of the small size of the receptor cell. A high correlation exists among these three types of response when examined in the same species.

Olfactory Bulbar Activity.   Infusion of odorous chemicals into the nares induces rhythmic, oscillatory responses (induced waves) in the olfactory bulb. The olfactory bulb also develops a slow potential shift superimposed with regular oscillation when recorded through a DC-coupled preamplifier. The slow potential is produced in the dendritic network within the glomeruli, and the induced waves are the result of synchronous activity of the secondary bulbar neurones.

Sensitivity and Specificity of Olfaction

The aquatic environment surrounding fishes makes their olfaction unique; the entire process takes place in water, and therefore volatility of odorants is less relevant than for those in air. Traditionally, however, volatile chemicals primarily odorous to humans had been widely utilised as stimuli for studying fish olfaction. Many of the early published detection thresholds such as 3.5×10^{-18} mol litre^{-1} β-phenethyl alcohol for *Anguilla anguilla,* 1×10^{-15} mol litre^{-1} butyl alcohol for *Ictalurus catus,* and 5.7×10^{-10} mol litre^{-1} morpholine for salmonids are not supported by studies employing modern electrophysiological techniques (cf. Hara 1975).

Since Sutterlin and Sutterlin (1971) and Suzuki and Tucker (1971) independently provided electrophysiological evidence that olfactory receptors of Atlantic salmon (*Salmo salar*) and white catfish (*Ictalurus catus*), respectively, are highly sensitive to amino acids, research on fish olfaction has centred on recordings of electrical responses to amino acids at different levels of the olfactory system. Amino acids are potent olfactory stimuli in a wide variety of fish species, and the threshold concentrations lie as low as 0.1 nmol litre^{-1} (cf. Hara 1975, 1982). Generally, unsubstituted L-α-amino acids containing unbranched and uncharged side chains are the most effective olfactory stimuli. Ionically charged α-amino and α-carboxyl groups appear essential; acylation of the former or esterification of the latter results in reduced activity. Peptides are not stimulatory. The amino acids effective as olfactory stimuli are thus characterised by being simple, short and straight-chained, with only certain attached groups. Although some species specificities exist, the populations of olfactory receptors in general are tuned primarily to the neutral portion of the amino acid spectrum (Caprio 1982; Hara 1982). The amino acid specificities found for fish olfaction are similar to those for bacterial chemotaxis and a neutral amino acid transport system in the mammalian ileum. At least two receptor sites or transduction mechanisms are likely to be involved in amino acid detection: (1) neutral amino acids, and (2) basic amino acids. These receptor sites are separate and mutually exclusive (Rhein and Cagan 1983). It is not, however, clear whether these two distinct receptor sites occur in separate receptor neurones.

Overwhelming effectiveness of amino acids as olfactory stimuli for fish does not of course rule out other types of molecules from playing important roles as chemical messengers. For example, bile acids, especially taurine conjugates, are also potent olfactory stimulants for salmonid fishes, with a threshold concentration ranging between 0.1-10 nmol litre^{-1} (Døving, Selset and Thommesen 1980). Ciliated receptor cells which are distributed more distally on each lamella are primarily responsible for bile acid stimulation, and more proximally distributed microvillous receptor cells are for amino acid stimulation (Thommesen 1983). Responses induced by bile acids project to the medial part of the olfactory bulb, and amino acids elicit responses in the lateral part of the bulb (Døving *et al.* 1980). However, this hypothesis is not necessarily supported by recent studies with rainbow trout embryos, in which ciliated receptor cells respond to both amino acids and bile acids prior to the maturation of microvillous receptor cells (B. Zielinski, unpublished). High sensitivity of the olfactory receptor of salmonids to bile acids is implicated in their role as specific chemical signals (pheromones) in homing migration.

Olfaction-Taste Distinction and Overlap

In fish, as discussed earlier, amino acids are important chemical signals eliciting various physiological and behavioural reactions through both the olfactory and gustatory systems. In the majority of species studied, olfactory receptors are stimulated by a wide spectrum of amino acids, with a tendency to tune more to its neutral portion. The taste system demonstrates more variabilities. At one extreme, taste receptors of species such as catfish generally respond to a wide variety of amino acids; at the other extreme, species such as rainbow trout respond to only a limited selection of amino acids; and the rest fall between these two extremes (Yoshii, Kamo, Kurihara and Kobatake 1979; Goh and Tamura 1980; Kiyohara, Yamashita and Harada 1981; Caprio 1982; Marui, Evans, Zielinski and Hara 1983). Thus functional separation between olfaction and taste is obvious for this group of chemicals. There are also other groups of chemicals such as bile acids which stimulate both olfactory and taste systems with equivalent thresholds (Hara *et al.* 1984). There has long been a belief that olfaction is the immediate mediator of chemical communication in fishes, simply because of its lower active thresholds. However, because the fish gustatory system is as sensitive to as many chemical stimuli as the olfactory system, we can no longer consider olfaction to be the sensory modality for detection of aquatic chemicals only on the basis of its lower effective concentration.

Olfaction and Behaviour

Homing Migration in Salmon

Homing migration is displayed by many fish species. These fish may spend their early life in a home territory, migrate to an ecologically entirely different area, then return to spawn in their home territory. For a general review of fish migration see McKeown (1984). The salmon are best known for their spectacular migrations. This impressive feat comprises three migratory phases: (1) downstream journey of the young to the ocean, (2) return of spawning adults to the coastal area near the entrance to their home stream, and (3) the upstream migration. Phases (1) and (2) are thought to rely primarily on non-chemosensory mechanisms such as drift, random movement, and celestial, magnetic and sun-compass orientation, and will not be discussed here. Homing of salmon is well documented (e.g. Hasler 1966; Harden Jones 1968), and a number of studies on salmonid migration have demonstrated the importance of chemical information in homestream discrimination (reviewed by Hara 1970; Cooper and Hirsch 1982; Hasler 1983). The recent monograph by Hasler and Scholz (1983) describes the historical development of the olfactory hypothesis for salmonid homing, endocrine control of the olfactory imprinting process, and other factors controlling migration.

Olfactory Hypothesis.   Over a hundred years ago, Buckland (1880) postulated an olfactory basis for salmonid homing, and his statement is often quoted today:

> When the salmon is coming in from the sea he smells about till he scents the water of his own river. This guides him in the right direction, and he has only to follow up the scent, in other words, to 'follow his nose', to get up into fresh water, i.e., if he is in a travelling humour. Thus a salmon coming up from the sea into the Bristol Channel would get a smell of water meeting him. 'I am a Wye salmon', he would say to himself. 'This is not the Wye water; It's the wrong tap, it's the Usk. I must go a few miles further on', and he gets up stream again.

The olfactory hypothesis, cast into modern terms by Hasler and his students, proposes that salmonids 'imprint' to certain distinctive odours of the home stream during the early period of residence, and as adults they use this information to locate the home stream, at least during the late stages of the homing migration. The results of olfactory impairment experiments repeated over 20 times using seven different salmonid species are remarkably consistent; in 16 experiments the olfactory sense appeared to be necessary for correct homing, and, in addition, two studies demonstrated that blind fish homed nearly as well as control fish. Thus vision is not essential for relocating the original stream, at least during upstream migration. At present, little is known about how olfactory imprinting might operate in salmon homing. Whatever the underlying mechanism of imprinting, the smolt stage is of critical importance, because it is during this period that salmon become indelibly imprinted to distinctive odours of their natal tributary. This ability appears to be acquired rather than inherited; young salmon transplanted from their natal tributary into a second stream before smolt transformations occur will return to that second stream. There is evidence that this process is rapid. Less than 10 days, or perhaps only several hours, appear sufficient for imprinting to take place (cf. Hasler and Scholz 1983). The nature of the olfactory cues in home streams has been variously characterised as volatile, non-volatile, organic and inorganic. If odours distinctive to streams are often-suggested mixtures, or 'bouquets', as opposed to a single chemical, and fish discriminate subtle differences of their chemical composites, the identification of the active ingredients by conventional methods appears extremely difficult.

Homing of Artificially Imprinted Salmon.   Studies on artificial imprinting using coho salmon (*Oncorhynchus kisutch*) in Lake Michigan have been well described (Cooper and Hirsch 1982; Hasler 1983; Hasler and Scholz 1983). The basic principle was to expose salmon to low levels of synthetic chemicals, either morpholine or phenethyl alchohol, in place of, or in addition to, natural homestream odours in order to determine if as adults they could be attracted to a stream scented with that chemical. The fish were exposed to the chemicals in a fish hatchery during the smolt stage and then released directly into the lake. During the spawning migration 18 months later, two separate streams were scented respectively with morpholine and phenethyl alcohol in order to simulate home streams for the experimental fish. Approximately 95 per cent of the morpholine fish were captured in the morpholine scented stream and 92 per cent of the phenethyl alcohol fish were recovered in the phenethyl alcohol scented stream. The general conclusions drawn are that the fish are decoyed to the streams by the synthetic chemicals, able to learn or imprint to them during a brief period of the smolt stage, and retain these cues for 18 months without being again exposed to the chemicals.

However, two fundamental questions remain unanswered in these studies. Do salmonids really smell morpholine and phenethyl alcohol? How do synthetic chemicals make stream water specific for fish? Studies with modern techniques (electrophysiological, cardiac conditioning, and behavioural) have been unable to demonstrate that salmon detect these chemicals at the concentrations used. Salmon do not live in distilled water. Besides morpholine and phenethyl alcohol, in natural water there exist hundreds of other chemicals, some of which may be more potent in terms of olfactory stimulatory effectiveness and capable of imprinting naturally. Alternatively, morpholine would alter the chemistry of the whole stream water to create a novel condition to which salmon imprint. So, what chemical characteristics of morpholine are responsible for such an incredible action?

Do the imprinting odours serve as a sign stimulus to release a stereotyped behaviour pattern in fish? Migrating salmon are known to make a wrong choice at a stream junction, or bypass the natal tributary (overshooting). These fish rectify their errors through 'backtracking', that is, they eventually swim back downstream and reach their natal tributary. Fish deprived of their sense of smell swim downstream but do not swim back upstream. Thus the presence of the homestream odour evokes a positive rheotaxis, and absence of the odour evokes a negative rheotaxis. This is clearly demonstrated in the following experiments in which salmon smolts were exposed to either morpholine or phenethyl alcohol, stocked in Lake Michigan, and captured as adults in streams scented with the chemicals (Hasler and Scholz 1983). The fish were then transported to and released in a different section of the river located upstream from the capture site. Morpholine-treated salmon migrated upstream when morpholine was present and downstream when it was absent. Phenethyl alcohol-exposed fish, on the other hand, swam downstream in both cases.

Continuous excitation of the olfactory system induces sensory adaptation or fatigue so that a given odour is no longer perceived. The behaviour of a salmon ascending the river system seems analogous to that of a dog following an odour track; the dog does not stay exactly on the track but progesses along it criss-crossing. When morpholine was introduced on either the right or left side of a stream, movements of morpholine-treated fish were confined to the morpholine sides. If a fish swam out of the odour trail, it swam downstream until encountering the odour again (Hasler and Scholz 1983).

Pheromone Hypothesis.   An idea that pheromones emitted by juvenile fish attract migrating adult salmon into their natal streams has existed for some time. For example, G.H. Parker suggested, 'it is barely possible that a certain race of fish may give off emanations that differ chemically from those of other races, and that one might attribute the return of individual races to home streams to their power to sense the familiar emanation' (Chidester 1924). This hypothesis has been further formulated by Nordeng (1977), who proposes: (1) populations or races of salmon in different streams emit pheromones that serve to identify fish distinctly from each other, (2) the memory of this population-specific pheromone is inherited, and (3) homing adults follow pheromone trails released by juveniles residing in the stream, that is, the juveniles provide a constant source of population odour. The pheromones are thought to be released from the skin mucus. The olfactory bulbar neurones of the Arctic charr (*Salvelinus alpinus*) responded differentially to mucus emanating from different populations of the same species (Døving, Nordeng and Oakley 1974). Further electrophysiological and chemical analyses have shown that the skin mucus of salmonids contains species-specific compositions of amino acids responsible for olfactory stimulation and provides a basis for the individual and/or sexual recognition (Hara *et al.* 1984). Some species of fish are known to recognise other members of their own species by mucus.

Hasler's experiments on homing of salmon treated with synthetic chemicals cannot, however, be explained by the pheromone hypothesis. In all these experiments, there were no young salmon present in the scented streams at the time the adults were attracted to them. In addition, all the fish used were from the same spawning stock and randomly separated into groups for exposing to different odours. Therefore, the fact that different experimental groups behaved distinctively would seem to rule out the possibility that the result of artificial imprinting experiments could be explained by the pheromone hypothesis. The olfactory and pheromone hypotheses may not necessarily be mutually exclusive; under natural conditions population-specific odour may be just one component in the chemical environment to which a salmon imprints as a juvenile and responds as an adult.

Hormonal Regulation of Homing and Olfactory Imprinting.   Evidently, sex hormones play important roles in regulating reproductive behaviour in a number of teleost species (Stacey 1983a,b). Oestradiol-17β and testosterone increase dramatically during the course of the spawning season in salmonids, reaching a peak at early stages (Scott and Sumpter 1983; Ueda *et al.* 1984). Increased sex-hormone levels in both males and females of migrating salmon are correlated with high sensitivity to or discrimination of homestream water (Hasler and Scholz 1983). Spawned-out salmon with low levels of sex hormones no longer respond to their home-stream odour. Administration of gonadotrophin into salmon at a non-migratory stage increases locomotor activity and upstream movement, as the levels of sex hormones increase. Generally, a relationship between nasal and genital function exists in vertebrates. In fish, injections of oestradiol or testosterone modify olfactory bulbar activity by altering thresholds of responses to chemical and electrical stimulation (reviewed by Hara 1970; Demski and Hornby 1982). However, its underlying physiological mechanisms have not been well understood.

Juvenile salmon undergo marked transitions in morphology, physiology and behaviour just prior to their seaward migration (smolt transformation). This process is complex, and appears to be under the control of the endocrine system, particularly thyroid hormones and cortisol (Hoar 1976). Hasler and Scholz (1983) demonstrated that: (1) serum concentrations of these hormones rise just before smoltification occurs, and (2) injection of thyroid-stimulating hormone (TSH) and/or adrenocorticotrophic hormone (ACTH) into presmolts induces these transitions. However, Bern (1982) warns, in his closing remarks at a symposium on salmon smoltification, that 'We do not have information on blood levels of thyroid hormones and cortisol at different stages in various salmonids. We still need information on the kinetics of hormone production and degradation. ... we really do not know the physiological roles of thyroid hormones in salmonids (indeed in fish generally).' There is also evidence that olfactory imprinting occurred owing to TSH treatment. Presmolts receiving TSH or TSH plus ACTH and exposed to either morpholine or phenethyl alcohol demonstrated the ability to track their respective odour upstream, whereas those receiving ACTH, saline or no injections did not (Hasler and Scholz 1983). Only gonadotrophin-injected smolts, treated with TSH or TSH plus ACTH, displayed this attraction. Thus TSH injections mimic the events that activate olfactory imprinting in natural smolts (see also Chapter 3 by Huntingford, this volume).

Feeding

The majority of species rely upon information received by all their senses for food detection, recognition and selection. Whatever the senses involved, feeding behaviour shows a stereotyped sequence of behavioural components (cf. Atema 1980 and see Chapter 8 by Hart, this volume). The first step in the feeding sequence is arousal to the presence of food (alert or arousal stage). Animals become alerted to the presence of food by changing their respiratory and swimming patterns or activity levels. This process is primarily mediated by olfaction. The following locating or searching phase shows a wide diversity and is under the control of various sensory modalities, including chemical senses, depending upon the eco-logical niche of the animals involved. Feeding behaviour is completed by the food uptake and ingestion phase, which is triggered by taste stimuli. Many of the earlier studies dealt with general behavioural reactions to prey organisms and food extracts, and consequently the distinction of senses involved is not always clear. The literature on the role of chemoreception in feeding behaviour has been reviewed on several occasions (e.g. Kleerekoper 1969; Hara 1971, 1975, 1982; Atema 1980).

Detection of Food.   The significance of olfaction in the procurement of food by fishes was established experimentally by Bateson as early as 1890. He lists more than 15 freshwater and marine species that seek and recog-nise their food by the sense of smell alone. One of the interesting features common to these species is, Bateson notes, that they are all more or less nocturnal and remain in hiding by day. Catfish, for example, populate the bottoms of ponds, lakes and stagnant parts of rivers, and feed mostly at night. Detection of live prey (concealed earthworms) by ictalurid catfish (*Ictalurus nebulosus*) was abolished when olfaction was blocked by sectioning the olfactory tract, but not when barbels were removed. The yellow bullhead (*Ictalurus natalis*), on the contrary, shows essentially normal feeding behaviour with its olfactory epithelium cauterised, and even feeds normally with its forebrain removed. In the latter species external taste serves to locate dead bait at a distance and to trigger the pick-up reflex, whereas internal taste serves as a second, more restrictive screen for food intake controlling swallowing (Atema 1980). Thus, in the yellow bull-head, the entire feeding behavioural sequence from alert to swallowing is controlled by taste. Unlike catfish, yellowfin tuna (*Thunnus albacares*), which are highly visual predators, constantly cruise in large schools in the upper water layers in search of prey such as anchovy and squid. When a prey school is encountered, hungry tuna go into a feeding frenzy. They rely on vision in distant-prey detection under normal circumstances. However, under laboratory conditions, yellowfin tuna can detect and distinguish between odours of intact prey organisms, responding to odour quality of various prey rinses (Atema 1980). Feeding experience with prey causes a gradaul shift in responses from one to another prey odour or artificial mixture of odours. Furthermore, the hunger stage of tuna strongly influ-ences intensity of responses. Localisation of the odour source does not usually occur. Thus, in nature, tuna may use their olfactory sense when they cross odour trails of prey. Responses to the odour appear to be followed by orientation using other, probably visual, cues (Atema, Holland

and Ikehara 1980). Under a two-choice situation, whitefish (*Coregonus clupeaformis*), which are midwater plankton-feeding coregonids (Salmoniformes), show preference (attraction) for food extract. The behavioural reaction is eliminated when the nares of whitefish are cauterised. The preference reaction is characterised by a nosing and exploratory behaviour followed by a stationary posture in the side containing food extract, consequently resulting in a gradual decrease in crossing a boundary (Hara, Brown and Evans 1983).

Feeding Stimulants. Although behavioural studies to identify the active ingredients in food materials are fraught with inconsistencies in the response criteria, preparations of food materials, and chemosensory modalities involved, there is now evidence that feeding behaviour in different fish species is stimulated by somewhat different chemical substances (cf. Hara 1982).

The following generalisation regarding the chemical nature of feeding stimulants can be made. All feeding stimulants, to date, identified for teleosts are: (1) of low molecular weight (< 1000), (2) non-volatile, (3) nitrogenous, and (4) amphoteric. This generalisation applies to most of the cases in which amino acids, betaine, other amino acid like substances and nucleotides, particularly inosine, have been implicated. The high sensitivity to a relatively broad spectrum of amino acids and the apparent species-specific arrays of relative acuities suggest that biologically meaningful feeding stimulants may consist of fingerprint-like mixtures or chemical images (Atema 1980). For most species studied, single compounds such as betaine, glycine, alanine and taurine have been identified as representing major effectiveness in these mixtures. However, mixtures are nearly always more effective than single compounds: two exceptions are specific compounds isolated from marine invertebrates, which elicit exploratory feeding behaviour (Sangster, Thomas and Tingling 1975).

Anatomical Separation of Feeding Behaviour. In the cod (*Gadus morhua*) electrical stimulation of the lateral part of the lateral olfactory tract elicits the head-down and backing movements, characteristic of the feeding behaviour in this species (Døving and Selset 1980). The lateral part of the olfactory bulb of salmonids responds to feeding stimulant amino acids, whereas the medial part responds to bile acids (Thommesen 1978). Also, in goldfish the feeding response induced by a food odour is abolished by sectioning of the lateral part of the olfactory tract, and sectioning of the medial part has little effect (Stacey and Kyle 1983). These functional differentiations are consistent with the structural organisation of the teleost olfactory fibre connections, and suggest that olfactory neural inputs relevant to feeding may be processed through a channel relatively independent of those for other behaviour patterns such as reproduction.

Reproduction

Olfaction exerts a functional role in every aspect of the reproductive process from initial attraction and recognition of sexual status to sexual development of the young. Considerable work on the role of chemical signals, or pheromones, in reproductive behaviour of fishes has been conducted; however, as is the case with feeding behaviour, the sensory system involved and the nature of pheromones have not been rigorously investigated. Consequently, most of what is known of the interface between olfaction and reproduction is derived from studies on laboratory animals such as rodents (cf. Stoddart 1980). For detailed accounts the reader is referred to recent reviews (Kleerekoper 1969; Hara 1975; Pfeiffer 1982; Colombo *et al.* 1982; Liley 1982; Liley and Stacey 1983; Stacey 1983b). In this section, some of the major developments emphasising interactions between olfaction and reproductive behaviour currently in progress will be discussed.

The most convincing demonstration of the role of olfaction in mating is that exposure of males to water holding a gravid female elicits courtship behaviour in the goby (*Bathygobius soporator*) (Tavolga 1956). Of various internal body fluids from gravid females, only ovarian fluid elicits the behaviour. However, anosmic males do not respond to the female odour. Whether the stimulating ovarian substance is secreted by the eggs or by the ovary itself is not clear from this experiment. Many studies point to the gonads as a source of sex pheromones (cf. Colombo *et al.* 1982). In *Gobius jozo*, urine is more effective than ovarian washing in triggering courtship behaviour, and fluids collected from gravid females are more effective than those of postvitellogenic ones (Colombo *et al.* 1982). This contrasts with the observations of Tavolga (1956) that urine is inactive in *B. soporator*. Shortly after ovulation, the female goldfish releases a pheromone from the ovary which attracts the male and elicits the persistent courtship that accompanies spawning (Partridge, Liley and Stacey 1976). However, these authors did not rule out the possibility that the pheromone is also produced at other regions, such as the body surface. Although the chemical nature of sexual pheromones has not been identified conclusively in any teleost, recent studies indicate that the steroid glucuronides are likely candidates. In *G. jozo*, 5β-reduced androgen conjugates, particularly etiocholone glucuronide produced by a male mesorchial gland, attracts gravid females (Colombo *et al.* 1982). The sex attractant secreted by the ovaries of female zebrafish (*Brachydanio rerio*) also consists of steroid glucuronides, most likely a mixture containing oestradiol-17β- and testosterone glucuronide (van den Hurk and Lambert 1983). Ovarian pheromones in both species not only elicit courtship behaviour, but also inhibit fighting; anosmic males show extremely aggressive behaviour towards females. Some of the sex steroid hormones are known to influence olfactory mechanisms, probably mediated by steroid feedback on brain areas involved in centrifugal control of the olfactory bulb (Demski and Hornby 1982). However, no direct stimulatory effects on the olfactory receptors of the sex steroids has been demonstrated in any teleost species.

Although the ovarian sex pheromones are essential for triggering a series of courtship actions by the males, a complete courtship repertoire is expressed only in the presence of other sensory cues (see Chapter 16 by Fitzgerald and Wootton, this volume). In the male three-spined stickleback (*Gasterosteus aculeatus*), for example, the functioning of the olfactory system during reproductive behaviour (nest building, zigzag dance, and fanning) is supplemented by the gustatory system (Segaar, de Bruin and van der Meché-Jacobi 1983). The olfactory neural inputs, albeit not absolutely necessary, support nest building and are essential for subsequent development of courtship behaviour as expressed by the zigzag dance. General olfactory stimuli, rather than a specific pheromone, are assumed to influence the development of nest building by inducing hormonal changes necessary to start the reproductive cycle. The olfactory inputs are essential for promoting the sexual tendency during the early reproductive cycle, and once the zigzag scores had reached an essential level, sectioning the olfactory nerve no longer affects the normal sexual activities, probably because the loss of chemical information via the olfactory system is compensated by functioning of the gustatory nerves (IXth and Xth). Stimuli from eggs and embryos are perceived by receptors located in the area of the pharyngeal roof between the first and second gill arches and excite sexual behaviour during most of the reproductive cycle.

Most experiments dealing with the role of olfaction in reproductive behaviour in teleosts have generally considered the olfactory system as a uniform whole. However, as discussed earlier, both anatomical and electro-physiological studies support the contention that the olfactory system may be composed of two relatively independent parts (cf. Satou *et al.* 1983). Sectioning of the medial olfactory tract of goldfish reduces male sexual behaviour to the level of fish with complete tract section, whereas cutting the lateral olfactory tract has no effect (Stacey and Kyle 1983). In contrast, the feeding response to food odours is more affected by section of the lateral olfactory tract than of the medial. Similarly, in salmonids, feeding stimulant amino acids elicits responses more in the lateral part of the olfactory bulb, and responses induced by steroid bile acids project to the medial part of the bulb (Døving *et al.* 1980). The medial olfactory tract contains fibres of the terminal nerve, a cranial nerve containing luteinizing hormone-releasing hormone-immunoreactive material (Münz, Stumpf and Jennes 1981; Kah, Chambolle, Dubourg and Dubois 1984). It is suggested that this system may play an integrative function in processing olfactory and optic information in teleosts and is an essential element of the anatomical substrates for the role of olfaction and vision in homing and reproduction.

Fright Reaction

von Frisch (1938) accidentally discovered that when an injured minnow (*Phoxinus phoxinus*) was introduced into a normal shoal, they became frightened and dispersed. 'In fact, the incident that first drew the attention of von Frisch to this problem was his observation of a kingfisher swooping down and attacking a school (= shoal, Ed.) of minnows. After successfully capturing one, as the bird was flying away it dropped the minnow and von Frisch noticed that the other minnows, which were still schooling, suddenly dispersed' (quoted from Hasler and Scholz 1983). The general characteristics of the phenomenon is that when the skin of a fish is damaged (e.g. by a predator), alarm-substance cells are broken and release alarm substance (*Schreckstoff*). Nearby conspecifics smell the alarm substance and show a fright reaction (*Schreckreaktion*). Their fright reaction may then be treated as a visual signal by other conspecifics, leading to rapid transmission of the signal through a group of fish. This alarm-signal system consists of two main components: a behavioural component, the fright reaction, and a morphological component, the alarm substance (cells). In view of the recent and comprehensive reviews by Pfeiffer (1974, 1982) and Smith (1977, 1982), the following will focus on recent developments, together with a brief summary of earlier reviews.

Fright Reaction Behaviours.   The fright reaction is not stereotyped and differs considerably from species to species, involving cover seeking, closer crowding, rapid swimming or immobility. Thus seven intensities of response are arbitrarily established when a skin extract of the minnow is introduced into an aquarium containing minnows. The most intense reaction involves sudden flight into the hiding place, emergence, rapid swimming around the tank, and avoiding the feeding place for a considerable length of time. In a typical experiment, an extract of 0.002 mg of chopped fish skin was sufficient to elicit fright reaction in minnows. In the common shiner (*Notropis cornutus*), a skin preparation increases shoal cohesion and polarisation and decreases the variability in overall shoal dimensions (Heczko and Seghers 1981). The reaction is generally accepted to be mediated by olfaction; however, as von Frisch (1941) himself noticed, involvement of gustatory and/or other senses is not entirely ruled out. The extract of the skin and accompanying mucus contains various organic compounds such as fatty acids, carbohydrates, proteins, phospholipids and free amino acids, many of which stimulate gustatory as well as olfactory systems. Although the fright reaction may be initiated by a chemical stimulus, it can be transmitted through a school of fish by vision. When two aquaria with a group of fish in each are placed close together, an alarm-substance-induced fright reaction in one tank would trigger a fright reaction in the fish in the adjacent tank without exchanging water or chemical stimuli.

Alarm Substance.   The alarm substance is contained in a large epidermal cell, the 'club cell'. Alarm-substance cells can be distinguished from mucous cells, the other main fish epidermal secretory cells, by their negative reaction to periodic acid Schiff's (PAS) reagent. The mucous cells secrete the mucus at the skin surface through a pore, but the alarm-substance cells lack an opening at the skin surface and release their contents only when the skin is injured. Alarm-substance cells are usually found closer to the basal layer than the mucous cells and autofluoresce under ultraviolet light. Skin extract from breeding males or testosterone-treated fathead minnows (*Pimephales promelas*), which lose their alarm-substance cells, does not induce a fright reaction. Skin extract from females or non-breeding males, which retain alarm-substance cells, does induce a fright reaction (Smith 1982). The timing of the loss of alarm substance cells in nature corresponds with the development of androgen-induced secondary sexual characters and with higher levels of testosterone. Seasonal loss of alarm substance in male fatheads is interpreted as an adaptation to their abrasive spawning habits, reducing the chance of the alarm system 'misfiring' and interfering with spawning or parental care (Smith 1977). The seasonal loss of alarm substance is found in males of seven cyprinid species, six of which have abrasive spawning behaviour.

Identification of the Alarm Substance.   The alarm substance was variously characterised as being purine- or a ptcrine-like, small-ringed or double-ringed, polypeptide-like, or histamine-like. A mixture of several compounds including amino acids and peptides was also suspected for channel catfish, and the skin extract of Atlantic salmon, which lack the alarm substance, was about equally effective (Tucker and Suzuki 1972). Recent chemical analyses suggest that hypoxanthine-3(N)-oxide is most likely to be the alarm substance identified in the minnows. This substance is colourless, non-fluorescent, slightly water soluble, and unstable in solution (Argentini 1976, cited in Pfeiffer 1982). Behavioural experiments have confirmed that hypoxanthine-3(N)-oxide has full biological activity. However, its olfactory stimulatory activity has not been determined.

Taxonomic Distribution of the Alarm-substance Fright Reaction.   The alarm-substance fright reaction system described above appears to be restricted to one phylogenetically related group of fish, the Ostariophysi. However, possession of alarm substance and fright reaction is not universal within the group. Several groups lack one or both components of the alarm system. The reaction is not associated with any particular habit or type of social behaviour within the groups. The fright reaction is not species specific, fish react to the alarm substance emitted by others, but the intensity of the response is related to the phylogenetic proximity of the species. The fright reaction is genetically determined, and appears at a certain stage of development regardless of prior experience. The reaction does not develop until some time after the alarm substance is formed in the skin. In minnows, it appears at 51 days after hatching. Shoaling behaviour, by contrast, appears much earlier than the fright reaction.

Biological Significance of the Fright Reaction.   It is conceivable that individuals are likely to gain protection as a result of their responses to alarm substance released by a conspecific upon attack by a predator. It is, however, difficult to understand how the release of alarm substance can provide protection or other benefits to the damaged individual, and how such a signalling system might have evolved. Smith (1977) discusses some of the hypotheses, such as kin selection and chemical repellent, regarding the adaptive significance of fish alarm substance. Further research is required to answer this intriguing question properly.

Summary

Fish detect chemical stimuli through two major chemosensory channels, olfaction and taste. Unlike terrestrial animals, in fish all chemical activators are mediated by water, and chemoreception occurs entirely in the aquatic environment. Therefore, a distinction between olfaction and taste is tenuous, and volatility of olfactory stimulants (odours) is less relevant. Because the fish gustatory system is as sensitive as the olfactory system, we can no longer consider olfactory to be *the* sensory modality for detection of aquatic chemicals only on the basis of its lower effective concentration. The olfactory system is not a uniform whole; at least two types of olfactory receptor cell detect and encode chemical signals. The encoded olfactory information is transmitted and integrated into behavioural patterns through spatially separated neuroanatomical substrates within the olfactory centre. Thus feeding behaviour elicited by substances such as amino acids may be processed via the *lateral olfactory system*, whereas other behaviour such as reproduction may be processed through the *medial olfactory system.*

Four principal areas of interface between olfaction and fish behaviour are described:

(1)   *Homing migration in salmon.* The olfactory sense seems essential for homing migration. Salmonids imprint to certain distinct odours of the home stream during the early period of residence, and as adults they use this olfactory information to locate the home stream — the 'olfactory hypothesis'. Experiments in which salmon artificially imprinted to morpholine homed to a stream scented with the chemical verified the olfactory hypothesis. However, the physiological basis for olfactory detection of morpholine has not been established. Homing may be an inherited response to population-specific pheromone trails released by descending smolts — the 'pheromone hypothesis'.

(2)   *Feeding*. Feeding behaviour, a stereotyped sequence of behavioural components, is diversely controlled by various sensory modalities, depending upon the ecological niche of the animal. The initial arousal or alert process is primarily mediated by olfaction. Feeding stimulants seem species-specific, and have the following chemical nature: (a) low molecular weight, (b) non-volatile, (c) nitrogenous, and (d) amphoteric. Biologically meaningful feeding stimulants may consist of fingerprint-like mixtures or chemical images.

(3)   *Reproduction*. Olfaction exerts a functional role in every aspect of the reproductive process. Investigations involving fish are still in a descriptive stage. Not a single pheromone has been identified chemically. Recent studies point to the gonads as a source of fish reproductive pheromones, and indicate that steroids are likely candidates. Evidence is also presented that loss of chemical information via the olfactory system is compensated by functioning of the gustatory system.

(4)   *Fright reaction*. When the skin of an ostariophysid fish is damaged, alarm-substance cells are broken and release alarm substance. Nearby conspecifics smell the alarm substance and show a fright reaction. Hypoxanthine-3(N)-oxide is likely to be the alarm substance identified in minnows. The adaptive significance of the fright reaction still remains an intriguing question.

References

Atema, J. (1980) 'Chemical Sense, Chemical Signals, and Feeding Behaviour in Fishes', in J.E. Bardach, J.J. Magnuson, R.C. May and J.M. Reinhart (eds), *Fish Behavior and its Use in the Capture and Culture of Fishes*, International Center for Living Aquatic Resources Management, Manila, pp. 57-101

Atema, J., Holland, K. and Ikehara, W. (1980) 'Olfactory Responses of Yellowfin Tuna (*Thunnus albacares*) to Prey Odors: Chemical Search Image', *Journal of Chemical Ecology, 6*, 457-65

Bateson, W. (1890) 'The Sense-organs and Perceptions of Fishes; with Remarks on the Supply of Bait', *Journal of the Marine Biological Association of the United Kingdom, 11*, 225-56

Bern, H.A. (1982) 'Epilog', in H.A. Bern and C.V.W. Mahnken (eds), *Salmonid Smoltification, Aquaculture, Special Issue, 28*, v-x and 1-270

Brown, S.B. and Hara, T.J. (1982) 'Biochemical Aspects of Amino Acid Receptors in Olfaction and Taste', in T.J. Hara (ed.), *Chemoreception in Fishes*, Elsevier, Amsterdam, pp. 159-80

Buckland, J. (1880) *Natural History of British Fishes*, Unwin, London

Caprio, J. (1982) 'High Sensitivity and Specificity of Olfactory and Gustatory Receptors of Catfish to Amino Acids', in T.J. Hara (ed.), *Chemoreception in Fishes*, Elsevier, Amsterdam, pp. 109-34

Chidester, F.E. (1924) 'A Critical Examination of the Evidence for Physical and Chemical Influences of Fish Migration', *Journal of Experimental Biology, 2*, 79-118

Colombo, L., Belvedere, P.C., Marconato, A. and Bentivegna, F. (1982) 'Pheromones in Teleost Fish', in C.J.J. Richter and H.J.Th. Goos (eds), *Reproductive Physiology in Fish*, Pudoc, Wageningen, pp. 84-94

Cooper, J.C. and Hirsch, P.J. (1982) 'The Role of Chemoreception in Salmonid Homing', in

T.J. Hara (ed.), *Chemoreception in Fishes*, Elsevier, Amstedam, pp. 343-62

Demski, L.S. and Hornby, P.J. (1982) 'Hormonal Control of Fish Reproductive Behavior: Brain-Gonadal Steroid Interactions', *Canadian Journal of Fisheries and Aquatic Sciences, 39*, 36-47

Demski, L.S. and Northcutt, R.G. (1983) 'The Terminal Nerve: a New Chemosensory System in Vertebrates?', *Science (Washington, DC), 220*, 435-7

Døving, K.B. and Selset, R. (1980) 'Behavior Patterns in Cod Released by Electrical Stimulation of Olfactory Tract Bundles', *Science (Washington, DC), 207*, 559-60

Døving, K.B., Nordeng, H. and Oakley, B. (1975) 'Single Unit Discrimination of Fish Odours Released by Char (*Salmo alpinus*) Populations', *Comparative Biochemistry and Physiology, 47A*, 1051-63

Døving, K.B., Selset, R. and Thommesen, G. (1980) 'Olfactory Sensitivity to Bile Acids in Salmonid Fishes', *Acta Physiologica Scandinavica, 108*, 123-31

Evans, R.E., Zielinski, B. and Hara, T.J. (1982) 'Development and Regeneration of the Olfactory Organ in Rainbow Trout', in T.J. Hara (ed.), *Chemoreception in Fishes*, Elsevier, Amsterdam, pp. 15-37

Fesenko, E.E., Novoselov, V.I., Krapivinskaya, L.D., Mjasoedov, N.F. and Zolotarev, J.A. (1983) 'Molecular Mechanisms of Odor Sensing VI. Some Biochemical Characteristics of a Possible Receptor for Amino Acids from the Olfactory Epithelium of the Skate *Dasyatis pastinaca* and carp *Cyprinus carpio*', *Biochimica Biophysica Acta, 759*, 250-6

Goy, Y. and Tamura, T. (1980) 'Olfactory and Gustatory Responses to Amino Acids in Two Marine Teleosts — Red Sea Bream and Mullet', *Comparative Biochemistry and Physiology, 66C*, 217-24

Hara, T.J. (1970) 'An Electrophysiological Basis for Olfactory Discrimination in Homing Salmon: a Review', *Journal of the Fisheries Research Board of Canada, 27*, 565-86

Hara, T.J. (1971) 'Chemoreception', in W.S. Hoar and D.J. Randall (eds), *Fish Physiology*, vol. 5, Academic Press, New York, pp. 79-120

Hara, T.J. (1975) 'Olfaction in Fishes', in G.A. Kerkut and J.W. Phillis (eds), *Progress in Neurobiology*, vol. 5, Pergamon Press, Oxford, pp. 271-335

Hara, T.J. (1982) (ed.) *Chemoreception in Fishes*, Elsevier, Amsterdam

Hara, T.J., Brown, S.B. and Evans, R.E. (1983) 'Pollutants and Chemoreception in Aquatic Organisms', in J.O. Nriagu (ed.), *Aquatic Toxicology*, Wiley, New York, pp. 247-306

Hara, T.J., Macdonald, S., Evans, R.E., Marui, T. and Arai, S. (1984) 'Morpholine, Bile Acids and Skin Mucus as Possible Chemical Cues in Salmonid Homing: Electrophysiological Re-evaluation' in J.D. McCleave, G.P. Arnold, J.D. Dodson and W.H. Neill (eds), *Mechanisms of Migration in Fishes*, Plenum, New York, pp. 363-78

Harden Jones, F.R. (1968) *Fish Migration*, Edward Arnold, London

Hasler, A.D. (1966) *Underwater Guideposts — Homing of Salmon*, University of Wisconsin Press, Madison

Hasler, A.D. (1983) 'Synthetic Chemicals and Pheromones in Homing Salmon', in J.C. Rankin, T.J. Pitcher and R.T. Duggan (eds), *Control Processes in Fish Physiology*, Croom Helm, London, pp.103-16

Hasler, A.D. and Scholz, A.T. (1983) *Olfactory Imprinting and Homing in Salmon. Investigations into the Mechanism of the Imprinting Process*, Springer-Verlag, Berlin

Heczko, E.J. and Seghers, B.H. (1981) 'Effects of Alarm Substance on Schooling in the Common Shiner (*Notropis cornutus*, Cyprinidae)', *Environmental Biology of Fishes, 6*, 25-9

Hoar, W.S. (1976) 'Smolt Transformation: Evolution, Behaviour, and Physiology', *Journal of the Fisheries Research Board of Canada, 33*, 1233-52

Kah, O., Chambolle, P., Dubourg, P. and Dubois, M.P. (1984) 'Immunocytochemical Localisation of Luteinizing Hormone-releasing Hormone in the Brain of the Goldfish *Carassias auratus*', *General and Comparative Endocrinology, 53*, 107-15

Kiyohara, S., Yamashita, S. and Harada, S. (1981) 'High Sensitivity of Minnow Gustatory Receptors to Amino Acids', *Physiology and Behavior, 26*, 1103-8

Kleerekoper, H. (1969) *Olfaction in Fishes*, Indiana University Press, Bloomington

Liley, N.R. (1982) 'Chemical Communication in Fish', *Canadian Journal of Fisheries and Aquatic Sciences, 39*, 22-35

Liley, N.R. and Stacey, N.E. (1983) 'Hormones, Pheromones, and Reproductive Behavior in Fish', in W.S. Hoar, D.J. Randall and E.M. Donaldson (eds), *Fish Physiology*, vol. IX, Academic Press, New York, pp. 1-63

McKeown, B.A. (1984) *Fish Migration*, Croom Helm, London, 224 pp.

Marshall, N.B. (1967) 'The Olfactory Organs of Bathypelagic Fishes', *Symposium of the Zoological Society of London, 19*, 57-70

Marui, T., Evans, R.E., Zielinski, B. and Hara, T.J. (1983) 'Gustatory Responses of the Rainbow Trout (*Salmo gairdneri*) Palate to Amino Acids and Derivatives', *Journal of Comparative Physiology, 153*, 423-33

Münz, H., Stumpf, W.E. and Jennes, L. (1981) 'LHRH Systems in the Brain of Platyfish', *Brain Research, 221*, 1-13

Nordeng, H. (1977) 'A Pheromone Hypothesis for Homeward Migration in Anadromous Salmonids', *Oikos, 28*, 155-9

Oka, Y., Ichikawa, M. and Ueda, K. (1982) 'Synaptic Organization of the Olfactory Bulb and Central Projection of the Olfactory Tract', in T.J. Hara (ed.), *Chemoreception in Fishes*, Elsevier, Amsterdam, pp. 61-75

Ottoson, D. (1971) 'The Electro-olfactogram', in L.M. Beidler (ed.), *Handbook of Sensory Physiology*, vol. 4, part 1, Springer-Verlag, Berlin, pp. 95-131

Partridge, B.L., Liley, N.R. and Stacey, N.E. (1976) 'The Role of Pheromones in the Sexual Behaviour of the Goldfish', *Animal Behaviour, 24*, 291-9

Pfeiffer, W. (1974) 'Pheromones in Fish and Amphibia' in M.C. Birch (ed.), *Pheromones*, Elsevier, North-Holland, Amsterdam, pp. 269-96

Pfeiffer, W. (1982) 'Chemical Signals in Communication' in T.J. Hara (ed.), *Chemoreception in Fishes*, Elsevier, Amsterdam, pp. 307-26

Rhein, L.D. and Cagan, R.H. (1983) 'Biochemical Studies of Olfaction: Binding Specificity of Odorants to a Cilia Preparation from Rainbow Trout Olfactory Rosettes', *Journal of Neurochemistry, 41*, 569-77

Sangster, A.W., Thomas, S.E. and Tingling, N.L. (1975) 'Fish Attractants from Marine Invertebrates. Arcamine from *Arca zebra* and Strombine from *Strombus gigas*', *Tetrahedron, 31*, 1135-7

Satou, M., Fujita, I., Ichikawa, M., Yamaguchi, K. and Ueda, K. (1983) 'Field Potential and Intracellular Potential Studies of the Olfactory Bulb in the Carp: Evidence for a Functional Separation of the Olfactory Bulb into Lateral and Medial Subdivisions', *Journal of Comparative Physiology, 152*, 319-33

Scott, A.P. and Sumpter, J.P. (1983) 'The Control of Trout Reproduction: Basic and Applied Research on Hormones', in J.C. Rankin, T.J. Pitcher and R.T. Duggan (eds), *Control Processes in Fish Physiology*, Croom Helm, London, pp. 200-20

Segaar, J., de Bruin, J.P.C. and van der Meché-Jacobi, M.E. (1983) 'Influence of Chemical Receptivity on Reproductive Behaviour of the Male Three-spined Stickleback (*Gasterosteus aculeatus* L.)', *Behaviour, 86*, 100-66

Smith, R.J.F. (1977) 'Chemical Communication as Adaptation: Alarm Substance of Fish', in D. Müller-Schwarze and M.M. Mozell (eds), *Chemical Signals in Vertebrates*, Plenum, New York, pp. 303-20

Smith, R.J.F. (1982) 'The Adaptive Significance of the Alarm Substance-Fright Reaction System' in T.J. Hara (ed.), *Chemoreception in Fishes*, Elsevier, Amsterdam, pp. 327-42

Stacey, N. (1983a) 'Hormones and Pheromones in Fish Sexual Behavior', *BioScience, 33*, 552-6

Stacey, N.E. (1983b) 'Hormones and Reproductive Behaviour in Teleosts', in J.C. Rankin, T.J. Pitcher and R.T. Duggan (eds), *Control Processes in Fish Physiology*, Croom Helm, London, pp. 117-29

Stacey, N.E. and Kyle, A.L. (1983) 'Effects of Olfactory Tract Lesions on Sexual and Feeding Behavior in the Goldfish', *Physiology and Behavior, 30*, 621-8

Stoddart, D.M. (1980) *The Ecology of Vertebrate Olfaction*, Chapman & Hall, London

Strieck, F. (1924) 'Untersuchungen über den Geruchs- und Geschmackssinn der Elritzen', *Zeitschrift für vergleichende Physiologie, 2*, 122-54

Sutterlin, A.M. and Sutterlin, N. (1971) 'Electrical Responses of the Olfactory Epithelium of Atlantic Salmon (*Salmo salar*), *Journal of the Fisheries Research Board of Canada, 28*, 565-72

Suzuki, N. and Tucker, D. (1971) 'Amino Acids as Olfactory Stimuli in Freshwater Catfish, *Ictalurus catus* (Linn.)', *Comparative Biochemistry and Physiology, 40A*, 399-404

Tavolga, W.N. (1956) 'Visual, Chemical and Sound Stimuli as Cues in the Sex Discriminatory Behavior of the Gobiid Fish, *Bathygobius soporator*', *Zoologica, 41*, 49-64

Thommesen, G. (1978) 'The Spacial Distribution of Odour Induced Potentials in the

Olfactory Bulb of Char and Trout (Salmonidae)', *Acta Physiologica Scandinavica, 102,* 205-17

Thommesen, G. (1983) 'Morphology, Distribution, and Specificity of Olfactory Receptor Cells in Salmonid Fishes', *Acta Physiologica Scandinavica, 117,* 241-9

Tucker, D. and Suzuki, N. (1972) 'Olfactory Responses to Schreckstoff of Catfish' in D. Schneider (ed.), *Olfaction and Taste IV,* Wissenschaftliche Verlagsgesellschaft, Stuttgart, pp. 121-7

Ueda, H., Hiroi, O., Hara, A., Yamauchi, K. and Nagahama, Y. (1984) 'Changes in Serum Concentrations of Steroid Hormones, Thyroxine, and Vitellogenin during Spawning Migration of the Chum Salmon, *Oncorhynchus keta'*, *General and Comparative Endocrinology, 53,* 203-11

van den Hurk, R. and Lambert, J.G.D. (1983) 'Ovarian Steroid Glucuronides Function as Sex Pheromones for Male Zebrafish, *Brachydanio rerio'*, *Canadian Journal of Zoology, 61,* 2381-7

von Frisch, K. (1938) 'Zur Psychologie des Fisch-Schwarmes', *Die Naturwissenschaften, 26,* 601-6

von Frisch, K. (1941) 'Ueber einen Schreckstoff der Fischhaut und seine biologische Bedeutung', *Zeitschrift für vergleichende Physiologie, 29,* 46-145

Yoshii, K., Kamo, N., Kurihara, K. and Kobatake, Y. (1979) 'Gustatory Responses of Eel Palatine Receptors to Amino Acids and Carboxylic Acids', *Journal of General Physiology, 74,* 301-17

7 ROLE OF THE LATERAL LINE IN FISH BEHAVIOUR

Horst Bleckmann

The lateral line of teleost fish is usually visible externally as a row of small pores along the trunk and the head. These pores lead into an underlying canal, the lateral line canal. According to Parker (1904) the first people to recognise the pores of the lateral line were Stenon in 1664, Lorenzini in 1678, and Rivinius in 1687. In the head of most fishes the lateral line canal has three main branches, one of which passes forwards and above the eye (supra-orbital canal), another forward and immediately below the eye (infra-orbital canal), and a third downwards and over the lower jaw (hyo-mandibular canal). Together with the 'free organs' described later in this chapter, these three head and one trunk canal with their 'canal organs' constitute the lateral line system of teleost fish (Figures 7.1A,B).

Figure 7.1: Distribution of Ordinary Lateral Line Organs in *Phoxinus phoxinus* (A) and the Blind Cave Fish *Anoptichthys jordani* (B): • Free Neuromasts; ○ Canal Pores (Enlarged for Clarity); (C) Distribution of Specialised Lateral Line Organs in the Weakly Electric Fish *Gnathonemus petersii*. The ordinary lateral line organ on the trunk is indicated by a dashed line. SOC, supra-orbital canal; IOC, infra-orbital canal; HMC, hyo-mandibular canal; TLL, trunk lateral line; and SLLO, specialised lateral line organs (redrawn from Dijkgraaf 1933; Grobbel and Hahn 1958; and Szabo 1974)

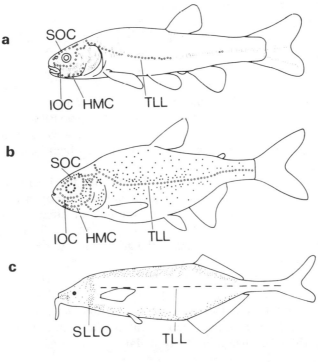

Most earlier investigators considered the lateral line as a system of glands for the production of the mucus seen so characteristically on the skin of many fishes (see Fuchs 1895), but it was Jakobson (1813) who first concluded from observing the extensive nerve supply, that the lateral line system is a sense organ which is probably stimulated mechanically. Knox (1825) and Leydig (1850, 1851) supported this idea and the lateral line was soon universally accepted as a mechanoreceptive system especially adapted for aquatic life.

Despite all these investigations the biologically relevant stimulus was heavily disputed. Some investigators presumed that the lateral line system was only an organ of touch (Merkel 1880; de Sede 1884); others supposed it to have a function between touch and hearing (Leydig 1850, 1851; Schulze 1870; Dercum 1880); to be an accessory auditory organ (Emery 1880); a pressure receptor (Fuchs 1895); or to serve rheotactic purposes (Hofer 1908). From behavioural experiments Richard (1896, quoted by Parker 1902) believed that the lateral line organ was connected with the production of gas in the air bladder, and Lee (1898) thought it gave a sense of equilibrium.

It was Hofer (1908) who convincingly demonstrated that both free neuromasts and canal organs are current-sensitive mechanoreceptors. He observed distinctive fin movements when weak water currents were impinging locally on the head or trunk of a pike (*Esox lucius*). Unilateral cauterisation of the lateral line organs rendered the operated part of the body insensitive, whereas all intact body areas were unaffected in sensitivity. From these experiments Hofer concluded that fish can 'feel at a distance', and his results have later been confirmed and extended with many different species, including elasmobranchs, other bony fishes and amphibians. All investigations report spontaneous or conditioned responses elicited by weak local water currents in blinded animals. Dijkgraaf (1963) finally demonstrated that local water currents around a swimming fish are the lateral line stimuli responsible for close-range obstacle detection. Schwartz (1965, 1971) has shown beyond doubt that the lateral line organs on the head enable surface-feeding fish to detect and locate sources of surface waves.

In 1969 Katsuki and Hashimoto noted that the lateral line organs of some teleost fish are sensitive to chemical stimuli. For instance, for some teleosts they found that the order of stimulatory effectiveness for monovalent cations is as follows: $Rb^+, K^+ > Na^+ > Cs^+, Li^+$. To give some further examples: the lateral line organs of the goby, *Gobius giurinus*, responds to divalent cations in the order: $Ba^{2+} > Ca^{2+} > Mg^{2+} > Sr^{2+}$. In addition, the lateral line system of bottom dwellers is often very sensitive to Na_4Cl (Kawamura and Yamashita 1983). Owing to the lack of any behavioural evidence, and to the fact that the lateral line nerves terminate in the octavo-lateralis area of the medulla and not in olfactory centres, it is very unlikely that the chemosensitivity of lateral line organs is more than an incidental by-product of receptor physiology.

Experimental results of the last three decades show that the lateral line organs can be split into two groups: (1) the 'ordinary' types, which are mechanoreceptive, and (2) the 'specialised' kinds, which are electro-receptive.

Ordinary Lateral Line Organs

Morphology and Physiology

The basic unit of the mechanosensitive ordinary lateral line system is the neuromast, which consists of a cluster of pear-shaped sensory cells surrounded by supporting cells (Figure 7.2A). These pear-shaped cells are called 'hair cells' because of a bundle of hairs on their upper (apical) cellular surface. This bundle projects into a jelly-like substance, the cupula. It was Schulze (1861) who first observed with the microscope that slight motions of the water made this cupula move and bend. Single neuromasts are distributed in a definite arrangement on the head and the trunk of teleost fish (Figures 7.1A,B). These organs, which are called free organs or superficial neuromasts, are often found at the bottom of a shallow pit or groove in the skin of cyclostomes and many bony fishes. In many quick- and constantly swimming fish a certain number of free neuromasts have evolved into canal organs, which communicate with the surrounding water by way of pores (Figure 7.2B). In canal organs the cupula is surrounded by sea water or by a special canal endolymph, a watery fluid. Canal organs may have developed to shield the neuromasts from the direct influence of the currents passing the body of a swimming fish. This point of view is sup- ported because in many fish which have secondarily returned to slow or intermittent swimming the lateral line canals have completely disappeared (Dijkgraaf 1963).

The hair cells of the lateral line (Figure 7.2C) are similar to those in the auditory and vestibular organs of vertebrates. The bundle of (30 to 150) hairs on the hair cells, called stereocilia or stereovilli, were first recognised as such by Schulze (1861). The stereocilia (diameter $0.2-0.8\,\mu m$) grow longer from one edge of the hair bundle to the other. Most hair cells also have a single membrane-bound kinocilium (diameter about $0.3\,\mu m$) with the true $9 + 2$ ciliary pattern. The kinocilium always occurs eccentrically at the tall edge of the bundle, thus all hair cells display a morphological polarisation. The adequate (natural) stimulus for a hair cell is a mechanical force applied to the distal end of the whole hair bundle. Thus a hair cell is essentially a mechanoreceptor sensitive to bundle deflections of the order of only thousandths of a degree (Hudspeth 1983). An individual hair cell itself lacks dendrites or an axon but is connected to the CNS by an afferent (sensory) and efferent (inhibitory) nerve fibre (Figure 7.2C).

Besides the morphological polarisation, hair cells also show a physio-

Figure 7.2: (A) Diagram of a Superficial Neuromast of Bony Fishes. Characteristic features are the pear-shaped sense cells, each bearing a hair bundle, and the jelly-like cupula ensheathing the hairs. (B) Part of a Fish's Body with Epidermal Neuromasts, Canal Pores and a Longitudinal Section Through a Lateral Line Canal with Canal Organs. (C) Schematic Diagram of a Vertebrate Hair Cell. The continuous plasma membrane envelops the individual stereocilia and the single kinocilium. Note the progression of stereocilia length, with the longest stereocilium at the edge of the hair bundle adjacent to the kinocilium. Ordinary lateral line organs display dual (afferent and efferent) innervation. (D) The Directional Sensitivity of the Hair Cell Approximates a Cosine Function of Stimulus Direction: the Output Varies as the Cosine of the Angle between the Direction of Maximum Sensitivity and the Applied Displacement. (E) The Directional Sensitivity of an Organ Illustrated in a Polar Coordinate System. At each direction of stimulation, the amplitude of the microphonic potential is plotted on the appropriate coordinate. The axis of the canal is parallel to the coordinate 0-180°. The dotted line indicates the outline of the organ, and the orientation of the two groups of hair cells within the sensory epithelium indicated by two sensory-hair bundles oriented with their kinocilia (large dots) pointing in opposite directions (redrawn from Flock 1965). (F) Schematic Illustration of Sensory Cells in Specialised (Electric) Lateral Line Organs. Example of an ampullary organ (top) and a tuberous organ (bottom) of teleost fishes. The apical cell membrane has lost the hair bundle typical for mechanosensitive hair cells. In contrast to hair cells, electroreceptive sense cells lack an efferent innervation (redrawn from Dijkgraaf 1963; Flock 1971; and Szabo 1974)

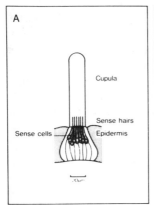

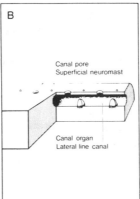

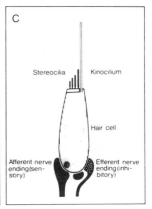

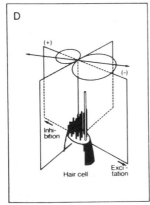

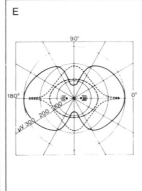

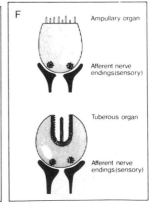

logical polarisation. Displacement of the sensory-hair bundle in the direction of the kinocilium is excitatory, i.e. there is a depolarisation of the hair cell which leads to an increased probability of nerve firing. Displacement in the opposite direction is inhibitory, i.e. there is a hyperpolarisation of the hair cell, which leads to a decreased probability of nerve firing (Flock 1971). Consequently the response of a single hair cell varies with the stimulus angle in a cosine fashion (Figure 7.2D). In single neuromasts or canal organs, adjacent hair cells are always oriented with their kinocilia facing in opposite directions. This means that displacement of the cupula will cause responses of opposite polarities from the two sets of cells, which work 180° out of phase, and that lateral line organs as a whole are directionally sensitive. The directional sensitivity can be demonstrated by recording the microphonic potential (summed potential of all hair cells of a single neuromast) and its dependence on stimulus angle. The microphonic potential is largest when the direction of stimulation is parallel to the major axis of the organ, and decreases with increasing deviation from this direction until it reaches a minimum when the direction of stimulation is perpendicular to it (Figure 7.2E). Single afferent fibres of the lateral line nerve couple only to hair cells that have the same orientation. A single afferent fibre therefore responds only (with a decrease or increase of the spontaneous discharge frequency) if the cupula is bent in one of the two possible directions with respect to the most sensitive axis of the neuromast. A fish may have hundreds of neuromasts, which may differ in their orientation and frequency response. Therefore, in theory at least, a fish should be able to determine the angle of a stimulus, its velocity, direction of movement, and shape by timing and spacing integration of the input from all neuromasts stimulated.

Sensitivity of the Ordinary Lateral Line Organs to Local Water Displacements

The appropriate stimuli for the ordinary lateral line organs are low-frequency water displacements or net flow. For the latter, threshold displacements as low as 0.025 mm water current per second have been reported (see Schwartz 1974). The maximum sensitivity (in terms of displacement) to vibrational stimuli in most fish species is in the frequency range 30-150 Hz, with threshold displacements (peak-to-peak) in the range 0.1-0.001 µm (e.g. Flock 1965; Kuiper 1967; Sand 1981; Bleckmann and Topp 1981; H. Münz, unpublished).

Free neuromasts respond best to water displacements directed parallel to the long axis of the cupula. Canal organs are most sensitive to water currents parallel to the axis of the canal, i.e. in this case the amplitude of the displacement of the canal fluid and the electrophysiological responses of single canal neuromasts is maximal (Sand 1981; Denton and Gray 1983; Gray 1984). From the work of Denton and Gray (1983) it was found that, for frequencies up to 80 Hz, the displacement of the canal fluid in the sprat

(*Sprattus sprattus*) is proportional to the velocity of the external medium, i.e. canal organs are relatively insensitive to slow changes of water flow along a moving fish.

There is some evidence that the sensitivity of ordinary lateral line organs can actively be reduced by means of efferent lateral line nerve fibres (Figure 7.2C). A reduction of the sensitivity of lateral line organs occurs especially during the power strokes of swimming movements associated with the contraction of white 'anaerobic' swimming muscles, i.e. those associated with rapid escape, turning or chasing. This may be essential for maintaining the sensitivity of the lateral line organs to water displacements the moment the movement stops (Russell 1974, 1976). In contrast to vigorous escape and attacking movements, many fish have evolved a style of locomotion and cruising using red muscles, which involves little body movement, and apparently may be ideal for lateral line detection. For instance, the surface-feeding fish *Aplocheilus lineatus* responds to a 140 Hz wave train of only 0.023 μm, even if its swimming speed is 10-14 cm s^{-1} (Bleckmann 1982). From schooling fish we know that during low-amplitude cruise swimming the water movements caused by the nearest neighbours provide information concerned with station-keeping (Partridge and Pitcher 1980).

Sensitivity of the Ordinary Lateral Line to Low-frequency Sound Waves

Water displacements and pressure waves are caused simultaneously by various 'sound' sources, i.e. by any kind of pulsating, vibrating or moving object. Let us consider the easiest case: a monopole (or its acoustic equivalent, a pulsating sphere of radius r). When $2\pi r < \lambda$ (λ = wavelength), the water displacement in the near field of a pulsating sphere, i.e. one that contracts and expands, is proportional to the volume increase (E) of the sphere and the square of its radius (r^2), and falls off by $1/s^2$ (s = distance from the centre of the sphere to the point in question) (Figure 7.3A). In the case of a vibrating sphere the near-field displacement is in direct proportion to the product of the sphere volume (r^3) and amplitude (d) of vibration (where $d << r$), and decreases with the cube of the distance ($1/s^3$) (Figure 7.3B). Conventionally one defines the near field of a monopole as the distance from the source at which the near-field velocity component Vn becomes equal to the velocity amplitude of water particles in the radiated far-field pressure wave, Vf. As can easily be calculated, the near-field component of a monopole reaches over a distance of approximately $\lambda/2\pi$ from the pulsating emitter. In water these are roughly 250, 25, 2.5, and 0.25 m for the vibration frequencies of 1, 10, 100, and 1000 Hz, respectively. To give a further numerical example: a ball of 1 cm diameter vibrating with an amplitude of 0.5 cm will cause maximal water displacements of 0.02 μm at a source distance of 30 cm.

In addition to near-field displacement, the motion of any object in water will cause pressure changes, which really consist of two types of pressure:

Figure 7.3: Calculation of Near-field Water Displacements, not Related to Compressional Waves, Caused by (A) a Pulsating (Monopole) and (B) a Vibrating Sphere (Dipole). In B at any given point the displacement of particle motion can be expressed as a radius vector d_s, where the particle velocity of the liquid is in the direction of the moving sphere and an angular vector d_φ, *where particle velocity is perpendicular to d_s.* The actual displacement is the vector sum of the angular and radial displacement. Consequently particle movements caused by a vibrating sphere have the shape of a figure eight with maximum values in the direction of vibration (redrawn from Harris and van Bergeijk 1962). (See also Figure 5.2 by Hawkins, this volume, for velocity/pressure ratio and phaselag of pulsating and vibrating spherical sound sources).

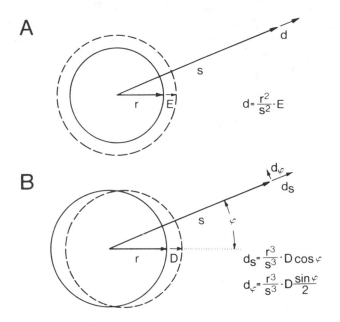

the near-field pressure associated with near-field motion, and the far-field pressure. The decay of the first can be described as $1/s^2$, the latter as $1/s$ (van Bergeijk 1967). Since near-field displacements attenuate markedly, the far field of a sound source consists of mainly pressure waves; the near field is a combination of pressure-waves and near-field displacements, with the latter being dominant. For example: the near-field displacement is 10^4 times greater than the displacement associated with pressure waves, when a ball with a radius of 1 cm, vibrating with a frequency of 100 Hz, is measured from a distance of 4 cm (Harris and van Bergeijk 1962). (See also Chapter 5 by Hawkins, this volume.)

There has long existed a controversy about what the proper stimulus to the lateral line really is. For instance, there are many behavioural and electrophysiological experiments indicating that lateral line organs respond to low-frequency pressure waves. In contrast there are a large number of studies leading to the opposite conclusion (for references see Sand 1981, 1984). One reason for this controversy may be the difficulty in separating the lateral line and inner ear functions. For instance, in many ablation experiments, only the trunk lateral line has been cut, leaving the head lateral line intact. Ablation of the whole lateral line can probably only be done chemically, taking care that such chemicals do not enter and interfere with the inner ear. In addition, in many studies a clear distinction between near and far field has not been done. Thus in such studies it is not clear whether the lateral line and the fish respectively, has responded to sound pressure or to near-field displacements. Harris and van Bergeijk (1962) have shown that the lateral line of the killifish (*Fundulus heteroclitus*) is sensitive to the near-field displacements of a sound source, and not to the pressure waves. Since that time the consensus of opinion among most investigators is that lateral line organs do not directly respond to sound pressure, but only to particle displacements that are associated with that pressure (see Chapter 5 by Hawkins, this volume). However, in the far field, particle oscillations caused by pressure variations are very small. For instance, if we assume a frequency of 100 Hz and a pressure variation of 1 μbar (which is 74 dB re. 2×10^{-5} N m^{-2} and thus considerable), the particle displacement in the far field will only be 0.001 μm, i.e. probably below the threshold of most lateral line organs (Harris and van Bergeijk 1962).

There is another reason why, under far-field conditions, the lateral line in most cases is not directly stimulated by sound sources. The tissue of fish has nearly the same acoustic properties (density) as water, therefore a fish of small size compared with the wavelength of the relevant sound frequencies (e.g. at 100 Hz 14.4 m) will be vibrated with the same phase and amplitude as the surrounding water particles when exposed to sound. As a result there is no net movement between sound-induced particle displacements and the fish.

However, close to the sound source the situation is more complex. Because of the strong attenuation of near-field displacements the movement of the medium adjacent to the sound source is greater than that of the fish itself. At the end of the fish furthest away from the source the reverse is true. In consequence there is a cross-over position along the length of the fish for which displacements of the fish and the medium are the same. The resulting complex stimulation pattern of the lateral line may provide the fish with information about the direction and the distance of the acoustic source (Denton and Gray 1983).

Indirect Stimulation of the Ordinary Lateral Line via Sound-induced Pressure Waves

Although pressure waves do not directly stimulate the lateral line of fish, they can — even under far-field conditions — induce surface displacements of the swimbladder. In this case the swimbladder acts as a pressure/displacement transducer which indirectly may stimulate the trunk lateral line canal organs via the fish's body (Sand 1981).

In clupeid fish, the swimbladder, inner ear and lateral line system are all linked. The central feature is a pair of pro-otic auditory bullae, which act as pressure-displacement converters. Each bulla is divided by a membrane into a gas-filled lower part and a liquid-filled upper part. Whereas the liquid-filled part is hydrodynamically connected via a fenestra in the upper wall of the bulla with the ear and the head lateral line, the gas-filled lower part is air-connected with the swimbladder via a precoelomic duct. The auditory bullae, because of the compliance of the gas, allow pressure changes (caused, for instance, by a sound source) to generate flows of liquid which, up to a frequency of 1000 Hz, stimulate the sense organs of the inner ear (mainly those of the utriculus) and the lateral line (Blaxter, Denton and Gray 1981). The time constant is such that at frequencies above 0.1 Hz there is little loss in the sensitivity of the bulla system, yet this system is insensitive to the maintained static pressure level or to very slow changes in that level, caused by vertical movements of the fish. However, displacements in the medium outside the fish have very little effect on the bullae and the utriculi, but they are accompanied by displacements in the lateral line canals (Blaxter *et al.* 1981).

Identification and Localisation of Stationary Objects

Fish can identify and localise stationary objects by means of their lateral line. Hofer (1908) was the first to observe that the predatory pike (*Esox lucius*), while swimming even with dimmed vision, kept itself at some distance from the wall of an experimental tank without touching it, and a ruler of 4 cm width held in the animal's way was perceived at about 0.5-1 cm distance. These avoidance reactions disappeared after cauterisation of the canal organs on the head. The perception of non-moving objects at a distance by blinded, quietly approaching fish has since been confirmed with many different species. Dijkgraaf (1963) tentatively explained reactions of this kind as due to the detection of water currents caused by the swimming movements and deflected by the obstacle. This means that in front of an object the fish will experience an 'unexpected' rise of water resistance.

Object identification via self-induced lateral line stimulation is most pronounced in the blind Mexican fish *Anoptichthys jordani* (Characidae), which lives in caves in constant darkness (see also Chapter 17 by Parzefall, this volume). If the cave fish is confronted with a new object, it swims around restlessly, avoiding collision with and maintaining a narrow gap between itself and the object. On approaching the object, the fish accelerates and then glides past it in close proximity without moving its tail. If the fish swims over the obstruction, it often rotates sideways shortly before reaching it so that one side of its body is directed towards the object while passing by, and it is probable that *A. jordani* gains information about these unknown objects in this way. Weissert and von Campenhausen (1981) and von Campenhausen, Reiss and Weissert (1981) provided evidence for this when they showed that *A. jordani* could no longer distinguish between objects if the fish were not allowed to pass close by: *A.*

jordani could be trained to differentiate between tunnels that contained vertical bars in various combinations of size, number and positions. *Anoptichthys jordani* can differentiate between vertically and horizontally oriented rectangular openings (4.3 cm × 2.3 cm width) and also between vertical bars with diameter of either 1, 2, 3 or 4 mm. In more crucial tests, *A. jordani* was trained successfully to discriminate either between two 'fences' (9.5 × 3.5 cm each) consisting of four or six bars (diameter of each bar 3 mm), or between two 'fences', each of which had six bars differing only slightly in respect to their relative position. However, when the fish were hindered in their gliding behaviour by walls, which were affixed at right angles to the left and right side of the tunnel openings, differentiation between different fences was no longer observed.

Anoptichthys jordani is richly supplied with thousands of neuromasts, which, together with the canal organs, form a characteristic pattern on the skin (Figure 7.1B). It is possible that for perceptual tasks the response of all neuronal channels of the lateral line organ is used simultaneously and processed with respect to the location of the organs on the skin and possibly also in the context of amplitude and phase relations among the channels. If so, the fish should be able to reconstruct its surroundings in three dimensions when swimming (Weissert and von Campenhausen 1981).

Identification and Localisation of Moving or Vibrating Inanimate and Animate objects

Fish with an intact lateral line system learn fairly quickly and easily to associate an approaching glass disc of varying diameter either with punishment or with food. For instance, a food-conditioned blinded minnow (*Phoxinus phoxinus*) about 5 cm in length locates a glass filament with a diameter of only 0.25 mm at a distance of up to 10 mm (Dijkgraaf 1963). In trained fish, tests with weak local water currents flowing from a pipette nozzle tend to produce the same reactions. It is obvious that such currents will stimulate the lateral line as soon as they reach the animal (Dijkgraaf 1963). Under natural conditions the perception of wave signals caused by conspecifics, prey or enemies is probably of great importance.

Prey Identification and Localisation by Surface Feeding Fish

For terrestrial insects the water/air interface forms a deadly clinging trap, resulting in an ideal two-dimensional surface for predators to roam and search from below. Surface-feeding fish with an intact head lateral line system are attracted by wave signals caused, for instance, by the struggling movements of a terrestrial insect trapped on the water surface (Schwartz 1965, 1971). Wave signals of this type have a maximal peak-to-peak (pp) amplitude which rarely exceeds 100 µm, and a frequency content which covers a range from about 5 Hz up to 140 Hz (Lang 1980). Abiotic wave signals caused by water drops or the falling of leaves and twigs are clearly distinguishable from prey waves in that they have an upper frequency limit

Figure 7.4: Examples of Relative Spectra (Relative Amplitude of the Signal versus Frequency) of Waves Caused by Biotic and Abiotic Sources (Redrawn from Lang 1980). Shaded area refers to frequencies found in wind-generated waves (Bleckmann and Rovner 1984)

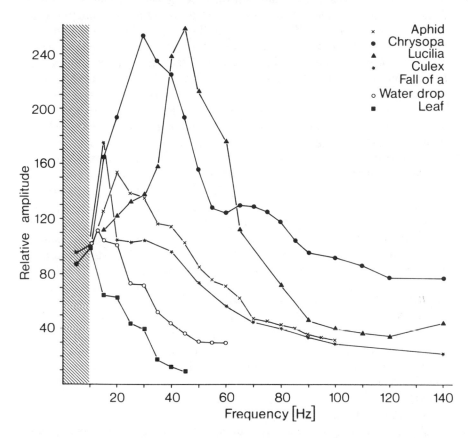

which rarely exceeds 40-50 Hz (for examples of wave spectra see Figure 7.4).

Surface-feeding fish show a remarkable sensitivity to water surface waves. For example, in its most sensitive frequency range of 100-140 Hz, the topminnow (*Aplocheilus lineatus*) and the African butterflyfish (*Pantodon buchholzi*) respond to wave signals which have a pp-amplitude of less than 0.02 μm (Bleckmann 1980; U. Müller, unpublished). This means that surface-feeding fish are very well adapted to perceive the low-amplitude high-frequency components that characterise prey signals in contrast to non-prey waves. The signal-to-noise ratio is further improved by a strong rise of threshold displacement for frequencies below 14 Hz. Both *A. lineatus* and *P. buchholzi* do not respond to stimuli of a frequency lower than 10 Hz, which under natural conditions are mainly caused by wind (Figure 7.4) or by the fishes' own body movements (Bleckmann 1982).

If rewarded with food, surface-feeding fish also respond to artificial wave signals (clicks) produced, for instance, by dipping a small rod once into the water surface (for examples of click signals see Figure 7.6A-C). Behavioural experiments have shown that even without vision these fish can determine both the direction and the distance of the wave source, if a single click of only 200-700 ms duration is presented (Schwartz 1965, 1971). In order to find out what factors of click signals might be relevant for distance determination, some knowledge of surface wave physics is necessary.

The Physical Properties of the Signal Transmission Channel Water Surface.   Particle movements induced by water surface waves are circular with a maximal diameter equal to that of wave height. Surface waves with amplitudes much smaller than the water depth h are radiated with dispersion, i.e. the propagation velocity has its minimum of 23.1 cm s^{-1} at 13 Hz and then rises with all other frequency values (Figures 7.5A).

In addition to being dispersed with propagation, water surface waves are also strongly attenuated with transmission distance, the slope of attenuation again being dependent on wave frequency. In Figure 7.5B the calculated attenuation of $10 \leq f \leq 140$ Hz circular water surface waves is plotted. As can be seen, attenuation is greatest within the vicinity of a wave source, and increases with increasing stimulus frequency (decreasing wave length) (Figure 7.5B). In deeper water, both the vertical and horizontal motion of water particles caused by surface waves decreases (Figure 7.5C). In summary, the reach of particle oscillation caused by disturbances of the water surface is very restricted, both in the horizontal and vertical plane. This is probably the reason why surface-feeding fish are closely restricted to the water/air interface.

Depending on the initial stimulus intensity, click signals consist of wave cycles of different amplitudes (about 0.1-20 μm) and frequencies (about 5 Hz up to 190 Hz) (Bleckmann 1980). Because of the filter properties of the water surface, all multifrequency wave signals more quickly lose their high-frequency components than others during signal propagation. As a result, the upper frequency limit of a multifrequency wave signal decreases with increasing source distance (see spectra in Figure 7.6A-C). In addition, click signals show the following characteristics:

(1)   According to their frequency content and the dispersion of water surface waves they are always frequency modulated downwards. In contrast to a click's amplitude and frequency range, this frequency modulation mainly depends on source distance but not on the kind of click production or initial stimulus intensity (Bleckmann and Schwartz 1982).

(2)   Like all other concentric wave signals, the curvature of the wave front of clicks is most pronounced close to the wave source and then

Figure 7.5: (A) Propagation Velocities of Surface Waves with Different Frequencies and Wavelengths. (B) Attenuation as a Function of Frequency and Distance from the Vibration Source. (C) Decrease of Particle Oscillation with Increasing Water Depth (Calculated According to Formulae given by Sommerfeld (1970), Bleckmann (1985) and Schuijf (1976). Explanations of symbols: *A*, pp-amplitude of wave signal, *Ao* pp-amplitude of wave signal at stimulus origin, C_{ph} phase velocity, λ wavelength, *g* gravitational acceleration, *h* depth (surface to bottom) of the water, *z* depth of observation point below the surface, *R* radius of surface wave, χ kinematic viscosity of water, η surface tension, ϱ density of water, ξ maximal vertical component of particle movement, *k* $\lambda/2\pi$.

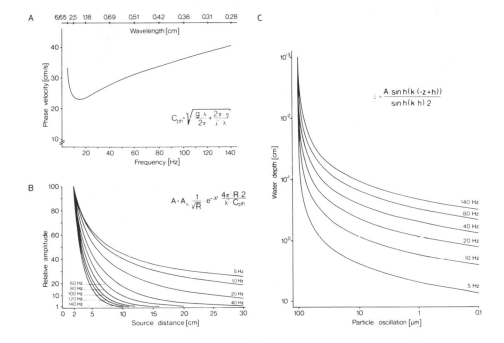

decreases continuously. Thus the upper frequency limit, the frequency modulation, the amplitude and the decline of stimulus amplitude per unit of propagation distance all decrease in a regular manner during click propagation.

Determination of Wave Source Distance.   Experiments with defined artificial wave signals have shown that distance estimation in surface-feeding fish is first of all based on the evaluation of the frequency modulation (time structure) of the wave signal (click). If, for example, these fish are confronted in a source distance of 7 cm with a wave signal whose frequency modulation closely resembles that of a click at 15 cm, they swim at a mean 3-5 cm beyond the wave source (Bleckmann and Schwartz 1982; Hoin-Radkovski, Bleckmann and Schwartz 1984) (Figure 7.6D). The use of frequency modulation for distance determination is further supported in experiments with fish (*Aplocheilus lineatus*) whose head lateral line organs have all been destroyed except one. These animals can still determine the

Figure 7.6: Left Side: Mean Swimming Distance (Arrow Length) of *A. lineatus* towards the Wave Source (Small Vertical Lines Drawn Below the Glass Tube, GT) which is Dependent on Stimulus (Insets) and Source Distance (x-axis). The stimulus amplitudes are not to scale. In all cases the surface waves were produced by blowing a defined air stream through a glass tube on to the water surface. Right Side: Power Spectra of the Corresponding Wave Signal. The y-axis in all cases is 0 to 60 dB. In A, B and C the highest power value of A was set equal to 0 dB. In D-G the highest power value of the corresponding spectrum was defined as 0 dB. The stimuli are: a Click in (A) 5, (B) 10, and (C) 15 cm source distance, a simulated click (D), an upward frequency-modulated wave signal (E), a 35 Hz wave signal (F), and a 100 Hz wave signal (G)

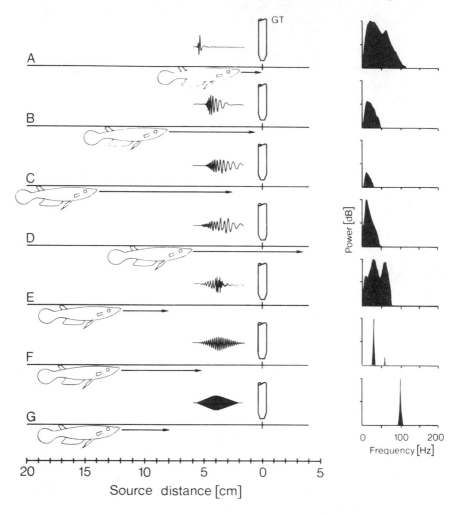

source distance if clicks are presented (Müller and Schwartz 1982). This gives clear evidence that the curvature of the wave front, which can only be measured with at least three neuromasts (Hoin-Radkovski *et al.* 1984), cannot be the main (or only) cue used for distance determination.

However, that the evaluation of both the curvature of the wave front and the amplitude spectrum is involved in distance determination can be seen in experiments in which single-frequency (sf) wave signals, which do not contain any frequency modulation at all, were presented (Figure 7.6F-G). Surface-feeding fish now tend to underestimate the source distance if it exceeds 6-8 cm. However, with low-frequency sf-signals, there is still a positive correlation between swimming distance and source distance. Summarising all results obtained with sf-signals we can learn that the relative localisation error at a given source distance: (a) rises with increasing stimulus frequency, and (b) at a given stimulus frequency it increases with source distance (Bleckmann 1980; Bleckmann and Schwartz 1982; Hoin-Radkovski *et al.* 1984).

Assuming that the curvature of the wave front is determined through the arrival-time differences of the wave front at different neuromasts, this is exactly what is to be expected. Theoretical calculations (Hoin-Radkovski *et al.* 1984) suggest that accuracy of distance determination to sf-signals should decrease both with increasing source distance (because the curvature becomes more and more flattened) and increasing frequency of the stimulus (because an increase in frequency results in an increase of phase velocity and thus in a decrease of arrival-time differences). In addition, the fact that at a given source distance surface fish swim progressively shorter in response to sf-signals at higher frequencies indicates that the amplitude spectrum is also evaluated for distance determination; i.e. if no other cues are available, a higher-frequency wave signal is 'expected' to have travelled a shorter distance than a low-frequency one. This point of view is further supported by results of experiments with artificial upward frequency modulated wave signals, which contain high-amplitude, high-frequency wave components. These signals also cause an underestimation of source distance (Bleckmann and Schwartz 1982; Hoin-Radkovski *et al.* 1984) (Figure 7.6E). Bearing in mind the filter properties of the water surface this probably reflects a good localisation strategy.

That the evaluation of wave signals is based on frequency and amplitude discrimination is further supported in experiments in which fish were trained to differentiate between signals that differed either in frequency or in amplitude. In the range 10-140 Hz, the frequency difference limen of *A. lineatus* was found to be 10 per cent, i.e. a wave signal of frequency F is clearly distinguishable from one of the frequency $F \pm 0.1\ F$ (Bleckmann, Waldner and Schwartz 1981). In contrast, amplitude discrimination is best if higher-frequency wave signals are presented. For example, *A. lineatus* can distinguish a 70 Hz wave signal of 2.3 μm pp-amplitude from one of the same frequency but with an amplitude of only 0.14 μm (I. Waldner, unpublished). It is very likely that surface-feeding fish use their wave-discriminating ability also for the differentiation between prey and non-prey waves.

Determination of Target Angle.   In theory, for the determination of target angle the following cues might be used:

(1)   the time differences that appear between the stimulation of different neuromasts and which depend on target angle. Such a mechanism in particular would be facilitated by the low propagation speed of water surface waves (see Figure 7.5A)

(2)   Single neuromasts of the lateral line organs show a bidirectional sensitivity (Figure 7.2E). As the long axis of different head neuro-masts in all surface-feeding fish so far investigated differs with respect to the long axis of the fish's body (Schwartz 1971), surface wave stimulation evokes a spatial and temporal neural response of a certain intensity pattern, which depends on target angle.

If these are the cues used for target-angle determination, we should get certain sensory deficits in fish whose head neuromasts in part have been removed. Indeed, surface-feeding fish with complete unilateral ablation of neuromasts show well-differentiated responses only if the stimulus approaches the intact body side. In contrast, stimuli impinging first upon the side deficient in neuromasts cause disoriented turns directed away from the side deficient in neuromasts cause disoriented turns directed away from the wave source. Most pronounced localization errors occur in fish with only one neuromast left intact. Although these fish (*A. lineatus*) remain responses depend solely on the topographic position of the intact neuro-mast. A more rostral position results in a smaller turn and a more caudal position in a larger one (Müller and Schwartz 1982). Thus in intact fish the turning response is probably calculated by integration of the information that reaches the brain from all head lateral line neuromasts.

Lateral Line and Intraspecific Communication

The use of 'water touch' in sexual display behaviour was first observed by Stahr (1897, quoted by Hofer 1908; and see Chapter 10 by Turner, this volume). Today many examples for the use of water touch in sexual display behaviour are known (see Dijkgraaf 1963). Commonly the two partners keep alongside each other at a distance of only a few centimetres, their heads often pointing in opposite directions. Sexual stimulation is done by pushing movements that create small (not yet quantified) flows of water against a companion. Even when both animals are blinded, this behaviour can be observed.

In fighting fish *Betta splendens*, both nest building and the hatching of the juveniles are done by the male. At an age of about 3 days, up to 150 juveniles leave the air-bubble nest and begin to investigate their surroundings actively. In doing so they always keep in contact with the surface film of the water since as anabantid fish they depend on breathing air. In the event of danger the male takes an oblique position with the head close to the water surface and begins to produce surface waves by trembling movements of its pectoral fins. Up to a distance of 40 cm, the juveniles, even in complete darkness, orientate themselves to the vibration source and then intermittently swim in the wave-source direction. Reaching the signalling male they are sucked up by him and transported back to the nesting site (Kühme 1963).

The response of juvenile *B. splendens* to water surface waves can also be released with artificial wave signals, probably perceived with head neuromasts which are already well developed at that time. Single-frequency signals between 8 and 10 Hz and with an amplitude of at least 13 μm turned out to be the most effective stimulus. Only one or two minutes after stimulus onset, more than 70 per cent of all fish have accumulated close to the wave centre (S. Kaus, unpublished). Taking into account the fact that juvenile *B. splendens* respond to a signalling male at a distance of up to 40 cm, the wave train produced by the male must have an initial pp-amplitude of at least 100 μm.

Lateral Line Function and Shoaling

Another striking behaviour of fishes which at least partly depends on lateral line function is schooling (shoal is a general term for a social group of fish, and the term school is restricted to a synchronously turning and accelerating fish group; Pitcher 1979 and see Chapter 12 by Pitcher, this volume). Fish schools are well known for their remarkable synchrony, which even persists if the school as a whole executes complicated evasive nanoeuvres. This requires individual fish to respond quickly to short-term changes in the velocity and direction of their neighbours. From analysis of video films we know that in normal saithe (*Pollachius virens*) schools there is a significant clustering of neighbours parallel to each other which is the position in which changes in velocity of the nearest fishes are most apparent visually. However, even temporarily blindfolded fish are capable of matching changes in velocity and heading of at least their first two neighbours. Only when the fish are both blinded and have their lateral line nerves cut at the opercula are they then unable to school (Pitcher, Partridge and Wardle 1976; Partridge and Pitcher 1980). Thus the information for schooling comes not only from vision but also from the posterior lateral line.

The importance of the lateral line in fright responses and sudden velocity changes within a school was demonstrated by an experiment in which an entire saithe school had lateral line nerves cut at the opercula. If such a school was startled by a dark object suddenly being thrust overhead, the latency with which lateralis-sectioned fish responded to the object was a function of the distance and direction of the object in question (Partridge and Pitcher 1980). In contrast, fish in control schools showed no significant relationship between latency to startle for either distance or bearing (angle in the horizontal plane to the nearest neighbour). That the startle response in herring (*Clupea harengus*) shoals is triggered by pressure waves alone was demonstrated by Blaxter and Hoss (1981). If the gas-filled pro-otic bullae, which act as part of the pressure-detecting system (see above) are destroyed, the threshold of the startle response increases remarkably (Blaxter and Hoss 1981).

The bream (*Abramis brama* (L.)) is a true social animal. Quantitative comparisons of its shoaling behaviour have shown that both lateral line and vision are employed simultaneously to maintain individual distance (Pitcher 1979). Paradoxically, bream with an intact lateral line cover more ground (as measured by swimming behaviour) and make more periodic forays, spending about 20 per cent of their time up to 12 body lengths from their fellow: The converse was observed when there was no lateral line information. Thus it seems possible that deprivation of lateral line information about neighbour fish who could nevertheless be seen makes bream in this experiment more 'anxious' and hence increases their motivation to be part of the group (Pitcher 1979).

Specialised Lateral Line Organs (Electroreceptors)

In 1917 Parker and van Heusen discovered that the catfish *Ictalurus nebulosus* responds to galvanic currents of less than 1μA. About four decades later Lissmann and Machin (1963) established an electrical sensitivity down to $0.75 \mu V cm^{-1}$ in several species of siluroid fishes, and in 1968 Roth noted that some electroreceptive fish not only can detect constant electric fields but in addition are able to sense the polarity of such a field. The first evidence for the biological use of electric sense organs in teleosts was given by Kalmijn and Adelmann (in Kalmijn 1974). They demonstrated that *I. nebulosus* uses its electric sense to locate living prey by the bioelectric fields (see below) emanating from it.

Electroreceptors are ordinal characteristics of the African freshwater teleost order Mormyriformes (Gymnarchidae: one species, and Mormyridae: many genera and species), the South American order Gymnotiformes (Gymnotidae: many genera and species), and the large and world-wide order Siluriformes (many families and species). In addition the teleost *Xenomystus nigri* (Osteoglossiformes: Notopteridae) has recently found to be electroreceptive (Bullock, Bodznick and Northcutt 1983). Electroreceptors of electroreceptive fish (Figure 7.1C) were discovered as such by Bullock, Hagivara, Kusano and Negishi (1961). It is very likely that they have developed from the mechanosensitive hair cells of the ordinary lateral line as the afferent fibres of teleost electroreceptors form part of the lateral line nerve. If so electroreceptors constitute an integral part of the lateral line system of fishes. All electroreceptors lack the cupula of ordinary lateral line organs. They are located in an invagination of the epidermal basement membrane. The apical membrane of the sensory cells is connected either by a jelly-filled canal, varying in length, to the epidermal surface or by way of specialised cells covering the sense organ (Szabo 1974). In teleosts two large main groups of electrosensitive cells are distinguishable (Figure 7.2F): (1) the 'tuberous organs', which are high-frequency receptors (passband *c.* 50-2000 Hz) and which are used in active electrolocation (see below), and (2) the 'ampullary organs' which are low-frequency receptors (passband *c.* 0.1-50 Hz) and which are used in passive electrolocation (see Bullock 1982). The specialised lateral line organs are not a substitute for the ordinary lateral line organs but an additional sensory system. Consequently, in all fish displaying a specialised lateral line system, the ordinary lateral line shows its usual distribution.

Possible Natural Electric Fields

The natural electric fields that may be detected by electrosensitive fish are: (1) inanimate electric fields (e.g. indirectly caused by the Earth's magnetic field), and (2) animate incidental electric fields, which are produced by all living organisms. Some fish, most belonging to the families Gymnarchidae, Mormyridae and Gymnotidae, possess electric organs which have derived from muscle or nerve tissue. In addition some siluriforms (genus *Malapterurus*) and perciforms (genus *Astroscopus*) are known to be electric (in the genus *Astroscopus* there is no evidence for electroreception, nor is there any obvious or recognised function of the electric discharge) (Bullock *et al.* 1983). The electric fields produced by the electric organs of these fish therefore constitute a third source of natural electricity. A few species, such as the electric eel (*Electrophorus electricus*) of the order Gymnotiformes, can produce strong electric discharges of up to several hundred volts. However, the majority of species are limited to weak electric discharges in the range of millivolts to a few volts. Many of these fish fire their electric organ in a rather sinusoidal manner (wave species) with a very stable discharge frequency of up to about 1700 Hz. In contrast, other electric fish have a pulse interval long compared to pulse duration (pulse species). Both wave and pulse species can use their electroreceptor system in a passive and also an active mode. For a more refined distinction between active and passive mode of electroreception see Kalmijn 1985. In the active mode, sensory information is obtained from distortions of the animal's own electric field, caused by various objects with conductivities lower or higher than that of the surrounding water (Figure 7.7). In this case the receptors involved are exclusively the high-frequency tuberous organs.

Passive Mode of Electroreception

Since the internal and external milieux of aquatic animals are electrochemically dissimilar, a potential difference of up to $200 \mu V \, cm^{-1}$ across their boundary layer occurs. These d.c. (direct current) bioelectric fields, which in wounded species locally display even higher values, are modulated by relative movements of body parts. For instance, substantially low-frequency potential fluctuations are in phase with the respiratory gill movements. The detection of voltage gradients induced by prey, or by motion in

Figure 7.7: The Principle of Active Electrolocation. The location of the electric organ is indicated by the black bar. The solid lines give the current flow associated with electric organ discharge. The electroreceptors, which monitor voltage changes across the epidermis, are found in pores of the anterior body surface. Each object with conductivity different from that of the surrounding water distorts current pattern and thus alters transepidermal voltage in the area of skin (*) nearest to the object (redrawn from Heiligenberg 1977)

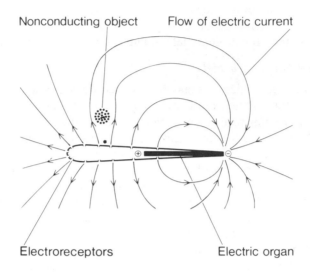

magnetic fields like the Earth's, is established both in elasmobranchs and teleosts.

It is well known that sharks and rays use their electric sense (ampullary organs) to detect flat fishes buried in the sand. They make well-aimed dives at the prey if they have approached it up to a distance of less than 50 cm, uncover it from the sand and devour it voraciously. Like sharks and rays, weakly electric teleosts and the catfish *Ictalurus* can also locate the position of a live prey fish, even when all but the bioelectric stimuli emanating from the prey are attenuated by a layer of agar. But *Ictalurus*, for instance, no longer responds to the prey fish when a very thin, electrically insulating film of polyethylene is added to the agar. That the electroreceptors and not visual or chemical cues are involved in this behaviour can be further seen in experiments in which the presence of a prey fish is simulated by passing low-frequency current (8-64 μV cm^{-1}) between two electrodes buried in the sand. In this case, weakly electric fish and *Ictalurus* display the same characteristic feeding response to the electrodes as they do to actual prey (Kalmijn 1974).

Long-distance Orientation

Electric fields, called motional electric fields, in the magnitude of 0.05-5.0 μV cm^{-1} are induced whenever water is moving or a fish is swimming through the Earth's magnetic field (Kalmijn 1974). The possible navigation of migratory fish by sensing indirectly the geomagnetic field is a fascinating idea. For instance, elasmobranchs respond to uniform electric fields with voltage gradients of only $0.005\,\mu V\,cm^{-1}$ (Kalmijn 1985). Since, under natural conditions, the effective field depends on the direction in which the animal is moving, motional electric fields may provide electrosensitive animals with orientational cues, i.e. they may indicate the fish's heading relative to the Earth, its velocity and its latitude (Kalmijn 1985). Catfish can also orientate in strictly uniform electric fields as low as $1.0\,\mu V\,cm^{-1}$. This can be inferred from the ability of the fish to select the correct (rewarded) direction from among 12 uniform electric fields arranged in a circle (Kalmijn *et al.* 1976, quoted by Bullock 1982).

The ability to detect motional electric fields has been of special interest in long-distance migrating teleosts such as the American eel (*Anguilla rostrata*), or the Atlantic salmon (*Salmo salar*). Rommel and McCleave (1973) have claimed electrosensitivity for *S. salar* and *A. rostrata.* However, repeated efforts to confirm electrosensitivity for these two species have failed (e.g. Enger, Kristensen and Sand 1976; Bullock *et al.* 1983). Thus, contrary to former suggestions, it is very unlikely that eel and salmon use motional electric fields for migratory orientation.

Active Mode of Electroreception and Electrocommunication

Lissmann (1951) was the first to propose the theory that a weakly electric fish may locate nearby objects by detecting the distortions which these produced in the fish's own electric organ field (Figure 7.7). Indeed, the weakly electric fish, *Gymnarchus niloticus* could be trained to respond to the presence of a glass rod 0.2 mm in diameter, even though it was placed in a porous pot so as to exclude mechanical and visual detection (Lissmann and Machin 1958). The threshold of *G. niloticus* was found to be as low as $0.15\,\mu V\,cm^{-1}$. Further behavioural experiments have shown that weakly electric fish depend greatly upon the active electric sense in manoeuvering in their muddy environments. For instance they persist in attempting to enter a hole through which they were previously trained to pass, even when the hole was mechanically closed by agar, a substance which can be made electrically 'transparent' (Kalmijn 1974). In this case the use made of ampullary organs resembles the task allotted to the ordinary lateral line organs except that the stimulus of mechanical forces is replaced by changing electric potentials. In view of the fact that, even in the ordinary lateral line organs, potential changes in the receptor cells serve as a link between mechanical deformation of the hair and the nervous response, the transformation of mechanosensitive hair cells to electrosensitive cells seems comprehensible.

Like the mechanosensitive ordinary lateral line, electroreceptors also have a communicating function. For example, electric fish can detect each other's presence by using their high-frequency electroreceptors (tuberous organs) to sense the electric-organ discharge of conspecifics. In some species electric fish get information not only about the presence of a neigh-bouring fish, but also about its position, distance, sex and state of activity. For instance, if the weakly electric fish *Gnathonemus petersii* is passively moved within a sheltering tube towards another specimen at a certain distance (about 30 cm), at least one of the fish lowers or ceases its electric-organ discharge activity (Möller and Bauer 1973). In another weakly electric fish, *Sternopygus macrurus*, the resting discharge frequency of sexually mature males differs from that of females. Daily injections of androgens have shown that steroid hormones influence the electric organ pacemaker frequency in *Sternopygus* (Heiligenberg and Bastian 1984), and that the tuning of electroreceptors follows hormone-induced frequency shifts even if the receptors do not experience the animal's electric-organ discharge (C. Keller, personal communication). During the breeding season, males of the genus *Sternopygus*, which are territorial, produce variations in their discharge when in the presence of females. These variations consist of 'rises' (an increase in frequency followed by decrease back to resting frequency) and 'interruptions' (temporary cessation in dis-charge) (Hopkins 1974).

Most electric fish, even within one species, have an individual electric-organ discharge frequency which often slowly changes over the course of weeks and months (Heiligenberg and Bastian 1984). However, if it happens that the frequencies of neighbouring fish are similar, this impairs the electrolocating performance. As an evolutionary response, wave and pulse species have developed mechanisms to minimise the effect of inter-fering signals. For example wave species protect their electrolocation ability by shifting the electric-organ discharge frequency away from interfering stimulus frequencies. Such behaviour, first discovered by Watanabe and Takeda (1963, quoted by Heiligenberg 1977) is called 'jamming avoidance response'. This response can be elicited at stimulus intensities 100 to 1000 times weaker than the intensity of the animal's own electric field. Since electrolocation performance will markedly suffer only when the stimulus intensity approaches the intensity of the animal's own near field, jamming avoidance response acts as an early warning system that shifts the electric-organ discharge frequency into a safer range long before an approaching conspecific might impair elecrolocation abilities (see Heiligenberg 1977).

Summary

The lateral line organs of teleost fishes can be divided into 'ordinary' (mechanosensitive) lateral line organs and 'specialised' (electrosensitive) lateral line organs. The mechanosensitive lateral line is found in all teleosts and comprises free superficial neuromasts which in quick and persistent swimmers are partly transformed into canal organs. All neuromasts are covered by a gelatinous cupula which encloses the sensory hairs from the underlying mechanosensitive hair cells. The appropriate stimuli for the mechanosensitive lateral line are minute water displacements which shear the cupula parallel to its base, involving a deformation of the sensory hairs.

The ordinary lateral line organs found throughout the teleosts are used as 'distant touch' receptors. They serve mainly to detect and locate moving animals, as well as inanimate mobile and immobile objects at short range on the basis of current-like water disturbances. The ordinary lateral line organs of teleosts are of special importance for the detection and localisation of prey, for enemy avoidance, for schooling, and for intraspecific communication.

The specialised electrosensitive lateral line organs mainly found in the African freshwater teleost families Gymnarchidae and Mormyridae, the South American family Gymnotidae, and the world-wide order Siluriformes comprise two broad classes of receptors: the ampullary organs, which are sensitive to weak low-frequency potential changes, and the tuberous organs (not found in siluriforms) which are sensitive to weak high-frequency potential changes occurring at their external openings. These specialised lateral line organs are used to detect electric stimuli produced by inanimate sources (e.g. electric fields caused by the Earth's magnetic field) or incidentally by all organisms. The former serves an orientation function, and the latter is engaged in prey capture behaviour.

Some fish such as *Eigenmannia virescens* and *Gymnarchus niloticus* produce their own electric field by means of electric organs. These fish can locate other animals or obstacles at some distance from the disturbance caused by bodies of different conductivity from water within their own electric field. In addition, electric fish can perceive the electric field of nearby conspecifics. With the aid of their electroreceptors, these fish not only get information about their neighbour's presence but also about its distance, sex and state of activity.

Acknowledgements

I thank J. Ruthven, Drs H. Münz, J. Schweitzer and W. Plassmann for helpful discussions and comments on the manuscript. I especially thank J. Ruthven for linguistic assistance, and K. Grommet for drawing the figures.

References

Blaxter, J.H.S. and Hoss, D.E. (1981) 'Startle Response in Herring: the Effect of Sound Stimulus Frequency, Size of Fish and Selective Interference with the Acoustico-lateralis system', *Journal of the Marine Bioacoustics Association, UK*, 61, 871-9.
Blaxter, J.H.S., Denton, E.J. and Gray, J.A.B. (1981) 'Acousticolateralis System in Clupeid Fishes', in W.N. Tavolga, A.N. Popper and R. Fay (eds), *Hearing and Sound Communication in Fishes*, Springer-Verlag, Heidelberg, pp. 39-59.

Bleckmann, H. (1980) 'Reaction Time and Stimulus Frequency in Prey Localization in the Surface-feeding Fish *Aplocheilus lineatus*', *Journal of Comparative Physiology, 140,* 163-72

Bleckmann, H. (1982) 'Reaction Time, Threshold Values and Localization of Prey in Stationary and Swimming Surface-feeding Fish *Aplocheilus lineatus* (Cyprinodontidae)', *Zoologisches Jahrbuch der Physiologie, 86,* 71-81

Bleckmann, H. (1985) 'Perception of Water Surface Waves: How Surface Waves are used for Prey Identification, Prey Localization and Intraspecific Communication', in H.J. Autrum and D. Ottoson (eds), *Progress in Sensory Physiology 5,* Springer-Verlag, New York, pp. 147-66

Bleckmann, H. and Rovner, J. (1984) 'Sensory Ecology of the Semiaquatic Spider *Dolomedes triton.* I. Roles of Vegetation and Wind-generated Waves in Site Selection', *Behavioral Ecology and Sociobiology 14,* 297-301

Bleckmann, H. and Schwartz, E. (1982) 'The Functional Significance of Frequency Modulation within a Wave Train for Prey Localization in the Surface-feeding Fish *Aplocheilus lineatus* (Cyprinodontidae)', *Journal of Comparative Physiology, 145,* 331-9

Bleckmann, H. and Topp, G. (1981) 'Surface Wave Sensitivity of the Lateral Line System of the Topminnow *Aplocheilus lineatus*', *Naturwissenschaften, 68,* 624-5

Bleckmann, H., Waldner, I. and Schwartz, E. (1981) 'Frequency Discrimination of the Surface-feeding Fish *Aplocheilus lineatus* — a Prerequisite for Prey Localization?', *Journal of Comparative Physiology, 143,* 485-90

Bullock, T.H. (1982) 'Electroreception', *Annual Review of Neuroscience, 5,* 121-70

Bullock, T.H., Hagiwara, S., Kusano, K. and Negishi, K. (1961) 'Evidence for a Category of Electroreceptors in the Lateral Line of Gymnotid Fishes', *Science, 134,* 1426-7

Bullock, T.H., Bodznick, D.A. and Northcutt, G. (1983) 'The Phylogenetic Distribution of Electroreception: Evidence for Convergent Evolution of a Primitive Vertebrate Sense Modality', *Brain Research Reviews, 6,* 25-46

Campenhausen, C. von, Riess, I. and Weissert, R. (1981) 'Detection of Stationary Objects by the Blind Cave Fish *Anoptichthys jordani* (Characidae)', *Journal of Comparative Physiology, 143,* 369-74

Denton, E.J. and Gray, J.A.B. (1983) 'Mechanical Factors in the Excitation of Clupeid Lateral Lines', *Proceedings of the Royal Society of London B, 218,* 1-26

Dercum, F. (1880) 'The Lateral Sensory Apparatus of Fishes', *Proceedings of the Academy of Natural Science, Philadelphia,* 152-4

de Sede, P. (1884) 'La Ligne Laterale des Poissons Osseux', *Revue Scientifique Serie de Tom, 7,* 467-70

Dijkgraaf, S. (1963) 'The Functioning and Significance of the Lateral Line Organs', *Biological Reviews, 38,* 51-106

Emery, C. (1880) 'Le Specie del Genera Fierasfer nel Golfo di Napoli e Regioni Limitrofe', *Fauna and Flora des Golfes von Neapel, Monographie,* pp. 1-76

Enger, P.S., Kristensen, L. and Sand, O. (1976) 'The Perception of Weak Electric D.C. Currents by the European Eel (*Anguilla anguilla*)', *Journal of Comparative Biochemistry and Physiology, 54A,* 101-3

Flock, A. (1965) 'Electron Microscopic and Electrophysiological Studies on the Lateral Line Canal Organ, *Acta Otolaryngologica, 199,* 1-90

Flock, A. (1971) 'Sensory Transduction in Hair Cells', in *Handbook of Sensory Physiology.* vol. I, *Principles of Receptor Physiology,* Springer-Verlag, Heidelberg, 396-441

Fuchs, S. (1895) 'Über die Funktion der unter der Hautliegenden Canalsysteme bei den Selachiern', *Archiv über die gesamte Physiologie, 59,* 454-78

Gray, J. (1984) 'Interaction of Sound Pressure and Particle Acceleration in the Excitation of the Lateral-line Neuromasts of Sprats', *Proceedings of the Royal Society of London B, 220,* 299-325

Grobbel, G. and Hahn, G. (1958) 'Morphologie und Histologie der Seitenorgane des augenlosen Höhlenfischs *Anoptichthys jordani* im Vergleich zu anderen Teleosteern', *Zeitschrift für Morphologie und Ökologie der Tiere, 47,* 249-566

Harris, G.G. and van Bergeijk, W.A. (1962) 'Evidence that the Lateral-line Organ responds to Near-field Displacements of Sound Sources in Water', *Journal of the Acoustic Society of America, 34,* 1831-41

Heiligenberg, W. (1977) 'Principles of Electrolocation and Jamming Avoidance in Electric Fish. A Neuroethological Approach', in V. Braitenberg (ed.), *Studies of Brain Function,*

Springer-Verlag, New York, 1-85

Heiligenberg, W. and Bastian, J. (1984) 'The Electric Sense of Weakly Electric Fish', *Annual Review of Physiology, 46*, 561-83

Hofer, B. (1908) 'Studien über die Hautsinnesorgane der Fische', *Berichte der Königlich Bayerischen Biologischen Versuchsstation München, 1*, 115-64

Hoin-Radkovski, I., Bleckmann, H. and Schwartz, E. (1984) 'Determination of Source Distance in the Surface-feeding Fish *Pantodon buchholzi* (Pantodontidae)', *Animal Behaviour, 32*, 840-51

Hopkins, C.D. (1974) 'Electric Communication: Functions in the Social Behaviour of *Eigenmannia virescens*', *Behaviour, 50*, 270-305

Hudspeth, A.J. (1983) 'Mechanoelectrical Transduction by Hair Cells in the Acousticolateralis Sensory System', *Annual Review of Neuroscience, 6*, 187-215

Jakobson, L. (1813) 'Extrait d'un Memoire sur une Organe Particuliere de Sens les Raies et les Squales', *Nouveau Bulletin des Sciences, par la Société philomatique de Paris, 3*, 332-7

Kalmijn, A.J. (1974) 'The Detection of Electric Fields from Inanimate and Animate Sources Other than Electric Organs' in A. Fessard (ed.), *Handbook of Sensory Physiology* vol. III/3, Springer-Verlag, New York, pp. 147-200

Kalmijn, A.J. (1985) 'Theory of Electromagnetic Orientation: a Further Analysis', in L. Bolis and R.D. Keynes (eds), *Comparative Physiology of Sensory Systems*, Press Syndicate, University of Cambridge, Cambridge, pp. 525-63

Katsuki, Y. and Hashimoto, T. (1969) 'Chemoreception in the lateral line system of the bony fish', *Proceedings of the Japan Academy, 45*, 209-14

Kawamura, T. and Yamashita S. (1983) 'Chemical Sensitivity of Lateral Line Organs in the Goby, *Gobius giurinus*', *Journal of Comparative Biochemistry and Physiology, 74A*, 253-7

Knox, R. (1825) 'On the Theory of the Existence of a Sixth Sense in Fishes; Supposed to Reside in Certain Peculiar Tubular Organs, Found Immediately under the Integuments of the Head in Sharks and Rays', *Edinburgh Journal of Science, 2*, 12-16

Kühme, W. (1963) 'Verhaltensstudien am maulbrütenden (*Betta anabatoides*) und nestbauenden Kampffisch (*Betta splendens*)', *Zeitschrift für Tierpsychologie, 18*, 33-55

Kuiper, J.W. (1967) 'Frequency Characteristics and Functional Significance of the Lateral Line Organ', in P. Cahn (ed.), *Lateral Line Detectors*, Indiana University Press, Bloomington, pp. 105-21

Lang, H.H. (1980) 'Surface Wave Discrimination between Prey and Nonprey by the Backswimmer *Notonecta glauca* L. (Hemiptera, Heteroptera)', *Behavioral Ecology and Sociobiology, 6*, 233-46

Lee, F.S. (1898) 'The Function of the Ear and the Lateral Line in Fishes', *American Journal of Physiology, 1*, 128-44

Leydig, F. (1850) 'Über die Schleimkanäle der Knochenfische', *Müllers Archiv für Anatomie und Physiologie*, 170-81

Leydig, F. (1851) 'Über die Nervenknöpfe in den Schleimkanälen von *Lepidoleprus, Umbrina* und *Corvina*', *Müllers Archiv für Anatomie und Physiologie*, 235-40

Lissmann, H.W. (1951) 'Continuous Electrical Signals from the Tail of a Fish, *Gymnarchus niloticus*', *Nature, 167*, 201-2

Lissmann, H.W. and Machin, K.E. (1958) 'The Mechanism of Object Location in *Gymnarchus niloticus* and Similar Fish' *Journal of Experimental Biology, 35*, 451-86

Lissmann, H.W. and Machin, K.E. (1963) 'Electric Receptors in a Non-electric Fish (*Clarias*)', *Nature, London, 199*, 88-9

Merkel, F. (1880) *Über die Endigungen der sensiblen Nerven in der Haut der Wirbeltiere*, Rostock

Möller, P. and Bauer, R. (1973) 'Communication in Weakly Electric Fish, *Gnathonemus petersii* (Mormyridae). II. Interaction of Electric Organ Discharge Activities of Two Fish', *Animal Behaviour, 21*, 501-12

Müller, U. and Schwartz, E. (1982) 'Influence of Single Neuromasts on Prey-localizing Behaviour of Surface-feeding Fish, *A. lineatus*', *Journal of Comparative Physiology, 149*, 399-408

Parker, G.H. (1902) 'Hearing and Allied Senses in Fishes', *Bulletin of the US Fish Commission, 22*, 45-64

Parker, G.H. (1904) 'The Function of the Lateral Line Organ in Fishes', *Bulletin of the US Bureau of Fisheries, 24*, 185-207

Parker, G.H. and van Heusen, A.P. (1917) 'The Responses of the Catfish, *Amiurus nebulosus*, to metallic and non-metallic rods', *American Journal of Physiology, 44*, 405-20

Partridge, B.L. and Pitcher, T.J. (1980) 'The Sensory Basis of Fish Schools: Relative Roles of Lateral Line and Vision', *Journal of Comparative Physiology, 135*, 315-25

Pitcher, T.J. (1979) 'Sensory Information and the Organisation of Behaviour in a Shoaling Fish', *Animal Behaviour, 27*, 126-49

Pitcher, T.J., Patridge, B.L. and Wardle, C.S. (1976) 'A Blind Fish Can School, *Science, 194*, 963-5

Rommel, S.A. and McCleave, J.D. (1973) 'Sensitivity of American Eel (*Anguilla rostrata*) and Atlantic Salmon (*Salmo salar*) to Weak Electric and Magnetic Fields', *Journal of the Fisheries Research Board of Canada, 30*, 657-63

Roth, A. (1968) 'Electroreception in the Catfish, *Amiurus nebulosus*', *Zeitschrift für vergleichende Physiologie, 61*, 196-202

Russell, I.J. (1974) 'Central and Peripheral Inhibition of Lateral Line Input During the Startle Response in Goldfish', *Brain Research, 80*, 517-22

Russell, I.J. (1976) 'Central Inhibition of Lateral Line Input in the Medulla of the Goldfish by Neurons which Control Active Body Movements', *Journal of Comparative Physiology, 111*, 335-58

Sand, O. (1981) 'The Lateral Line and Sound Reception', in W.N. Tavolga, A.N. Popper and R. Fay (eds), *Hearing and Sound Communication in Fishes*, Springer Verlag, Heidelberg, pp. 459-78

Sand, O. (1984) 'Lateral-line Systems', in L. Bolois, R.D. Keynes and S.H.P. Madrell (eds), *Comparative Physiology of Sensory Systems*, Cambridge University Press, London, pp. 3-32

Schuijf, A. (1976) 'Variation of Hydrodynamic Parameters with Depth in Capillary Gravity Waves', in A. Schuijf and A.D. Hawkins (eds), *Sound Reception in Fish*, Elsevier, Oxford, pp. 183-4

Schulze, F.E. (1861) 'Uber die Nervenendigung in den sogenannten Schlundkanälen der Fische und über entsprechende Organe der durch Kiemen atmenden Amphibien', *Archiv für Anatomie, 1861*, 759

Schulze, F.E. (1870) 'Über die Sinnesorgane der Seitenlinie bei Fischen und Amphibien', *Archiv für Mikriskopische Anatomie, 6*, 62-88

Schwartz, E. (1965) 'Bau und Funktion der Seitenlinie des Streifenhechtlings *Aplocheilus lineatus*', *Zeitschrift für vergleichende Physiologie, 50*, 55-87

Schwartz, E. (1971) 'Die Ortung von Wasserwellen durch Oberflächenfische', *Zeitschrift für vergleichende Physiologie, 74*, 64-80

Schwartz, E. (1974) 'Lateral-line Mechano-receptors in Fishes and Amphibians', in A. Fessard (ed.), *Handbook of Sensory Physiology*, vol. III/3, Springer Verlag, New York, pp. 257-78

Sommerfeld, A. (1970) *Vorlesung über theoretische Physik. Band II. Mechanik der deformierbaren Medien*, Dietrichsche Verlagsbuchhandlung, Wiesbaden

Szabo, T. (1974) 'Anatomy of the Specialized Lateral Line Organs of Electroreception', in A. Fessard (ed.), *Handbook of Sensory Physiology*, vol. III/3, Springer Verlag, New York, pp. 13-58

von Bergeijk, W.A. (1967) 'Introductory Comments on Lateral-line Function', in P. Cahn (ed.), *Lateral Line Detectors*, Indiana University Press, Bloomington, pp. 73-81

Weissert, R. and Campenhausen, C. von (1981) 'Discrimination between Stationary Objects by the Blind Cave Fish *Anoptichthys jordani* (Characidae)', *Journal of Comparative Physiology, 143*, 375-81

PART THREE: BEHAVIOURAL ECOLOGY

INTRODUCTION

T.J. Pitcher

Behavioural ecology is the study of the ways in which behaviour is influenced by natural selection in relation to ecological conditions. Along with its sister discipline of sociobiology, defined as the study of factors shaping the evolution of social behaviours, this new approach has stimulated an explosion of fresh work and insights in the past decade. An introduction to the field is given by Krebs and Davies (1981), and state-of-the-art reviews are provided in Krebs and Davies (1984). The search for optimal and evolutionarily stable solutions to animals' problems in search, choice and conflict entails attempts to match theory with experimental evidence, so that the testing of alternative theories forms a core of hard science in the discipline. Behavioural ecology is therefore a particularly exciting area to be working in at present, but the teleost fishes have received much less attention than birds and mammals. This section of the book is an attempt to redress the balance.

Four major topics are reviewed: foraging, mating, social behaviour and the effects of space and time, each covered by two chapters. In addition, two case studies in the behavioural ecology of fishes conclude this section. In the first three pairs of chapters, the first member of each pair gives a synoptic review of the topic, and the second highlights one significant aspect in more detail. A review of foraging is followed by the way in which it is constrained by predators; mating by the evolutionary basis of parental care; social behaviour (shoaling) by the importance of individual differences. The next pair of chapters, on spatio-temporal patterning, are based on two important examples of the effects of the 'template' of habitat. First, a chapter covers how the daily regime of illumination, including twilight periods, shapes the behaviour of predators and prey in fish communities on reefs and kelp forests. Secondly, a chapter discusses how the tidal regime shapes the behaviour of marine littoral fishes. This behavioural-ecology section of the book closes with examples of these major principles applied to two case studies: the first a review of the sticklebacks, a traditionally popular group with fish ethologists, and the second an account of the behavioural ecology of the specialised and fascinating cave-dwelling fishes.

One major topic of behavioural ecology not covered in this book is that of fish migration: readers will find a recent review in McKeown (1984).

In Chapter 8, Paul Hart opens the behavioural ecology section with a review of the decision rules which may enable us to understand and predict how fishes feed. He points out that the global problem is the optimal allocation of time and resources to feeding, reproduction and defence which maximises the animal's lifetime reproductive success. Hart discusses how, once fishes are in a feeding mode, they deal with patchy and variable food, alternative prey types and sizes, the presence of other fishes, predators, morphological constraints and learning, demonstrating that the strategies used represent trade-offs between benefits from energy gained and the costs or risks involved. Many of the earlier theories on the subject, such as classical 'optimal foraging theory', fail to account for feeding behaviour seen in groups of fish.

Manfred Milinski, in the following chapter, analyses one of the foraging tradeoffs in detail, using many results from his own extensive series of carefully designed experiments on predators and foraging fishes. Usually, both conflicting goals of avoiding death and eating are fulfilled less efficiently. Feeding fishes take greater risks when hungry or parasitised, or for better-quality food. Smaller vulnerable fishes will settle for suboptimal food, and shoaling enhances feeding through lowered timidity, increased predator detection and monitoring. Milinski concludes by nothing that the subtlety of fish responses may make it difficult to predict the precise details of behaviour in particular cases.

Chapter 10, by George Turner, reviews mating strategies in teleosts. A conflict of interest between the sexes is reflected in the number of sexual partners and the degree of parental care. There are many solutions to the conflict, so that the stable breeding systems which have evolved vary greatly between species. Turner discusses the theory and evidence for mate choice and sexual selection in teleost fishes, followed by an analysis of breeding systems. 'Promiscuous' mating is probably the primitive arrangement in the group. The chapter covers nest-site defence, monogamy, how the evidence supports competing theories of polygyny and polyandry, alternative male strategies, leks and the 'hotspot' hypothesis, brood-care helpers, asexual reproduction in clones of poecilids, and some complex and unusual strategies in wrasse, Antarctic plunderfish and others. Turner concludes by noting that some of these phenomena defy classification as 'systems', and, in the light of current theoretical thinking in this field, advises that it may be more profitable to consider the mating strategies of individuals.

Following the review of mating, Robert Sargent and Mart Gross put forward a novel analysis in Chapter 11 to explain the evolution of parental care in fishes. The concept that present reproduction bears a cost limiting future reproduction is named as 'Williams' principle', after G.C. Williams. The authors develop a quantitative model of this tradeoff between breeding now or later, and use it to analyse the prevalence of male broodcare in teleosts, changes in parental behaviour with offspring age, brood size, and territory defence, using examples from sticklebacks, medaka, sunfish, guppies, sculpins and cichlids. Sargent and Gross finish by putting forward a number of suggestions for extensions and tests of their 'Williams' principle' model.

Social groups of fishes are termed 'shoals', and my own chapter in the book comes next, using much of the recent work carried out in the laboratory at Bangor. The main theme is that the most useful analysis of shoaling is based upon the rule governing an individual fish's decision to join, stay with or leave the group. The older conceptual framework, emphasising group behaviours and advantages, not surprisingly reflected the impressive co-ordination seen in travelling fish shoals but was prone to invalid group-selectionist arguments. After briefly reviewing shoal definitions, functional aspects of school structure and the lack of evidence for a hydrodynamic function in shoals, I argue that predator defence and foraging benefits are the key to understanding shoaling behaviour, and discuss how individuals may take advantage of these benefits and trade them off with costs. In addition to reviewing previous work on shoals, the chapter briefly summarises some new findings in the form of the attack abatement effect, predator inspection visits, conflict and exploitation during predator attack on shoals, and forage-area copying. Chapter 12 proceeds with sections on migration, mixed-species shoals, and optimal shoal size, and closes with a suggestion that it may be worth while looking for shoal-limiting behaviours.

In Chapter 13, Anne Magurran looks in detail at individual differences in fish behaviour, reinforcing the points made by Turner, Hart and Pitcher in previous chapters. At one time, animals that failed to perform what was considered species-typical behaviour, for example standing idly by during courtship ceremonies, were labelled as aberrant or sick, or were even omitted from data altogether. This chapter shows how misguided that view was, and reviews individual differences in foraging, predator avoidance, mating, habitat use and dominance hierarchies, using experimental evidence from the recent literature. Alternative behavioural strategies are often an evolutionarily stable solution to the problems of optimising individual performance contingent upon what others are doing, or to patchiness of resources in space and time. In some cases the alternatives represent genetic differences, and in others fish are endowed with the behavioural plasticity to opt for the best solution. Magurran concludes by warning against describing behaviour in terms of the average individual.

We now come to two chapters on spatial and temporal patterning. In Chapter 14, Gene Helfman shows how the behaviour of coral-reef, kelp-forest and lake fishes is shaped by the daily cycle of light, twilight and dark. Few fishes can be optimally adapted to all these light conditions, and so in each of these communities, daytime sets of zooplanktivores, invertebrate-eaters and piscivores are replaced by alternative sets of nocturnal species, although herbivores and cleaners tend to be diurnal. At night, killer molluscs and killer isopods threaten refuging fishes. Twilight has a profound influence on the evolution, physiology and behaviour of fishes in both these sets, since many ambush-piscivores, striking from below, have the advantage at low light levels during the 'changeover' period. Helfman finishes with a discussion of differences between the role of twilight periods in the tropics and temperate habitats.

In Chapter 15, Robin Gibson reviews the behavioural ecology of intertidal fishes, where the normal locomotory medium is removed for hours at a time. Some intertidal fishes avoid the problem by migrating to and fro, but residents shelter, burrow or have evolved amphibious traits; about five families of teleosts are specialised for seashore life. Gibson describes the physical changes which shore fishes experience, along with the morphological, physiological and behavioural adaptations evolved to minimise them. He proceeds to discuss behavioural adaptations in locomotion, habitat selection, maintenance, fidelity, reproduction and the shore fishes' impressive homing ability. Gibson concludes with an account of the synchronisation of behaviour with the tidal cycle, and warns against the lack of appropriate tidal cues in laboratory experiments on these fishes.

The final two chapters in this section are case studies, the authors revisiting many of the concepts put forward in earlier chapters. First, Gerard FitzGerald and Bob Wooton review the behavioural ecology of the sticklebacks, since Tinbergen's pioneering experiments one of the most-studied freshwater fishes. Research on sticklebacks, which — unusually for fish — have the double advantage of being easy to maintain in captivity and easy to observe in the wild, has produced significant advances in our understanding of behaviour in relation to habitat, foraging, predator defence and breeding. In the field, important insight of the forces shaping behaviour has come from a comparison of stickleback communities with differing competition, food and predation, although the full stories have yet to be elucidated. FitzGerald and Wooton leave us with the prediction that stickleback studies will contribute as much to the young science of behavioural ecology as they did to the young science of ethology.

Jakob Parzefall reviews the behavioural ecology of cave-dwelling fishes in the final chapter in this section: members of 12 teleost families have become such 'troglobionts'. He begins by comparing feeding and reproductive behaviour in cave-dwellers and their epigean relatives, including experiments that reveal the genetic basis of some of these differences. He discusses how cave fishes, most of which are blind, deal with food, mates and rivals: in many cases behaviour appears to generalised and reduced. Competing theories to account for these reductions, including vision, are analysed. Pheromone-mediated sexual behaviour and nocturnal traits may be 'pre-adaptations' for cave life. Parzefall argues that cave fishes could be very important for the study of evolution, since he feels that changes in many of their traits (with exceptions) owe more to the accumulation of neutral mutations than to stabilising natural selection.

References

Krebs, J.R. and Davies, N.B. (1981) *Introduction to Behavioural Ecology*, Blackwell, Oxford, pp. 292

Krebs, J.R. and Davies, N.B. (1984) (eds) *Behavioural Ecology*, 2nd ed, Blackwell, Oxford, pp. 493

McKeown, B.S. (1984) *Fish Migration*, Croom Helm, London/Timber Press Beaverton, Oregon, pp. 224

8 FORAGING IN TELEOST FISHES

Paul J.B. Hart

A short while observing a feeding fish might give the impression that it would be impossible to understand why behavioural changes occur as they do. A foraging roach (*Rutilus rutilus*), for example, cruises about its habitat digging into the bottom, snapping at a drifting insect in midwater and plucking at a piece of leaf at the water's edge. How can these apparently disparate actions be explained by a unified foraging theory? The introduction of evolutionary concepts has helped to provide theories that predict decision rules for foraging animals, the principal concept being that efficient feeders will be favoured by natural selection. By minimising the time spent feeding, a fish will have more time to carry out other activities; alternatively, by minimising the energy spent to capture its prey, a fish will have more food energy left for metabolism and growth. So, time minimisation or energy maximisation are the two ways in which the fish can be efficient as a forager. Following on from this assumption, a number of decision rules can be derived which make predictions about how a fish should behave.

The global problem for fish is how to allocate time to feeding, reproduction and territorial behaviour in a way that will maximise representation in the next generation. There are two levels at which this distribution can be discussed. Using the lifespan of a fish as the time scale, one can ask how the individual should schedule growth and reproduction to maximise fitness. This leads to the analysis of the evolution of life histories and is an area of active research (Charlesworth 1980; Roff 1983). Allotting time to growth implies feeding, and, at a lower level of organisation, the problem is how to account for the moment-to-moment assignment of time to feeding in a pattern which maximises lifetime reproductive success. On the scale of a lifespan, the broad distributions are the summation of the small-scale allocations.

Once it has decided to feed (see Chapter 2 by Colgan, this volume), a fish is faced with a number of further decisions. Very often it will be necesssary to look for food, and the fish might choose a search path to maximise the energy gain per unit of search time. Food is usually patchily distributed, with different patches having higher or lower profitabilities. The fish has to decide how long to stay feeding in a particular patch, and, within the patch, which size or species of prey it would be appropriate to take.

The principal consideration in this chapter will be with foraging behaviour, but as this is constrained by the physiology and morphology of the fish (see Bond 1979; Brett 1979; Brett and Groves 1979), they too should be mentioned. Similarly the behaviour of an individual will be influenced by the presence of conspecifics and by predators. The resulting strategies of food gathering are a compromise between the benefits derived from food gathered and the costs associated with the strategy. Costs and benefits need to be expressed in terms of a common currency. As a fish gathers food, it takes in and expends energy and nutrients. Most models assume that costs and benefits can be expressed in units of energy, which is regarded as the common currency. It is often assumed further that maximising net energy gain is an optimal strategy but this is not so unless the animal is also maximising its fitness (Sih 1982), defined as the animal's lifetime reproductive success. This has never been measured in a fish, and in only a few animals from other groups (for example red deer (*Cervas elephas*): Clutton-Brock, Guiness and Albon 1982). Very often, however, energy and fitness maximisation are correlated (Hart and Connellan 1984).

There are two ways in which efficient feeding strategies can originate; they can be acquired over evolutionary time, the results of natural selection being inbuilt into each individual, or they can be learned during the lifetime of the fish. All vertebrates have complex neurophysiology and learning capacities (MacPhail 1982); consequently one would expect learning to have an effect on feeding behaviour, modifying the simple strategies characteristic of most of optimal foraging theory (OFT) (Dill 1983). This chapter will therefore also examine how learning can modify feeding strategies, and how foraging fish use their learning abilities to respond to environmental variability.

Behavioural Feeding Strategies

When to Start Feeding

A fish that has not fed for a long time will take food very readily, after which its feeding activity begins to decline. This is illustrated by pike feeding on different-sized prey. If a 20 cm pike has just eaten a minnow weighing around 0.5 g, it will take several more in the same feeding session (Hart and Connellan 1984). When the same-sized pike eats a minnow weighing around 4.5 g, it is unlikely to eat again for another 24 h. It is convenient to say that the motivational state of the pike is changed by feeding, and that the size of prey determines the degree to which the motivational state is altered. Assuming that stomach contents are reduced by digestion in an exponential way (Pitcher and Hart 1982), it can be shown that pike always start to feed again when they have between 0.3 and 0.7 g of food left in their stomachs, and this is constant over all prey sizes eaten. This implies that the motivational state of the pike must return to some level before the pike will again attack and catch prey.

Modern approaches to the study of motivation, which are discussed in detail by Colgan in Chapter 2 of this volume (and see also McFarland and

Houston 1981), argue that motivational state is determined by the combined effects of internal and external causal factors. The degree of stomach fulness and other physiological parameters (Holmgren, Grove and Fletcher 1983) are internal factors and the presence of prey of the appropriate size and species is an external factor. The internal state influences the way in which a fish responds to external stimuli. In pike, I. Clarke (unpublished data) has studied the response of semi-satiated and hungry individuals to a range of prey-like stimuli. A series of trials were carried out in which a recently fed or a hungry fish was presented with a silhouette of a minnow model which only moved in the horizontal plane. The recently fed pike was unresponsive to the prey for a median of 211.1 s each trial, whereas the hungry pike was unresponsive for a median of only 23.1 s. Further evidence shows that the way a pike responds to a prey-like stimulus is strongly influenced by its internal state.

A deeper understanding of the factors initiating feeding behaviour requires knowledge of gustatory physiology (see Holmgren *et al.* 1983). A summary of possible controlling factors is given in Figure 8.1. This suggests mechanisms by which feeding behaviour is switched on and off, but does not explain how feeding is integrated with other behaviours into a fitness-maximising sequence. This is because the model in Figure 8.1 does not account for the dynamic interactions that will occur between activities. At present, knowledge of physiological factors controlling appetite in fish is poor (Holmgren *et al.*, 1983) and the problem awaits further study. Equally, more work is required before we understand how behaviours associated with different functions are integrated.

Searching for Food

Once a fish has decided to start feeding, its first task is to find suitable food items. Many will be clumped and occur in patches so that the foraging fish has to travel. Assuming that fish are maximising their fitness, one is led to two considerations; the first concerns the route the fish should take to maximise the net gain of food, the second is concerned with the speed at which the animal should move. Optimising the route has been studied quite extensively (eg. Smith 1974; Pyke 1978; Jansen 1982). Optimal speed has been studied mainly theoretically by Ware (1975, 1978, 1982) and Pyke (1981).

The search paths of the three-spined stickleback (*Gasterosteus aculeatus*) looking for *Tubifex* worms in a grid of containers on the floor of a 9 cm × 27 cm aquarium showed marked responses to the presence of prey (Thomas 1974). If food was found and eaten, the linear distance travelled after discovery decreased significantly (Figure 8.2). Also, after a fish had approached and rejected a food item, the linear distance travelled increased significantly. This 'area-avoiding search' moves the fish fast and directly away from an area where a rejected item has been found, so increasing the likelihood of discovering a new area with desirable food.

Similarly, Beukema (1968) showed that, through time, searching stickle-

Figure 8.1: A Model of Feeding Behaviour in Fish, Illustrating How Internal and External Factors Influence the Need to Feed (Hunger). (Modified from Colgan 1972.)

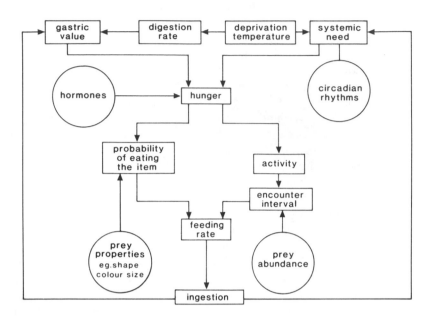

backs increased their encounter rate with prey by decreasing their rate of turning and by avoiding those parts of the experimental tank that never contained prey. After a while the fish did consistently better than the performance predicted from a model assuming random search. By increasing their search efficiency over time (higher encounter rate, avoidance of non-productive areas), sticklebacks were increasing their rate of food capture. This is an example of an animal learning an efficient strategy, a feature which will be discussed in more detail later.

Do the search paths chosen maximise net energy gain and fitness? Pyke (1978) examined theoretically the 'basic hypothesis that ... the pattern of movements of an animal is such that the foraging efficiency of the animal is maximised'. The model developed simulates an animal moving over a square grid with food items located only at grid points. The searching predator does not respond to prey from a distance and eats every item encountered. These last two properties mean that Pyke's animal is more of a harvester than a forager.

Pyke used the model to examine the way in which the size of the searched area, the length of the foraging bout, the behaviour of the animal at the boundary of the area, the degree of symmetry of movements, the memory of the animal, and the distribution of food affected the optimal pattern of movements. A key measure of the nature of the search path is its directionality, which measures the degree to which each successive move is

Figure 8.2: Sticklebacks Searching for *Tubifex* Set in a Rectangular Grid on the Floor of an Aquarium. Each point on the curves shows the linear distance (LD) travelled between 5s intervals both before and after discovering a worm (signified by the squiggle at the intersection of the two axes). One curve (o---o) shows the resulting LDs before and after fish had found and rejected a worm. The second (●——●) is for fish that found and ate the worm

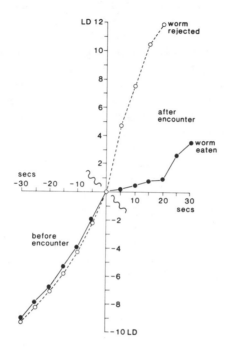

Source: Thomas (1974).

in the same direction as the last. Directionality can vary from zero when the animal moves randomly, to 1 when the animal moves forwards in a straight line. A given grid size has an optimal directionality at which the highest number of grid points is visited once or more. In the simulations, directionality varied between 0.6 and 0.8, depending on grid size, with a trend towards straighter paths as the grid enlarged. The calculated values agree reasonably with the data on finch flock movements in the Mohave desert (Cody 1971, 1974), but for data on bumblebees the predicted values were lower than those observed. Similarly, Lawrence (1979), studying sticklebacks searching for *Tubifex* on the floor of an aquarium, estimated from the model that directionality should have been 0.67 but this was observed to be between 0.43 and 0.53. For both the bumblebees and the sticklebacks, the searching animal could see flowers or prey from a distance and turn to travel straight to them. It is most likely for this reason that the directionality values fall below those predicted. The model assumes that animals move around their habitats in the manner of automated senseless vacuum cleaners, never detecting a prey until it is engulfed. The theory of search paths has lagged behind that for other aspects of foraging behaviour, perhaps because the problems to be solved are greater (Krebs, Stephens and Sutherland 1983).

A second variable of locomotion that could influence fitness during foraging is the speed of movement between patches or items. An analysis of speed during foraging is complicated by the range of constraints that will bear on the animal (Pyke 1981). In moving from one point to another, a fish might be approaching food, but at the same time it might be exposing itself to predators. Consequently it is hard to choose the optimality criteria. Ware (1978) made the assumption for pelagic fish that an appropriate criterion for optimisation would be growth-rate maximisation. Growth rate was made a function of the net food intake per unit time, the standard metabolic rate and the cost of swimming, each being expressed as functions of body weight. A simulation of the energy budget of fish of different sizes showed that the optimal swimming speed during foraging was a function of food concentration and fish size (Figure 8.3). At about $6 \times 10^{-4} \mathrm{cal\,cm^{-3}}$ the maximum foraging speed is predicted to be about $117\,\mathrm{cm\,s^{-1}}$ (2.9 body lengths $(\mathrm{BL})\mathrm{s^{-1}}$) for a 40 cm fish at 15°C. This compares well with the maximum sustained swimming speed of 3 $\mathrm{BL\,s^{-1}}$ at 15°C for a 42 cm salmon. Ware observes that this agreement indicates that the complete range of foraging speeds is contained within the fish's metabolic scope for activity (see Brett and Groves 1979).

A more detailed analysis of swimming-speed optimisation will be found in Pyke (1981).

Exploiting Patchy Resources

Biologists interested in animal feeding behaviour have worked extensively on the problems of how long an individual should stay in a food patch. In

Figure 8.3: Theoretical relation between the optimal foraging speed ($U_{f,opt}$) and food concentration (ϱ) for three different sizes of fish; 10, 30, and 60 cm

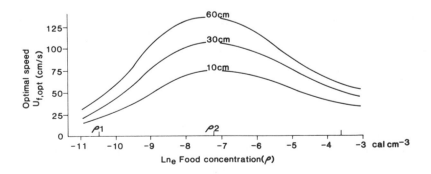

Source: Ware (1978).

addition, patches will be varying distances apart, requiring more or less travel between them, an activity which costs energy but gains none. In this section I consider rules for an animal feeding on a patchily distributed food source.

An early study of the problem was by Krebs, Ryan and Charnov (1974) who tested black-capped chickadees (*Parus atricapillus*) foraging for pieces of mealworm in artificial pine cones arranged in patches. The work was stimulated by the idea (Gibb 1962) that titmice (*Parus* spp.) looking for insect larvae searched cones until an expected number had been caught. This was labelled 'hunting by expectation'. Krebs *et al.* also tested the idea that the titmice hunted at each cone for a fixed time after which they left. Finally Krebs and his co-workers tested a model developed by Charnov (1973), which assumed that foraging birds would be maximising their net energy gain. The outcome of the work favoured the maximisation model, which was later elaborated by Charnov (1976). This will be described in detail as it provides a clear insight into the costs and benefits of moving from patch to patch.

The model provides a rule for allowing the animal to decide when to leave a patch, assuming that patches are entered at random. It further assumes that, while in the patch, consumption by the forager reduces food abundance so that the intake rate decreases steadily. This has the consequence that the cumulative food intake climbs to an asymptote as residence time in the patch increases (Figure 8.4a). It is imagined that the habitat contains food patches of varying quality, mixed in a random manner, and that the animal visits many patches during a single foraging bout, avoiding patches already visited (Figure 8.4b). No food is found outside the patches. Each patch type has its own curve of cumulative gain (as in Figure 8.4a). The following equation describes the net rate of energy intake , E_n.

$$E_n = \frac{\sum P_i g_i(T_i) - eE_T}{t + \sum P_i T_i} \tag{8.1}$$

where P_i = proportion of visited patches that are of type i; E_T = energy cost per unit time whilse travelling between patches; E_{si} = energy costs while searching in patch of type i; $h_i(T)$ = assimilated energy from hunting for T time units in a patch of type i minus all but the energy costs of searching; $g_i(T) = h_i(T) - E_{si}T$ = assimilated energy gain corrected for the cost of searching; t = travel time between patches.

The optimal solution is that the animal should choose values of T so that it leaves the patch when the marginal capture rate (i.e. the rate just before leaving) is equal to the average capture rate for the food patches in this habitat. This prediction is dependent on the animal's knowing the distribution of patch types and the average capture rate, each assumed to be learnt instantaneously.

The model predicts a response which varies with the abundance of food

Figure 8.4: (A) Cumulative Energy Intake from Two Patch Types, High and Low Quality, as Time in the Patch Increases. (B) A Hypothetical Environment with Two Patch Types (H and T) and the Movements of a Foraging Animal Both within and between Patches. Food is Available Only in the Patches

(A)

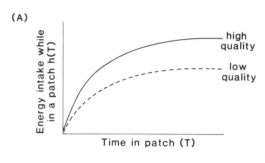

(B)

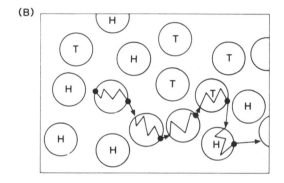

Source: Charnov (1976).

items in a patch and the travel time between patches. Their influences can best be illustrated by a diagram (Figure 8.5). Consider a habitat with two types of food patch, one with a high density and one with a low density of food items. In each the animal will experience a particular cumulative gain of food for a given residence time (curves H and L in Figure 8.5A), and there will be an average gain for the habitat (curve Av in the figure). The average net gain for the habitat is given when the line AB in Figure 8.5A, which takes account of travel time between patches, touches the cumulative gain curve. In the two patch types, net gain will be maximised when the rate of gain equals the average for the environment. This occurs where lines parallel to AB graze the cumulative gain curves for the patches. As shown in Figure 8.5A, the animal should leave a patch of low density sooner than a high-density patch ($T_{opt}L < T_{opt}H$). In Figure 8.5A a fixed travel time is assumed. Figure 8.5B illustrates the way the residence time changes as travel time increases ($T_{opt}STT < T_{opt}LTT$). This can also be understood in terms of equation 8.1. As t gets bigger the denominator increases, so reducing E_n. The animal can compensate for this by gathering more food

Figure 8.5: (A) A Graphical Interpretation of the Marginal Value Theorem. The curves show the cumulative net gain from different times spent in the two patch types of the habitat (H and L). Also shown is the average cumulative net gain (Av). The horizontal axis shows the mean travel time between patches. $T_{opt}L$ — optimal time in low-quality patch; $T_{opt}Av$ — average optimal time in a patch; $T_{opt}H$ — optimal time in a high-quality patch. See text for further details. (B) The Effects of Travel Time on the Optimal Time Spent in the Patch. LTT — long travel time; STT — short travel time; $T_{opt}STT$ — optimal time in patch when travel time is short; $T_{opt}LTT$ — optimal time in patch when travel time is long. See text for further details

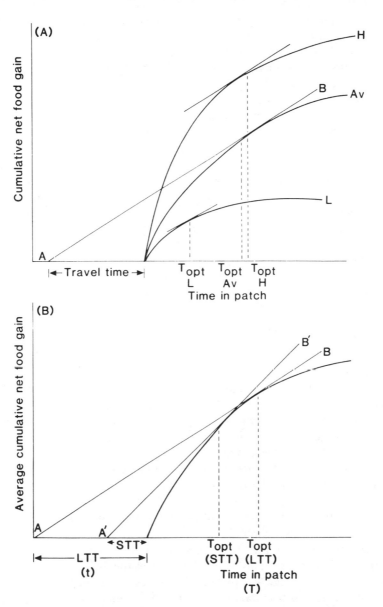

from the patch, although there comes a point where the benefit of staying longer is outweighed by the paucity of new food items to be found.

The marginal value theorem (MVT), as Charnov named it, has not been tested on fish, but has been applied widely to birds and insects (see Krebs *et al.* 1983 for a summary and full reference list). Many of the tests have resulted in qualitative agreement with predictions from the theory: in other words the animal's residence times have increased or decreased as predicted. Other tests gave quantitative agreement (for example Cowie 1977) whereas yet others gave results that differed qualitatively and quantitatively from MVT predictions (Howell and Hartl 1980). Nevertheless, there has been sufficient agreement to show that the theory is worth pursuing, but the original theory was not designed to explain all the behaviour one might observe when animals are collecting food from a patchy environment.

Krebs *et al.* (1974) argued that the marginal rate in a patch could be estimated by a feature they called 'giving-up time' (GUT). This is the time between finding the last item   and leaving the patch, and should be inversely proportional to the gathering rate, decreasing in rich habitats. If animals have perfect information about the amount of food within patches, GUTs should be the same for all patch qualities within a habitat. McNair (1982) shows that the MVT was not designed to make predictions about GUT and is a residence-time (RT) theory. As such, predictions about GUTs cannot be derived directly from equation 8.1. MacNair develops a model based on GUT which predicts that it should be positively correlated with patch quality within a particular habitat. He also shows that the MVT is only a special case of his own model, which is of the form

$$H_e = \frac{\sum p_i E(Y_i)}{\sum p_i E(R_i) + \tau} \tag{8.2}$$

where H_e = the habitat rate of energy intake; p_i = the probability of encountering patch type i; $E(Y_i)$ = the expected value of the cumulative net yield from patch of type i; $E(R_i)$ = the expected patch residence time for patch type i, selected by the forager by altering t_i, the GUT; τ = the mean travel time between patches.

Both $E(Y_i)$ and $E(R_i)$ depend on t_i, the GUT for the patch, which is the characteristic used to decide when to leave. Alternatively, t_i could be the residence time, in which case equation 8.2 would be a version of the MVT. Equally, other decision rules could be employed, such as the hunting-by-expectation idea of Gibb (1962); in each case similar predictions would follow, showing that the exact decision rule used is not critical to the behavioural outcome. This is highlighted by the fact that both NcNair's GUT strategy and the Charnov MVT (a residence-time strategy) predict that on average, foragers will spend more time, and a greater proportion of their time, in good patches than poor ones. Both predict that a greater energy yield should come from good patches and that residence time and yield should increase as travel time increases.

The theoretical approaches that have been discussed so far make predictions about behaviour constrained only by the characteristics of the food-gathering task. This approach makes it possible to understand the principal constraints on observed strategies. No theory can hope to incorporate every constraint, but those deriving from interactions with conspecifics or predators are so important that they cannot be omitted.

The group foraging behaviour of fish may be the reason for there being little work with fish to test the MVT and its successors. In a shoal, behaviour of an individual depends on what others in the group are doing (see Chapter 12 by Pitcher, this volume), a single foraging fish would be constrained only by the properties of the food distribution. Godin and Keenleyside (1984) studied the way six individuals of the cichlid species *Aequidens curviceps* distributed themselves between two patches which had either the same or different profitabilities. It was expected that the fish would distribute themselves between the patches in the proportions predicted by the ideal free theory (Fretwell and Lucas 1970): that is, in proportion to patch profitability. The theory requires that each individual goes to the place where its chances of food gain are highest, that the food patches are homogeneous with respect to features that influence survival, that the arrival of more individuals on a patch decreases its suitability, that there is no resource defence and that all individuals in the group should have the same needs and abilities (Milinski 1984).

The fish did distribute themselves according to the patch profitability ratio, but this was despite the unequal foraging performace of each fish. This was thought to be a result of individual differences in perceptual abilities and the speed with which fish could learn the different patch profitabilities. A similar experiment, with similar results, is reported by Milinski (1984), but he was in addition testing a theory about the learning rule used by the fish to decide which patch to be in. This will be discussed below.

The results of Godin and Keenleyside (1984) illustrate how the foraging behaviour of a fish usually reflects a compromise between conflicting pressures. The theories of Charnov (1976) and McNair (1982) are only applicable without modification to solitary foragers at little risk from predation. The influence of conspecifics on foraging behaviour is considerable, and is discussed further in Chapters 12 and 13, by Pitcher and by Magurran, this volume.

Prey Selection

Patches of prey are unlikely to contain items uniform in size, taxonomic category or nutritional value. Most field studies of prey preference by fish show selection for certain sizes and species. For example Mann (1982) studied the diet of pike in the River Frome, Dorset, UK, and found that pike older than 1 year select dace (*Leuciscus leuciscus*) preferentially and ate fewer of the small fish species, such as bullheads (*Cottus gobio*) and stoneloach (*Noemacheilus barbatulus*). The apparent selectivity from field studies like this often confounds two processes: the availability of different prey and selection by the predator.

Field workers have used 'electivity indices' to relate the quantities of each species eaten to their abundance in the habitat. One of the most common was devised by Ivlev, and is expressed as

$$E_i = \frac{r_i - p_i}{r_i + p_i} \tag{8.3}$$

where r_i is the proportion of prey i in the diet, and p_i the proportion in the habitat. This index will show whether a prey species is or is not being eaten in proportion to its abundance in the habitat, but does not distinguish the mechanism behind any selection revealed. This knowledge can only be gained through experiment.

The problem of prey choice has been studied extensively, both theoretically and experimentally (Pyke, Pulliam and Charnov 1977; Krebs 1978; Hughes 1980). Theoretical studies have developed rules for the decisions animals make when they are choosing prey. When collecting food, a particular item will contain a given amount of energy and nutrients and will cost the forager energy to capture, subdue and swallow. Prey will vary in size, and it is usually true that larger items have more value, in terms of energy or nutrients, than smaller items. In turn, smaller items will probably cost less, energetically, to subdue and eat. The relation between the costs and benefits can be made more explicit in the following way, considering a simplified environment with only two prey types, small, P_s, and large, P_l. Let the energy content of a small item be E_s and the costs in energy of subduing and handling it h_s. Similarly E_l and h_l represent the same elements of the large prey. In the natural environment, the abundance of large and small prey will determine the frequency with which the predator encounters them. Let λ_s be the encounter rate for small prey, in terms of items per unit time, and λ_l the rate for large prey. It is now possible to write a relation between the various components to calculate the energy gained per unit effort.

Assume that the animal will be foraging for T_f time units. During T_f it will collect

$$(\lambda_s E_s + \lambda_l E_l) T_f \tag{8.4}$$

units of energy, which will cost the forager

$$T_f + T_f(\lambda_s h_s + \lambda_l h_l) \tag{8.5}$$

units of time to collect. The net gain is given by

$$\frac{T_f(\lambda_s E_s + \lambda_l E_l)}{T_f + T_f(\lambda_s h_s + \lambda_l h_l)} \qquad (8.6)$$

which can be divided through by T_f to give

$$\frac{\lambda_s E_s + \lambda_l E_l}{1 + \lambda_s h_s + \lambda_l h_l} \qquad (8.7)$$

The problem for the forager is whether to eat both P_s and P_l whenever they are encountered, or to be selective and ignore small prey. Selectivity will pay when the net gain from eating P_l only is greater than the net gain from a diet containing both prey types. In terms of the simple model the condition for selection is

$$\frac{\lambda_l}{1 + \lambda_l h_l} > \frac{\lambda_s E_s + \lambda_l E_l}{1 + \lambda_s h_s + \lambda_l h_l} \qquad (8.8)$$

which can be reduced to

$$\frac{1}{\lambda_l} < \frac{E_l}{E_s} h_s - h_l \qquad (8.9)$$

This model can easily be extended to provide rules for a predator faced with a range of prey sizes (see Werner and Hall 1974 for a mathematically more rigorous discussion).

Inequality (8.9) implies two things about the way a selective forager should behave. It should eat either large prey or a mixture of both, switching sharply from one mode to another. In other words, so long as inequality (8.9) holds, small prey should never be taken. Another feature of the model is that the choice made does not depend on the absolute abundance of the different prey types. Even if small prey are very abundant, they should not be taken so long as $1/\lambda_l < (E_l/E_s) h_s - h_l$.

This model of diet selection has been tested on a wide range of animals, from ants and crabs to sunfish and howler monkeys (Krebs *et al.* 1983). Many of the tests show qualitative, if not quantitative, agreement with predictions, although there is one deviation which is common. In many instances the inequality is not strictly obeyed so that foragers eat unprofitable prey even when conditions predict that they should not. This disagreement is consistent with the idea that as animals cannot have complete knowledge of their environment they must continue to sample non-profitable elements, so updating what they know (see also p. 323).

A detailed study of diet selection and its consequences by the sunfishes (*Lepomis* spp.) has been made by Werner and his co-workers (Werner 1974, 1977; Werner and Hall 1974; Werner and Mittelbach 1981; Werner, Mittlebach and Hall 1982). The times taken to chew and swallow natural and artificial prey of known size (handling time) by the bluegill sunfish (*Lepomis macrochirus*) and green sunfish (*L. cyanellus*) were determined by Werner (1974). Energetic costs of handling were assumed to be proportional to handling time. As prey size increased, handling time increased, slowly at first and then exponentially until an upper limit, determined by morphology, was reached. Werner calculated handling time per unit of weight eaten, plotting the result for different-sized fish against prey weight (Figure 8.6). This showed that larger fish have greater latitude in the sizes of prey that they can take profitably.

Werner and Hall (1974) used a model of prey choice to make predictions about diet composition in bluegills which were tested by experiment. Bluegills were allowed to feed on *Daphnia magna*, 3.6 mm, 2.5 mm and 1.4 mm in length at six different densities. The densities were adjusted to take account of the different availabilities of each size, small *Daphnia*

Figure 8.6: The Cost of Prey Capture, Expressed as Handling Time per Unit Capture, Plotted Against Prey Weight for Five Different Sizes of Bluegill Sunfish. The dotted line represents the average handling time for the environment. Only those items with a handling time less than the average should be included in the diet of energy maximisation is to occur. The 4s line therefore defines diet width for the different sizes of fish, showing that larger fish can take a wider selection of prey

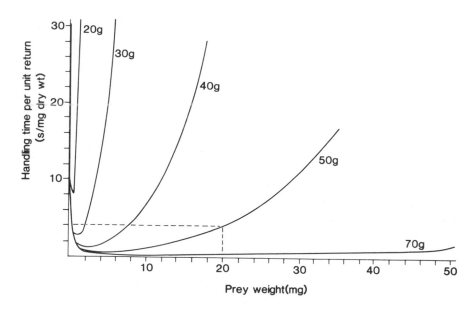

Source: Pitcher and Hart (1982) after Werner (1974).

being less visible than large ones. At the lowest prey density (20 of the largest prey with the abundances of the other two adjusted accordingly), all three prey sizes were taken with equal frequency. As abundance increased to a maximum of 350 per size class, the bluegills concentrated more and more on the large prey (Figure 8.7). The main effect of increasing prey density was to reduce the search time and thereby increase the encounter rate (λ in inequality (8.8)). Using the model, Werner and Hall (1974) calculated the search times after which the fish should switch from one of the feeding patterns in Figure 8.7 to another. The result is reproduced in Figure 8.8 showing that the theory predicts well what the fish actually do.

In principle, then, optimal diet choice has been successful in predicting how foraging animals should choose prey. As with the MVT and its derivatives, the early models have failed to consider many aspects of diet selection, such as the effects of vigilance. The next generation of models is trying to provide a more realistic view of foraging.

Constraints on Feeding strategies

The constraints on feeding behaviour imposed by other activities have already been mentioned. A fish in the reproductive state is most likely to stop feeding. Avoidance of predation and social interactions will interfere with feeding, which can change the strategy employed. The detailed functions of shoaling behaviour in foraging are described in this volume in Chapter 12 by Pitcher and the influence of predation on foraging in Chapter 9 by Milinski.

Behavioural strategies for foraging are also constrained by morphology. The largemouth bass, the green sunfish and the bluegill, living in lakes in Michigan, USA, exploit different size ranges of prey, which are related to mouth shapes and positions and to body form. The bluegill is very good at locating and snapping up small prey requiring fine manoeuvrability for their capture, whereas largemouth bass are best at capturing faster-moving prey which require rapid pursuit (Werner 1977). Their differences in diet are reflected in the relation between gain per unit effort and prey size.

The effect body shape has on manoeuvrability is discussed in detail by Webb (1982). There have been four major trends in the evolution of body form that affect the pattern of locomotion. In all fish groups there is a fusiform carnivore type which has a large tail and caudal area (e.g. the pike), a form which is often primitive. This type has formed the basis for the evolution of elongate forms (e.g. eels, *Anguilla* spp) and of benthic forms (e.g. plaice, *Pleuronectes platessa*; angler fish, *Lophius piscatorius*). A further type is specialised for continuous high-speed swimming and has a forked tail on the end of a narrow caudal peduncle (e.g. mackerel, *Scomber scombrus*). Finally, a type which has appeared several times is a deep-bodied form in which forward movement is driven by the dorsal, ventral

Figure 8.7: A Test of Optimal Foraging in Bluegill Sunfish. (a) The mean number of *Daphnia* of three different size classes, eaten per fish at 20 prey per size class (I — largest, IV — smallest). (b) The number of *Daphnia* of different sizes eaten per fish at (from left to right), densities of 50, 75 and 200 prey per size class. (c) The numbers eaten with prey at 300 and 350 per size class. In each histogram the stippled area represents the expected numbers in the stomachs if items were eaten as encountered (taking into account differential visibility)

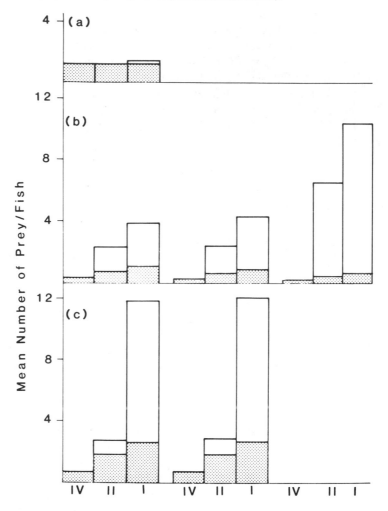

Source: Werner and Hall (1974)

and lateral fins. These forms, like the John Dory (*Zeus faber*), have sacrificed high-speed forward locomotion for the ability to manoeuvre finely. The largemouth bass and the bluegill illustrate well the two main locomotor strategies and their consequences for foraging. Most deep-bodied fish have flexible fins and neutral buoyancy so that it is not necessary to keep swimming to maintain position in the water. This allows the feeding fish to

Figure 8.8: The Way in Which Bluegill Sunfish Switch their Feeding Habits as Prey Density Changes: Predictions and Tests. The selection patterns for different prey sizes, as shown in Figures 8.7a, b and c, are plotted against search time in seconds. A mean ($\pm$SE) is plotted for the experiments at the lowest density, otherwise individual experiments are plotted. The dashed lines are the points at which the selection pattern should change as determined theoretically. If $T_s < 29$ s, only the largest prey (size I) should be taken, and if $T_s > 29$ s, the two largest prey types should be eaten. If $T_s > 295$ s, any selection is suboptimal and prey should be taken as encountered

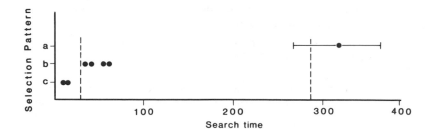

Source: Werner and Hall (1974).

aim precisely at small prey. Many fish can use both caudal- and pelvic-fin propulsion but specialisation in one can lead to decreased efficiency in the other (Webb 1982).

Capture of fast-moving prey requires a well-integrated sequence of jaw movements, which have been closely studied by Rand and Lauder (1981) in the pickerel, *Esox niger.* Earlier work had led to the conclusion that the nervous output to the jaw mechanism was highly stereotyped and uninfluenced by the movements or properties of the prey. New data (Rand and Lauder 1981) showed that when attacking prey, the pickerel employed two types of fast start and two types of jaw movement. The start used depended on how far the pickerel was from the prey, and the jaw movements varied with the position of the prey. Prey in corners were caught with a technique that increased the suction created by the opening mouth, so making it harder for the prey to escape. The results show how intimate is the relation between the behaviour employed and the morphology of the prey-catching equipment. They also illustrate the ways in which the behavioural repertoire, and as a consequence the prey types taken, will be limited by the structure of the fish. Morphology alone confines the diet to a subset of what is available; behaviour allows the fish to further refine its selection in response to features that will change from day to day.

Learning and Foraging Behaviour

The models that have been described assume that a foraging fish responds in a constant way to certain variables of the environment. Few foraging environments are simple enough to make it possible for a predator to survive for long by using unvarying rules. Within the domain frequented by a foraging fish, the positions of patches and the types of prey will change with the seasons so that a predator must be constantly modifying its behaviour if it is to stay alive. An example would be the way the size composition of a river-fish community changes over a year. In early summer there will be large numbers of fish fry available, which will disappear as the season advances. As a result a pike, for example, would eat different sizes and species of fish at different times of the year (Mann 1982). This change in diet could be a result of differences in the encounter rate, but it is also possible that the predator learns about changes in availability (see also Werner and Mittelbach 1981).

Decision making is the essential task performed by the foraging animal (Figure 8.9). Early OFT models underplayed the role of memory and internal states. Learning makes it possible for the forager to adapt to systematic variation in its environment, and can take two forms (Krebs *et al.* 1983): 'learing how' and 'learning about'. In both cases the evolutionary strategy has been to invest in learning capacity which will cost the developing animal both time and energy. The capacity will only evolve if the benefits to be gained outweigh the costs (Orians 1981), and in a wide

Figure 8.9: A Conceptual Scheme of the Principal Processes Contributing towards Decision Making in Animals. Learning occurs when the animal monitors the consequences of its own decisions made in the face of changing external and internal conditions

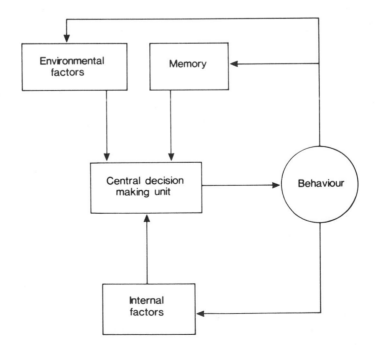

range of organisms they are too small or life is too short for the benefit of learning to be large (Staddon 1983), a restriction that does not apply to most fish.

'Learning how' mostly changes the costs or benefits of foraging. For example, Werner *et al.* (1981) found that the capture rate of bluegills feeding on *Chironomus plumosus* increased with successive experiments so that at the end of a series the rate was four times greater. Several other fish species have been shown to increase their feeding efficiency with experience (see Dill 1983 for a summary). The important point is that changes in components of efficiency with time could differentially alter the profitabilities of different foods (Hughes 1979). If experience of a prey lowers pursuit and handling costs, the net return will increase and the prey could change its rank in the diet.

It has been suggested that some foraging animals learn to detect a cryptic prey, forming what has been called a 'search image' (Curio 1976; Barnard 1983). This may assist the predator towards greater efficiency at finding profitable items. The idea is familiar in everyday life as we are often told to look for some object which merges in with its background and we only 'see' it after we have discovered the clues that give it away. In a similar manner we can miss an object if it has a form different from that expected. A person who normally shaves with an electric razor might miss a wet razor if asked to 'look for a razor' among a range of objects. In this instance the searcher would bear in mind an image of the object sought, attending only to certain features. This selective attention is part of the 'search image' phenomenon. Saying that an animal has a 'search image' for some prey type implies that it has learnt to see certain diagnostic clues which are used in the search. A central perceptual change is implied.

Deciding whether foraging animals use a 'search image' has proved difficult (Krebs 1973; Lawrence and Allen 1983). An animal using a search image cannot easily be distinguished from one that has learnt to forage in certain areas, has learnt to handle prey more effectively with experience, or simply prefers one type of prey over another. The term 'search image' is still incorrectly used (Lawrence and Allen 1983) by most authors, and there is no good evidence to show that any animal uses search images under natural conditions, although careful laboratory experiments have demonstrated the effect (Dawkins 1971).

Assessing profitabilities of food sources comes under the heading of 'learning about'. An example would be bluegill sunfish foraging in turbid water where only one prey can be seen at a time (Gardner 1981). To be able to judge the abundance of large and small prey in these conditions, the predator must retain a memory of past encounter rates.

Foraging from a patchily distributed food source presents a different problem. Many animals continue to take prey from unprofitable patches, and an explanation of this is that individuals must continue to sample all patches to update information about them. The critical question is how much time the predator should devote to sampling. A principal conclusion from recent work is that the time required to learn the difference in gain from two patches is inversely related to the difference between them (Kamil and Yoerg 1982). For foraging animals it is only significant to be able to detect large differences in patch profitability (Krebs *et al.* 1983), where the problem is relatively simple. Complex neural apparatus, or more time, would be required to discern the difference between patches with close profitabilities. It then bcomes unlikely that fitness will be increased sufficiently to outweigh the costs of extra neural equipment.

If an animal samples patches before exploitation, then it needs to have the ability to learn the patch profitabilities and remember them. When other conspecifics are involved, it will be necessary to take account of what they too are doing (see also Chapter 12 by Pitcher, this volume). In these circumstances the strategy adopted should be an evolutionarily stable strategy (ESS) (Maynard Smith 1982), which means that once a strategy is adopted by each individual in the population, an alternative strategy would not be able to invade and replace it. When resources have variable payoffs, a genetically determined ESS would be too inflexible, but an alternative is to have an ESS achieved through learning (Milinski 1984). A learning rule for several conspecifics foraging from a patchily distributed resource was devised by Harley (1981) and is called the relative payoff sum (RPS) learning rule. It has been extended by Regelmann (1984) to include unequal foraging abilities of individuals and travel time. This rule, once adopted by all the population, cannot be bettered by any other, so leading to an ESS. The rule assumes that a fish faced with two patches for the first time will assign arbitrary payoffs (r_1 and r_2) to each, and will then choose one on the basis of the probability f given by $r_1/(r_1+r_2)$. r_1 and r_2 reflect an inherent bias towards one or the other patch bearing no relation to known profitabilities. Assume that the fish chooses patch 1 first and takes the first prey that is available. The payoff sums S_1 and S_2 for the two patches are then updated to give $S_1=r_1+p_1$ and $S_2=r_2$, where p_1 is the payoff received from eating the prey. The fish now decides again which patch to go for, using a new probability f', given by $S_1/(S_1+S_2)$. If this dictates patch 1 again, then the new payoff sums will be $S_1=r_1+xp_1+p_1$ and $S_2=r_2$. The new term $0<x<1$ is the factor giving the degree to which the first payoff should be discounted as a result of memory loss. Clearly, when x is close to 0, the fish has a very short memory so that each decision is based entirely on the size of the last payoff received. The next probability of $f''=S_1/(S_1+S_2)$ is evaluated and the fish acts by choosing one of the two patches. In reality p_1 and p_2 may change with time.

Intuitively, the RPS model assumes on first encountering the two patches a fish chooses one arbitrarily. In response to the reward received from the patch, the individual reassesses its choice, staying with the patch chosen if it proved profitable and moving to the next if it did not. At each return the fish updates its assessment and compares it with its memory of past profitabilities. The model allows for varying rates of memory decay and only applies to fish selecting prey presented at intervals from two separate sources.

Regelmann (1984) modified this RPS learning rule to allow fish to have unequal foraging abilities and to account for travel time between food sources. The new theory was tested by Milinski (1984). The RPS rule predicts that the fish will distribute themselves according to the patch profitabilities, that good competitors will decide earlier where to feed and will switch less between patches, that the distribution of bad competitors will be determined by patch profitabilities and by the distribution of good competitors, and that travel costs will decrease the amount of switching. Six individual sticklebacks in an aquarium were offered *Daphnia magna* in two patches at profitability ratios of 2:1 or 5:1 and their distributions were observed. Milinski (1984) found that the predictions accurately described the outcome of the experiments, indicating that the sticklebacks were using some form of RPS learning rule to regulate their behaviour. The fish all showed some sampling behaviour in the early part of each trial, but soon settled to a steady exploitation frequency.

The degree to which learning is involved in foraging is only now being investigated extensively and there is much work to be done. Fish are well able to learn, so there is every reason to expect significant changes in our understanding of the rules governing their behaviour as new results become available.

Risk and Foraging Behaviour

The problem of foraging in the face of risk is illustrated by an example from the management of salmon. Most attempts to manage the Pacific salmon have assumed that it is sufficient to be able to maximise the mean yield that will be available to the fishing fleet over a period of time. For many years the fishery was regulated on this basis. To achieve the desired maximum, the fishery was regulated to allow a fixed number of adults to escape the fishing gear and to run up the rivers to spawn. This escapement was calculated on the basis of a relation between spawning stock size and number of recruits 2 to 4 years later. The flaw with this method is that the realised recruitment from a given spawning stock is very variable so that the year-to-year yield to the fisherman fluctuates widely around the mean. For many years it was assumed that the parameter of interest was the mean yield, but to the fisherman the annual variability could be more critical (Walters 1975). Poor years mean many underpaid men, and in good years each fisherman gets a good reward but many usable fish go uncaught. Walters (1975) developed a method of evaluating the costs and benefits of variable mean yields, making the point that it could be desirable to accept a lower mean if this was associated with less variability.

The example from fishing illustrates a point that is so far neglected in most foraging models. It is acknowledged (Pyke *et al.* 1977) that animals operate in a stochastic environment so that rewards obtained from foraging are uncertain. It is assumed that it is permissible to substitute maximisation of net gain by maximisation of mean net gain, implying, as in the case of the fishery, that the variability that would exist around that mean was not important to the animal. As with the fisherman, variability could be critical to an animal near to starvation. For many diurnal birds it is necessary for sufficient food to be gathered through the day to last the animal over night. If it is an hour before dusk and the bird has some way to go before it has accumulated sufficient energy for the night, then the variability of the food resources being collected is critical.

The problem of how animals respond to variable resources was tackled by Caraco, Martindale and Whittham (1980), who studied the response of yellow-eyed juncos (*Junco phaeonotus*) to variability in their food supply. They found that birds with a positive energy budget avoided an uncertain source, and others that had a negative energy budget chose the risky source where there was a chance that their energy deficit could be cleared quickly. In this condition, going for the sure supply would have given no chance of clearing the deficit.

The responses of fish to risk have not been studied. As fish are cold-blooded, it might be assumed that they would be less vulnerable to starvation. Ivlev (1961) discussed the degree to which different freshwater fish species can tolerate starvation and his conclusion shows that some fish are perhaps not that different from birds and mammals. The bleak (*Alburnus alburnus*) cannot go without food for more than a few days. This species feeds on small planktonic organisms and surface insects, and can be expected to feed in the same manner as a small passerine bird, that is, many small items will be taken throughout the day. Pike are able to go without food for months, and under natural circumstances they take prey only once every one to three days (Diana 1979). These differences could have implications for the way in which the different species of fish react to variability when foraging, and some comparative studies would provide a valuable contribution.

Summary

This chapter has taken the view that fish, like all animals, are problem-solving devices with the goal of maximising the production of repro-ductively viable offspring. One of the principal tasks fish have in achieving this goal is to find sufficient energy and nutrients. The discussion has con-centrated on the solution of problems specific to food acquisition. Problems considered are: when to start feeding, how and at what speed food should be searched for, how a forager should deal with a patchy resource, and what species or sizes of prey should be selected. The way solutions to these problems are constrained by morphology, social relations and predation is discussed. The final two sections consider the way learning is used in foraging, and the way in which animals deal with resource unpredictability. These aspects are now being incorporated into theory (e.g. Clark and Mangel 1984 and see Chapter 12 by Pitcher, this volume), but still need experimental testing.

Acknowledgements

I would like to thank Ian Clarke for commenting upon an earlier draft of the manuscript.

References

Barnard, C.J. (1983) *Animal Behaviour: Ecology and Evolution*, Croom Helm, London

Beukema, J.J. (1968) 'Predation by the three-spined stickleback. (*Gasterosteus aculeatus* L.)', *Behaviour, 31*, 1-126

Bond, C.E. (1979) *The Biology of Fishes*, Saunders, Philadelphia

Brett, J.R. (1979) 'Environmental Factors in Growth', in W.S. Hoar, D.J. Randall and J.R. Brett (eds), *Fish Physiology*, vol. VIII: *Bioenergetics and Growth*. Academic Press, New York, pp. 599-675

Brett, J.R. and Groves, T.D.D. (1979) 'Physiological Energetics', in W.S. Hoar, D.J. Randall and J.R. Brett (eds), *Fish Physiology*, vol. VIII: *Bioenergetics and Growth*, Academic Press, New York, pp. 279-352

Caraco, T., Martindale, S. and Whittham, T.S. (1980) 'An Empirical Demonstration of Risk-sensitive Foraging Preferences', *Animal Behaviour, 28*, 820-30

Charlesworth, B. (1980) *Evolution in Age-structured Populations*, Cambridge University Press, Cambridge

Charnov, E.L. (1973) 'Optimal Foraging: Some Theoretical Explorations', unpublished PhD thesis, University of Washington

Charnov, E.L. (1976) 'Optimal Foraging Theory: the Marginal Value Theorem', *Theoretical Population Biology, 9*, 129-36

Clark, C.W. and Mangel, M. (1984) 'Foraging and Flocking Strategies: Information in an Uncertain Environment', *American Naturalist, 123*, 626-41

Clutton-Brock, T.H., Guiness, F.E. and Albon, S.D. (1982) *Red Deer: Behaviour and Ecology of Two Sexes*, University of Chicago Press, Chicago

Cody, M.L. (1971) 'Finch Flocks in the Mohave Desert', *Theoretical Population Biology, 21*, 142-58.

Cody, M.L. (1974) 'Optimization in Ecology', *Science, 183*, 1156-64

Colgan, P.W. (1972) 'Motivational Analysis of Fish Feeding', *Behaviour, 45*, 38-65

Cowie, R.J. (1977) 'Optimal Foraging in Great Tits (*Parus major*)', *Nature (London), 268*, 137-9

Curio, E. (1976) *The Ethology of Predation*, Springer-Verlag, Berlin

Dawkins, M. (1981) Perceptual Changes in Chicks, Another Look at The "Search Image" Concept', *Animal Behaviour, 19*, 566-74

Diana, J.S. (1979) 'The Feeding Pattern and Daily Ration of a top Carnivore, the Northern Pike (*Esox lucius*)', *Canadian Journal of Zoology, 57*, 2121-7

Dill, L.M. (1983) 'Adaptive Flexibility in the Foraging Behaviour of Fishes', *Canadian Journal of Fisheries and Aquatic Sciences, 40*, 398-408

Fretwell, S.D. and Lucas, J.R. (1970) 'On Territorial Behaviour and Other Factors Influencing Habitat Distribution in Birds. I. Theoretical Development', *Acta Biotheoretica, 19*, 16-36

Gardner, M.B. (1981) 'Mechanisms of Size Selectivity by Planktivorous Fish: a Test of Hypotheses', *Ecology, 62,* 571-8

Gibb, J.A. (1962) 'L. Tinbergen's Hypothesis of the Role of Specific Search Images', *Ibis, 104,* 106-11

Godin, J-G.J. and Keenleyside, M.H.A. (1984) 'Foraging on Patchily Distributed Prey by a Cichlid Fish (Teleostei, Cichlidae): a Test of the Ideal Free Distribution Theory', *Animal Behaviour, 32,* 120-31

Harley, C.B. (1981) 'Learning the Evolutionarily Stable Strategy', *Journal of Theoretical Biology, 89,* 611-33

Hart, P.J.B. and Connellan, B. (1984) 'The Cost of Prey Capture, Growth Rate and Ration Size in Pike (*Esox lucius*) as Functions of Prey Weight', *Journal of Fish Biology, 25,* 279-91

Holmgren, S., Grove, D.J. and Fletcher, D.J. (1983) 'Digestion and the Conrol of Gastro-intestinal Motility', in J.C. Rankin, T.J. Pitcher and R. Duggan (eds), *Control Processes in Fish Physiology,* Croom Helm, London, pp. 23-40

Howell, D.J. and Hartl, D.L. (1980) 'Optimal Foraging in Glossophagine Bats: When to Give Up', *American Naturalist, 115,* 696-704

Hughes, R.N. (1979) 'Optimal Diets under the Energy Maximisation Premise: the Effects of Recognition Time and Learning', *American Naturalist, 113,* 209-21

Hughes, R.N. (1980) 'Optimal Foraging Theory in the Marine Context', *Oceanography and Marine Biology, Annual Review, 18,* 423-81

Ivlev, V.S. (1961) 'Experimental Ecology of the Feeding of Fishes', Yale University Press, New Haven

Jansen, J. (1982) 'Comparison of Searching Behaviour for Zooplankton in an Obligate Planktivore, Blueback Herring (*Alosa aestivalis*) and a facultative planktivore, bluegill (*Lepomis macrochirus*)', *Canadian Journal of Fisheries and Aquatic Sciences, 39,* 1649-54

Kamil, A.C. and Yoerg, S.J. (1982) 'Learning and Foraging Behaviour', in P.P.G. Bateson and P.H. Klopfer (eds), *Perspectives in Ethology* vol. 5, Plenum Press, New York, pp. 325-46

Krebs, J.R. (1973), 'Behavioural Aspects of Predation', in P.P.G. Bateson and P.H. Klopfer (eds), *Perspectives in Ethology,* vol. 3, Plenum Press, New York, pp. 73-101

Krebs, J.R. (1978) 'Optimal Foraging: Decision Rules for Predators', in J.R. Krebs and N.B. Davies (eds), *Behavioural Ecology: an Evolutionary Approach',* Blackwell, Oxford, pp. 23-63

Krebs, J.R. Ryan, J.C. and Charnov, E.L. (1974) 'Hunting by Expectation or Optimal Foraging? A Study of Patch Use by Chickadees', *Animal Behaviour,, 22,* 953-64

Krebs, J.R. Stephens, D.W. and Sutherland, W.J. (1983) 'Perspectives in Optimal Foraging Theory', in G.A. Clark and A.H. Bush (eds), *Perspectives in Ornithology,* Cambridge University Press, New York, pp. 165-221

Lawrence, E.S. (1979) 'Optimal Foraging: Search Patterns during Feeding in the Stickleback (*Gasterosteus aculeatus* L.)'. Unpublished BSc thesis, University of Leicester

Lawrence, E.S. and Allen, J.A. (1983) 'On the Term "Search image"', *Oikos, 40,* 313-14

McFarland, D.J. and Houston, A. (1981) *Quantitative Ethology,* Pitman, London

McNair, J.N. (1982) 'Optimal Giving-up Times and the Marginal Value Theorem', *American Naturalist, 119,* 511-29

MacPhail, E.M. (1982) *Brain and Intelligence in Vertebrates* Oxford University Press, Oxford

Mann, R.H.K. (1982) 'The Annual Food Consumption and Prey Preferences of Pike (*Esox lucius*) in the River Frome, Dorset', *Journal of Animal Ecology, 51,* 81-95

Maynard Smith, J. (1982) *Evolution and the Theory of Games,* Cambridge University Press, Cambridge

Milinski, M. (1984) 'Competitive Resource Sharing: an Experimental Test of a Learning Rule for ESSs', *Animal Behaviour, 32,* 233-42

Orians, G.H, (1981) 'Foraging Behaviour and the Evolution of Discriminatory Abilities', in A.C. Kamil and T.D. Sargent (eds), *Foraging Behaviour, Ecological, Ethological and Psychological Approaches,* Garland STPM Press, New York, pp. 389-405

Pitcher, T.J. and Hart, P.J.B. (1982) *Fisheries Ecology,* Croom Helm, London

Pyke, G.H. (1978) 'Are animals efficient foragers?', *Animal Behavior, 26,* 241-50

Pyke, G.H. (1981) 'Optimal Travel Speeds of Animals', *American Naturalist, 118,* 475-87

Pyke, G.H., Pulliam, H.R. and Charnov, E.L. (1977) 'Optimal Foraging: a Selective Review of Theory and Tests', *Quarterly Review of Biology, 52,* 137-54

Rand, D.M. and Lauder, G.V. (1981) 'Prey Capture in the Chain Pickerel, *Esox niger.* Correlations between Feeding and Locomotion Behaviour', *Canadian Journal of Zoology, 59,* 1072-8

Regelmann, K. (1984) 'Competitive Resource Sharing: a Simulation Model', *Animal Behaviour, 32,* 226-32

Roff, D.A. (1983) 'An Allocation Model of Growth and Reproduction in Fish', *Canadian Journal of Fisheries and Aquatic Sciences, 40,* 1395-1404

Sih, A. (1982) 'Optimal Patch Use: Variation in Selective Pressure for Efficient Foraging', *American Naturalist, 120,* 666-85

Smith, J.N.M. (1974) 'The Food Searching Behaviour of two European Thrushes, I. Description and Analysis of Search Paths', *Behaviour, 48,* 276-302

Staddon, J.E.R. (1983) *Adaptive Behaviour and Learning,* Cambridge University Press, Cambridge

Thomas, G. (1974) 'The Influence of Encountering a Food Object on the Subsequent Searching Behaviour in *Gasterosteus aculeatus* L.', *Animal Behaviour, 22,* 941-52

Walters, C.J. (1975) 'Optimal Harvest Strategies for Salmon in Relation to Environmental Variability and Uncertain Production Parameters', *Journal of the Fisheries Research Board of Canada, 32,* 1777-84

Ware, D.M. (1975) 'Growth, Metabolism and Optimal Swimming Speed of a Pelagic Fish', *Journal of the Fisheries Research Board of Canada, 32,* 33-41

Ware, D.M. (1978) 'Bioenergetics of Pelagic Fish: Theoretical Change in Swimming Speed and Relation with Body Size', *Journal of the Fisheries Research Board of Canada', 35,* 220-8

Ware, D.M. (1982) 'Power and Evolutionary Fitness of Teleosts', *Canadian Journal of Fisheries and Aquatic Sciences, 39,* 3-13

Webb, P.W. (1982) 'Locomotor Patterns in the Evolution of Actinopterygian Fishes', *American Zoologist, 22,* 329-42

Werner, E.E. (1974) 'The Fish Size, Prey Size, Handling Time Relation in Several Sunfishes and Some Implications', *Journal of the Fisheries Research Board of Canada, 31,* 1531-6

Werner, E.E. (1977) 'Species Packing and Niche Complementarity in Three Sunfishes', *American Naturalist, 111,* 553-78

Werner, E.E. and Hall, D.J. (1974) 'Optimal Foraging and the Size Selection of Prey by the Bluegill Sunfish (*Lepomis macrochirus*)', *Ecology, 55,* 1042-52

Werner, E.E. and Mittelbach, G.G. (1981) 'Optimal Foraging: Field Tests of Diet Choice and Habitat Switching', *American Zoologist, 21,* 813-29

Werner, E.E., Mittelbach, G.G. and Hall, D.J. (1981) 'The Role of Foraging Profitability and Experience in Habitat Use by the Bluegill Sunfish', *Ecology, 62,* 116-25

9 CONSTRAINTS PLACED BY PREDATORS ON FEEDING BEHAVIOUR

Manfred Milinski

Teleost fish are frequently in danger of being preyed upon. Predators of many taxa are specialised piscivores and there are more of them that attack small fish. Therefore, young fish normally live under a high risk of predation, which decreases as they grow older and bigger. Growing fast is not only a good strategy for escaping the prey spectrum of many predators, but also for increasing reproductive success; bigger teleosts are generally able to produce more numerous offspring than smaller ones because bigger females produce more eggs, and bigger males can defend breeding sites better and have a higher social rank; for example, dominant male three-spined sticklebacks (*Gasterosteus aculeatus*) maintain larger territories and have priority of access to females (Li and Owings 1978). Thus, there is a high selection pressure to feed most efficiently in order to grow quickly.

Optimal foraging theory predicts how an animal should proceed to achieve a maximum rate of energy intake, given its potential types of food and their distribution. Several investigations have shown that animals probably forage using the optimality rules predicted. These include decisions on where to forage, how long to sample, which food items to select and when to leave a food patch in order to find another one (see Chapter 8 by Hart, this volume). Finding food in the most economic way is only one of the many sub-goals for maximising fitness. Another one is the avoidance of being preyed upon. Unfortunately, an animal is mostly not able to fulfil the tasks sequentially, for example foraging when hungry and avoiding predators when satiated. Often there is some interaction between foraging and risk of predation: (a) the animal is more conspicuous by its inevitable movements while foraging; (b) it has to concentrate on finding food at the expense of vigilance for predators; (c) food can be plentiful in places of high risk of predation. Therefore we expect to find a tradeoff between energetic returns from foraging and the risk of predation accepted by the animal.

Balancing Feeding and Predation Risk

Many predators adopt an ambush mode of hunting; they wait motionless and often concealed until the prey is within range and then strike suddenly. Several birds, such as kingfishers and herons (Figure 9.1), and some fish, for example the pike (*Esox lucius*), adopt this strategy. So also during foraging a small fish has to be vigilant in case of a sudden attack. If in some

Figure 9.1: When Feeding, a Small Fish Takes the Risk of Being Attacked by Birds and Fish Predators

places the risk of predation is higher than in others, the fish could avoid these places, especially when the predator is known to be present.

A problem arises when the predator is detected in the place where food is most plentiful. By feeding there, the fish could ostensibly maximise its rate of food intake, but with a low probability of surviving the foraging trip. Here the best strategy depends on the specific circumstances of the situation and on the fish's behavioural options (= strategy set). Let us assume that it does not pay the fish to wait and come out at night (but see Chapter 14 by Helfman, this volume) or even to starve for a while (see Chapter 8 by Hart, this volume). Hence the best tactic might be a compromise, i.e. feeding near the predator but not closer than at a point where the risk of predation would override the payoff for feeding most efficiently. The stickleback in Figure 9.1 has to approach the big perch when it tries to pick up the row of *Tubifex* worms, which are a prey of high quality. How close would the stickleback dare to approach the predator in order to have a good meal? In an experiment (Milinski 1985) three-spined sticklebacks were allowed singly to feed upon *Tubifex* worms which lay on the bottom of plexiglass cylinders. These were of a height such that the fish while feeding lost sight of a big cichlid (*Oreochromis mariae*) which observed each movement of the stickleback from behind a glass partition. On first sight of the cichlid, the sticklebacks usually raised their spines and moved backwards slightly, showing that they regarded it as a threat. If the sticklebacks dared to feed at all in the presence of the predator, we would expect them to avoid the prey next to the cichlid. The fish indeed preferred the worms that were furthest away from the predator (Figure 9.2), and they fed more slowly than without the cichlid nearby. Thus, the sticklebacks arrived at a compromise between feeding efficiently and avoiding the predator. We cannot decide whether this compromise is quantitatively the best solution.

Under natural conditions there are many different influences possibly determining the optimum of the compromise. If we could detect and measure all of them, including the animal's constraints, optimality theory would be a suitable tool to predict the best compromise. A great problem for this approach consists in the need to sum up different short-term currencies of fitness, i.e. we have to convert net energy gain from a foraging strategy into probability of death from the risk of predation incurred. In principle, this should be possible to solve. As a first step we should determine qualitatively which influences affect the balance between foraging and avoiding the risk of predation.

(1)   The animal should be more disposed to take the risk of predation with decreasing value of alternative food in a safer place.
(2)   If the risk of predation is increased, the animal's behaviour should change to compensate for this; for example, it should reduce its feeding rate in order to become more vigilant and ready to take evasive actions, or it should simply avoid the risky food.
(3)   An increased need for food should render an animal more willing to accept risky food. The probability of starvation should be pro-

Figure 9.2: Mean Number of *Tubifex* Worms on which Sticklebacks Fed at Different Distances from a Fish Predator; each Fish was Allowed to Consume Eight Worms (after Milinski 1985)

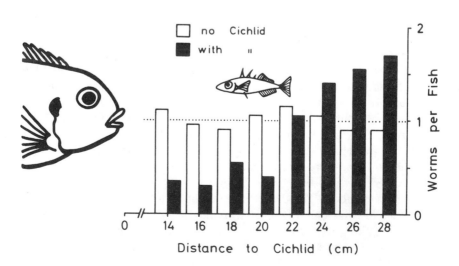

portional to the risk of predation taken in order to feed efficiently. Such a motivational influence is a dynamic one because the state of hunger changes during feeding and so does the amount of risk which should be accepted.

In the following I will consider the evidence for these influences that is available from studies on teleost fish.

Value of Alternative Food and the Risk of Predation

Whether fish are willing to run a higher risk of predation if the patch containing a predator provides much more food than an alternative patch without a predator was investigated by Cerri and Fraser (1983). Minnows (*Rhinichthys atratulus*) were confined to an artificial stream system consisting of several compartments, some of which contained predatory fish (*Semotilus atromaculatus*). The prey fish could easily switch between compartments whereas the larger predators could not. The availability of food was low in all compartments in one treatment of the experiment, but there was additional food in the predator compartments in another treatment. During 6 days of observation, prey fish were found in predator compartments more frequently when there was additional food. This result appears to show that the minnows spent more time close to the predators and therefore accepted a higher risk when patches with predators allowed a higher feeding rate than alternative places. Nevertheless, the minnows took into account that there was a risk of predation because fewer fish were found in the compartments containing additional food with predators being present than when there were no predators in a control treatment.

In Cerri and Fraser's experiment, the value of the alternative food patches was very low, so that the minnows probably had to forage in the risky patches. This was different in a study of several species of armoured catfish (Loricariidae) by Power (1984). The larger size classes of these fish, having outgrown their vulnerability to piscivorous fishes, restricted their foraging to deeper water and did not exploit standing crops of algae in shallow water, even in the dry season when food was most limited. Only in shallow water were they susceptible to several species of bird predators. The large loricariids could afford to avoid the risky habitat, for there were no seasonal changes in mortality rates although somatic growth rates decreased when food was scarce.

From these examples we can conclude that the food value of alternative patches without predators determines the degree to which fishes are disposed to forage in risky places.

Increased Risk of Predation

We might expect compromise between feeding and avoiding predation to be shifted to more pronounced predator avoidance when the risk of predation is increased. Werner, Gilliam, Hall and Mittelbach (1983) investigated whether three size classes of bluegill sunfish (*Lepomis macrochirus*), of which the smallest class was very vulnerable to predation by largemouth bass (*Micropterus salmoides*), responded to both increased predation risk and habitat profitability in choosing habitats in which to feed. The bluegills were stocked on both sides of a divided pond which contained largemouth bass only on one side. Each side of the pond consisted of three different habitats, of which only one offered refuges from predation by dense vegetation but had much less food than the others. It turned out that only fish of the smallest size were preyed upon, although bluegills of the medium size class were assumed to be chased more often than the largest fish. The smallest fish therefore had an increased risk of predation as compared with the other size classes. Without predators, bluegills of all three size classes preferred the benthos, where food was most plentiful (Figure 9.3). With predators present, only the smallest fish had a less pronounced preference for the benthos, and foraged nearly as often in the less profitable vegetation. Thus, the size class that had an increased risk of predation balanced the two conflicting demands in that they suffered only a little from predation, but grew more slowly than the other size classes. It could not be decided whether all small fish divided their time between different habitats in the same way, or whether some individuals were more willing to take the risk of foraging in the benthos than others.

Figure 9.3: Average Percentage Composition of the Diet in Two of Three Habitats for Three Size Classes of Bluegill Sunfish. (Modified after Werner *et al.* 1983.)

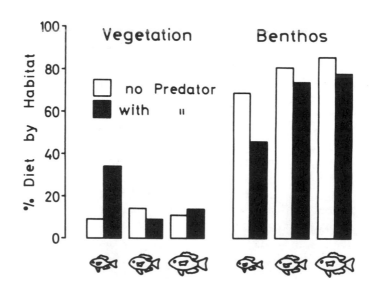

We have seen that increased risk of predation in one habitat can alter the choice of places in which to forage. It is also possible that *within* a patch the risk of predation is different for feeding upon one type of food as compared with others; this may alter the choice of items by a forager within a given habitat.

In a study of juvenile coho salmon (*Oncorhynchus kisutch*), Dill (1983) investigated the influence of a predator on ,the distance a coho would swim upstream from its holding station to intercept drifting prey. Bigger prey items were detected and attacked from a greater distance than smaller ones, suggesting that the fish maximised their rate of success for each prey size. On the other hand, by swimming a greater distance the fish becomes more visible, and hence presumably more vulnerable, to predators. Thus, the fish should swim shorter distances for a given prey item when it perceives risk of predation to be high. Dill tested this hypothesis by measuring the attack distance of the fish in response to three sizes of surface-drifting flies in an artificial stream channal. In some trials fish were presented singly with a photograph of a big rainbow trout, resembling a predator for young coho, and in other trials the coho could feed undisturbed. Attack distances were reduced in the presence of the model predator, particularly in response to the largest prey (Figure 9.4). These results support the hypothesis that juvenile coho salmon are less willing to travel long distances, or to spend long periods of time moving, to obtain prey when risk of predation is increased. The risk seems to grow exponentially with distance travelled, as can be seen from the marked effect when the prey were large. The extent of the reduction in attack distance was greater at higher frequencies of predator presentation, suggesting that coho are able to judge the level of risk and adjust their behaviour accordingly (Dill and Fraser 1984).

Increased risk of predation reduced the attack distance most for the large profitable prey. Thus, in the presence of predators, coho should capture fewer and smaller prey items than in their absence and therefore suffer from a reduced energy intake as did the small bluegills when choosing to forage in a safer but less profitable habitat (Werner *et al.* 1983). In the next section we will see that fish can change their social behaviour to compensate for increased risk of predation.

Predator Detectability and Feeding Efficiency

When foraging, a fish is at risk — especially from a suddenly attacking predator — because handling food can impair a full view around the fish for some seconds. If there were a way of improving the probability of detecting a predator so that it could be observed from a greater distance, the fish could evaluate the actual risk and sometimes continue feeding for longer. There might be no net benefit if predator detection were improved

Figure 9.4: The Effect of the Presentation of a Model Predator on the Distance Swum to Attack Flies of Different Size by Juvenile Coho Salmon. (Modified after Dill 1983.)

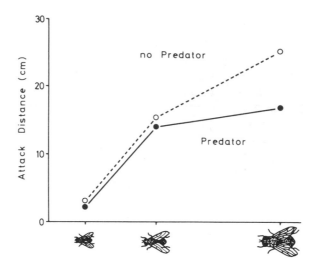

by investing more time in scanning than in feeding. A more profitable strategy consists in enlarging the number of eyes looking around, which has been shown to work in bird flocks (e.g. Powell 1974; Lazarus 1979). Whether the members of a fish shoal have a similar advantage has recently been investigated by Magurran, Oulton and Pitcher (1985). They observed shoals of 3, 6, 12 or 20 minnows (*Phoxinus phoxinus*) foraging on an artificial food patch during the simulated stalking approach of a model pike. Minnows in large shoals reduced their foraging sooner but remained feeding on the patch for longer (Figure 9.5). The relatively late reaction of small shoals to the model and the rapid cessation of feeding provide good evidence that they detected the predator later than larger shoals. Several specific anti-predator behaviours were observed which proved this point (see also Chapter 12 by Pitcher, this volume.)

Why did smaller shoals leave the patch earlier than larger ones? There could be several functional explanations for this result.

(1)   Fish in small shoals could not observe the approaching pike and therefore had no possibility of calculating the actual risk. A good strategy would then be to assume the highest risk and flee immediately upon detection of the predator.

(2)   Fish in small shoals must be more certain that the predator will successfully attack because they cannot rely on the confusion effect (see Chapter 12).

(3)   If the predator strikes successfully, a member of a small shoal has a

Figure 9.5: Median and Interquartile Time that Shoals of 20, 12, 6 or 3 Minnows Spent on a Food Patch during the Approach of a Model Pike. (Modified after Magurran *et al.* 1985.)

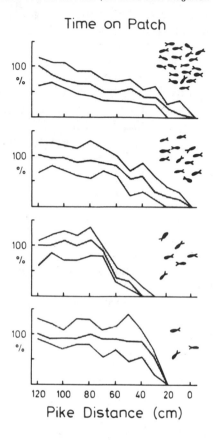

Time on Patch

Pike Distance (cm)

higher chance of being the victim than each member of a large shoal (but see Chapter 12).

Although these effects reduce the risk for members of a large shoal, they do not provide complete safety. By continuously reducing their feeding effort during the predator's approach, the fish in large shoals shifted the compromise more and more from feeding to vigilance in monitoring the pike's approach. Magurran *et al.* (1985) showed that the minnows in larger shoals did not suffer from increased competition for food, which could play a role under different conditions and might also influence the compromise achieved. For example, when apparent competition was increased for young coho salmon, they reduced the weighting given to risk of predation and travelled further to prey than when they were tested in visual isolation (Dill and Fraser 1984). There should be an optimal shoal size minimising the cost/benefit ratio caused by competition and risk of predation. (See Chapter 12 by Pitcher, this volume.)

Hunger and Avoidance of Predation

In this chapter we will see how altering the need for food, i.e. the risk of starvation, changes the compromise between feeding and anti-predator behaviour. Since hunger decreases during feeding, we would expect that the compromise would change as a function of satiation. This will be illustrated by a study on sticklebacks which have to scan for predators while feeding upon dense swarms of waterfleas.

Aggregations of tiny prey animals in the zooplankton are a valuable food source, especially for small fish. Big fish with large mouths can suck in many items from the swarm of prey in one gulp if the swarm density is high enough (McNaught and Hasler 1961). From less-dense swarms, fish have to catch one item after another. Small fish can economically adopt this latter tactic (Janssen 1976) even if prey items are very small. Such particulate feeders, too, can increase their feeding rate when they choose a swarm of a high density (Heller and Milinski 1979) because the distances between prey items are shorter there. Many predators, however, suffer from a so-called confusion effect when they attack dense swarms of prey (see Chapter 12 by Pitcher, this volume). I will show in the following that a stickleback's decision to attack in a high-density swarm of waterfleas not only increases the fish's hunting success, but also affects its own risk of predation.

During the summer, three-spined sticklebacks live mostly on planktonic prey (Wootton 1976, and Chapter 16 by Fitzgerald and Wootton, this volume). Whether these fish preferentially attack swarms of a high density was investigated in the laboratory (Milinski 1977a). The fish were given singly the choice between two test tubes, one containing a swarm of 40 waterfleas (*Daphnia magna*), the other one only two waterfleas. One group of sticklebacks was hungry whereas another one had become nearly satiated before the trial by feeding upon *Tubifex* worms. The results were rather puzzling because most of the hungry fish attacked the swarm, whereas most less-hungry fish chose the single items (Figure 9.6). Further experiments confirmed that the hungry fish did not simply overlook the singletons but really chose to attack the swarm (Milinski 1977b; Heller and Milinski 1979). Does this mean that hunting in a high density is most profitable for hungry fish, and attacking a low-density swarm provides the highest energy intake for more-satiated fish? This was tested in another series of experiments in which hungry sticklebacks and more-satiated ones were allowed to feed upon either a high or a low prey density kept constant during a trial (Heller and Milinski 1979). The hungry fish fed significantly faster in the high prey density whereas the less-hungry ones achieved a higher feeding rate in the low density. Therefore, it pays a hungry fish to choose a high-density swarm and a less-hungry one to attack a lower density.

If only the risk of starvation is important, optimal foraging theory does

Figure 9.6: Choice of Either a Low or a High Density of Prey by Less-hungry Sticklebacks and by Hungry Ones. (Modified after Milinski 1977a.)

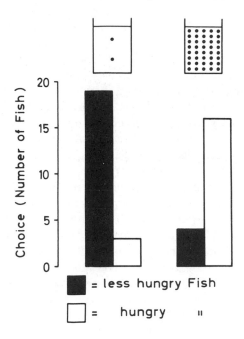

not explain these results. If hunting in place A is more profitable than in B, there is no reason for a less-hungry predator to hunt in B. Thus, there must be other costs at stake, which increase with the density of the prey swarm attacked.

A candidate for this could be the confusion effect from which three-spined sticklebacks have been shown to suffer (Ohguchi 1981). Hungry sticklebacks, however, must be able and willing to overcome the confusion imposed by a high-density swarm, otherwise they could not achieve their highest feeding rate there. Since it seems to be far-fetched to assume that the *ability* to overcome the confusion is dependent on hunger, it must be concluded that the less-hungry fish were not willing to pay the costs for overcoming the confusion.

What could these costs consist of? If the fish has to concentrate on the difficult task of tracking one out of many similarly looking targets of a dense swarm, it probably cannot pay sufficient attention to other events, e.g. a suddenly approaching predator of its own (see Figure 9.1). There is always a certain probability of a sudden attack by an ambush predator. If this risk is lower than the costs of not reducing its energy deficit as quickly as possible, the stickleback should overcome the confusion and attack the high-density swarm.

To test this hypothesis, the stickleback's expected risk of predation was increased. At the new level it should prefer a lower prey density than with the baseline risk level (Milinski and Heller 1978). A model of a European kingfisher (*Alcedo atthis*), an efficient predator of sticklebacks (Kniprath 1965), was moved over the tank before the stickleback was allowed to choose between different prey densities. Frightened fish attacked the lowest prey density whereas unfrightened ones preferred the highest density (Figure 9.7). Now, one question remains to be answered. Is a hungry stickleback when attacking a high-density swarm less likely to detect an approaching predator than a similarly hungry one attacking a low density? To test this, hungry sticklebacks were allowed to prey upon either a high density of waterfleas or a low one. After they had consumed the fourth waterflea, a conspicuous model of a kingfisher, or a cryptic model or no model, was flown over the tank (Milinski 1984a). If the fish detected the predator, they should hesitate before attacking the fifth waterflea. Figure 9.8 shows that in both prey densities the sticklebacks hesitated significantly longer after the conspicuous bird had been shown than in the control experiment. After the cryptic model had been flown over the tank, only the fish in the low prey density hesitated for longer than in the control. This shows that at least some of the fish attacking waterfleas in the high density must have overlooked the cryptic predator.

The unfrightened fish did not feed more quickly in the high-density swarm than in the low-density one as would be expected. This probably resulted from the high density being twice as high as in the experiment discussed earlier (Heller and Milinski 1979), providing too much confusion for the fish to overcome.

Figure 9.7: Choice of Different Swarm Densities by Sticklebacks which were Either Undisturbed or Frightened by a Model Kingfisher. (After Milinski and Heller 1978.)

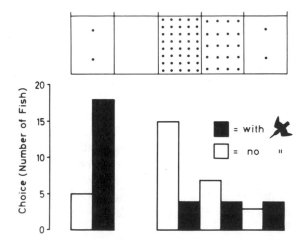

Figure 9.8: Consumption Time for One Prey Item by Sticklebacks Feeding Either upon a High or a Low Prey Density. While feeding, fish had been shown either a conspicuous model predator or a cryptic one or no predator. NS, not significant. (Modified after Milinski 1984a.)

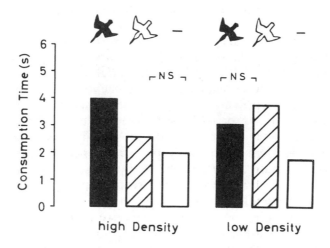

In conclusion, when the sticklebacks were very hungry, they took the risk of overlooking a suddenly approaching predator in order to feed most efficiently. When the risk of predation was increased, or when the fish were less hungry, they gave higher priority to predator detection, thereby sacrificing feeding efficiency.

A similar dynamic motivational influence on the tradeoff between risk of predation and food intake was found in coho salmon (*Oncorhynchus kisutch*) (Dill and Fraser 1984). With increasing satiation the fishes' attack distance was progressively reduced, but only when predation risk was high. Thus, hungry coho did not appear to weight risk as heavily as they did when more satiated.

Parasitisation and the Risk of Predation

We have seen that when fish have to balance two conflicting demands, i.e. to feed efficiently and to avoid a predator of their own, they arrive at a compromise which is influenced by their state of hunger and by the actual risk of predation. Hunger can be increased either by starvation or by enhanced needs, e.g. because parasites use up a great proportion of the energy the food contains. Three-spined sticklebacks are frequently parasitised by the cestode *Schistocephalus solidus* which can exceed the weight of its host (Arme and Owen 1967). Parasitised sticklebacks are hungrier than uninfested ones, not only because they have to nourish their parasites and swimming is energetically more demanding for them (Lester

1971), but also because they are poor competitors and have therefore to feed upon less-profitable prey (Milinski 1984b).

We would expect that parasitised fish give less weight to predator avoidance in order to find enough food. This was recently investigated by Giles (1983). Three-spined sticklebacks infested by *S. solidus* and unparasitised ones were allowed to settle singly in a tank before they were frightened by suddenly lowering a model heron's head over the test tank, so that the tip of the bill hit the water surface. The fish either stopped swimming immediately or jumped away from the model in alarm to remain still somewhere else. The mean recovery time, however, was twice as long for unparasitised fish as for parasitised ones. Moreover, the recovery time was significantly negatively correlated with parasitic load. Since feeding occurred mostly after the recovery period, the parasitised sticklebacks resumed feeding much earlier than the healthy ones.

We would predict that heavily parasitised sticklebacks should approach a predator less cautiously than healthy ones in order to find food. When the experiment shown in Figure 9.2 was repeated with sticklebacks infested by *S. solidus* (Milinski 1985), there was no obvious influence of the cichlid's presence on the sticklebacks' foraging behaviour (Figure 9.9). They fed as quickly and as close to the place of the cichlid as with no predator present. This risky behaviour was especially profitable concerning energy gain when a parasitised stickleback was competing for food with a healthy one in the presence of the predator. The parasitised fish clearly outcompeted the healthy fish by consuming three times as many worms as the latter dared to catch.

Figure 9.9: Mean Number of *Tubifex* Worms on which Parasitised Sticklebacks Fed at Different Distances to a Fish Predator; Each Fish was Allowed to Consume Eight Worms. (After Milinski 1985.)

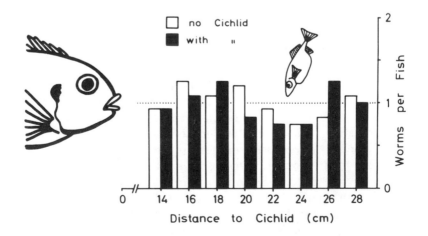

It has not been investigated whether parasitised sticklebacks are preyed upon more frequently than unparasitised ones. Van Dobben (1952), however, was able to show that roach (*Rutilus rutilus*) infested by the cestode *Ligula intestinalis* were caught by cormorants (*Phalacrocorax carbo*) at a much higher frequency than in the population. It could not be determined whether parasitised fish were preferentially preyed upon because of their conspicuousness, their lowered ability to flee or because they took a higher risk while feeding.

Three-spined sticklebacks infested by *S. solidus*, and dace (*Leuciscus leuciscus*) parasitised by the eyefluke *Diplostomum spathaceum*, swim preferentially close to the water surface where they are highly susceptible to bird predators (Arme and Owen 1967; Crowden and Broom 1980). Infested sticklebacks are supposed to do this because of their enhanced demand for oxygen (Lester 1971).

Parasites influence their teleost hosts in various ways that increase the fishes' risk of being preyed upon. The higher need for food forces the infested fish to give less weight to avoidance of predation than to feeding, although they are slower when fleeing.

Conclusions

Natural selection favours genetically determined programs allowing individuals to behave in any situation in such a way that they maximise their lifetime reproductive success. Whether animal behaviour approximates this expectation is difficult to investigate. Normally, there should be more than one demand at a time; under natural conditions an animal is seldom completely satiated and absolutely safe from predators and competitors. If these demands are conflicting, i.e. fulfilling one of them maximally can only be achieved at the expense of fulfilling the others, there should be a behavioural compromise optimising all costs and benefits in a way which maximises fitness.

For the most efficient avoidance of being preyed upon, a teleost should be continuously vigilant, stay far away from places where predators are supposed to be, hide and avoid any conspicuous movement. To maximise its rate of food intake the fish should pay attention only to finding food, forage where food is most plentiful — as also may be the predators, and move while searching and handling food. Obviously these demands are always conflicting, the degree depending on the specific situation. A number of investigations have shown that fish do balance the conflicting demands in an adaptive way. We have seen that fish avoid feeding close to predators, prefer less risky alternative patches, especially those providing not much less food, feed also close to a predator when vigilance is enhanced by shoaling, accept less-profitable alternative patches and/or types of food when the risk is increased, but give higher priority to feeding when energetic needs are increased either by starvation or parasitic load.

The dynamic influence of the decreasing need for energy with satiation on the behavioural compromise can be shown also by the widespread phenomenon that feeding rate decreases with satiation (McCleery 1978; Heller and Milinski 1979; Krebs 1980). Such satiation curves can be explained by the increasing proportion of time spent vigilant while deficit costs decrease. Similarly, with satiation, animals increasingly concentrate on the most profitable types of food, i.e. those that cost least handling time per calorie ingested (see references in Heller and Milinski 1979). This altered choice possibly saves time for vigilance, although the net rate of energy intake is not maximised to any further extent.

Great progress would be made if we could measure all the costs and benefits a situation provides for an individual and then predict the best compromise quantitatively. One difficulty is that we must sum up different short-term currencies of fitness; e.g. how much energy gained by feeding compensates for a risk of predation increased by 20 per cent? (see Chapter 7 by Bleckmann, this volume). A promising attempt has been made to solve this problem theoretically (Gilliam 1982).

If we can determine costs and benefits of a specific situation, the optimal behaviour predicted could differ from the one observed because the animal is either not willing or not able to use all of the information and instead adopts a simple 'rule of thumb' (e.g. Janetos and Cole 1981; Krebs, Stephens and Sutherland 1983). This can be the optimal solution because constraints and costs of sampling have to be considered for calculating the optimum.

Sometimes animals use obviously more information than we can detect, as suggested by the subtlety with which they recognise tiny changes of costs and benefits. For example, the direction in which the fish predator is facing determines how close the stickleback approaches in the experiment shown in Figure 9.2. In spite of this, it may well be that the adaptive flexibility and subtlety of the foragers' response to small changes in costs and benefits may prevent our ever being able to predict exactly how fish will behave when faced with a conflict between the need to get food and the need to avoid predators.

Summary

Often fish must avoid being preyed upon when they are feeding. When both demands are conflicting (i.e. maximising food intake is achieved only at the expense of efficiently avoiding predation and vice versa), it has been shown experimentally that fish make a compromise by fulfilling both needs less efficiently. They take a greater risk in order to feed more efficiently when the need for food is increased by starvation or by parasites, or when feeding is much more rewarding in places with predators. Smaller fish which have an increased risk of predation accept less rewarding food in order to avoid the predator than do bigger fish. Diet selection within a patch can be altered if one type of food is riskier to feed upon than others. By foraging in large shoals, fish can detect and monitor an approaching predator more easily and can continue feeding for longer than fish in smaller shoals.

The examples discussed provide qualitative evidence for teleosts changing their behaviour adaptively when costs and benefits of feeding and avoidance of predation vary. Because of the difficulty in measuring all costs and benefits and knowing how accurately the fish may do this sum, it is difficult to predict the compromise behaviour quantitatively.

References

Arme, C. and Owen, R.W. (1967) 'Infections of the Three-spined Stickleback, *Gasterosteus aculeatus* L., with the Plerocercoid larvae of *Schistocephalus solidus* (Müller 1776), with Special Reference to Pathological Effects', *Parasitology, 57*, 301-14

Cerri, R.D. and Fraser, D.F. (1983) 'Predation and Risk in Foraging Minnows: Balancing Conflicting Demands', *American Naturalist, 121*, 552-61

Crowden, A.E. and Broom, D.M. (1980) 'Effects of the Eyefluke, *Diplostomum spathaceum*, on the Behaviour of Dace (*Leuciscus leuciscus*)', *Animal Behaviour, 28*, 287-94

Dill, L.M. (1983) 'Adaptive Flexibility in the Foraging Behaviour of Fishes', *Canadian Journal of Fishery and Aquatic Sciences, 40*, 398-408

Dill, L.M. and Fraser, A.H.G. (1984) 'Risk of Predation and the Feeding Behaviour of Juvenile Coho Salmon (*Oncorhynchus kisutch*)', *Behavioural Ecology and Sociobiology, 16*, 65-72

Giles, N. (1983) 'Behavioural Effects of the Parasite *Schistocephalus solidus* (Cestoda) on an Intermediate Host, the Three-spined Stickleback, *Gasterosteus aculeatus* L', *Animal Behaviour, 31*, 1192-4

Gilliam, J.F. (1982) 'Habitat Use and Competitive Bottlenecks in Size-structured Fish Populations', unpublished PhD dissertation, Michigan State University, East Lansing

Heller, R. and Milinski, M. (1979) 'Optimal Foraging of Sticklebacks on Swarming Prey', *Animal Behaviour, 27*, 1127-41

Janetos, A.C. and Cole, B.J. (1981) 'Imperfectly Optimal Animals', *Behavioural Ecology and Sociobiology, 9*, 203-9

Janssen, J. (1976) 'Feeding Modes and Prey Size Selection in the Alewife (*Alosa pseudoharengus*)', *Journal of the Fisheries Research Board of Canada, 33*, 1972-5

Kniprath, E. (1965) 'Stichlinge als Nahrung des Eisvogels und des Teichhuhns', *Ornithologischer Beobachter, 62*, 190-2

Krebs, J.R. (1980) 'Optimal Foraging, Predation Risk and Territory Defence', *Ardea, 68*, 83-90

Krebs, J.R., Stephens, D.W. and Sutherland, W.J. (1983) 'Perspectives in Optimal Foraging', *Perspectives in Ornithology*, Essays Presented for the Centennial of the American Ornithologists' Union, Cambridge University Press, Cambridge, pp. 165-221

Lazarus, J. (1979) 'The Early Warning Function of Flocking in Birds: an Experimental Study with Captive *Quelea*', *Animal Behaviour, 27*, 855-65

Lester, R.J.G. (1971) 'The Influence of *Schistocephalus* Plerocercoids on the Respiration of *Gasterosteus* and a Possible Resulting Effect on the Behavior of the Fish', *Canadian Journal of Zoology, 49*, 361-6

Li, S.K. and Owings, D.H. (1978) 'Sexual Selection in the Three-spined Stickleback. I. Normative Observations', *Zeitschrift für Tierpsychologie, 46*, 359-71

McCleery, R.H. (1978) 'Optimal Behaviour Sequences and Decision Making', in J.R. Krebs and N.B. Davies (eds), *Behavioural Ecology: an Evolutionary Approach*, Blackwell, Oxford, pp. 337-410

McNaught, D.C. and Hasler, A.D. (1961) 'Surface Schooling and Feeding Behavior in the White Bass, *Roccus chrysops* (Rafinesque), in Lake Mendota', *Limnology and Oceanography, 6*, 53-60

Magurran, A.E., Oulton, W.J. and Pitcher, T.J. (1985) 'Vigilant Behaviour and Shoal Size in Minnows', *Zeitschrift für Tierpsychologie, 67*, 167-78

Milinski, M. (1977a) 'Experiments on the Selection by Predators against Spatial Oddity of their Prey', *Zeitschrift für Tierpsychologie, 43*, 311-25

Milinski, M. (1977b) 'Do All Members of a Swarm Suffer the Same Predation?', *Zeitschrift für Tierpsychologie, 45*, 373-88

Milinski, M. (1984a) 'A Predator's Costs of Overcoming the Confusion Effect of Swarming Prey', *Animal Behaviour, 32*, 1157-62

Milinski, M. (1984b) 'Parasites Determine a Predator's Optimal Feeding Strategy', *Behavioural Ecology and Sociobiology, 15*, 35-7

Milinski, M. (1985) 'Risk of Predation Taken by Parasitised Sticklebacks under Competition for Food', *Behaviour, 93*, 203-16

Milinski, M. and Heller, R. (1978) 'Influence of a Predator on the Optimal Foraging Behaviour of Sticklebacks (*Gasterosteus aculeatus* L.)', *Nature, London, 275*, 642-4

Ohguchi, O. (1981) 'Prey Density and Selection against Oddity by Three-spined Sticklebacks', Supplement No. 23 to *Zeitschrift für Tierpsychologie*

Powell, G.V.N. (1974) 'Experimental Analysis of the Social Value of Flocking by Starlings (*Sturnus vulgaris*) in Relation to Predation and Foraging', *Animal Behaviour, 22*, 501-5

Power, M.E. (1984) 'Depth Distributions of Armored Catfish: Predator-induced Resource Avoidance', *Ecology, 65*, 523-8

Van Dobben, W.H. (1952) 'The Food of the Cormorants in The Netherlands', *Ardea, 40*, 1-63

Werner, E.E., Gilliam, J.F., Hall, D.J., and Mittelbach, G.G. (1983) 'An Experimental Test of the Effects of Predation Risk on Habitat Use in Fish', *Ecology, 64*, 1540-8

Wootton, R.J. (1976) *The Biology of the Sticklebacks*, Academic Press, London

10 TELEOST MATING SYSTEMS AND STRATEGIES

George Turner

The mating system of an animal is defined by the number of members of the opposite sex with which an individual mates. This will be intimately related to the extent and type of parental care because care-giving may interfere with further opportunities for mating. Since these two factors cannot entirely be divorced, I will refer to the combination of these systems as the breeding system of the animal.

The teleost fishes are the largest extant vertebrate group, and so it is scarcely surprising that they show a great range in their systems for mating and parental care. Many species appear to mate indiscriminately, and others require a protracted period of courtship before choosing a partner. In some species a male and female may remain together for years; in others males may have many mates; and often females mate with many males. Some species may be self-fertilising; others are certainly simultaneous hermaphrodites, producing eggs and sperm at the same time, and releasing either when spawning. In many species sex change occurs: males may change into females (protandrous hermaphrodites) or females into males (protogynous hermaphrodites). In some species there are no males at all.

Bearing in mind that the problems inherent in observing fast-moving aquatic species have meant that fewer detailed field studies have been carried out on fishes than on terrestrial vertebrates, the range of breeding systems of the teleosts can be analysed in the light of current evolutionary biology. This chapter will review the types of mating strategy found in teleost fishes, and discuss theories that attempt to explain why species have developed particular systems. In addition, several examples will be considered in more detail, and some unusual and aberrent strategies will be discussed. Before this, it is necessary to discuss theories of mate choice, since selection by one sex is likely to be an important force in shaping the mating strategies of individuals of the opposite sex. A section will also be devoted to the evolution of breeding systems in teleosts, which will demonstrate the interrelatedness of mating system and parental care (see Chapter 11 by Sargent and Gross, this volume). The importance of individual differences in mating strategies within a species will be emphasised; the topic of individual differences is dealt with in more depth elsewhere in this book (see Chapter 13 by Magurran, this volume).

Mate Choice and Sexual Selection

Suppose a female's entire reproductive output for a year is to be entrusted to the care of a single male. It is likely that it would be beneficial to females to be very discriminating in the choice of their mate. In this and other ways it is inevitable that sexual selection will influence the mating strategies of fish.

In most cases there will be a markedly unequal pressure on the sexes to exercise care in the selection of mates. There is a difference between males and females in regard to the cost of gamete production: eggs are more expensive to make (or at least take longer to replace) than sperm. This means that, all other things being equal, females can best increase their fitness by being choosy about their mates, and males should concentrate on increasing the number of mates which they have. Obviously this is a generalisation, not an inviolable rule: where males care for the offspring and where this interferes with further mating, it is to be expected that males will also be selective. These criteria are often met in teleost fishes and male selectivity has been demonstrated in several species (*Cottus bairdi*: Downhower and Brown 1981; *Cyprinodon macularis*: Loiselle 1982). In species where a long-lasting pair bond is essential for the survival of the offspring, it is necessary for each partner to be selective about the other: in the event of failure of a brood at a late stage, a great deal of time and energy will be necessary to raise another brood to the same stage. Larger individuals will be better able to defend their young, and so the requirement of each sex to have the largest possible partner may lead to assortative mating, with males and females of approximately the same size pairing off. Where there is a division of labour between the sexes, the need to choose a larger partner will be most strongly felt by the sex which is not required to perform the major role in brood defence. If this is the case, a typical male/female size ratio of pairs may develop (Barlow and Green 1970).

Mate choice in females has been more extensively studied. It has been suggested (as reviewed by Halliday 1983) that females could choose males on the basis of two distinct classes of beneficial characteristic. These are immediate or material qualities, which may be contrasted with genetical qualities ('good genes'), which, if they have any advantage at all, ensure that the female's offspring are fitter types. Into the former category would come characteristics such as male size, aggressiveness or nest-site quality, which would increase brood or female survival. There is no doubting the selective benefits for females which make a choice on this basis.

Male dragonets (*Callionymus lyra*) have spectacular colours and greatly extended fins. Male lionhead cichlids (*Steatocranus casuarius*) have an enormous fatty lump on the forehead. The male Mexican swordtail (*Xiphophorus helleri*) has the lower part of its tail prolonged into a 'sword' (Figure 10.1). In bristle-nosed catfish (*Ancistrus* sp.), males have branched fleshy tentacles on their heads. None of these features seems to be related to the immediate survival of the individual, to the mechanics of reproduction, or to brood care. Nor do they appear to have any value as weapons for use in combat between males. They are much more developed than would be required to distinguish the sexes, they all appear to be expensive to make, and in some cases increase the fish's vulnerability to predators. Similar characteristics are found in other animal groups, the most celebrated being the peacock's tail. To explain these 'epigamic' features, the theory of sexual selection by female choice was proposed by Darwin (1871) and developed by Fisher (1958). It seems reasonable that males should be chosen for characteristics that increase the survival or competitiveness of their offspring, but in order for female choice to produce the exaggerated male characteristics, females must choose males on the basis of attributes whose only advantage is that other females also choose on this basis. In other words, it will benefit a female to mate with males which will produce attractive sons (often called Fisher's 'sexy son' hypothesis). It is possible that such a preference could arise randomly; perhaps attractiveness might initially mean merely conspicuousness to females. Once started, Fisher suggested that such a process would be self-reinforcing and lead to a runaway increase in the expression of the character until it became such a disadvantage that increased predation, or excessive cost of making the character, balanced the benefit from female choice. In these terms, being attractive to a female is an adaptive character for a male, so that the dichotomy between 'natural' and 'sexual' selection is an unhelpful one (See Halliday 1983), and it may be more helpful to regard sexual selection as a category of natural selection [a definition based on set theory rather than exclusive categories: Ed.]

Simulations have been developed to test the compatibility of the 'runaway' prediction with the workings of natural selection as they are currently understood. These have produced differing results, either supporting (Lande 1980), or rejecting (O'Donald 1983) Fisher's theory. In my opinion, there are no other convincing explanations for the evolution of peacocks' tails (see Krebs and Davies 1981). The theory is made more plausible if you allow for the possibilities of constraints and limitations on mechanisms of female choice. For example, many birds prefer to incubate dummy eggs which are vastly larger than their own; a similar supernormal stimulus effect could be built into the design of some mate-choice mechanisms. It should also be noted that in some fish species, the female is more brightly coloured than the male, as in the cichlids *Cichlasoma nigrofasciatum, Crenicichla* sp., and *Chromidotilapia guentheri.* In some species the colours are short-lasting and may indicate willingness to mate, and in other cases it may be that the female is the sexually selected sex.

The Evolution of Breeding Systems in the Teleosts

It is likely that the ancestral method of reproduction in bony fishes is a promiscuous mating system with no courtship or mate choice and no

Figure 10.1: Some Characteristics of Male Teleosts Probably Resulting from Sexual Selection. (A) Bright colours and exaggerated finnage in the dragonet, *Callionymus lyra*: (B) the extended 'sword' on the tail of the male Mexican swordtail, *Xiphophorus helleri*; (C) the fatty lump on the forehead of the lionhead cichlid, *Steatocranus casuarius*. In all cases the male is in the foreground, with the female behind

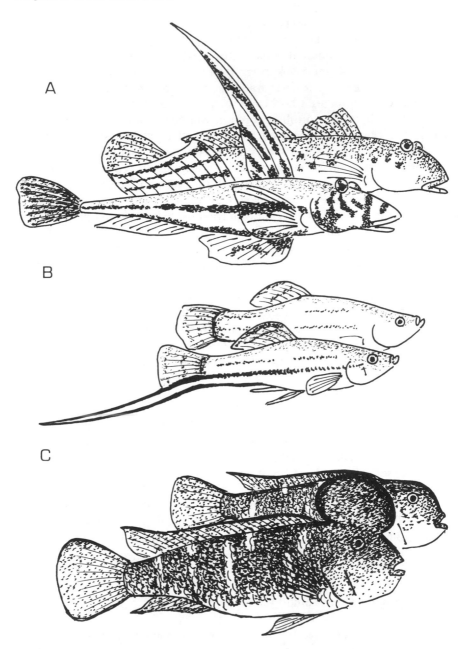

parental care (Balon 1981). For species that do not spawn in large groups some courtship is necessary: a fish needs to mate with an individual of the same species and a different sex. In some habitats, survival of the young can be enhanced by the provision of increased resources to each of a smaller number of offspring: in its simplest form this can entail the production of large yolky eggs. Often it is beneficial to spend time choosing a suitable site for the eggs: to conceal them from predators, or to protect them from the physical environment. Thus salmon and most killifish bury their eggs; many cyprinids lay them in dense mats of weeds; darters choose sheltered rocks. In such circumstances, it is possible that suitable sites will be in short supply, the guarding of such a site prior to spawning will help ensure that a fish has a spawning site when it finds a mate. Clearly if a male is able to control a prime site, he may be able to mate with a succession of females (Barlow 1962) and so male competition and polygamy will develop. This stage of development is shown by most gobies, blennies and sculpins. In this event it may involve little extra cost for the male to guard and clean the eggs in his territory, and so males may be preadapted for brood care. To increase brood survival further, males of many species are specially adapted behaviourally or morphologically: sunfish dig pits for their eggs; sticklebacks build nests of plant fragments; Siamese fighting fish contruct floating rafts of mucus bubbles. Splashing tetras lay their eggs out of water on overhanging leaves, the male having an extended tail which it uses to splash water on to the developing brood. Seahorses brood the young in a pouch, and many cichlids and catfish are mouth-brooders. In the majority of teleosts where parental care is provided, it is provided by the male.

In circumstances where the fry are particularly vulnerable to predation, two guarding parents may be so much better than one that it would be beneficial to the female also to participate in caring. This situation is found in the angelfish, some other cichlids, and also in the shore-living butterfish. Division of labour may become unequal, and here the larger parent would be best employed guarding the territory and the smaller parent keeping the eggs free of dirt and fungus. Sometimes this division of labour will lead to the larger parent having the opportunity to expand its territory and control several members of the opposite sex, perhaps eventually becoming free of parental duties. In this way female care could develop. Thus, in teleosts, the probable direction of evolution was: no care; male care; biparental care; female care (Baylis 1981). There will always be the possibility that circumstances will favour a reversal in direction.

This theory, then, is that males will be preadapted for parental care through the advantages of polygamy and territoriality. There are alternative explanations for this high incidence of male care; for example, it has been pointed out that in externally fertilising species, the female has the opportunity to desert before the eggs are fertilised, whereas in internally fertilising species, the male can obviously get away first (Trivers 1972). It would then be necessary for the remaining parent to guard, otherwise brood survival will be low.

Monogamy

If each individual mates with only one member of the opposite sex, then that system is said to be monogamous. However, this term is also applied to a social system in which a long-lasting association between individuals exists. This is probably easier to observe in the field, but monogamous matings may occur between individuals which do not remain together outside the breeding season. For a discussion of these problems see Wickler and Seibt (1983). For the purposes of this chapter monogamous mating can be considered to be a state where reproduction results from the exclusive mating of the same two individuals, either once or repeatedly. Where the male and female remain together to co-operate in the defence of their young or of a common territory, this will be called monogamous pairing.

Monogamy is rare in fishes: it occurs in many of the cichlids (Fryer and Iles 1972; Barlow 1974); in some catfish (Blumer 1983); and in some chaetodontids (butterflyfish: Reese 1975) and pomacentrid species (damselfish: Fricke and Fricke 1977). In the chaetodontids, pairs may stay together for several years (Fricke 1973: cited by Keenleyside 1979), feeding and spawning in a large home-range. The eggs are planktonic and not cared for by the parents. Long-term pair bonds are also established in the reef-dwelling *Serranus tigrinus*, which also show no parental care. Such social organisation is similar to that of the monogamous chaetodontids, but with one notable difference: this species is a simultaneous hermaphrodite. As with many reef fish laying pelagic eggs, spawning takes place almost every evening; in this case one individual produces eggs and exhibits a barred pattern, the other produces sperm and lacks the dark stripes. After spawning they exchange roles (and colour patterns) and spawn again. In these two groups parental care can be ruled out as an explanation for pair-bonding. In the case of the serranid, it appears that co-operative foraging is important (Pressley 1981), and co-operative defence of a feeding territory or shelter may occur in some species. *Serranus tigrinus* is not strictly a monogamously mating species (although monogamously pairing), as they often mate with individuals from neighbouring territories.

In the monogamous cichlids, spawning is typically preceded by extensive courtship, and pairs often stay together for several days before choosing a nesting site (Meral 1973). Eggs are laid on a solid substrate and guarded, as are the fry, by both parents (Baerends and Baerends-van Roon 1950; Meral 1973; McKaye 1977). Predators are often abundant in areas where cichlids raise their broods, and thus provision of care is probably the most important criterion in the maintenance of a monogamous system in these species. There is frequently a tendency towards a division of labour in the parental roles of these fishes. Whereas some species (such as *Etroplus maculatus*: Cole and Ward 1969; *Pterophyllum scalare*: Chien and Salmon 1972) show little differentiation in male and female roles, in many cichlids the female provides most of the intimate care of the offspring, and the male remains at the periphery of the territory (e.g. *Cichlasoma* spp: Barlow 1974, *Lamprologus brichardi*: Limberger 1983). There is no evidence that males are any more aggressive than females in the defence of the young; rather, the marked size dimorphism renders perimeter defence more appropriate and probably makes cleaning of the eggs, etc. more difficult. It has also been suggested (Rechten 1980) that the larger partner is able to adopt the least energetically expensive or risky parental tasks and is able to feed more often when employed in perimeter defence. Meral (1973) has observed that males are more inclined to leave the fry unguarded than females, and when alarmed they are more likely to seek shelter (see Rechten 1980). This lessening of parental duties in males also manifests itself in a tendency to desert the brood in the later stages of care, if there is an opportunity to spawn with another female (Meral 1973; Keenleyside 1983). Interestingly, in the catfish *Ictalurus nebulosus*, the situation is reversed; the male looks after the brood at close range, and the female defends the perimeter and often leaves the male alone for long periods (Blumer 1983). In the cichlids, it is easy to see how this occasional polygamy and division of labour could result in evolution towards a polygynous system (see below) with female care.

Polygamy

A polygamous mating system is found in a great many teleost species. This may take the form of polygyny, where one male fertilises the ova of several females; polyandry, where one female mates with several males; or 'promiscuity', where both sexes typically have several partners. (Some workers consider the latter term to have unfortunate connotations.) Again there are problems of definition: there may be alternative mating strategies within a species, for example some males adopt an aggressive displaying strategy and attempt to gain exclusive access to a number of females, and others (sneakers) try to fertilise some of the eggs of females mating with these dominant individuals. If the females are actively choosing the displaying males, then the females and aggressive males are acting according to a polygynous system, whereas the sneaking males are behaving in a promiscuous manner: what, then, do you call the system? There is also a considerable difference between a system where males compete for and may mate with a number of females, while females choose and mate with a number of males (confusingly called meshed polygyny-polyandry by Brown 1975), and a system where no apparent choice occurs. Both can be categorised as promiscuous systems, so this category tends to become a dustbin for systems that do not fit other categories. The confusion could be reduced by considering individual mating strategies within a species, rather than trying to construct a complex classification of systems.

Resource-based Polygamy (the Verner-Orians Model)

Several models have been proposed to investigate the selective forces that may influence the adoption or maintenance of polygamous mating systems. The Verner-Orians model deals with the evolution of resource-based polygyny (Orians 1969, based on Verner and Willson's (1966) work on birds).

The model is designed for territorial species and assumes that males control resources necessary to the females, for example in the rearing of their young. Different males will hold resources of varying quality and are assumed to be indiscriminate in the choice of mates, which may not be the case if sperm production is costly, as in the lemon tetra (Nakatsuru and Kramer 1982). Males will always benefit from polygyny, which will not be true if male parental care is necessary for the successful rearing of the progeny, and females will always choose their mates with care, no cost of choosiness being envisaged (see Parker 1983 for a discussion). The females will initially choose the males that possess resources superior at the time when the female settles in a territory, but as these males become paired the females will be left with the alternatives of choosing males holding inferior resources or opting to become the second female in the territory of a superior male. It is assumed that this will lead to a reduced resource availability to the newcomer and perhaps to the resident female also (neither need always be the case). It is predicted that females will switch from monogamy to polygamy when the fitness of the second female in a good territory exceeds that of the first female in a poor-quality territory (Figure 10.2).

Vehrencamp and Bradbury (1984) have reviewed attempts to test the model and concluded that it has not been rigorously applied (although Plezczynska and Hansell 1980 may be an exception). The model is constructed from the standpoint of a choice between monogamy and polygyny: this is likely to be the case for bird species; less so for the teleosts. The model does not take into account the aggressive behaviour of the females concerned; this could easily prevent polygamy. Meral (1973) reports that females of the monogamous convict cichlids (*Cichlasoma nigrofasciatum*) are particularly aggressive to other females prior to spawning; this could evolve to prevent males choosing to pair monogamously with another female, but would also inhibit polygamy.

Species where the male alone cares for the offspring may be suitable for analysis in the light of this model. In one such species, the mottled sculpin, *Cottus bairdi*, it has been shown that larger males are more capable of protecting the eggs and that females prefer to spawn with such fish

Figure 10.2: the Verner-Orians Polygyny Threshold Model. the fitness of a female joining a male on his territory is plotted against the amount of resource available at the time of settlement. The size of the male symbol is proportional to the amount of resource which that male controls. Different males are indicated by the numbers. Animals in a monogamous pair are indicated by open symbols, those in a harem by closed symbols. Polygyny will occur when the resources available to a female that becomes the second to arrive on the territory of the best male exceed those available to it if it becomes the first on the territory of the best unmated male

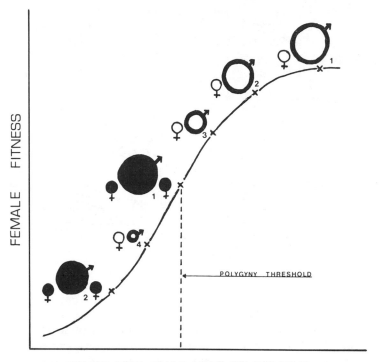

(Downhower, Brown, Pederson and Staples 1983). There is great variation in the success of the males, as assumed by the model. Smaller males are chosen more as the season progresses, as might be expected if sharing proved costly, but a closer scrutiny reveals that smaller females generally spawn later than the larger ones. Smaller females avoid large males because of the danger of being eaten: sculpins are notorious cannibals — a 40 mm size difference is likely to be fatal! (Figure 10.3). Details like this cannot be anticipated in general models which are formulated from consideration of a few selective factors.

A General Theory of Polygamy: The Emlen-Oring Model

Another more general model for the adoption of polygamy was developed

Figure 10.3: Part of the Courtship behaviour of *Cottus* sp. (After Morris 1956.) The displaying male often envelops the head of the female; for a large hungry male which is already guarding several batches of eggs (which the female may try to eat), courtship may turn into cannibalism. This is probably the reason why female *Cottus bairdi* do not like to mate with males much larger than themselves

by Emlen and Oring (1977), again specifically formulated with avian systems in mind. This rests on more general assumptions, and considers the polygamy potential: the opportunity for members of one sex to control, directly or through control of an essential resource, a group of members of the opposite sex (Figure 10.4). This potential cannot be realised if polygamy detracts from parental care to such an extent that monogamy would yield more offspring from an individual of the potentially polygamous sex. Unlike the Verner-Orians model, in this case the occurrence or otherwise of polygamy is assumed to be as a result of benefit or disadvantage to the multiply mating sex. Opportunities for polygamy will arise when a resource or group of individuals of the opposite sex is clumped in distribution in such a way that it is economically defensible. This might be the case in the sculpins if good-quality nest sites were in short supply and females were compelled to mate with the males that controlled these, but it appears that this is not the case (Downhower *et al.* 1983).

In the Lake Tanganyika cichlid *Lamprologus brichardi*, males are occasionally seen to mate polygamously, although monogamy is more usual. From Limberger's (1983) laboratory studies it appears that the only

Figure 10.4: The Emlen-Oring Theory of Polygamy, Applied to Resource-based Polyandry. Females are represented by dark-coloured fish, males by light-coloured ones. The resources they defend are sea anemones. When resources are scattered (top), then males are scattered and females are unable to control more than one (functional) male, and monogamy results. When resources are clumped (bottom), males are clumped in distribution, providing females with the opportunity to control a harem. This is a simplified account: see text for more details

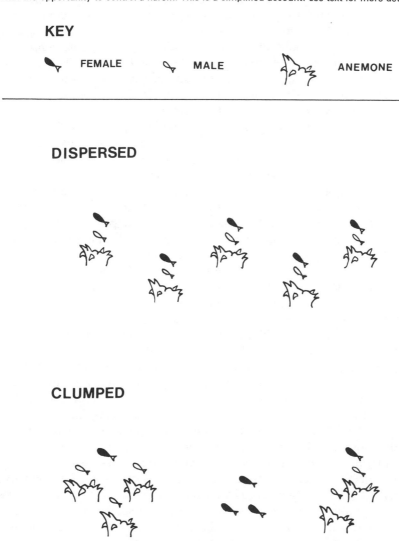

factor influencing polygyny is the opportunity for males of sufficient competitive ability to monopolise more than one female. There is no apparent cost to harem formation, as the male contributes very little care to the offspring, and the tendency of the offspring of previous broods to assist in the care of their siblings (Taborsky and Limberger 1981) is likely to reduce further the importance of the male's contribution. Thus a male can control two separate territories, each containing a female and the young of several broods, and the reduction in his contribution to territorial defence will have minimal effect on juvenile survival. There appears to be no increase in female activity nor reduction in growth rate as a result of sharing a male. The rarity of polygyny in the field would seem to be caused by the difficulty of maintaining control of two females in the face of intense competition with other males. In terms of the Emlen-Oring model, it appears that there is rarely an adequate polygamy potential, although there are no conflicting pressures to oppose the tendency when the potential exists. The Verner-Orians model is not appropriate in this case because there is no evidence that males initially hold the territories; rather, females choose the territory and the males then compete to control them.

In the cleaner wrasse (*Labroides dimidiatus*), sociality is perhaps encouraged by the shortage of the appropriate cleaning stations. Whatever the explanation, this species lives in groups of 5 to 20 and a single dominant male spawns with several females. The eggs are pelagic and no parental care is provided. *Labroides dimidiatus* is a hermaphrodite species: individuals begin reproductive life as females, but can transform into males upon removal of the dominant male from the group. It is typically the largest and most dominant female which does this, starting to behave and look like a male within 30 minutes of the male's removal and becoming fully functional in 2 weeks (Robertson 1972). Despite the aggressive behaviour of the females, their territories are small enough to permit them to be accommodated within the range of a single male. Here the potential for polygyny is high, and there is no conflicting requirement for parental care. Male competition will be reduced by hermaphroditism, the subordinates all being reproductively active females.

It would also appear that there is considerable potential for polygamy in the anemone fishes, *Amphiprion* spp., where a number of individuals shelter within and defend a common territory based around a patchily distributed resource, namely a suitable sea anemone. Again, sex reversal can occur, but in this case it is a protandrous one (males change to females) and the larger females are socially dominant to the males (Fricke 1974). Polyandry does not appear to occur very often since, within a group sharing an anemone, the dominant male suppresses the reproductive activity of the subordinate males by means of aggression. However, in cases where several anemones are grouped together, a single female is able to control more than one anemone, and the males are not, each suitable host containing a separate group of males and juveniles. The female then lays eggs with each dominant male in turn (Moyer and Sawyers 1973).

Thus, although the potential for polyandry generally exists in this species, as one female is certainly able to monopolise access to several males, it is prevented by the behaviour of the males, except in the situation of clumped anemones. It is not clear what benefits the female derives from polyandry, and it would appear that polygyny would be more favourable to the species as a whole, as one dominant male breeding with several females would produce many more eggs than would one female mating with one or more males. However, selection at the level of the species is unlikely to be important when compared with selection at the level of the individual, and a mating system once evolved may have an inherent stability which is difficult to break. If living and reproducing in groups is advantageous, the evolution of sex reversal in either direction would initially be favoured, as then any group of two fishes would be able to reproduce without the need to leave the protection of the anemone. If protandry arose first, then a switch to protogyny might be difficult. If small protogynous females were to develop, they would be prevented from breeding by larger females, and if they became males when large, they would still only be able to fertilise the eggs of one female as the females are too aggressive to be amenable to living in a harem.

When trying to explain what has caused the evolution of an animal's social system, the idea of adaptiveness and the correlation with ecological parameters is often not enough; a social strategy may be stable once evolved, even though another strategy might be better adapted to the animal's environment. The initial system used by the population may determine which of several possible systems will be the one to evolve, simply because the best tactic to adopt depends on what everyone else is doing. This is the type of thinking characteristic of the game-theory approach, and a strategy which is stable once evolved is known as an evolutionarily stable strategy, of ESS. For a more detailed explanation of this approach, see Maynard Smith (1982) and Parker (1984).

The Complex Mating System of the Blue-headed Wrasse

A protogynous system with dominant polygynous males, sex reversal and no parental care is not confined to *L. dimidiatus*, but is found in a great many species of wrasse and parrotfish (Labridae and Scaridae). Within some of these species there is the further complication that males may adopt different reproductive strategies (Warner and Hoffman 1980; Warner 1982). In the blue-headed wrasse, *Thalassoma bifasciatum*, there are two distinct types of male: primary males are small and look identical to females; secondary males are large and totally different in appearance. Secondary males result from the metamorphosis of primary males or of females. A few large females may adopt secondary male coloration but remain functional females: a similar phenomenon occurs in the minnow, *Phoxinus*, a non-sex-changing species (Pitcher 1971). There is no known functional explanation in either species. In the wrasse, the secondary males actively display towards females and aggressively defend a territory in which they spawn. The primary males do not indulge in display or territorial defence but attempt to sneak in and fertilise some of the eggs laid by a spawning pair. This sneaking behaviour, often with males mimicking the behaviour and appearance of females, is common in fishes (see Magurran, Chap 13, this volume). However, when there are large numbers of these primary males, they can adopt more successful tactics; they band together and attempt group spawnings with females. They also harass the primary males by invading their territories *en masse*, not only to participate in the spawnings but often also to chase or entice the female away before spawning commences. As the population size increases, the secondary males are forced away from the favoured spawning areas on the reef by this interference and are proportionately less successful than in smaller populations, simply because the invading groups are larger (Figure 10.5). This is shown in comparisons between populations of *Thalassoma bifasciatum* and also by looking at *T. lucanascum*, which although capable of sex change and of producing secondaries (Warner 1982) rarely does so. This species exists at much higher population densities than *T. bifasciatum*. In *T. lucanascum* the ability to change sex is probably selectively neutral when it exists in large populations, and advantageous in small populations. See Charnov (1982) for a fuller discussion of these species.

Promiscuity

A system in which each male mates with several females and each female with several males is generally described as a 'promiscuous' mating system (see above). Strictly speaking, the mating system of the blue-headed wrasse would have to come into this category. So would that of the herring (*Clupea harengus*), a species in which apparently no mate choice or competition occurs. Enormous numbers of these fish congregate at regular breeding grounds and spawn simultaneously, and apparently without courtship. Observers of these spawnings (e.g. Aneer *et al.* 1983) have been unable to detect any behavioural differences between the sexes. Spawning appears to be stimulated by pheromones in the milt of the males (Hourston, Rosenthal and Stacey 1977). Selectivity and courtship impose an energetic cost as well as being time-consuming and possibly increasing vulnerability to predators. The systems of the wrasse and herring are considered promiscuous, although the processes of male competition and female choice will be strikingly different in each case.

Leks

A lek is an aggregation of males, each of which defends a territory and courts any arriving females. The males provide no parental care. There is a surprising lack of clear evidence for what really goes on in leks. It is known that a few males always achieve the vast majority of matings (see Davies

Figure 10.5: The Mating System of *Thalassoma bifasciatum*, the Rainbow Wrasse. The size of the box indicates the relative numbers of each form in the population; the thickness of the arrow the strength of the behavioural effect of one form on the actions of the other. Secondary males are territorial and competitive, these tendencies are mediated through aggression (AGG). Primary males are co-operative and indulge in mass spawnings; they attempt to interfere (INT) with the primary males through mass incursions to drive or coax females out of the secondaries' territories. They also indulge in sneaking fertilisations (SN). Females appear to actively choose (choice) secondary males when they can. At low population densities (top), there is room for a relatively large proportion of the population to be territorial secondary males, and in these cases they are able to exclude successfully the relatively small number of primary males. When the population density is high (bottom), the secondary males have to contend with a much larger number of co-operating primary males, whose interference is now of a much greater effect. The result of this is that at high density the primary males and females delay their metamorphosis to secondary males, since only the most successful competitors among them will achieve more matings than the primaries and females

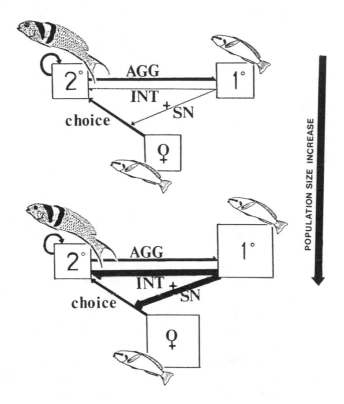

1978; Bradbury and Gibson 1983 for references). In some cases male success appears to be related to the position of the male's territory within the lek (as in the lek-like system of the wrasse *Halichoeres melanochir.* Moyer and Yogo 1982). Male dominance, size, age and display vigour have all been implicated as cues which females may use to select males from within leks. It is also possible that females may copy one another's preferences. Teasing apart these suggestions and their interactions has not, so far, proved possible.

Why do males clump together in leks? Bradbury and Gibson (1983) list a number of suggestions. Different ones may be applicable to different species, and more than one selective pressure may be acting at a time. It has been suggested that aggregation may increase the visibility of males to females. This is difficult to test: it needs to be shown that the increase in attraction of females is enough to raise the success of each male, and when one male is not achieving many matings, he would appear to be better off moving to another place and avoiding the competition. Males may aggregate to reduce predation risk. Perhaps it is the females that are trying to minimise predation risk: possibly this is what makes centrally placed territories attractive. Males may require a certain habitat for their display grounds, and if this were in short supply, they would be forced into close contact. This may be the case where male mouth-brooding cichlids dig a spawning pit: only a soft substrate would be suitable (e.g. *Haplochromis burtoni*: Fernald and Hirata 1977). Possibly, however, the males are being compelled to choose these areas because the females benefit from reducing competition for food. In *H. burtoni* the females and non-reproductives remain in the immediate vicinity of the males and often attempt to feed in their territories, which appear to be rich in food. On the other hand, *Oreochromis mossambicus* females and juveniles inhabit areas of hard substrate and are generally in much better physical condition than the males, suggesting that the males are indeed confined to a suboptimal habitat (Bowen 1984). However, laboratory observations suggest that males that control territories show slower growth than non-territorial fishes (G. Turner, in preparation), and the loss of condition may be due to the energetic cost of territorial defence and reproduction. Again there is no conclusive evidence: the females and juveniles may be forcibly excluded from the breeding areas by the dominant males.

It has been suggested that females compel the males to gather together in order to facilitate mate choice. This presumes that the male's genetic contribution (his only contribution to the progeny) is important to the fitness of the offspring, and that there is considerable variance in male quality, which can be assessed by the females either directly or as a result of male competition.

The so-called hotspot theory suggests that males gather at areas where a large number of females are likely to pass. Like the conspicuousness theory, there is the problem of unsuccessful males; unless the females are choosing to mate only with lekking males, why don't the males that are unable to achieve mating while in the lek move to quieter areas?

Other Breeding Systems

These then are the main breeding systems found in the teleosts: promiscuous, polygynous, polyandrous and monogamous mating systems;

care by both, either or neither of the parents. There are also a number of unusual systems, or unusual specialistions of the above systems, to be found.

An extreme development of the pair bond occurs in some deep-sea angler fishes (Ceratiidae: Norman and Greenwood 1975). The male attaches himself on to the female and becomes a virtual parasite: almost all the organs degenerate: the female even provides nourishment through a placenta-like arrangement of blood vessels. Several males may attach to one female. The system is beneficial to sexually reproducing species living at low population density, and where location of conspecifics is difficult.

Care of young is not always carried out by the parents: in *Lamprologus brichardi*, the young of previous broods remain with the parents and assist in the care of subsequent batches of their siblings. These helpers will clearly increase their inclusive fitness (Hamilton 1964) by helping kin, and it appears that suitable breeding territories are in short supply anyway (Taborsky and Limberger 1981). The helpers also benefit by having access to shelter in the parental territory (Taborsky 1984). Many cichlid species are known to include fry of other broods or other species in the group of young they protect (McKaye and McKaye 1977). This may be to dilute the effect of predation on their own offspring or it may simply be that the parents cannot tell the difference between their own and other fry. Sometimes broods coalesce and are reared by both pairs of fish: there is even a report of cichlids assisting a pair of another species in the care of its brood.

A particularly strange case is that of the Antarctic plunderfish (*Harpagifer bispinis*: Daniels 1979). In this species eggs are normally cared for by the females; if one is removed, another fish takes over the care duties. Even non-breeding fish will care for, and not eat, unfamiliar eggs introduced into their aquarium. The adult fish may not be closely related to the eggs and it seems that this is altruistic behaviour. Although the larvae are planktonic, it appears that these fish exist as largely isolated races (Wheeler 1975) and so it is not inconceivable that selection between groups is occurring.

That is not the only example of a fish caring for unrelated eggs: the tesselated darter *Etheostoma olmstedi* (Constantz 1979) is a small stream-living species. The males guard stones which are kept clean to facilitate spawning. The dominant males are able to increase their success in finding mates by leaving the eggs and searching for females: the eggs harden after a few days and become immune to cannibalism; prior to hardening they are guarded by the father. In the absence of the dominant male, subordinates move into the abandoned caves and keep the surface free of dirt, algae, etc., since females prefer to lay their eggs on clean surfaces. They inadvertently clean the dominant's eggs (which they cannot remove from the rock) and thus help the progeny to survive. In both these species, care of the unrelated eggs has very little cost and in the darter considerable beneficial side-effects, as spawning sites are in short supply.

Another unusual mating system is shown by some of the viviparous fishes of the family Poeciliidae: these always mate with fish of a different species! In fact they exist as all-female populations, derived, it is believed, from the hybridisation of two related species (Schultz 1969). When a female *Poeciliopsis monacha* mates with a male *P. lucida*, only female offspring are produced: these are morphologically intermediate and contain characteristics of both species. However, they produce eggs which apparently contain almost entirely maternal genes. These fishes are typically found in areas where *P. lucida* is the most abundant species and the hybrid females usually mate with these males. This maintains the population of hybrids. If the hybrids mate with *P. monacha* males, normal *P. monacha* would result. Occasionally the hybrids produce diploid eggs with paternal as well as maternal genes, and when these are fertilised, triploid individuals result. These apparently produce triploid eggs which are not fertilised, but are merely stimulated into development by sperm. Hybrid diploids and triploids are extremely abundant in nature, although it seems that they are less well adapted than the sexual species (Lanza 1982). The advantage of producing only females means that the rate of population increase should be much faster for an asexual species than a sexual species. A similar situation exists in another poeciliid fish *Poecilia formosa* (Schultz and Kallman 1968). Internal self-fertilisation is reported to occur in a cyprinodont fish, *Rivulus marmoratus* (Harrington 1961).

Concluding Remarks

Vehrencamp and Bradbury (1984) suggest that in the search for a few ecological factors to explain the occurrence of particular mating systems, it has been easy to overlook other components. They provide a list of all the ways in which they can conceive that an organism can improve its fitness by alteration of its breeding system, and suggest that investigation of all these is really required to permit a full consideration of the selective forces which brought about a certain system in a certain species. Huntingford (1984) suggests that the current classification of mating systems is an oversimplification and that it really only describes extremes of what is actually a multidimensional continuum, where the axes describe features of the system; some systems will fall conveniently in a cluster and others will be intermediate. Vehrencamp and Bradbury (1984) consider the feasibility of basing a classification of mating systems on such an approach, but it remains a task of considerable practical difficulty. It may perhaps never be practical to construct such a classification, nor to analyse all the relevant fitness components, but an appreciation of the limitations of the current approach should guide future work in the complex field of mating systems.

The recent emphasis on individual (as well as kin) selection rather than group selection has greatly improved the understanding of mating systems.

Group- or species-selective explanations were formerly sought, not only for apparently altruistic acts, but even for behaviour which has obvious adaptive value. When Jones (1959) discussed the alternative mating strategies of the Atlantic salmon, he found it difficult to explain the precocious maturation of male parr and their subsequent participation in spawning. He suggested that they acted as an insurance against some of the eggs not being fertilised in a fast-flowing stream. This would be required to explain why the females tolerated the small precocious males; but Jones himself had noted that they tried to chase away the sneaking males. The advantage to the young males is clear: they get a chance to fertilise some eggs. What is more, they get a chance to do it now, without having to wait several years and without the hazards of sea life and migration. Provided there is no great cost to them, this will be a favourable strategy for many males (see Magurran, Chapter 13, this volume). The problem with Jones's argument is that it tries to deal with the adaptive significance of a system, but there is no system. The animals do not develop a mating system for the mututal benefit of all members of the species. What is observed is the result of conflict between individuals, whether of the same or of different sexes. The appropriate method of considering these situations is game theory, and in discussng these questions we might perhaps be better off without the word 'system' and deal with the 'strategies' adopted by individuals within a species.

Summary

Mating systems are intimately bound together with the form of parental care in the breeding system. Teleosts exhibit a wide variety of both mating-system and parental-care tactics. 'Promiscuity' is frequent, and in many cases probably primitive. Polyandry, polygyny and more complex polygamy are found with varying frequency. Monogamy occurs less often and may or may not be associated with a need for biparental care. In normally nonogamous species, polygamy may occur where conditions permit, in accordance with the expectations of Emlen and Oring's model. The more specific conditions assumed for the Verner-Orians model are apparently not met in any teleost species so far studied in detail.

The existence of alternative strategies, conditional mating strategies for individuals, and conflicts of interest between the sexes, means that there is rarely likely to be a single unalterable system for a species. Consideration of individual mating strategies is more appropriate for the modern game-theoretical approach and is less likely to lead to erroneous explanations based on systems for the good of the species as a whole.

References

Aneer, G., Florell, G., Kuatsky, U., Nellbring, S. and Sjoestedt, L. (1983) 'In-situ Observations of Baltic Herring (*Clupea harengus membras*) Spawning Behaviour in the Askoe-Landsort Area, Northern Baltic proper', *Marine Biology*, 74, 105-10

Baerends, G.P. and Baerends-van Roon, J.M. (1950) 'An Introduction to the Ethology of Cichlid Fishes', *Behaviour, Supplements*, 6, 1-243

Balan, E.K. (1981) 'About Processes which Cause the Evolution of Guilds and Species', *Environmental Biology of Fishes*, 1, 129-38

Barlow, G.W. (1962) 'Evolution of Parental Behaviour in Teleost Fishes', Abstract, *American Zoologist*, 2, 504

Barlow, G.W. (1974) 'Contrasts in the Social Behaviour between Central American Cichlid Fishes and Coral Reef Surgeon Fishes', *American Zoologist*, 14, 9-34

Barlow, G.W. and Green, R.F. (1970) 'The Problems of Appeasement and of Sexual Roles in the Courtship Behaviour of the Blackchin Mouthbrooder *Tilapia melanotheron* (Pisces: Cichlidae)', *Behaviour*, 14, 84-115

Baylis, J.R (1981) 'The Evolution of Parental Care in Fishes with Reference to Darwin's Rule of Male Sexual Selection', *Evironmental Biology of Fishes*, 6, 223-52

Blumer, L.S. (1983) 'Parental Care and Reproductive Ecology of the North American Catfish *Ictalurus nebulosus*', *Dissertation Abstracts International (B)*, 43, 10

Bowen, S.H. (1984) 'Differential Habitat Utilisation by Sexes of *Sarotherodon mossambicus* in Lake Valencia: Significance for Fitness', *Journal of Fish Biology*, 24, 115-21

Bradbury, J.W. and Gibson, R.M. (1983) 'Leks and Mate Choice', in P.P.G. Bateson (ed.), *Mate Choice*, Cambridge University Press, Cambridge, pp. 109-38

Brown, J.L. (1975) *The Evolution of Behaviour*, W.W. Norton, New York

Charnov, E.L. (1982) 'The Theory of Sex Allocation', *Monographs in Population Biology*, 18, 1-355

Chien, A.K. and Salmon, M. (1972) 'Reproductive Behaviour of the Angelfish *Pterophyllum scalare*. I. A Quantitative Analysis of Spawning and Parental Behaviour', *Forma Functio*, 5, 45-74

Cole, J.E. and Ward, J.A. (1969) 'The Communicative Function of Fin-flickering in *Etroplus maculatus* (Pisces: Cichlidae)', *Behaviour*, 35, 179-99

Constantz, G.D. (1979) 'Social Dynamics and Parental Care in the Tesselated Darter (Pisces: Percidae)', *Proceedings of the Academy of Natural Sciences of Philadelphia*, 131, 131-8

Daniels, R.A. (1979) 'Nest Guard Replacement in the Antarctic Fish *Harpagifer bispinis*: Possible Altruistic Behaviour', *Science*, 205, 831-3

Darwin, C. (1871) *The Descent of Man and Selection in Relation to Sex*, John Murray, London

Davies, N.B. (1978) 'Ecological Questions about Territorial Behaviour', in J.R. Krebs and N.B.Davies (eds), *Behavioural Ecology: an Evolutionary Approach*, Blackwell, Oxford, pp. 317-50

Downhower, J.F. and Brown, L. (1981) 'The Timing of Reproduction and its Behavioral Consequences for Mottled Sculpins, *Cottus bairdi*', in R.D. Alexander and D.W. Tinkle (eds), *Natural Selection and Social Behaviour*, Chiron Press, New York, pp. 78-85

Downhower, J.F., Brown, L., Pederson, R. and Staples, G. (1983) 'Sexual Selection and Sexual Dimorphism in Mottled Sculpins', *Evolution*, 37, 96-103

Emlen, S.T. and Oring, L.W. (1977) 'Ecology, Sexual Selection and the Evolution of Mating Systems', *Science*, 197, 215-23

Fernald, R.D. and Hirata, N.R. (1977) 'Field Study of *Haplochromis burtoni*: Quantitative Behavioural Observations', *Animal Behaviour*, 25, 643-53

Fisher, R.A. (1958) *The Genetical Theory of Natural Selection*, Dover, New York

Fricke, H.W. (1973) 'Behaviour as Part of Ecological Adaptation', *Helgolander Wissenschaftliche Meeresuntersuchungen*, 24, 120-44

Fricke, H.W. (1974) 'Oeko-Ethologie de Monogame Anemonenfisches *Amphiprion bicinctus* (Freiwasseruntersuchung aus Roten Meer)', *Zeitschrift für Tierpsychologie*, 36, 429-512

Fricke, H. and Fricke, S. (1977) 'Monogamy and Sex-change by Aggressive Dominance in Coral Reef Fish', *Nature, London*, 266, 830-2

Fryer, G. and Iles, T.D. (1972) *The Cichlid Fishes of the Great Lakes of Africa*, T.F.H. Publications, New Jersey

Hamilton, W.D. (1964) 'The Genetical Theory of Social Behaviour', *Journal of Theoretical Biology*, 7, 1-52

Halliday, T.R. (1983) 'The Study of Mate Choice', in P.P.G. Bateson (ed), *Mate Choice*, Cambridge University Press, Cambridge, pp. 3-32

Harrington, R.W. (1961) 'Oviparous Hermaphrodite Fish with Internal Fertilisation', *Science*, 134, 1749-50

Hourston, A.S., Rosenthal, H. and Stacey, N. (1977) 'Observations on Spawning Behaviour of Pacific Herring in Captivity', *Meeresforschung*, 25, 156-176

Huntingford, F.A. (1984) *The Study of Animal Behaviour*, Chapman & Hall, London

Jones, J.W. (1959) *The Salmon*, Collins New Naturalist, London

Keenleyside, M.H.A. (1979) *Diversity and Adaptation in Fish Behaviour*, Springer-Verlag, Berlin

Keenleyside, M.H.A. (1983) 'Mate Desertion in Relation to Adult Sex-ratio in the Biparental Cichlid Fish *Herotilapia multispinosa*', *Animal Behaviour*, 31, 683-88

Krebs, J.R. and Davies, N.B. (1981) *Introduction to Behavioural Ecology*, Blackwell, Oxford, 292 pp.

Lande, R. (1980) 'Sexual Dimorphism, Sexual Selection and Adaptation in Polygenic Characters', *Evolution*, 34, 292-305

Lanza, J. (1982) 'Population Control of an All-female Fish *Poeciliopsis monacha-lucida*', *Biotropica*, 15, 100-7

Limberger, D. (1983) 'Pairs and Harems in a Cichlid Fish, *Lamprologus brichardi*', *Zeitschrift für Tierpsychologie*, 62, 115-44

Loiselle, P.V. (1982) 'Male Spawning-partner Preference in an Arena Breeding Teleost, *Cyprinodon macularis californiensis* Girard (Atherinomorpha: Cyprinodontidae)', *American Naturalist*, 120, 721-32

Maynard Smith, J. (1982) *Evolution and the Theory of Games*, Cambridge University Press, Cambridge

McKaye, K.R. (1977) 'Competition for Breeding Sites between the Cichlid Fishes of Lake Jiloa, Nicaragua', *Ecology*, 58, 291-302

McKaye, K.R. and McKaye, N.M. (1977) 'Communal Care and Kidnapping of Young by Parental Cichlids', *Evolution*, 31, 674-81

Meral, G.H. (1973) 'The Adaptive Significance of Territoriality in New World Cichlidae', unpublished PhD thesis, University of California, Berkeley

Morris, D. (1956) 'The Function and Causation of Courtship Ceremonies', in *Colloque Internationale sur l'Instinct*, Fondation Sugar-Polignac, Paris, pp. 261-6

Moyer, J.T. and Sawyers, C.E. (1973) 'Territorial Behaviour of the Anemonefish *Amphiprion xanthurus* with Notes on the Life History', *Japanese Journal of Ichthyology*, 20, 85-93

Moyer, J.T. and Yogo, Y. (1982) 'The Lek-like Mating System of *Halichoeres melanochir* (Pisces: Labridae) at Miyoke-juma, Japan', *Zeitschrift für Tierpsychologie*, 60, 209-26

Nakatsuru, K. and Kramer, D.L. (1982) 'Is Sperm Cheap? Limited Male Fertility and Female Choice in the Lemon Tetra (Pisces, Characidae)', *Science*, 216, 753-5

Norman, J.R. and Greenwood, P.H. (1975) *A History of Fishes*, 3rd edn, Benn, London

O'Donald, P. (1983) 'Sexual Selection by Female Choice', in P.P.G. Bateson (ed.), *Mate Choice*, Cambridge University Press, Cambridge, pp. 53-66

Orians, G.H. (1969) 'On the Evolution of Mating Systems in Birds and Mammals', *American Naturalist*, 103, 589-603

Parker, G.A. (1983) 'Mate Quality and Mating Decisions', in P.P.G. Bateson (ed.), *Mate Choice*, Cambridge University Press, Cambridge, pp. 141-66

Parker, G.A. (1984) 'Evolutionarily Stable Strategies', in J.R. Krebs and N.B. Davies (eds), *Behavioural Ecology*, 2nd edn, Blackwell, Oxford, Chapter 2

Pitcher, T.J. (1971) 'Population Dynamics and Schooling Behaviour in the Minnow, *Phoxinus phoxinus* (L)' (Chapter 6), unpublished D.Phil. thesis, University of Oxford

Plezczynska, W. and Hansell, R.I.C. (1980) 'Polygyny and Decision Theory: Testing a Model in Lark Buntings', *American Naturalist*, 116, 821-30

Pressley, P.H. (1981) 'Pair Formation and Joint Territoriality in a Simultaneous Hermaphrodite: the Coral Reef Fish *Serranus tigrinus*', *Zeitschrift für Tierpsychologie*, 51, 33-46

Rechten, C. (1980) 'Brood Relief Behaviour of the Cichlid Fish *Etroplus maculatus*', *Zeitschrift für Tierpsychologie*, 52, 77-102

Reese, E.S. (1975) 'A Comparative Field Study of the Social Behaviour and Related Ecology of Reef Fishes of the Family Chaetodontidae', *Zeitschrift für Tierpsychologie, 37*, 33-61

Robertson, D.R. (1972) 'Social Control of Sex-reversal in a Coral-reef Fish', *Science, 117*, 1007-9

Schultz, R.J. (1969) 'Hybridisation, Unisexuality and Polyploidy in the Teleost *Poeciliopsis* (Poeciliidae) and other Vertebrates', *American Naturalist, 103*, 605-19

Schultz, R.J. and Kallman, K.D. (1968) 'Triploid Hybrids between the All Female Teleosts *Poecilia formosa* and *Poecilia sphenops*', *Nature, London, 219*, 280-2

Taborsky, M. (1984) 'Broodcare Helpers in *Lamprologus brichardi*: their Costs and Benefits', *Animal Behaviour, 32*, 1236-52

Taborsky, M. and Limberger, D. (1981) 'Helpers in Fish', *Behavioural Ecology and Sociobiology, 8*, 143-5

Trivers, R.L. (1972) 'Parental Investment and Sexual Selection', in B. Campbell (ed.), *Sexual Selection and the Descent of Man 1871-1971*, Chicago University Press, Chicago, pp. 136-79

Vehrencamp, S.L. and Bradbury, J.W. (1984) 'Mating Systems and Ecology', in J.R. Krebs and N.B. Davies (eds), *Behavioural Ecology: an Evolutionary Approach*, 2nd edn, Blackwell, London, pp. 251-78

Verner, J. and Willson, M.L. (1966) 'The Influence of Habitats on Mating Systems of North American Passerine Birds', *Ecology, 47*, 143-7

Warner, R.R. (1982) 'Mating Systems, Sex Change, and Social Demography in the Rainbow Wrasse, *Thalassoma lucanascum*', *Copeia, 1982*(3), 653-61

Warner, R.R. and Hoffman, S.G. (1980) 'Local Population Size as a Determinant of Mating Systems and Sexual Composition in Two Tropical Marine Fishes, *Thalassoma* sp.', *Evolution, 34*, 508-18

Wheeler, A. (1975) *Fishes of the World: an Illustrated Dictionary*, Ferndale Editions, London

Wickler, W. and Seibt, U. (1983) 'Monogamy: an Ambiguous Concept', in P.P.G. Bateson (ed.), *Mate Choice*, Cambridge University Press, Cambridge, pp. 33-50

11 WILLIAMS' PRINCIPLE: AN EXPLANATION OF PARENTAL CARE IN TELEOST FISHES

Robert Craig Sargent and Mart R. Gross

Parental care may be defined as an association between parent and off-spring after fertilisation that enhances offspring survivorship. This phenomenon has attracted the attention of evolutionary biologists since Darwin; however, it was not until the recent insurgence of behavioural ecology and sociobiology (e.g. Williams 1966a; Trivers 1972; Alexander 1974; Wilson 1975) that the variety of parental care patterns in animals has attracted such rigorous study. Perhaps because we ourselves are mammals, we tend to think of parental care as being the principal occupation of females, possibly with some help from males. A survey of the vertebrates, however, reveals that mammals are merely at one end of the spectrum, with predominantly female care, and fishes are at the other end, with predominantly male care (Table 11.1; see also Chapter 10 by Turner, this volume). Within teleost fishes with external fertilisation (about 85 per cent of all teleost families), one finds that the four states of parental care, ranked in descending order of their frequencies, are: no care, male care, biparental care, and female care. This seemingly peculiar trend has attracted considerable attention from evolutionary biologists, who have proposed several hypotheses about the origins of parental care in fishes (see reviews by Maynard Smith 1977; Blumer 1979; Perrone and Zaret 1979; Baylis 1981; Gross and Shine 1981; Gross and Sargent 1985).

Fishes have several attributes that make them ideal subjects for the study of the behavioural dynamics of parental care within the context of life-history evolution. First, they exhibit considerable phylogenetic diversity of the states of parental care. Secondly, many species are easily studied in the field. Thirdly, many species adapt readily to the laboratory where one can conveniently control or manipulate variables of interest. Fourthly, guarding the eggs or brood on the substratum is the predominant form of parental care in fishes (Table 11.2). Guarding, unlike feeding, is a divisible resource (Wittenberger 1981), in the sense that a unit of parental resource may be given to one or several offspring. Thus, a fish may be able to guard large clutches as easily as small ones (Williams 1975: p. 135). This assumption makes the dynamics of parental behaviour in fishes relatively easy to model mathematically.

Rather than catalogue the remarkable diversity of kinds of parental care exhibited by fishes (see Breder and Rosen 1966 for a review), we focus in this chapter on an important evolutionary principle and show how it leads to a better understanding of parental behaviour. We name this principle after its major progenitor, G.C. Williams. The principle assumes that reproduction has a cost, which can be modelled into a general framework for

Table 11.1: Approximate Phylogenetic Distribution of the States of Parental Care (Expressed in Percentage Terms), by Family, over the Major Vertebrate Classes. Within each class, if no species within a family is known to exhibit parental care, then that family is classified as 'No care'. If at least one species within a family exhibits one of the other states of parental care, then that family is classified under that state. Families that exhibit more than one of the other states of parental care are counted more than once. The data are presented at the family level to increase the likelihood that the states of parental care in the table are independent of phylogenetic relationships. The data were compiled from the following sources: mammals (Kleiman 1977; Kleiman and Malcolm 1981), birds (Lack 1968), reptiles (Porter 1972), amphibians (Salthe and Meacham 1974; Gross and Shine 1981), and fishes (Breder and Rosen 1966; Blumer 1982, Gross and Sargent 1985).

Taxon	Male care	Female care	Biparental care	No care
Mammals	0	90	10	0
Birds	2	8	90	0
Reptiles	0	15	Rare	85
Amphibians	24	32	14	30
Non-teleost fishes	6	66	0	28
Teleost fishes	11	7	4	78

predicting and studying a variety of problems on the evolution and behavioural ecology of fish parental care. We illustrate the model with selected examples from fishes, and discuss its implications for future research.

The Cost of Reproduction — Williams' Principle

Williams (1966a,b) proposed that a parent that continues to invest in its offspring does so at the expense of its potential future reproduction. If we assume that a parent, at any point in time, is selected to maximise its remaining lifetime reproductive success, we can model the tradeoff between present and future reproduction as follows:

$$R = P(RE) + F(SE) \tag{11.1}$$

where R = the parent's expected remaining lifetime reproductive success; P = present reproduction (the number of offspring at stake times their expected survivorship); F = the parent's expected future reproduction; RE = reproductive effort (*sensu* Williams 1966a,b) — the proportion of parental resources (energy, time, risk) devoted to fecundity and offspring

Table 11.2: Approximate Distribution of Teleost Families with External Fertilisation over the Four Most Common Forms of Parental Care as Classified by Blumer (1982). These categories are not necessarily mutually exclusive; some families contain species with more than one form of parental care. The data were compiled from Blumer (1982), with additions noted in Gross and Sargent (1985)

	Male	Male or biparental	Male or female	Biparental	Male, female or biparental	Biparental or female	female
Substratum guarding	28	6	2	10	5	2	1
Mouth-brooding	2	1	1	0	2	0	1
External egg-carrying	3	1	0	0	0	0	2
Brood pouch	1	0	0	0	0	0	1

survivorship; and SE = somatic effort (*sensu* Williams 1966b) — the proportion of resources devoted to the parent's own growth, maintenance and survival. SE = 1 − RE.

The sum of RE and SE represents the total resources available to a parent (RE + SE = 1); thus, reproductive effort (e.g. parental care) is at the expense of somatic effort and potential future reproduction of the parent. Thus, Williams' principle is that natural selection favours animals that maximise their remaining lifetime reproductive success, subject to this constraint. We make the common biological assumption that P and F both increase with diminishing returns (Real 1980) with increasing RE and SE, respectively (see Figure 11.1a). In other words, we assume that small investments into P or F yield higher returns per investment than do large investments into P or F. The optimal reproductive effort (RE*), which maximises R, will occur where dR/dRE = dP/dRE + dF/dRE = 0, or:

$$dP/dRE = dF/dRE \qquad\qquad (11.2)$$

The assumption of diminishing returns (Figure 11.1a) means that d^2P/dRE^2, d^2F/dRE^2, and thus d^2R/dRE^2 are always negative. Because d^2R/dRE^2 is negative for all values of RE, and because we assume that there exists an RE* for which dR/dRE = 0, then this RE* must maximise R. At this level of RE, the parent obtains equal rates of return on its investments into present and future reproduction (equation 11.2; Figure 11.1b). In our model the cost of reproduction (CR) is the loss in expected future reproduction (F) that is attributable to present reproduction (P). We will use our CR model to analyse parental investment; however, first we discuss some controversy surrounding Williams' principle, and present evidence of its applicability for fishes.

There has been considerable discussion as to how to measure reproductive effort (RE) (Hirshfield and Tinkle 1975; Pianka and Parker 1975; Hirshfield 1980), and whether or not a cost of reproduction (CR) even exists (Lynch 1980; Bell 1984a,b). Bell, for example, has suggested that a zero or positive correlation between present and future reproduction (P and F) among individuals within a population refutes the idea of a CR. We believe that such controversy stems from a misunderstanding of Williams' original model. In particular, it is important to note that the cost of reproduction is based on RE being a proportion of an individual's resource budget, and that the dynamics of the tradeoff are assumed to be independent of resource-budget size. Although two adults may have the same optimal RE (RE*), one may have higher fecundity and higher growth because it has a larger resource budget. Thus, lack of an observed negative correlation between P and F among individuals within a population cannot by itself refute the idea of a CR.

Several studies with fishes indicate that a CR tradeoff does in fact exist.

Figure 11.1: (a) Present Reproduction (P) is Assumed to Increase with Diminishing Returns with Increasing Reproductive Effort (RE), and Future Reproduction (F) is Assumed to Increase with Diminishing Returns with Increasing Somatic Effort (1—RE). (b) The Optimal Reproductive Effort (RE*) Occurs where the Rates of Return on Investment into Present and Future Reproduction are Equal (dP/dRE = —dF/dRE)

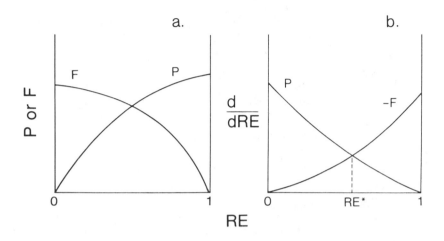

Perhaps the best-known experiment is that of Hirshfield (1980), who studied the Japanese medaka, (*Oryzias latipes*) among several temperature- and ration-controlled environments. Holding ration constant, total energy devoted to eggs and to growth (total production) did not differ among temperatures; however, as temperature increased among treatments, fecundity increased, and growth and survival both decreased. Although we do not know how temperature affected RE*, these results indicate a CR (Figure 11.2), because of the negative correlation between P and F among treatments, while total production remained constant. Further evidence is supplied by Reznick (1983), who found a similar negative correlation between reproduction and growth among five stocks of guppies (*Poecilia reticulata*), all of which had similar values for total production. In species with male parental care, it is known that males lose weight as a function of residence time on the territory, and at the end of a reproductive bout or breeding season, males suffer heavy mortality (*Lepomis gibbosus*: Gross 1980; *Pimephales promelas*: Unger 1983). Thus Williams' assumption of a cost of reproduction seems to hold for fishes.

If optimal RE* is independent of resource-budget size, as assumed by our model, then P and F should both increase as ration increases. This pattern has been reported for several species of fishes (e.g. *Oryzias latipes*: Hirshfield 1980; *Poecilia reticulata*: Reznick 1983; *Cichlasoma nigrofasciatum* Townshend and Wootton 1984).

Figure 11.2: The Cost of Reproduction over Three Temperature Regimes in *Oryzias latipes* (from Hirshfield 1980). (a) Adult female growth decreases as fecundity increases. (b) Adult female survival (percentage not sick or dead) decreases as fecundity increases

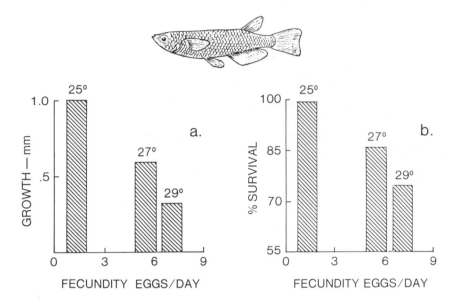

Parental Care Evolution in Fishes: Why is Male Care So Common?

Traditionally, models for the evolution of parental care have contained a tradeoff between parental care and the number of offspring in the brood (e.g. Trivers 1972; Maynard Smith 1977); however, there may be a more important tradeoff for fishes. Unlike some other vertebrates, fishes continue to grow after they have become reproductive. In fishes, both female fecundity and number of eggs fertilised by males tend to increase with body size (Gross and Sargent 1985). Any energy spent on reproduction is at the expense of growth; thus, a fish's expected future reproduction depends on how much it invests in growth, and on how big it gets (e.g. Reznick 1983). Our model states that parental care in fishes will be sensitive to this trade-off between reproduction and growth.

How might the costs of parental care to growth affect which sex is most likely to evolve parental care? Imagine a species of fish with external fertilisation, no parental care, and males who hold territories where they obtain all of their spawnings. Assume that if either sex shows parental care it loses growth. Then the sex most likely to evolve care is the one with the lower relative costs to future fertility, because it will receive the higher relative benefits from care. It is well known in fishes that female fecundity increases with accelerating returns with increasing body length (Bagenal and Braum 1978). If a male's ability to obtain matings were to increase linearly, or with diminishing returns with body length, then females would have the higher costs to future fertility, and thus would be less likely to evolve parental care. We illustrate this below.

Consider the following example from Gross and Sargent (1985; and see Figure 11.3) Assume that two fish have just spawned and that each one has present length s. If a parent guards its offspring, it can expect to be length g in the next breeding season. If it deserts, however, it will attain a larger size, d, because it has not invested energy into parental care. A fish's relative cost of parental care to future fertility is the ratio of its expected gain in fertility if it deserts to its expected gain in fertility if it guards (D/G). This ratio, which is always greater than one, is ranked in descending order of F against body length curve types: accelerating, linear, and diminishing (Figure 11.3). Gross and Sargent (1985) examined the shapes of these curves for five species of fishes with male care: the bluegill sunfish (*Lepomis macrochirus*); the pumpkinseed sunfish (*L. gibbosus*); the rock bass (*Ambloplites rupestris*); the three-spined stickleback (*Gasterosteus aculeatus*); and the mottled sculpin (*Cottus bairdi*). In each species the female curves increased with accelerating returns with body length, but the male curves increased either linearly, or with diminishing returns. Thus, for these five species, females would have the higher relative costs to their future fertility if they showed parental care. This is presumably why male care rather than female care evolved in these species, which we offer as an explanation for the prevalence of male parental care in fishes. Gross and Sargent (1985) review this and competing hypotheses for the evolution of male parental care in fishes (see also Chapter 10 by Turner, this volume).

Figure 11.3: The Relationship between Expected Future Reproductive Success and Body Length May Be Accelerating (1), Linear (2), or Diminishing (3). At the next breeding season, a fish that is presently of length *s* will attain length *g* if it guards, or *d* if it deserts. If the expected gain of a fish that deserts is the same for each type of curve, then the relative cost of guarding, D/G, to expected future fertility is ranked in descending order among curve types: accelerating (1), linear (2), and diminishing (3)

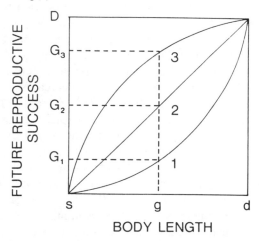

Behavioural Ecology of Parental Care: the Cost of Reproduction and the Dynamics of Parental Behaviour

Our CR model suggests that levels of parental care are unlikely to be static within an individual over time, or among individuals under different conditions. We look in particular at four phenomena that are common to species with parental care and ask:

(1)   How should parental behaviour vary with age of the offspring?
(2)   How should parental behaviour vary with brood size?
(3)   How should parental behaviour vary with the probability of breeding again?
(4)   How should reproductive effort, and specifically parental behaviour, vary with increased investment in territory defence?

Our approach to answering these questions is through a simple graphical analysis of the model. We consider parents who have already allocated energy to mating and who are now defending eggs or offspring; thus, any pending expenditures of RE are in the form of parental care. Recall from expression (11.2) that the optimal reproductive effort, RE*, is the one that yields equal rates of return on investment into present and future reproduction (P versus F), which is a very useful property of the model. For example, if the rate of return on investment into P, dP/dRE, were raised, then RE* would increase so as to balance the two rates of return, dP/dRE = −dF/dRE (Figure 11.4). Similarly, if the rate of return on investment into future reproduction, dF/dSE (= −dF/dRE), were raised, then the optimal somatic effort would increase and RE* would decrease (Figure 11.5). Thus, we can study the dynamics of parental care by perturbing (i.e. manipulating) the tradeoff between present and future reproduction, and looking for evidence of the predicted shift in RE*. Because we cannot measure RE* directly, we look for a negative correlation between present and future reproduction among treatments.

Parental Behaviour and Offspring Age

In fishes, it has been observed that, at first, parental care increases as the offspring get older, but later, parental care decreases as the offspring approach independence (e.g. van Iersel 1953; Barlow 1964). Although these parental-behaviour dynamics can be explained as reflecting the physical needs of the offspring, they may instead reflect the parent's resolution of a tradeoff between P and F. We can describe this process with a progression of CR models whose tradeoffs change with time, or, more precisely, with offspring age. Let us consider four developmental stages of fish offspring: (1) newly fertilised eggs (i.e. zygotes); (2) eggs that are about to hatch; (3) wrigglers; and (4) free-swimming fry (see Keenleyside 1979). The hypothetical P versus F tradeoffs for these developmental states

Figure 11.4: Raising the Rate of Return of Investment into Present Reproduction (e.g. by Increasing Brood Size) Raises the Optimal Reproductive Effort. (a) A parent with a large brood has a higher expected present reproduction than a parent with a small brood (P2 > P1). (b) As dP/dRE increases, RE* increases

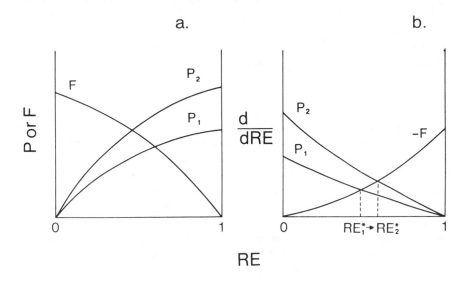

Figure 11.5: Raising the Rate of Return on Investment into Future Reproduction (e.g. by Increasing the Likelihood of Remating) Raises the Optimal Somatic Effort, which Lowers the Optimal Reproductive Effort. (a) A parent who experiences F2 has a higher expected future reproduction than a parent who experiences F1. (b) As −dF/dRE increases, RE* decreases

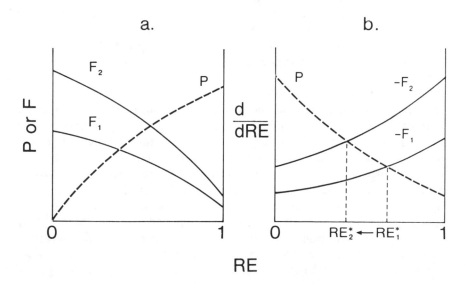

are presented in Figure 11.6. We assume that during a brood cycle a parent's resource budget remains roughly constant, and that changes in RE* over time produce changes in the net levels of observed parental behaviour. We also assume that, as the brood cycle progresses, changes in a parent's ability to improve offspring survival far outweigh any decreases in its ability to affect its own survival (and thus F). Therefore, for simplicity, we treat the F(RE) curve as not changing with offspring age; however, the P(RE) curve depends on offspring age (Figure 11.6). Let

$$P(RE) = bl(RE) \tag{11.3}$$

where b is brood size, which is assumed to be a constant, and l is the probability of the offspring surviving to reproduce, which increases with diminishing returns with increasing RE. At spawning (Figure 11.6a), offspring survivorship is zero if the parent offers no care, but improves somewhat with maximum parental care (RE = 1). Just before hatching (Figure 11.6b), the offspring are still completely dependent on parental care (P = 0 if RE = 0), but now the parent can achieve higher juvenile survivorship with maximum parental care (RE = 1), simply because the offspring are older and thus closer to reproductive maturity. Because the rate of return on investment into P (dP/dRE) has increased over that for newly spawned eggs, RE* also increases. When the offspring are at the wriggler stage (Figure 11.6c), they have some finite probability of surviving on their own (P > 0 if RE = 0), and the difference in offspring survivorship between RE = 0 and RE = 1 is now less than when the offspring were just hatching. Thus the parent now enjoys a lower dP/dRE than for hatching eggs, and so should lower its RE. Finally, when the fry are free swimming (Figure 11.6d), the difference in offspring survivorship between RE = 0 and RE = 1 is very small, and is always less than −dF/dRE. Because the parent has higher expected future reproduction when RE = 0 than it can gain if it invested anything in its offspring (RE > 0), the optimal reproductive effort is zero (RE* = 0), and the parent deserts its offspring.

Thus, we have extended our basic CR model to one that changes with time over a brood cycle, and with it we can explain why the dynamics of parental care change with offspring age. Although the dynamics we have described in Figure 11.6 will vary among species, they are likely to represent the general pattern in fishes with parental care. To illustrate these dynamics we consider three-spined stickleback (*Gasterosteus aculeatus*).

The three-spined stickleback is a species characterised by males who aggregate in shallow water, establish territories, build nests, court females, and care for the developing eggs and newly hatched fry (see Chapter 16 by FitzGerald and Wootton, this volume). A single male may undergo several brood cycles during the breeding season. It is well known for the three-spined stickleback that paternal fanning increases with age of the eggs up

Figure 11.6: Hypothetical Parental Reproductive Tradeoffs for Four Different Stages of Offspring Development: (a) Zygote, (b) Late Embryo, (c) Wriggler, and (d) Free-swimming Fry. The top row shows present and future reproduction versus reproductive effort; the second row shows the rates of return on investment into present and future reproduction versus reproductive effort. Under this model, RE* increases between stages a-b, and decreases between stages b-c and c-d. See text for assumptions and explanation

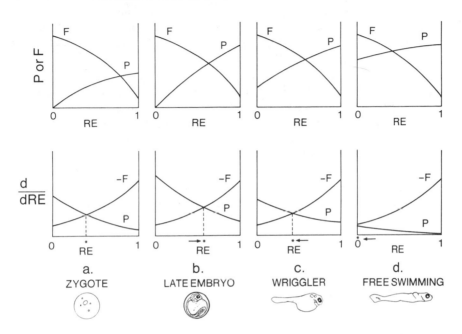

until about a day before hatching and then continues to decline as the offspring approach independence (van Iersel 1953; van den Assem 1967). Although this result is predicted by our model, the pattern in fanning over the brood cycle might also be explained by the eggs requiring more oxygen as they get older, and less oxygen after they hatch. Thus fanning may not corroborate our model. Another form of parental care in the three-spined stickleback is defence of the brood from predators. Unlike fanning, a brood's requirements in defence from predators are likely to be independent of age. In sticklebacks the major brood predators are conspecifics of either sex (Kynard 1978). Motivation for brood defence can be measured by presenting a persistent threat to the brood on a male's territory, and then recording the resident male's behaviour. One such threat is another male stickleback in a glass cylinder next to a resident male's nest (Wootton 1976). Over a brood cycle such tests show that as the eggs approach hatching, brood defence increases (bites per minute towards an intruder). After hatching, however, brood defence wanes as the fry approach independence (Huntingford 1977; Sargent and Gebler 1980; Sargent 1981). Similar brood defence dynamics were found in the pumpkinseed sunfish, *Lepomis gibbosus* (Colgan and Gross 1977). Thus, both fanning and brood defence exhibit the kinds of dynamics described by our model (Figure 11.7a).

Parental Behaviour and Brood Size

Several laboratory and field studies have shown considerable variation in the levels of parental behaviour among parents, and much of this variation can be explained by a positive correlation between parental behaviour and brood size (e.g. van Iersel 1953; Pressley 1981). Under our model, there may be two components to this correlation:

(1)   Bigger parents may have larger resource budgets, which allow them to accumulate bigger broods and thus to show higher brood defence. Several studies have documented a positive correlation between parental body size and parental behaviour (Downhower and Brown 1980; Gross 1980).

(2)   Parents with larger broods experience a higher rate of return on investment into brood defence (dP/dRE, see Figure 11.4), and thus should show higher brood defence in order to maximise remaining lifetime reproductive success.

Figure 11.7: Parental Behaviour versus Offspring Age in *Gasterosteus aculeatus* (from Sargent 1981). (a) Both fanning and brood defence parallel the RE*s given in figure 11.6. (b) Males with three clutches show higher brood defence than males with one clutch at each stage of offspring development

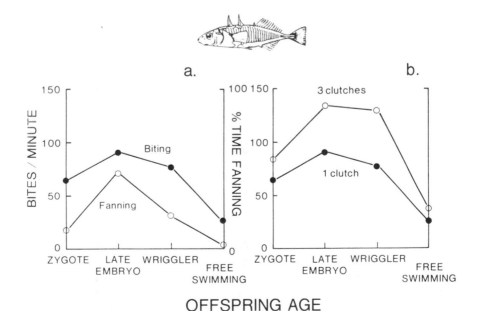

To test these hypotheses, Sargent (1981) randomly assigned competing male sticklebacks (*Gasterosteus aculeatus*) either one or three clutches. Males with three clutches consistently showed higher brood defence than males with one clutch (Figure 11.7b), during each stage of offspring development. Larger males may have had higher brood defence than smaller males; however, this difference was not statistically significant. Thus, although resource-budget size may affect the level of parental behaviour, brood size has an additional effect, because parents with larger broods enjoy a higher rate of return on brood defence.

Parental Behaviour and Expected Future Reproduction

A parent's expected future reproduction will vary with a number of factors, both intrinsic and extrinsic. Examples of intrinsic factors may include age, health, and social status; extrinsic factors may include sex ratio, food availability, and predators. Our model predicts that changes in expected future reproduction will produce changes in RE*. For example, raising the rate of return on investment into future reproduction (-dF/dRE) lowers RE* (Figure 11.5), or the optimal investment in P.

An example of this is found in the parental behaviour of males of the biparental cichlid, *Herotilapia multispinosa* (Keenleyside 1983). In this species both parents guard the eggs on the substratum, and guard the fry after they hatch. Each guarding male is defending only one clutch, and either must complete the brood cycle or desert his offspring before he can mate again. Keenleyside studied these fish in ponds in which he manipulated the adult sex ratio (7:5, 1:1, and 5:7 females to males). Note that as the proportion of females increases among ponds, so does a guarding male's expected future reproduction (but not that of the female). According to our model, as a male's chances of remating increases, his RE should decrease. One would expect, therefore, that as the proportion of females increases among ponds, so should the proportion of guarding males who desert their offspring. This is precisely what Keenleyside found (Figure 11.8). His experiment corroborates our model and demonstrates that the optimal reproductive effort is sensitive not only to a change in the rate of return on investment into the offspring at stake, but also to a change in the rate of return on investment into future reproduction.

Parental Behaviour and Territory Defence

Almost all species of fishes that guard their eggs and fry exhibit territory defence of the spawning site. For males, energy spent on territory defence may be at the expense of other forms of parental care (e.g. aeration of the eggs). Thus, males that vary in their territory defence may vary in the efficiency with which they raise offspring. It thus becomes interesting to ask, 'Can variation in territory defence affect optimal reproductive effort?'

R.C. Sargent (unpublished)[1] divided male sticklebacks (*Gasterosteus aculeatus*) into two treatments: competitive and solitary. Replicate 76 -litre

Figure 11.8: Male Desertion versus Adult Sex Ratio in *Herotilapia multispinosa* (from Keenleyside 1983). As the adult sex ratio (proportion of females) increases, the proportion of males who desert their offspring increases

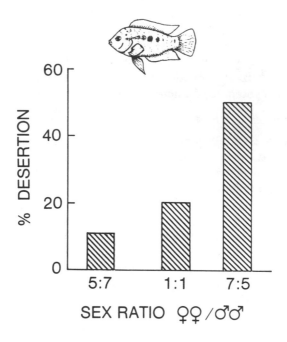

aquaria were bisected using a partition with one male nesting in each compartment (30 × 30 cm²). Competitive males were separated by transparent partitions; solitary males by opaque partitions. Each male was allowed to spawn with one female per brood cycle, and was allowed as many brood cycles as he could achieve within a breeding season. All males died by the end of the season (about 16 weeks). The competitive males spent considerable time fighting and displaying at the transparent partition, which resulted in their spending less time fanning than solitary males (Table 11.3). These behavioural differences between the two treatments resulted in several direct (and indirect) costs of being a competitive male, which are summarised in Table 11.3. Competitive males lost weight faster than solitary males, thus territory defence imposed an energetic cost on top of parental care. Competitive males took longer to build their nests because they spent time fighting. Competitive males also took longer to hatch their eggs, because they spent less time fanning than solitary males. (The length of time between spawning and hatching increases as fanning decreases (van Iersel 1953); this difference in hatching time between competitive and solitary males has been reported previously by van den Assem (1967)). These temporal costs of territoriality resulted in competitive males having longer brood cycles than solitary males. Both the energetic and temporal costs of territoriality contributed to solitary males averaging more brood cycles per breeding season than the competitive males.

Thus, a CR may have two components: energy and time. Competitive males appear to pay higher costs for both components. Because of these costs, territoriality may result in a lower rate of return on investment into future reproduction for competitive males. One can also argue, however, that because competitive males have longer brood cycles and spend more energy per brood cycle, territoriality may lower the rate of return of investment into present reproduction. This suggests that being competitive is a poor strategy for a stickleback, yet aggressive territoriality is characteristic of the species! Why, then, are sticklebacks territorial? In the field, the denser the aggregation of nesting males, the higher the average number of eggs per male (Kynard 1978), which suggests that dense aggregations of competitive males attract more females per male than solitary males. Gross (1980) found a similar increase in females per male, with increasing male density for the pumpkinseed sunfish (*Lepomis gibbosus*). Competitive males, therefore, may experience a *higher* rate of return on investment into present reproduction, because they get more clutches for their efforts. As an indirect test of this hypothesis, Sargent presented males in the two treatments additional gravid females the day before the eggs were due to hatch. By this time, most males with one clutch should have switched to the parental phase (*sensu* van Iersel 1953), and should not court additional females (van Iersel 1953; Sargent 1981). In Sargent's experiment, however, most competitive males courted these new females, indicating that they were still trying to increase their brood size, and thus their present reproduction. Most solitary males did not court these additional females (Table 11.3).

Again, there appears to be a positive relationship between P, F, and resource-budget size. Within the competitive treatment, there was a positive correlation between a male's condition (weight/length^3), which reflected his energy reserves, and the number of brood cycles that he

Table 11.3: A Summary of Sargent's (unpublished)[1] Results for Solitary versus Competitive Male *Gasterosteus aculeatus*. All comparisons are statistically significant at the 0.05 level

Comparison	Solitary	Competitive
Percentage of time fighting the day before the eggs hatched	0	40
Percentage of time fanning the day before the eggs hatched	44	21
Weight loss, grams/day	0.0017	0.0063
Time to nest completion, days	3.2	4.1
Hatching time, days	6.0	7.1
Brood cycles per season	3.5	2.5
Percentage of males who courted new females the day before the eggs hatched	17	92

enjoyed ($r = 0.59$, $p < 0.05$). Moreover, males in better condition lost weight faster than males in poor condition ($r = 0.61$, $p < 0.05$). Thus it appears that present and future reproduction will both increase with condition in sticklebacks. Although there are measurable costs of territoriality when manipulating males into being competitive versus solitary, among competitive males, energy devoted to present reproduction is positively correlated with future reproduction.

In summary, territorial males may incur higher temporal and energetic costs than solitary males. These higher costs will lower their expected number of future brood cycles. However, dense aggregations of nesting males may attract more females per male, thus territorial males may enjoy more matings per brood cycle, and presumably a higher lifetime reproductive success results. Note that males may face two simultaneous tradeoffs: (1) P versus F; and within P (2) quantity versus quality of offspring. Such simultaneous tradeoffs would increase the complexity of parental investment dynamics, and must be considered by future researchers.

Summary and Suggestions for Research

Williams' principle assumes that reproduction always has a cost. As animals attempt to maximise their remaining lifetime reproductive success, the cost of reproduction will limit the amount that they can invest into future reproduction. Animals who maximise this tradeoff obtain equal rates of return on investments into present and future reproduction. This tradeoff, when modelled, provides a powerful technique for analysing parental behaviour in fishes. With it we can explain the prevalence of male parental care, the changes in parental behaviour over a brood cycle, the positive correlation between parental behaviour and brood size, the negative correlation between parental behaviour and expected future reproduction, and the dynamics of parental behaviour with territory defence. Future research might well be directed into the following areas:

(1)   *The effects of social status on optimal reproductive effort.* Throughout we have assumed that the optimal reproductive effort (RE*) is independent of resource-budget size, which we supported with four species of fishes in which both present and future reproduction increase with the size of the resource budget. There may be an interesting exception to this assumption, however. Specifically, in many species of fishes, dominant individuals, who presumably have the largest resource budgets, can inhibit or even prevent the reproduction of subordinate individuals (e.g. Jones and Thompson 1980). Thus an individual's rate of return on investment into present reproduction may depend on social status, which in turn will affect RE*.

(2)   *Indivisible resources.* For mathematical ease, we limited our model

and its interpretations to those parental behaviours that represent divisible resources to the offspring (Wittenberger 1981). Guarding, a divisible resource, is the most common parental behaviour in fishes; however, other parental behaviours, such as aeration of the eggs and feeding the young, may represent indivisible resources (i.e. what is given to one offspring cannot be share by other offspring). If the eggs or fry are under competition for parental care, then our model would have to be made more complex to describe parental behaviour. Unfortunately, there has been little research on competition among siblings for parental resources in fishes.

(3)   *Simultaneous tradeoffs.* In the final section above, we presented evidence that parental male fishes may be maximising remaining lifetime reproductive success subject to two simultaneous tradeoffs: (a) present versus future reproduction; and (b) within present reproduction, quantity versus quality of offspring. More empirical studies are needed to determine whether the phenomenon of simultaneous tradeoffs is widespread, and, if so, how they interact with each other. Should these simultaneous tradeoffs be important, they will have to be incorporated into future models.

(4)   *Dynamic models.* For mathematical ease, we limited our attention to static situations. Essentially, we described parents with similar backgrounds and suddenly subjected some of them to different conditions. Then we asked 'How does this perturbation affect RE* among treatments?' In a more comprehensive model, the tradeoff between present and future reproduction would change with the age of the parent and with the age of the offspring. Not only would such a model be very complex mathematically, but documentation and understanding the temporal dynamics of this tradeoff would be very difficult. Nevertheless, such models are being used in optimal foraging theory, and probably will soon become a part of parental investment theory.

Acknowledgements

This research was funded by an NSERC of Canada Operating Grant (UO244) to M.R.G. A NATO Postdoctoral Fellowship (National Science Foundation, USA) supported R.C.S., and M.R.G. was supported by NSERC of Canada University Research Fellowship.

References

Alexander, R.D. (1974) 'The Evolution of Social Behavior', *Annual Review of Ecology and Systematics*, 5, 325-83

Bagenal, T.B. and Braum, E. (1978) 'Eggs and Early Life History', in T.B. Bagenal (ed.), *Fish Production in Fresh Waters*, Blackwell, Oxford, pp. 165-201

Barlow, G.W. (1964) 'Ethology of the Asian Teleost *Badis badis*. V. Dynamics of Fanning and Other Parental Activities, with Comments on the Behavior of the Larvae and Postlarvae', *Zeitschrift für Tierpsychologie, 21*, 99-123

Baylis, J.R. (1981) 'The Evolution of Parental Care in Fishes, with Reference to Darwin's Rule of Male Sexual Selection', *Environmental Biology of Fishes, 6*, 223-51

Bell, G. (1984a) 'Measuring the Cost of Reproduction. I. The Correlation Structure of the Life Table of a Plankton Rotifer', *Evolution, 38*, 300-13

Bell, G. (1984b) 'Measuring the Cost of Reproduction. II. The Correlation Structure of the Life Tables of Five Freshwater Invertebrates', *Evolution, 38*, 314-26

Blumer, L.S. (1979) 'Male Parental Care in the Bony Fishes', *Quarterly Review of Biology, 54*, 149-61

Blumer, L.S. (1982) 'A Bibliography and Categorization of Bony Fishes Exhibiting Parental Care', *Zoological Journal of the Linnean Society, 76*, 1-22

Breder, C.M., Jr and Rosen D.E. (1966) *Modes of Reproduction in Fishes*, T.F.H. Publications, Neptune City, NJ

Colgan, P.W. and Gross, M.R. (1977) 'Dynamics of Aggression in Male Pumpkinseed Sunfish (*Lepomis gibbosus*) over the Reproductive Phase', *Zeitschrift für Tierpsychologie, 43*, 139-51

Downhower, J.F. and Brown, L. (1980) 'Mate Preferences of Female Mottled Sculpins, *Cottus bairdi*', *Animal Behaviour, 28*, 728-34

Gross, M.R. (1980) 'Sexual Selection and the Evolution of Reproductive Strategies in Sunfishes (*Lepomis*: Centrarchidae)', PhD thesis, Unversity of Utah, pp. 1-319, University Microfilms International No. 8017132, Ann Arbor, Michigan

Gross, M.R. and Sargent, R.C. (1985) 'The Evolution of Male and Female Parental Care in Fishes', *American Zoologist*, in press

Gross, M.R. and Shine R. (1981) 'Parental Care and Mode of Fertilization in Ectothermic Vertebrates', *Evolution, 35*, 775-93

Hirshfield, M.F. (1980) 'An Experimental Analysis of Reproductive Effort and Cost in the Japanese Medaka, *Oryzias latipes*', *Ecology, 61*, 282-92

Hirshfield, M.F. and Tinkle, D.W. (1975) 'Natural Selection and the Evolution of Reproductive Effort', *Proceedings of the National Academy of Sciences, 72*, 2227-31

Huntingford, F.A. (1977) 'Inter- and Intraspecific Aggression in Male Sticklebacks', *Copeia, 1977*, 158-9

Jones, G.P. and Thompson, S.M. (1980) 'Social Inhibition of Maturation in Females of the Temperate Wrasse *Pseudolabrus celidotus* and a Comparison with the Blennoid *Tripterygion varium*', *Marine Biology, 59*, 247-56

Keenleyside, M.H.A. (1979) *Diversity and Adaptation in Fish Behaviour*, Springer-Verlag, Berlin

Keenleyside, M.H.A. (1983) 'Mate Desertion in Relation to Adult Sex Ratio in the Biparental Cichlid Fish *Herotilapia multispinosa*', *Animal Behaviour, 31*, 683-8

Kleiman, D.G. (1977) 'Monogamy in Mammals', *Quarterly Review of Biology, 52*, 39-69

Kleiman, D.G. and Malcolm, J.R. (1981) 'The Evolution of Male Parental Investment in Mammals', in D.J. Gubernick and P.H. Klopfer (eds), *Parental Care in Mammals*, Plenum, New York, pp. 347-87

Kynard, B.E. (1978) 'Breeding Behavior of a Lacustrine Population of Threespine Sticklebacks (*Gasterosteus aculeatus* L.)', *Behaviour, 67*, 178-207

Lack, D. (1968) *Ecological Adaptations for Breeding in Birds*, Methuen, London

Lynch, M. (1980) 'The Evolution of Cladoceran Life Histories', *Quarterly Review of Biology, 55*, 23-42

Maynard Smith, J. (1977) 'Parental Investment: a Prospective Analysis', *Animal Behaviour, 25*, 1-9

Perrone, M., Jr and Zaret, T.M. 'Parental Care Patterns of Fishes', *American Naturalist, 85*, 493-506

Pianka, E. and Parker, W. (1975) 'Age-specific Reproductive Tactics', *American Naturalist, 109*, 453-64

Porter, K.R. (1972) *Herpetology*, W.B. Saunders, Toronto

Pressley, P.H. (1981) 'Parental Effort and the Evolution of Nest-guarding Tactics in the Threespine Stickleback, *Gasterosteus aculeatus* L.', *Evolution, 35*, 282-95

Real, L.A. (1980) 'On Uncertainty and the Law of Diminishing Returns in Evolution and Behavior', in J.E.R. Staddon (ed.), *Limits to Action: the Allocation of Individual Behavior*, Academic Press, New York, pp. 37-64

Reznick, D. (1983) 'The Structure of Guppy Life Histories: the Tradeoff between Growth and Reproduction', *Ecology*, *64*, 862-73

Salthe, S.N. and Meacham, J.S. (1974) 'Reproductive and Courtship Patterns', in B. Lofts (ed.), *Physiology of the Amphibia*, vol. 2, Academic Press, New York, pp. 309-521

Sargent, R.C. (1981) 'Sexual Selection and Reproductive Effort in the Threespine Stickleback, *Gasterosteus aculeatus*', PhD thesis, State University of New York, Stony Brook, pp. 1-85, University Microfilms International No DA 8211237, Ann Arbor, Michigan

Sargent, R.C. and Gebler, J.B. (1980) 'Effects of Nest Site Concealment on Hatching Success, Reproductive Success, and Paternal Behavior of the Threespine Stickleback, *Gasterosteus aculeatus*', *Behavioral Ecology and Sociobiology*, *71*, 137-42

Townshend, T.J. and Wootton, R.J. (1984) 'Effects of Food Supply on the Reproduction of the Convict Cichlid, *Cichlasoma nigrofasciatum*', *Journal of Fish Biology*, *23*, 91-104

Trivers, R.L. (1972) 'Parental Investment and Sexual Selection', in B. Campbell (ed.), *Sexual Selection and the Descent of Man*, Aldine-Atherton, Chicago, pp. 136-79

Unger, L.M. (1983) 'Nest Defense by Deceit in the Fathead Minnow, *Pimephales promelas*', *Behavioral Ecology and Sociobiology*, *13*, 125-30

van den Assem, J. (1967) 'Territory in the Three-spined Stickleback, *Gasterosteus aculeatus* L. An Experimental Study in Intra-specific Competition', *Behaviour Supplements*, *16*, 1-167

van Iersel, J.J.A. (1953) 'An Analysis of Parental Behaviour of the Male Three-spined Stickleback (*Gasterosteus aculeatus* L.)', *Behaviour Supplements*, *3*, 1-159

Williams, G.C. (1966a) 'Natural Selection, the Costs of Reproduction, and a Refinement of Lack's Principle', *American Naturalist*, *100*, 687-90

Williams, G.C. (1966b) *Adaptation and Natural Selection*, Princeton University Press, Princeton

Williams, G.C. (1975) *Sex and Evolution*, Princeton University Press, Princeton

Wilson, E.O. (1975) *Sociobiology: the New Synthesis*, Harvard University Press, Cambridge

Wittenberger, J.F. (1981) *Animal Social Behavior*, Duxbury Press, Boston

Wootton, R.J. (1976) *The Biology of the Sticklebacks*, Academic Press, London

Note

1. R.C. Sargent's unpublished experiment (Table 11.3) has been analysed as the resolution of simultaneous trade offs, and is in press in *Behaviour*

12 FUNCTIONS OF SHOALING BEHAVIOUR IN TELEOSTS

Tony J. Pitcher

Predators and food are the keys to understanding fish shoals; synchronised co-operation defeats predators, and optimal food gathering in shoals reflects a shifting balance between joining, competing in, or leaving the group. In the wild, predators may arrive while shoaling fish are feeding, and so vigilance is a crucial behaviour. Once detected, predator defence takes precedence over feeding, since an animal's life is worth more than today's dinner.

Travelling fish schools display impressive co-ordination and were once viewed as egalitarian leaderless societies (Breder 1954; Shaw 1962; Radakov 1973) in which co-operation preserved the species. In contrast to such classical group-selectionist views, contemporary ethology reveals social behaviour to be nothing more than animals co-operating only when it pays. Distinct coexisting behavioural strategies of sneaking or scrounging are often evolutionarily stable (Barnard 1984; Parker 1984). In fish shoals, homogeneity and synchrony have been overemphasised; recent work reveals that individuals constantly reappraise the costs and benefits of being social. Reappraisal is reflected in decisions to join, stay with or leave groups, and observed behaviour allows us a glimpse of these underlying tensions. In teleosts, a major constraint is swimming; fish physiologically and morphologically adapted to cruise fast, such as mackerel, break ranks less often to avoid the alternative of rapid dispersal. Under some circumstances, however, the underlying tensions between individuals may be uncovered in even the most phalanx-like of cruising fish shoals.

Shoaling behaviour has attracted much speculation about function (e.g. Shaw 1978; Partridge 1982a), but until recently few critical experiments have been performed. The aim of this chapter is to review in the light of current theory the areas where such evidence of function has been gathered. The arguments presented in this chapter do not support the views of Hamilton (1971) or Williams (1964) that shoaling is primarily a matter of cover-seeking, or Breder's (1976) view that hydrodynamics is the major factor. Furthermore, I will present arguments that, for fish shoals, simple attack avoidance and attack dilution have been incorrectly associated with selection for grouping behaviour.

Definitions of Shoaling

A clarification of terms will facilitate discussion of function, since shoaling behaviour continues to suffer from the semantic confusion of the 1960s

294

(Keenleyside 1955; Hemmings, 1966; Radakov 1973). Groups of fish which remain together for social reasons are best termed 'shoals' (Kennedy and Pitcher 1975; Pitcher 1983), in an analogous way to the term 'flock' for birds. Defined as a social group of fish, 'shoal' has no implications for structure or function.

Synchronised and polarised swimming groups are termed 'schools'. Schooling is therefore one of the behaviours exhibited by fish in shoals (Figure 12.1), and schools have a structure measured in polarity and synchrony. Shoals which travel are almost bound to school on the way. Other action patterns of fish in social groups, such as foraging or anti-predator manoeuvres, can be accommodated within the taxonomy outlined in Figure 12.1 (Pitcher 1983).

In North America the term 'school' is still used to cover both phenomena distinguished in this chapter. The dichotomy between 'facultative' and 'obligate' schooling fish introduced by Breder (1959) has, however, been largely rejected in favour of direct estimates of the proportion of time which shoaling fish spend travelling in polarised groups

Figure 12.1: Relationship of Definitions of Shoaling and Schooling Behaviour Illustrated by a Venn Set-theory Diagram. Criteria for the two behaviours are indicated. Three other behaviours are superimposed as examples of how other functional categories relate to these terms. Some further specific examples of anti-predator tactics are shown: spawning and foraging behaviours might be similarly elaborated. Unlike previous definitions, this scheme can augment descriptions of observed behaviours (from Pitcher 1983)

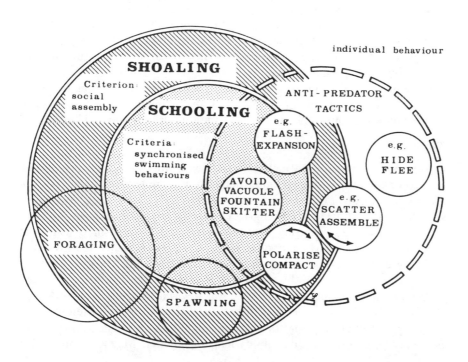

(Partridge 1982b; Pitcher 1983). The reasoning behind this set of concepts is discussed in full in Pitcher (1983).

Perspective on Structure and Function in Shoals

Much of the long search for structure in fish schools has been sterile for the explanation of function, not least because of the failure to distinguish between shoaling and schooling. Measurements of position and movements for individual fish, crucial to tests of current theory, have generally been lost in pooled values for the structural parameters of nearest neighbour distances, bearings and polarity (Cullen, Shaw and Baldwin 1965; Hunter 1966; Pitcher 1973; Inagaki, Sakamoto, Aoki and Kuroki 1979; Partridge, Pitcher, Cullen and Wilson 1980).

Individuals have been recognised only rarely; only two rigorous accounts of individual positions having been published. Healey and Prieston (1973) identified peripheral and central fish in coho salmon (*Oncorhynchus kisutch*) schools, and Pitcher, Wyche and Magurran (1982) measured individual position preferences in mackerel (*Scomber scombrus*) schools. Muscialwicz and Cullen (unpublished experiments) have demonstrated consistent individual 'initiators' and 'followers' in zebra danio (*Brachydanio rerio*) shoals, and Pitcher (1979b) presented similar evidence in bream (*Abramis brama*) shoals, but did not recognise consistent individuals. Work on school structure has now begun to assess the costs and benefits to individuals, even if they are difficult to translate into the measures of lifetime fitness which strict interpretations of theory demand (Krebs and Davies 1978). This is not least because of the difficulty of expressing anti-predator, feeding and hydrodynamic advantages in equivalent units. Recent findings on anti-predator tactics, choice of neighbour, sorting by size, joining schools, and the acquisition and transmission of information about food and predators are described in detail below.

Elective Group Size in Shoals

In the looser organisation of the shoal, behaviour for spatial positioning and co-ordination with swimming neighbours is not needed, but we still require a measure of cohesion (packing) and synchrony since, as we will see, these are adjusted according to circumstances. Cohesion in terms of the distances and synchrony between fish could be measured using the same three-dimensional methods as used for schools (Pitcher 1975; Pitcher and Lawrence 1984) but can usually be adequately scored by eye, as in the aggregation scores used by John (1964), Andorfer (1980) and Cerri (1983). A more quantitative measure was devised by Helfman (1984) and Wolf (1983), who presented a distribution of fish groups of different sizes.

Using a stricter criterion for a group, including only fish considered to be behaving together within 4 body lengths, Pitcher, Magurran and Allan

(1983) introduced the term 'elective group size' (EGS). ('Elective' distinguishes the measure from group sizes deliberately created by the experimenter, as in Pitcher, Magurran and Winfield 1982.) EGS turns out to be a sensitive measure of the fish's perception of the current balance of costs and benefits of shoaling, and can be measured either for individuals or for groups. For example, Figure 12.2 shows the distribution of EGS values in undisturbed shoals of minnows (*Phoxinus phoxinus*) and dace (*Leuciscus leuciscus*) foraging in semi-natural conditions in a fluvarium. In these shoals, many small fish find it of benefit to forage individually or in small groups, whereas larger fish are in somewhat larger shoals.

Figure 12.2: Elective Group Size (EGS) for Two Length Classes of Dace and Minnows Shoaling on Separate Occasions in a Large 10m Fluvarium Tank. Twenty-five small and large fish of each type were present (mean lengths, dace: 175 and 97mm, minnows: 56 and 39mm). Histograms give the percentage of fish observations for each observed group size: total observations is shown as *n*. Groups were scored as fish within 5 body lengths, recorded at a standard time once per day over a 10 day period. Median group sizes are indicated by the solid circles, 90 percentiles by the bars (for further details see Pitcher *et al.* 1983)

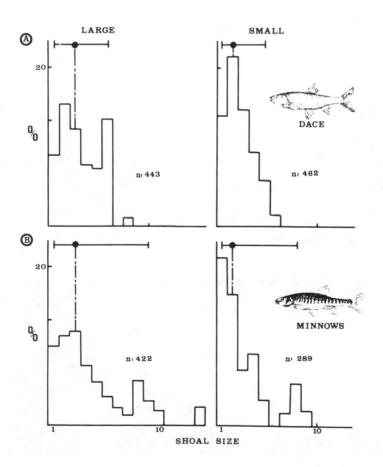

Similarly, McNicol and Noakes (1984) have demonstrated a very sensitive response of juvenile brook trout (*Salvelinus fontinalis*) to crowding, group numbers, food and water flow. Behaviour changes rapidly from shoaling through individual territory-holding to dominance hierarchies.

Hydrodynamic Advantage

The beating tail of a swimming fish generates a wake of counter-rotating spinning vortices, which can be visualised using special techniques (Figure 12.3). It is reasonable to assume that schooling fishes avoid swimming into the wake of those in front and this constrains their positions in the school. Teleost fishes exhibit a wide diversity of tail structure, including tails specialised for acceleration (e.g. barracuda), fast cruising (e.g. tuna), or manoeuvrability (e.g. perch) (Yates, 1983), but it is not yet known if these differences influence fish positions in schools as would be expected.

Figure 12.3: Stationary Thrust Vortices Seen behind a 50cm-long Swimming Cod Shown on a Still Frame from Videotape. The cod swam in an annular track within a 10m diameter tank. Distance between the vortices is about 0.5m, which were generated by a cod 20cm below the water surface which has left the field of view. Vortices, which are in the positions predicted by hydrodynamic theory, were made visible by shining red light at a shallow angle so that slight disturbances of the water surface cast clear shadows. At the top of the figure is a shadow cast by a fence used to keep the fish in the annular track. (Further details are given in Partridge and Pitcher 1979.)

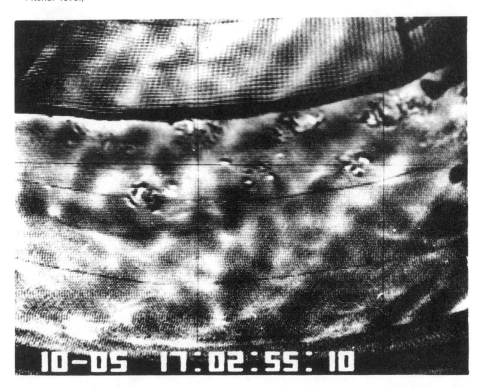

The idea of positive hydrodynamic advantage, rather than passive avoidance of wakes, has been put forward several times (Breder 1965; Belyayev and Zurev 1969). Weihs (1973, 1975) formalised the theory, making five specific predictions about spacing and behaviour in travelling schools which would enable fish to make energy savings up to 65 per cent. These were:

(1)   Lateral neighbours should be 0.4 body lengths apart.
(2)   Their tailbeats should be in antiphase.
(3)   Fish in the next row should be centred between the leaders, 5 body lengths behind.
(4)   Lateral neighbours should swim 0.3 body lengths apart to maximise benefit from a lateral 'push-off' effect.
(5)   Uplift from rigid pectorals in negatively buoyant fish could aid fish behind.

Fish gaining benefits 1 to 3 should swim in a diamond lattice.

None of the extensive work on the three-dimensional structure of fish schools gives evidence of the rigid diamond lattice envisaged by Weihs, but there remains the possibility that advantage could be gained through a statistical tendency to adopt the advantageous positions. Using a very large amount of three-dimensional data on cruising herring (*Clupea harengus*), cod (*Gadus morhua*), and saithe (*Pollachius virens*) schools ($n \simeq 20000$), Partridge and Pitcher (1979) demonstrated that predictions 1 to 3 were not met. For prediction 3, even when four out of only 659 fish found centred behind two leaders were 5 body lengths behind, they did not maintain these putatively beneficial positions as might be expected. In fact, the snouts of following fish were very often ahead of the region where the swimming vortices stabilise, and often ahead of the tail itself, so no hydrodynamic benefit could accrue on this theory. Since tailbeat frequency does not alter vortex spacing as the fish swim faster, Weihs' hydrodynamic model predicts no change in positions with swimming speed. Yet in fact school structure alters and becomes more compact with speed (Pitcher and Partridge 1979). MacCullum and Cullen (in preparation) have also shown that the tailbeat prediction, '2' above, does not hold for schooling yellowtail scad (*Trachurus mccullochi* see index under Carangidae).

Very few teleost fish hold their pectorals rigidly out when swimming (indeed the fast cruising, negatively buoyant scombrids have evolved precise grooves in which to tuck them away). Elasmobranchs which swim with rigid pectorals to generate lift rarely school in the same cohesive way as teleosts, so there is no evidence for prediction 5. (Sharks do seem to shoal, however.)

A much more serious criticism of the hydrodynamic hypothesis can be put forward on theoretical grounds. Weihs' benefits 1 to 3 accrue only to alternate rows in the school, and therefore are evolutionarily unstable in the absence of altruism. There is no hint of the likely scramble for good positions by fish finding themselves in non-benefit rows. It has been suggested that if advantageous fish positions are shuffled by turns, hydrodynamic benefit might be shared in time (Blake 1983), but this does not overcome the fundamental objection that no evidence of instability in poor positions is seen.

Benefit 4, the lateral 'push-off' effect, is fundamentally different in this respect since individual fish gain from their own behaviour, but no unequivocal evidence in support has been produced yet. The saithe studied by Partridge and Pitcher (1979) swam about 0.9 body lengths apart and would have gained about 35 per cent of the maximum benefit from this lateral effect had their neighbours been on the same horizontal level. Saithe neighbours, however, tended to be about 10° above or below. For the lateral advantage to be maximised, fish should choose to swim next to neighbours of the same size. Evidence of a statistically significant tendency towards such choice of neighbour size has been found in herring and mackerel schools (Pitcher, Magurran and Edwards 1985). Additional evidence supporting the lateral-neighbour effect has been put forward for bluefin tuna schools (Partridge, Johansson and Kalish 1983).

Abrams and Colgan (1985) showed that shiner (*Notropis heterodon*) shoals became deeper vertically and more compact when attacked by largemouth bass. They argued that the fish sacrificed a flat, hydrodynamically efficient structure in favour of an unobstructed visual field in the presence of a predator. The assumption that a flat school is hydrodynamically efficient is based upon a misreading of the hydrodynamical theory: in particular Weihs' analysis is based upon one layer out of a potentially multi-layered school. In fact, the presence of fish in adjacent vertical layers enhances the 'push-off' effect in Weihs' theory. The shiners in Abrams and Colgans' experiment behaved very like other cyprinids, described below, whose change in shoal structure on encountering a predator has little to do with hydrodynamics.

There have been many reports of fish using less oxygen in groups (Abrams and Colgan 1985; see Parker 1973 for earlier references), and hydrodynamic advantage is often claimed as the cause of the lower energy demands. Unfortunately, not one of these experiments has contained an adequate control for the motivational effects of group size. Group size has a profound effect upon the behaviour of individual fish. As will be discussed later, fish in smaller groups are more timid and nervous and consequently have higher respiratory rates (Itazawa, Matsumoto and Kanda 1978). Since small shoals are less likely to detect a predator early and are less effective in escaping from an attack, the higher respiratory rate of nervous fish can be interpreted functionally as a greater readiness for careful, considered evasion manoeuvres using red muscles. (Emergency flight in teleosts employs the white muscle, which is anaerobically fuelled.)

Drag-reducing fish mucus has been linked to hydrodynamic advantage in schools. By adding Polyox, a high molecular weight polymer of ethylene oxide, to the water, Breder (1976) showed that menhaden (*Brevoortia patronus*), a fast-cruising schooling species, beat their tails almost twice as fast as controls and swam proportionally faster. Polyox, which was shown not to be toxic, reduced drag on the fish's beating tails. Breder argued that the Polyox simulated the effect of fish mucus in reducing drag in schools, but his experiment did not control for slime production nor for group size and cannot therefore be taken as conclusive.

In conclusion, no valid evidence of hydrodynamic advantage in travelling schools has been produced, and existing evidence contradicts most aspects of the only quantitatively testable theory published. It is certainly surprising that no evidence has been put forward that fish positions in schools are, at the very least, constrained by the thrust vortices shown in Figure 12.3. The idea of hydrodynamic advantage remains attractive and plausible: it seems possible that the theory needs more attention. On the other hand, hydrodynamics could not help fish in unstructured shoaling, and is therefore unlikely to have been a primary reason for the evolution of the loose social groups we term shoals.

Countering Predators

Behaviour in fish shoals has been shaped by an evolutionary arms race (Dawkins and Krebs 1979) with predators; what we currently observe is behaviour reflecting the shoaling fish's tendency to stay one fin ahead in this race. The behaviour has, of course, also been shaped by foraging, as dealt with in the latter part of this chapter: predator attack must have evolutionary priority though (Dawkins and Krebs' life/dinner principle). This section aims to review critically how shoaling fish may counter predator attack by avoidance, dilution, abatement, evasion, confusion, detection, mitigation, inspection, inhibition and prediction. In particular, it aims to dispel three fallacies which appear widely in the literature:

(1)   Avoidance is enhanced by grouping.
(2)   Probability of capture by a predator is diluted by grouping *per se*.
(3)   Fish in shoals behave homogeneously.

Predator Avoidance

When visual range is low in relation to the relative speed of predator and prey, grouping of prey has been considered to be favoured over dispersal since the predator has less chance of coming within the detection envelope of a group than the many such envelopes of scattered fish (Treisman 1975). Although implicitly group-selectionist since the detection probability for any particular fish is identical in both cases, this wheel has been re-invented several times (Brock and Riffenburgh 1960; Olsen 1964; Cushing and Harden-Jones 1968; Wilson 1975; Partridge 1982a) and, as first pointed out by Williams (1964), is profoundly mistaken. In one variant of the theory (see Taylor 1984) predators who encounter fewer prey as a consequence of overlapping detection envelopes of grouped prey are supposed to go away somewhere else. All of these authors confuse the probability of detection of a shoal with that of an individual fish (see Figure 12.4a). The average group member is protected only if the consumption rate of the predator is less when feeding on groups: the consumption rate must be considered over many encounters and there is no evidence that this is less for grouped prey because of detection differences, rather than more immediate factors which will be considered below. Furthermore, the standard argument implies that a group member is protected only on detection by a predator, so avoidance can hardly be the important factor.

The argument about detectability of shoals remains valid, however, and under good visibility (clear environment and/or high visual acuity) the reverse situation holds; for example, flocks of birds are generally more conspicuous than individuals. The predator-avoidance hypothesis is usually put forward in the context of marine pelagic schooling species. It is important to note that the hypothesis makes no predictions about the behaviours to be adopted by fish once within the predator's detection envelope.

Two additional arguments lead me to believe that the passive-avoidance hypothesis is not supported by evidence, and has not therefore had a major influence on the evolution of shoaling.

Figure 12.4A: The Search Avoidance Effect is a Fallacy. Diagram illustrating the fallacy of reduced probability of detection. Shaded circles represent spheres of detection around potential fish prey for a searching predator. The chances of the predator finding the shoal in the right-hand diagram are less than for finding any solitary fish in the left-hand diagram, but for any particular individual prey fish they are identical in both cases (solid bars). Note that the search and consumption rates of the predator are assumed to be the same in both cases. The search rate is in fact irrelevant to this model. For a given search rate the predator may consume slower in the right-hand diagram and will therefore become hungrier. Some authors assume that this means the predator leaves the area (or model?) protecting the shoaling fish, but in practice a hungry predator may hunt harder, selecting against grouping

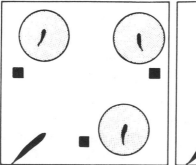

First, like herds of mammals, many shoals have been shown to be intimately accompanied by their fish predators (Pitcher 1980). For example, roach (*Rutilus rutilus*) in a British river were never out of range of predatory pike (*Esox lucius*); barracuda (*Sphrynaenidae*) accompany shoals of grunts (*Pomadasyidae*) on coral reefs. For fish living near cover in some form, such as weeds or substrate, dispersal and refuging would meet the end of predator avoidance more effectively than grouping. When there are few conspecifics around, this is precisely what shoaling minnows do (Marrugan and Pitcher 1983, 1986). Similarly, Savino and Stein (1982) found that bluegill sunfish exposed to largemouth bass predators schooled less when cover was increased.

Secondly, fish shoals near the surface are more visible *from the air* than individuals. This takes care of the argument about pelagic shoalers, since their diving avian predators (gannets, terns, pelicans, auks) certainly detect shoals at long range. Avian predation is a major source of mortality in such fish, and so, by virtue of this conspicuousness, shoaling cannot have been selected to minimise detection. Again dispersal would be more effective than huge marine shoals for simple avoidance of these predators.

In conclusion, like social groups in other vertebrates, I think that it is unlikely that shoaling behaviour evolved to avoid detection by predators. This end is better served by crypsis or refuging, an evolutionary route taken by the many teleosts, such as blennies and lurking predators, which rarely shoal.

There are, however, several observations which suggest that minimising detection is important in shaping the behaviour of shoals that already exist. The oblate spheroid shape of many shoals is likely to be an adaptation in this respect, which minimises the detection envelope of the shoal under water (Pitcher and Partridge 1979). Elongated and flattened shoals produce a spherical envelope of visibility, a smaller detection volume than the ellipsoid envelope of a spherical object (see Chapter 4 by Guthrie, this volume).

Helfman (Chaper 14, this volume) shows how the behaviour of shoals on coral reefs may be temporally and spatially patterned by minimising contact with predators. Daily migration routes are perpetuated culturally in grunt shoals (Helfman and Schultz 1984).

Dilution of Attack and the Abatement Effect

The often-quoted 'dilution effect' appears to be widely misunderstood. The theory states that an individual group member gains advantage simply through a reduced probability of being the one attacked in an encounter with a predator (Bertram 1978; Foster and Treherne 1981). This probability reduces as the reciprocal of group size, giving the prediction that the log attack rate per individual will plot linearly with a slope of -1 against log group size. Such relationships have been unequivocally demonstrated in several experiments and in Foster and Treherne's paper (see below). The point, however, is that simple dilution of attack cannot be a selection pressure favouring grouping because the effect applies only to members of a particular group being attacked.

The fallacy can be demonstrated using a simple example. Assume that a predator has a fixed attack rate and picks an individual prey at random, and that all prey are within detection range. Consider the analogy of predator with perfect vision armed with a rifle. For a fixed total number of individuals distributed among several groups, the average probability of being the one attacked is the same whatever the size of the groups. This may be explained as follows. When in small groups, the member of an attacked group has a high probability of being the one picked, but this group will be attacked seldom: when this particular fish is in a larger group, he will suffer attacks more often even though the dilution effect reduces his individual risk per attack (see Figure 12.4b). The point is further illustrated in naval strategy: unescorted convoys should disperse once found and attacked. The purpose of the convoy is mutual protection and as an aid to armed escort, not attack dilution, as exemplified by the effects of Exocet cruise missiles during the Falklands war.

Foster and Treherne (1981) claim to have distinguished a dilution effect from predator confusion (see below) in clupeids attacking flotillas of surface-living insects. Fish are not detected by the insects until an attack occurs, and so the picture is not confounded by vigilance or social alarm signals. Almost exactly as predicted by the dilution law, individual insects in larger groups suffered a lower attack rate. The confusion effect (see below) was distinguished since attacks launched on larger groups of the constantly moving insects were less successful. The results support dilution

Figure 12.4B: The Dilution Effect is a Fallacy. Diagram illustrating the fallacy of the simple dilution effect as represented in the literature (see text). Open bars indicate the probability of an individual being in a group which is attacked. Solid bars show the probability of being the one picked from the group attacked. The overall probability of being killed by the predator is the same for both solitary (left-hand diagram) and group (right) strategies. Note that this model assumes that all prey are within attack range (shaded zone). Search is not included in this model, and solitary/shoaling behaviours are considered mutually exclusive strategies

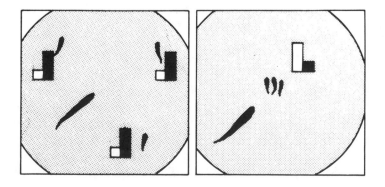

of attack on particular groups, but, as argued above, this does not prove dilution to be a selection pressure leading to grouping. Foster and Treherne's results could be explained by an alternative hypothesis since they ignore possible reluctance by the fish to attack larger groups, a behaviour which can be seen in many fish predators. In fact, larger insect groups suffered significantly fewer attacks than individuals, an observation supporting such predator reluctance. In support of the dilution theory, both of these authors cite Hamilton's (1971) theoretical analysis of an unseen snake attacking frogs in a pond, but in fact Hamilton's argument is not concerned with dilution in the sense used by these authors. Hamilton analyses the dangers of individuals finding themselves on the margins of groups, leading to behaviour in which animals seek cover behind others. Marginal individuals are proportionally fewer in larger groups. Hamilton's snake is very like our predator with a rifle since all frogs are detected and within range.

A form of attack abatement can operate, provided that attack is considered along with search or encounter. Previously we assumed that all prey were in equal-sized groups, but we will now allow groups of different sizes to coexist, and endow individuals with the freedom to join or leave any group. A searching predator encounters groups of prey rather than individuals. When a prey group is encountered, one individual is singled out for attack. Unlike our predator with perfect vision and a rifle, this time we draw an analogy with a nocturnal predator searching through the habitat equipped with low-resolution radar and an old firearm accurate only at close range. Under these circumstances, grouping pays: for example, an individual leaving a group of 100 reduces its chance of encounter from one to a half, but its chance of death given an attack goes from one-hundredth to one (J.R. Krebs, personal communication). I term this the 'attack abatement' effect to distinguish it from previous analyses (see Figure 12.4c). In a further exploration of this argument, Turner and Pitcher (1985) show that the attack abatement ESS is for all prey to group, even though this may not maximise benefit for all individuals.

The simple dilution effect may be a fallacy, but the attack dilution in a larger group could confer benefits indirectly in a more subtle way, through the wider spread of information that a hungry predator was in the vicinity. More fish gain this information in larger groups once an attack is launched, and the analogy with ships under radio silence in a naval convoy knowing of the vicinity of a U-boat is still valid. Because of dilution, individuals are able to experience an attack and gain knowledge about the attacking predator at no extra cost. The benefits of this knowledge would not always be immediate, but may be delayed until fish were able to join in group evasion manoeuvres, as described below. Most individuals in small scattered groups would not have this advantage, although the precise level at which the effect operated might vary with the nature of the predator and the scale of the grouping. The value of knowledge about the proximity of a predator

Figure 12.4C: The Attack Abatement Model. Diagram showing how search and dilution acting together may favour grouping, termed here the attack abatement effect to distinguish it from previous analyses. Shifts between grouping and solitary behaviour are allowed in this model (open prey fish shape). The predator searches for groups rather than individuals, and so the chances of being in the group attacked depend upon the number of groups. Left-hand diagram illustrates two groups, a group of two and a solitary prey fish: right-hand diagram shows the situation after the solitary fish has joined the pair to form a single group of three. Solid bars represent the probability of being the one eaten if the group is attacked; open bars the probability of being in the group found and attacked. Shaded areas are detection zones. In this diagram the solitary fish does better by joining to become a member of a group of three, but the original pair now does worse: however, the ESS in this model is to join (Turner and Pitcher 1985). Note that the predator search for groups out of range is an essential feature of the abatement model

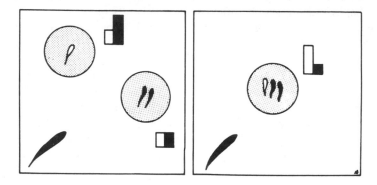

could help select for grouping behaviour, but it is difficult to see how this 'cognitive dilution benefit' from attack dilution could be tested directly by experiment.

Barnard (1984) suggests a related benefit through time-wasting: once a kill is made, the rest of a group can escape while the predator handles the prey. This benefit would help any group animals so that, balanced against the probability of being the one attacked, a clear benefit of attack dilution emerges. Taylor (1984) derives a mathematical model equivalent to this 'time-wasting' hypothesis.

Predator Evasion

Experiments have confirmed that the attack success of aquatic piscivores declines with prey group size. This has been shown for cephalopod, pike and perch predators (Neil and Cullen 1974), sticklebacks (Milinski 1979a), pirhanas (Tremblay and FitzGerald 1979), and mink (Poole and Dunstone 1975) and seems to be widely applicable. Minnows who find themselves apart from a group are more likely to be eaten by pike (Magurran and Pitcher 1986), and pike have reached a stage in the arms race where their attack behaviour aims to split individuals from the shoal. How, then, does shoaling protect fish under attack?

When a potential underwater predator is detected, shoals become more cohesive (e.g. Andorfer 1980; Rüppell and Gösswein 1972). The dramatic effect of a predator on the elective group size of a minnow shoal is illustrated in Figure 12.5; compaction persists for some considerable time after the predator has left. Two experiments have examined mixed-species shoals. Under threat of predation, mixed-species shoals of cyprinids sort into conspecific groups (Allan and Pitcher 1985). Odd fish with few conspecifics seek refuge, but shoal if enough of their own species are present (Wolf 1985). Small groups which become separated from a larger shoal may rejoin through a narrow 'neck' or 'pseudopodium' of rapidly moving fish, two or three fish wide: the smaller shoal appearing to deflate like a balloon as its members transfer to the main body (see Figure 12.6; Radakov 1973; Pitcher and Wyche 1983). Clearly, behaviour has evolved that puts a premium on preserving the integrity of the group under threat.

The reason for shoaling fish adhering to conspecifics when alarmed is to take advantage of co-operative escape tactics such as those illustrated in Figure 12.6. Some tactics with a wide occurrence are discussed in detail below. Individuals co-operate because these behaviours only work when fish behave as a co-ordinated group: detailed examination of many of the tactics reveals the tensions between individuals in the form of exploitation (or 'sneaking') benefit produced by others, such as is described for bird flocks under predation (Charnov and Krebs 1975); however, this study is still in its infancy in fish shoals.

(1) Evade and the Minimum Approach Distance. Through judiciously unhurried movements, fish attempt to maintain a distance of 5 to 15 body lengths between themselves and the predator (e.g. Jakobssen and Jarvik 1978; Pitcher and Wyche 1983; Magurran and Pitcher 1985). Weihs and Webb (1983) have shown a clear benefit in maintaining such a minimum approach distance in order to preserve an advantage in relative manoeuvrability and acceleration by the smaller prey. In itself, though, this applies no more to shoals than to individuals.

(2) Fountain Effect. This manoeuvre has been widely reported (Potts 1970; Nursall 1973; Radakov 1973; Pitcher and Wyche 1983; Magurran and Pitcher 1986). Velocity relative to the predator is maximised and the shoal reassembles behind the stalking predator. Wardle (Chapter 18, this volume) shows how this behaviour, which is also observed around divers or fishing gear, can come about through a simple visual response in groups of fish.

(3) Trafalgar Effect. Webb (1980) proved that the average latency between sudden stimulus and fast acceleration to escape was less in shoals than for individuals. Webb's result was almost certainly brought about by the rapid transfer of information across a shoal by observation of and reaction to the startles of fellow fish, as much as to external frightening stimuli. Treherne and Foster (1981) likened this to the rapid transfer by

Figure 12.5: The Dramatic Effect of Exposure to a Predator on the Elective Group Size (EGS) of Shoals of 20 minnows. Measures were made in a large arena tank. Three separate replicates are shown from left to right. Each histogram gives the relative frequency of records of groups within 4 body lengths scored under standard conditions. Medians, quartiles and numbers of observations are given above each histogram. (From Pitcher *et al.* 1983.) (a) Baseline measurements for undisturbed minnows; (b) 2h after a 1h exposure to a hunting pike, which was then removed; (c) one day after the pike exposure

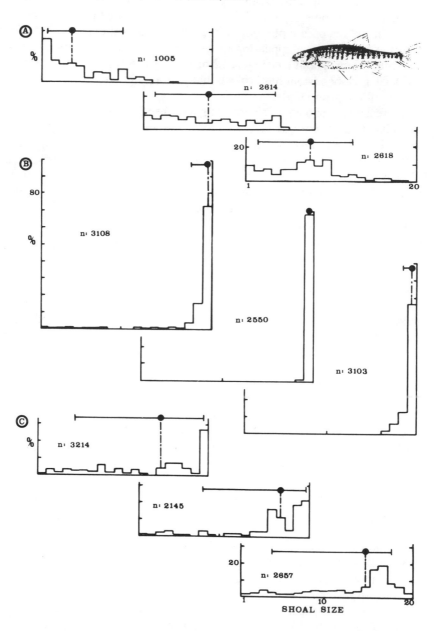

Figure 12.6: Illustration of the Repertoire of Anti-predator Tactics in Sandeel (US, Sand Lance; Ammodytidae) Schools under Threat from Hunting Mackerel. Tactics are drawn in plan view: a small turning school of sandeel is seen from the side in the foreground. Observations were made in a 12 m arena tank. (Further details are given in Pitcher and Wyche 1983.)

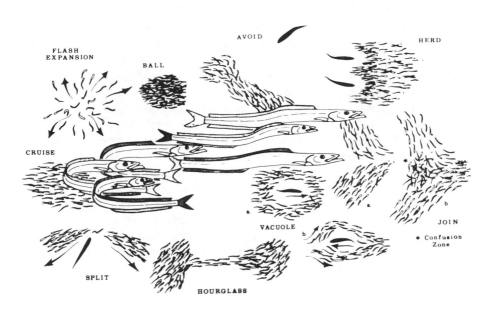

flags of battle signals through Nelson's fleet at Trafalgar, and estimated that a wave of reaction to an attack could propagate through a flotilla of surface-living insects seven times faster than the approach speed of the predator. Evasion turns are thought to spread through bird flocks in similar fashion (Potts 1984), although in this case birds synchronised their turns by anticipating the approaching wave of reaction. Unrehearsed high kicks in lines of human chorus girls spread faster than the human reaction time through a similar anticipation. Potts' 'chorus-line' hypothesis is therefore a subset of the Trafalgar effect. There is some evidence for an effect of this kind in saithe schools (Pitcher 1979a), but more detailed examination would be worth while.

(4) Ball. Pelagic shoalers, out of reach of cover, may take refuge in a tightly packed ball (Breder 1951; Pitcher and Wyche 1983). For example, over 500 sandeels (mean length 140 mm) threatened by auks were concentrated into a 50 cm ball measured by Grover and Olla (1983). For fish shoals, such jostling throngs provide the only clear example of Hamilton's (1971) 'selfish-herd' effect, in which animals avoid dangerous margins by seeking shelter in the centre of a group (see also Wilson 1975). In such a case, average individuals would probably be better off dispersed than in a conspicuous ball, but the benefits of getting the central locations must out-

weigh dispersal behaviour. Hamilton's selfish-herd theory does not illuminate other synchronised manoeuvres seen in fish shoals, or sorting into conspecifics in threatened heterospecific shoals.

Predator Confusion

Overt predator attack may be made less likely or less effective through confusion generated in the predator by multiple potential targets. It is sometimes implied that this 'confusion effect' is the main way in which a predator's effectiveness is reduced by shoaling (e.g. Milinski 1979b), but the effect is in reality only one of the categories surveyed here. Individuals who join a group all benefit equally from the confusion effect in its simple form, where predator attacks are spoilt by the multiple targets. As may be expected, however, detailed examination of the behaviour reveals conflicts as some individuals attempt to gain more of the benefit.

A great deal about the confusion effect remains to be elucidated: confusion may be perceptual, through overloading of the visual-analysis channel (either peripherally or centrally) (Broadbent 1965). Alternatively, confusion could be cognitive, the 'embarrassment of riches' effect, like a dog unable to choose between several juicy bones. Humans watching radar screens suffer from visual channel confusion, and Milinski (1985) quotes studies that suggest that humans suffer from perceptual confusion of targets similar to that shown by sticklebacks. All fish-behaviour workers are familiar with the difficulty of catching many small fish from a large tank. Because visual predators rely on complex movement and trajectory analysis (Guthrie 1980), it is difficult for them to overcome the perceptual-confusion effect. Large piscivores which lunge and gobble rapidly and randomly are one evolutionary solution to this problem. Smaller piscivores appear to have evolved a visual 'lock-on' in an attempt to minimise confusion, and, as we will see, this too has had consequences in the shoaling arms race.

Whatever the mechanism, several tactics performed by shoaling fish appear to enhance predator confusion beyond that engendered merely by multiple targets. This occurs mainly through increasing the relative movements between individuals. It is here that the opportunity to cheat on the equitably-shared standard confusion arises. Two analyses based on my recent work on minnow(*Phoxinus phoxinus*) shoals are described below.

In European minnows a sequence of behaviours based on 'skittering', a Mauthner-driven startle acceleration (Webb 1976; Eaton, Bombardieri and Mayer 1977; Guthrie 1980), reflects an evolutionary race between those who dodge and those who preserve shoal integrity (Magurran and Pitcher 1986). Minnows that skitter accelerate rapidly for 5 to 10 body lengths, brake, rise in the water 3 or 4 body widths, and then resume co-ordinated behaviour with the group, almost as if they have rapidly slotted into a new position in the shoal. Although skittering fish increase the confusion effect, thereby benefiting all shoal members, the important point is that an individual that skitters gains greater benefit by confounding a predator which may have attempted to 'lock on' to that individual. Neighbours to the original location are then placed at greater risk. This explains why the frequency of skittering increases as an attack progresses (Magurran and Pitcher 1986), and why the behaviour often occurs in bursts, appearing to spread through adjacent fish. Frequent slow and deliberate passings of one fish by another seen in polarised minnow schools under attack (see Figure 12.7a) may represent attempts to get neighbours to optomotor smoothly (Shaw 1965; Shaw and Tucker 1965) and inhibit skittering. (The optomotor response is to maintain station with a moving stimulus using visual cues; see Wardle, Chapter 18, this volume.)

Synchrony of skittering avoids the penalty while maintaining predator confusion. In minnows two behaviours achieve this end. The 'group jump' is a synchronised and polarised skitter by the majority of shoal members, which in effect relocates the whole shoal rapidly. It is not performed very often relative to normal skitters, which is not surprising as a predator could easily counter by anticipating the new location of the group.

The second behaviour is the 'flash expansion', a behaviour which has been widely reported from teleost shoals (Potts 1970; Nursall 1973; Pitcher 1979b; Pitcher and Wyche 1983). Here, a polarised group of fish, often stationary, explodes in all directions like a grenade, the fish rapidly reassembling after swimming 10 to 15 body lengths. In effect, the predator's group of potential targets vanish, to reappear somewhere else. Although a most effective confusion manoeuvre, the danger in this behaviour lies in reassembly, since, after exploding, individual fish are scattered and some may end up near enough to the predator to get attacked. For this reason, the tactic is relatively rare. There is clearly a premium on reassembly, although in minnows this can take as long as 30 s if the predator presses home an attack. Rapid reassembly is best, and recently discovered details of such group startles give clues as to how this is done. Blaxter, Gray and Denton (1981) showed that fish must retain an awareness of their relative positions while accelerating since they do not collide. The lateral line may mediate this effect, since fish with severed lateral-line nerves did occasionally collide when startled (Partridge and Pitcher 1980). Blaxter's research also showed that most startles were in a direction away from a sound (non-visual) stimulus, although the physiological basis of the directionality was not understood.

The Dangers of Joining: Confusion Zone in Schools

Poor co-ordination in travelling sub-schools which join may provide evolutionary pressure to avoid splitting and increase cohesion in schooling pelagic fish. Pitcher and Wyche (1983) described a zone of confusion at the rear of coalescing sandeel (Ammodytidae) schools, where the lack of co-ordination may attract predator attack (Figure 12.8) (Hobson 1968). Pitcher and Wyche also suggested that selection favouring individuals that

Figure 12.7A: A Polarised Minnow School under Threat from a Pike. Fish in foreground are out of focus, but other images (arrows) are blurred by movement during passing manoeuvres within the school. Passing probably enhances predator confusion, but may also represent a conflict between school neighbours (see text). (Photograph taken by flash in an arena tank with the aid of S. Smyth and B. Partridge.)

Figure 12.7B: Minnows Leaving a Feeding Patch when Approached by a Model Pike. Ice-cube tray feeding patch at left, pike model approaching from behind weeds, five 55mm-long minnows above and behind the model. The model predator is detected further away by minnows in larger shoals, but the fish allow it to approach closer before leaving the patch. (Photograph taken with the aid of E. Pritchard. See Magurran *et al.* (1985) for more details.

minimise the chances of being caught in a confusion zone may work through an enhancement of behaviour for synchronous turning, which shuffles fish positions. Furthermore, the precise location of the danger zone varies greatly with the angle of incidence of the joining sub-schools, and is not predictable, so inhibiting a melée to get out of the dangerous rear stations. Confusion zones therefore provide selection pressure for high co-ordination and synchrony but, paradoxically, this behaviour itself generates the zones. The cost for individuals of occasionally being caught randomly in a confusion zone could actually be greater than for less cohesive shoaling, but all would bear this cost and the situation could be an ESS.

Early Predator Detection

Minnows in larger shoals detect an approaching predator sooner (Magurran *et al.* 1985) (Figure 12.7b and see Chapter 9 by Milinski, this volume); birds in flocks have a similar advantage (Powell 1974; Kenward 1978; Lazarus 1979). There are two reports of the converse of this effect. Seghers (1981) reported that shiner shoals reacted to an approaching

Figure 12.8: Polarity of a Sandeel School Plotted across a Zone where Two Sub-schools were Joining (see Figure 12.6). The number of fish on widely different headings increase progressively from front to rear of the joining region: shading indicates the 'confusion zone' (Figure 12.6) formed as a consequence. Polarity was calculated as the standard deviation of headings of individual fish relative to the new common track of the school. Number of observations given above each data point: the relationship is statistically significant. (For further details see Pitcher and Wyche 1983.)

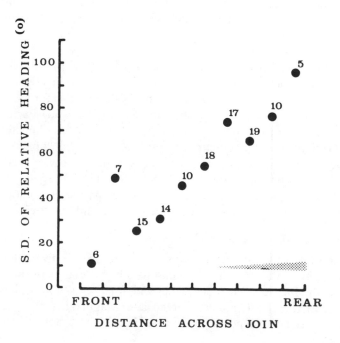

model later than individuals, but failed to detect any consistent effect with shoal numbers. This was almost certainly because he did not score the behaviour of individual fish in sufficient detail to show when the fish first detected the predator. Goodey and Liley (1985) reported that escape responses of guppies to a swooping aerial predator model were no faster in groups than for solitary fish, perhaps because all fish need to respond to such a rapidly-appearing threat without reference to other group members. This is not the same situation as detection of a gradually approaching threat from a stalking predator.

The minnow experiment is support for the 'many eyes' hypothesis (Bertram 1978; Barnard 1983) that the average probability of detecting a predator is higher in groups. An individual in the group therefore clearly has advantage through shoaling provided that the information about a putative predator is shared, as in bird alarm calls. The implications of this greater vigilance in the group for the optimal allocation of time to foraging and other behaviours is discussed in the later section dealing with foraging.

Warning Skitters — an Alarm Signal?   Minnows that detect an approaching predator (real or model) perform a low-intensity skitter, often returning to near their original position in a curved path. It is this behaviour which was scored in Magurran's work (Magurran *et al.* 1985). The movement does not appear to be fast enough to be a Mauthner fast-start response, unlike most skitters at later stages of attack (see above), although the difference is probably only one of degree. The behaviour is probably still fast enough to minimise the chance of a predator 'locking-on' to the performer. The skitter may act as a warning to shoal fellows, but is not precisely analogous to warning calls in bird flocks since it may be repeated by increasing numbers of individuals as the predator approaches, and indeed several fish may perform the behaviour together at a later stage. (It is this more conspicuous group behaviour which Seghers seems to have scored.) The cost of a warning skitter in our experiments was small, the fish ceasing to forage for a few seconds, although this might not be the case in the wild. Performing a skitter may also cost by attracting the predator's attention, but this would be countered by rapidly rejoining the group, when the confusion effect would operate again. It is possible that there is a small individual advantage in the skitter on a few occasions when a predator may be targeting that particular fish. From these pressures it is easy to see how the evolution of the warning beneficial to all individuals in the group could evolve.

Predator Inspection Behaviour

During the early stages of an attack by a stalking pike, individual minnows leave the shoal and approach to within 4 to 6 body lengths of the predator (Magurran and Pitcher 1986). There they wait for a few seconds, slowly turn, and return to the shoal. Pitcher, Green and Magurran (1986) term this behaviour a 'predator inspection visit' (see Figure 12.9). Sometimes a small

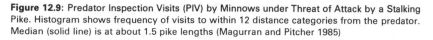

Figure 12.9: Predator Inspection Visits (PIV) by Minnows under Threat of Attack by a Stalking Pike. Histogram shows frequency of visits to within 12 distance categories from the predator. Median (solid line) is at about 1.5 pike lengths (Magurran and Pitcher 1985)

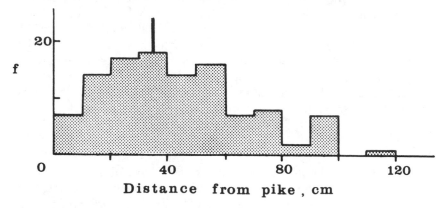

group will perform the behaviour together. Under more disturbed conditions in a fluvarium, the behaviour is performed more rapidly, with no obvious pause (Allan and Pitcher 1985). The same behaviour is shown to models and, initially, to other intruders such as large non-predatory fish; it has also been described for sticklebacks (Reist 1983; Giles and Huntingford 1984), and bluegill sunfish (G.S. Helfman, personal communication).

Approaching a predator as an individual clearly carries a danger of inciting or exacerbating an attack, the apparent leisurely manner of the behaviour perhaps being an attempt to minimise this cost. Predator inspection declines in frequency as a pike attack proceeds, reflecting the increasing likelihood of being selected as a target. At present we can only speculate about the benefits. Minnows in shoals will stay close to much larger cyprinids such as bream without fright; indeed they habituate more slowly to realistic predator models (Magurran and Girling 1985), so it is possible that one function of the behaviour is to distinguish friend from foe. Furthermore, fish that perform predator inspection visits gather knowledge about the predator's precise location and current state, which may be of advantage if sudden flight is needed from a predator that had previously only been in the vicinity but was now launching an overt attack. It is not known if any of this information is shared with others when the inspector returns to the shoal. G.S. Helfman (personal communication) speculates that the behaviour may inform the predator that it has been spotted, a suggestion that might be tested with judicious experiments.

Predator Prediction

Magurran and Pitcher (1986) noted that minnows sometimes appeared to be able to predict that a strike by a stalking pike was imminent, anticipating the attack by swift synchronised evasion. Human observers are able to make similar predictions, either live or when watching videotape. How this is achieved remains unclear, although it may relate to an 'unnatural-looking' stillness and indication of tension in the pike's body. Sometimes pike quivered before a strike, but the effect was not consistently linked to this. For small pike, only careful unobtrusive stalking had a payoff in prey capture, Magurran and Pitcher's results suggesting that shoaling cyprinids are currently ahead in the arms race. This effect was, however, strongly dependent upon size: for most experiments pike encountered minnows of approximately optimal size (55 mm long; based on handling time in relation to energy value: Hart and Connellan, 1984). But when larger pike were employed, shoaling was little protection. It is interesting that minnows would be sub-optimal-sized prey for such larger pike in the wild.

Predator Inhibition

Breder (1959), Hobson (1968), and others have described several instances of apparent predator inhibition by mass displays of shoaling fish. Motta (1983) describes several instances of displays by coral-reef prey to their predators which appeared to be analogous to mobbing by birds (Bertram 1978), although some of these instances may have been unrecognised inspection visits. Synchronously turning silvery fish shoals have frightened scuba divers (Pitcher 1979a) but the inhibition phenomenon remains to be investigated in detail.

Genes and Evolution in Shoal Behaviour

In three-spined sticklebacks (Giles and Huntingford 1984), and in guppies (Seghers 1974), predators have shaped interpopulation differences, and so we may speculate that the basic repertoire of shoaling responses to predators is likely to be innate. In support of this, minnows from provenances currently with and without pike exhibited a similar repertoire of anti-predator tactics, but those sympatric with the predator performed and integrated the behaviours more effectively (Magurran and Pitcher 1986). Further evidence was provided by Jakobssen and Jarvik (1978), who showed that naive salmon smolt shoals reacted selectively to pike predators, but not to non-predator fish. However, we are ignorant of the precise way in which genetic and environmental determinants interact in the normal development of shoaling fish to produce effective anti-predator tactics, so controlled experiments to tease out these factors would be timely (see Chapter 3 by Huntingford, this volume.)

It is possible that, in some fish, altruistic shoal behaviours may have evolved through kin selection. There is certainly no other satisfactory explanation for the evolution of fright substance in cyprinids (see Chapter 6 by Hara, this volume). Evidence for kinship in shoals of the freshwater shiner, *Notropis cornutus*, has been produced by Ferguson and Noakes (1981). Quinn and Busack (1985) have demonstrated chemosensory recognition of siblings in shoaling coho salmon juveniles, and Loekle, Madison and Christian have similar evidence for poecilid fish (Loekle *et al.*, 1982). Traditionally, it has been argued that kinship could only affect selection in some freshwater or reef fish, which do not travel far from a restricted habitat and spawning area, and that such factors could not impinge upon marine species with pelagic larvae. I feel that it may be worth while to question this assumption, if chemical kin recognition is considered alongside the growing evidence for natal-site reproductive homing in a wide range of teleosts. Recent work suggests close genetic affinity and high shoal fidelity among such pelagic schooling fish as tuna (J.M. Cullen, personal communication). The latest techniques of investigating close genetic affinity, such as sequencing mitochondrial DNA (A. Ferguson, personal communication) may be expected to provide some fascinating data on shoaling in the near future. If close kinship along with the means of kin recognition can be demonstrated in fish shoals, we may have to revise our ideas of the evolution of the behaviour as exclusively to the advantage of individual fish.

Successful Fish Predators

Despite the effective shoal tactics outlined above, some piscivores appear to be ahead in the arms race and specialise in preying on shoals. Most of them are large relative to their shoaling prey, or have the advantage of long-range detection (visual, olfactory or acoustic) and fast swimming. The list of detectors includes diving birds such as gannets, auks, terns and pelicans, which can spot shoals from the air at a distance. Once under water, most diving birds seem to pursue individual targets and rely on the sheer speed of the dive and their long sharp beaks for success. Pelicans acting as a group can herd their shoals of prey, catching alarmed fish at random from the shoal with their scooping beaks. Seals and dolphins are also said to herd fish prey and beat the confusion effect by multiple rapid strikes. Sawfish and teleosts with rostra, like marlin and swordfish, swim through shoals stunning fish randomly with rapid swings of their saws or swords. Thresher sharks also do this with their tails. Humpback whales herd shoaling capelin with foaming clouds of bubbles released in a huge ring beneath the fish. Shoals that seek shady cover (Helfman 1981) are exploited by a heron thoughtfully extending a wing to provide such a shady area!

Shoaling and Foraging

Visibility is restricted under water; even in the clearest water, absorption and scattering of light mean that objects are visible only over a few tens of metres. The visual range in most water extends only a few metres (see Guthrie, Chapter 4, this volume). Under these conditions, foraging for patchy food in a social group has immense benefits, although this aspect of shoaling behaviour was virtually ignored in the classical literature. Radakov

(1973) reviews early Russian observations which, unlike much contemporary western research, considered enhancements to foraging through social observation, but, as with the anti-predator function of shoaling, these were put forward mostly with no experimental proof and with arguments which were essentially group selectionist.

Many costs and benefits in foraging shoals have been recently explored quantitatively and are reviewed in detail below, but, surprisingly, theoretical behavioural ecology has yet to make testable predictions in this area. Conventional optimal-foraging theory (OFT) assumes that animals will maximise their energy intake through diet and food-patch selection (and see Hart, Chapter 8, this volume: Krebs 1978). Similarly, the marginal-value theorem predicts when foragers should leave a food patch (Charnov 1976; Hart, Chapter 8, this volume). Social behaviour is not taken into account in the simple classical versions of these theories, so the only group-related prediction of OFT or the marginal-value theorem is that individuals will tend to congregate on good patches (Krebs 1978) until depletion makes alternative sites more attractive. Unfortuntely, this is not how foraging groups of fish behave (Pitcher *et al.* 1983).

Another optimality theory, that of the 'ideal free distribution', predicts that individuals in foraging populations will distribute themselves among available resource, such as food patches, in proportion to the rewards encountered (Fretwell and Lucas 1970; Parker 1978; Milinski 1979a; Sutherland 1983). In this way each individual will gain the same food intake; optimal foraging for all in an alternative sense of OFT. Godin and Keenleyside (1984) timed six cichlids feeding on two patches as appearing to fit the ideal free distribution. Milinski (1979a) reported an apparently similar finding for sticklebacks foraging on daphnia. However, more detailed investigation by Milinski (1984b) included careful measurements of individual food intake and behaviour and revealed large consistent differences between fish. The basic ideal free distribution was not an appropriate model in these circumstances, although an alternative optimality theory could be modified *post hoc* to take account of the social interaction (the relative payoff sum rule; see Hart, Chapter 8, this volume.) 'Switchers' were poorer competitors who changed patches more frequently and had lower food intakes. Milinski's work shows that even though the gross picture may ostensibly fit an ideal free distribution prediction, the precise behaviour of individuals depends upon the amount of social competition for food. For example, Godin and Keenleyside may have obtained a different result by using a different group size in relation to the availability of food.

The ideal free distribution's prediction of equality of food intake is not likely to be met in practice since individuals vary in their ability to compete or scrounge food located by others, as has been clearly demonstrated in bird flocks (e.g. Barnard, Thompson and Stephens 1982). Theoretical ideal free distribution individuals are 'ideal' and 'free' because of the absence of constraints on where they should go or how they should perform. In the real world, shoaling fish are not free of such constraints and therefore basic ideal free distribution theory does not make such precise predictions as might be hoped, although estimates of the absolute numbers of animals to be found on food patches may be sufficient for some ecological purposes. Some indication of the factors to be taken into account by a more comprehensive theory are discussed below.

Fish foraging in shoals gain benefits through faster location of food, more time for feeding, more effective sampling, information transfer and opportunity for copying. Increased costs with group size are perceived through greater competition in various forms. Experiments which have investigated these costs and benefits, usually through manipulation of shoal size, are discussed in the next section.

Finding Food Faster

Faster location of patchy food in larger shoals has been clearly demonstrated by manipulating shoal size in two cyprinid species, minnows and goldfish (Pitcher *et al.* 1982a), as illustrated in Figure 12.10. A simple model based on a random food searcher who can detect when other searchers find food can explain this result. As the density of searchers rises, randomly located food is found sooner by someone since the aggregate search rate is higher for each additional fish searching. Non-finders then benefit by moving to the food. This model suggests strong selection pressure to be able to tell when others have found food.

Goldfish searchers watch for the 'nose-down' feeding posture of a food finder (Magurran 1984), minnows react to 'wriggling' foragers, and mackerel also appear to use visual cues from excited food finders. Krebs, MacRoberts and Cullen (1972) reported a similar finding for flocking tits in experiments in which flock size was manipulated. Faster food location in larger groups of stone loach (*Noemacheilus barbatulus*) a nocturnally feeding cobitid (Street and Hart 1985) is more likely to be the result of water turbulence set up by food locators than vision or olfaction. In bird flocks some individuals are consistently better food locators (Barnard 1984), but such differences have not yet been investigated for fish.

At first sight it is surprising that behaviours like 'nose down' or 'wriggling' which enable food locators to be detected are not strongly countered by behaviour to conceal or even bluff a find. Concealment would keep a valuable patch to the finder, whereas bluff might throw others off the scent while the bluffer moves off to fresh search areas. Bluffing would be a viable strategy only at low frequencies, but it is difficult to see why concealment behaviour is not more obvious when foraging fish groups are observed. One possible explanation is that of reciprocal altruism if individuals have fidelity to particular groups. Helfman (1984) has provided some evidence of group fidelity in yellow perch (*Perca flavescens*) shoals. Another reason may be that the morphophysiological constraints on

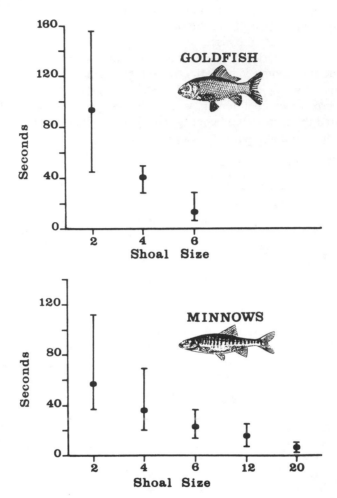

Figure 12.10: Goldfish and Minnows in Larger Shoals Find Food Faster. Graph shows the relationship between shoal size and time spent before a randomly located food item was found by a focal individual. Marked individuals were followed through the range of shoal sizes, but there were no strong individual differences, unlike the situation in bird flocks. Points show the medians and quartile for 16 replicates (Pitcher *et al.* 1982a.)

capturing and subduing prey make eating difficult to conceal: the 'nose-down' goldfish is using efficient suction from its protrusible cyprinid jaw mechanism (Lauder 1983), and piscivores have to shake prey in their jaws to incapacitate and swallow them.

Time Allocated to Feeding

In larger groups, cyprinids feeding on patchy food allocate more of their time budget to foraging (Magurran and Pitcher 1983) because they are less timid, spending a smaller proportion of their time in weed cover (Figure

12.11). Even when an approaching predator is detected, fish in larger shoals allow it to approach more closely before stopping feeding (see Figure 12.6b; Magurran *et al.* 1985), or will approach a dangerous area away from cover more readily (Milinski 1985 and Chapter 9, this volume). Individuals are less timid, spend less time in exclusively vigilant behaviour, and forage longer in larger shoals, all of which are direct benefits to fish who join small groups of conspecifics. Much of this particular benefit may accrue to fish in multi-species shoals even where the fish are not feeding on precisely the same food (Allan in prep.).

The conflicting demands of vigilance and foraging, exemplified by the work of Caraco (1980), Sih (1980) and Lendrem (1983) are described in Chapter 8 by Hart (this volume): the conflict appears to be resolved more in favour of feeding in larger fish shoals.

Sampling Feeding Sites

Experiments have now confirmed that fish in larger shoals gain greater foraging benefit through sampling behaviour. When feeding in an environment containing several food patches of different quality, although feeding most on the best patch, fish make visits to feed on poorer patches. Such behaviour is not predicted by classical OFT, and continues in goldfish even when the location of the best patch is completely stable over many trials (Pitcher and Magurran 1983). In larger goldfish shoals, individuals make

Figure 12.11: Shifts of Behaviour towards Increased Foraging and Lower Timidity with Shoal Numbers in Minnows and Goldfish. Shaded areas represent proportion of time spent in different behaviour categories for an experiment carried out in large tanks (medians for 8 replicates). Both species of cyprinids shoal, but only minnows spend an appreciable amount of time in polarised synchronised schools. Although goldfish spend less time in weed, and minnows less time in midwater, the trends are remarkably similar in the two species (data from Magurran and Pitcher 1983)

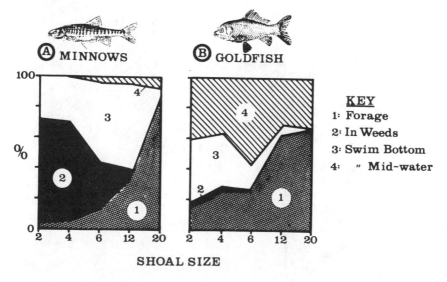

more visits to poor patches, partly because more time is budgeted to foraging (Figure 12.12). Pitcher and Magurran demonstrated that information gathered on visits to other patches was used to change foraging tactics when the location of the best feeding patch was switched: the visits therefore represented genuine sampling behaviour. More sampling took place in larger shoals. (Note that all experiments controlled for patch depletion.)

Cowie and Krebs (1979) reported sampling of poor feeding sites in great tits. In goldfish, Lester (1981, 1984) showed that foraging time on two food patches departed from OFT predictions the more the location of the better patch was altered. This could be interpreted as reflecting greater

Figure 12.12: Time Spent Foraging on Three Patches of Different Food Density in Shoals of Two and Five Goldfish. Total time shows that fish in the larger shoal spend more time foraging. Foraging time shows that fish in the smaller shoal spend more time feeding at the best site and are therefore closer to OFT predictions. Means are based on 8 replicates and are significantly different. In the larger shoal, goldfish sample poor sites more often. Further experiments, described in Pitcher and Magurran (1983), prove that such sampling is important for effective information transfer about good feeding sites

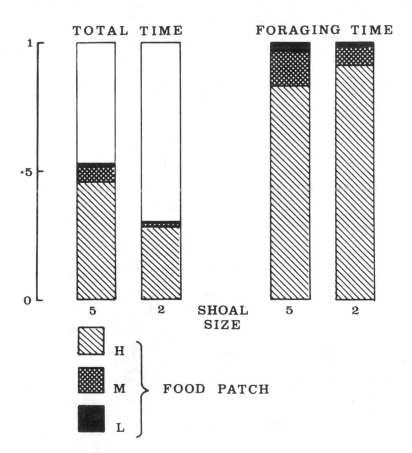

sampling behaviour in the face of more unpredictability of patch quality.

For any given food distribution in space and time, there is likely to be an optimal sampling regime which produces maximum food intake while allowing for unpredictability: unfortunately the theory of optimal sampling is not yet sufficiently developed to make general predictions (Krebs, Kacelnik and Taylor 1978; Orians 1981; Shettleworth 1984), and has not considered the social constraints we see in fish shoals. A simple reason for observing more sampling in larger shoals is that individuals have more foraging time at their disposal. Theory, however, may in the future be able to show how individuals benefit from doing more sampling rather than just spending their larger foraging budget on the best patch. One recent pioneering development in this respect is a novel approach to modelling social foraging by Clark and Mangel (1984), based on the value of information rather than classical OFT's emphasis on energy/time budgets. Clark and Mangel's model predicts that individuals should sample before taking decisions to stay with, leave or join foragers on a patch, precisely the kind of behaviour which we have clearly observed in cyprinid shoals.

Passive Information Transfer

Active information transfer (AIT) (Pitcher *et al.* 1982a) occurs when distinct overt behaviour communicates the presence and location of food, as in the famous honey-bee dance or in flocking nectar bats (Howell 1979). Passive information transfer (PIT) (Magurran 1984) occurs when behavioural cues about food are used by other animals. Recent insight into the manipulative nature of animal signals (Krebs and Dawkins 1984) suggests that, in evolutionary terms, there may be a continuous spectrum between these two extremes, but when observing animals now, behaviour is likely to be optimised for either AIT or PIT. AIT may be found when a signaller benefits from giving information about food to others through kinship or manipulation; PIT may be seen where behavioural detection of feeding is currently the winning strategy in the evolutionary race, perhaps because feeding is hard to conceal.

In fish shoals, PIT allows the faster food-finding effect discussed above. In addition, by reciprocal transfer of single goldfish that had knowledge about the location of good food patches ('informed' fish) between goldfish shoals with different food patch arrangements, Pitcher and Magurran (1983) demonstrated that PIT enabled fish in larger shoals to recover more rapidly from changes in the location of the good patch. In these experiments, the effect occurred even when the misinformed fish were in the majority, although the precise relative values of the patches and shoal size might influence this point. Using evidence largely from bird flocks, Giraldeau (1984) has suggested that a 'skill pool' effect enhances foraging when individual differences in foraging specialisations are shared through social learning. In a broad sense, the differences in information about good foraging sites in this experiment fit the 'skill pool' definition.

Partridge *et al.* (1983) consider that the concave leading edge of bluefin tuna schools implies group co-operation for hunting similar to that familiar in seals and whales, but there is no direct evidence for this speculation.

There is as yet no evidence for AIT in fish shoals, nor for Ward and Zahavi's 'information centre' (1973) hypothesis in which details of the location and quality of food are shared among groups. Information centres exist in the social insects and in many mammals, and have been experimentally demonstrated in quail (de Groot 1980).

Competition for Food

As the benefits of being in a group go up, costs due to increased intraspecific competition for food also increase (Bertram 1978). In goldfish shoals this increase in competition is reflected in a decrease in handling time for each food item (Street, Magurran and Pitcher 1984). Presumably, goldfish speeded up handling to avoid interference from others. A similar effect was noted for house sparrows by Barnard (1984), and Japanese medaka (*Oryzias latipes*) ate faster in larger groups (Uematsu and Takamori 1976). Bolting food in this way may bear digestive costs in both animals.

The effects of competition do not appear to become pronounced until a group size of about a dozen individuals is reached. The growth rate of Japanese medaka under standard densities and feeding regimes giving equal food intake increased with group size from 2 to 6 fish, but did not increase further as shoal numbers went up to 12 fish (Kanda and Itazawa 1978).

Competition between two sizes of fish was scored in minnow shoals (Pitcher, Magurran and Allan 1985b) and might be expected to be more severe in larger groups. Small minnows were forced out of feeding pots by large intruders, so small fish were forced to move about on the patch more. Milinski (1984b; see above) has demonstrated alternative coexisting feeding strategies depending upon competitive ability in sticklebacks. Fish that switched feeding patches more performed poorly in a test of competitive ability.

Such movers and stayers do not seem to be equivalent to the 'producers' and scrounging kleptoparasites seen in bird flocks (Brockmann and Barnard 1979). Giraldeau (1984) considers 'joiners' and 'discoverers' in relation to new food sites. Barnard and Sibly's model (1981) predicts that group size will have a strong effect upon the stable ratio of producers and scroungers, but, despite attempts, consistent individual strategies like these have not yet been observed in fish shoals (T.J. Pitcher, unpublished).

Forage Area Copying

Before individuals can benefit from foraging in a group, they have to join others feeding. Fish may join from a distance, but the same behaviour can be observed on a smaller scale within loosely structured shoals. In minnow and goldfish shoals, small groups with low EGS remain in visual contact. Individuals will soon join a group in which much 'nose-down' foraging is observed. This joining behaviour has been termed 'forage area copying' (FAC) by Barnard and Sibly (1981), following Krebs *et al.* (1972) who first observed the phenomenon in bird flocks.

FAC does not necessarily entail moving to an actual food find: joining a group of foragers is sufficient. Moreover, sometimes this behaviour may be difficult to distinguish from aggregation driven by other motivations such as anti-predator defence. For example, herons joined polystyrene models erected on mudflats (Krebs 1974), and rudd joined conspecifics held in jars (Keenleyside 1955), but food or defence reasons for joining were not disentangled. This difficulty has been overcome in some recent work on fish shoals which will be described next.

FAC has been tested experimentally for goldfish by Pitcher and House (in prep.). A shoal of test goldfish was placed next to a trained shoal feeding on the better of two patches behind a transparent barrier. The test fish had a choice between two food patches of equal food density. FAC was demonstrated when the test fish fed on the patch next to the trained shoal. In contrast, classical OFT predicts an equal distribution of test fish on the patches.

Using a similar protocol, Pitcher (in prep.) showed that FAC depends critically upon individual reward rate. Minnows in shoals of 20 were trained to feed upon three patches of different food density, and soon devoted most of their time to the best patch. Twenty minnows in a test shoal were placed in front of the transparent barrier with three equivalently arranged but equally rewarding food patches. Like the goldfish, test minnows chose to feed on the patch adjacent to trained fish, but this result was strongly dependent upon the level of food on their patches. When food was plentiful, or completely absent from their patches, little FAC was observed, even right at the start of a trial. This shows that the success rate experienced by foragers shapes behaviour on a very short time scale. Very low or very high rates of food finding override FAC in these cyprinid shoals.

In this experiment, FAC reappeared when the test fish were given no patches on which to search. (They foraged on the aquarium gravel.) Since all minnows in the experiment had prior experience of the particular design of patch, we can conclude that subtle cognitive cues are used in fish foraging behaviour, as pointed out by Tinbergen (1981) for starling flocks.

Controls in the latter experiment proved that the minnows were not aggregating through timidity for anti-predator advantage in any simple fashion; for example, fish moved freely and, on average, distributed equally among equally-rewarded patches when both trained and test fish experienced this arrangement. This control overcomes the objection that minnows may have aggregated to feed to take advantage of the higher feeding time in larger groups. It is worth emphasising that neither classical OFT nor the ideal free distribution predicted the behaviour of the fish in these cyprinid shoals.

Migration Advantages

Shoaling could increase the accuracy of homing on migration, since the mean direction or route is likely to be a more accurate estimate of the correct destination than any individual's choice (Larkin and Walton 1969). A similar effect may operate in bird flocks, but there appears to have been no critical test of this idea, which ought to be strongly dependent upon group size. Two pieces of work lending indirect support to the migration-enhancement hypothesis have been published.

First, adult coho salmon home more accurately to their natal river at higher densities (Quinn and Fresh 1984). The evidence is based on statistical analysis of tagging returns along with population size estimates, and the figures are subject to considerable variance. The point could be tested directly using the techniques pioneered by Hasler (1983) in proving the home-stream hypothesis.

Secondly, social transmission of information about diurnal migration routes between refuge and feeding sites has been demonstrated in grunt shoals (Helfman and Schultz 1984). Transplant experiments proved that individuals rapidly acquired knowledge from local residents about routes and shoaling sites. After only a short period with their new, conspecifics, isolated transplantees travelled the residents' routes to the correct locations. Such cultural transmission had not been shown in fish before, and is clearly a variation on the information centre idea.

Wijffels, Thines, Dijkgraaf and Verheijen (1967) showed that groups of eight *Barbus tincto* swam more quickly through a maze than individuals, but careful controls did not confirm Welty's (1934) conclusion that fish in groups learn routes faster than isolates. The fish in Wijffels *et al.*'s study performed better in a group if they had learned the maze in a group, and better alone if they had learned it alone, a general confirmation of the 'context' learning phenomenon analysed by Hinde and Stevenson-Hinde (1973).

Mixed-species Shoals

Most of the factors favouring larger group size in single-species shoals can, in the simplest analysis, apply equally to mixed-species shoals. These factors are more effective foraging by social observation, better vigilance for predators, and greater economy of time budgeting. In mixed shoals, these benefits are likely to be greater the more similar the fish morphologies and diet. In single-species groups, the benefits are countered by increased food competition, but this could be lower for a given group size in mixed shoals. Differences between species allow for exploitation or genuine symbiosis if food items wanted by one species of fish are flushed by the other. Some behaviours, such as predator-evasion tactics, may be more efficiently performed with conspecifics because of similar morphology. Furthermore, small numbers of one species in a mixed shoal may suffer through being conspicuous to predators. Many of these points have been investigated in detail for bird flocks (see Krebs and Barnard 1980) but there has been relatively little detailed reserch on mixed-species fish shoals.

In the open pelagic environment, mixed-species shoaling appears to be less common, the great commercially important shoals of clupeids, scombrids and gadoids usually being found in groups of their own kind. Mixed-species hake (Merlucciidae) shoals appear to be more common though. In more structured habitats, for example coral reefs, kelp forests or rivers, mixed shoals appear to be more common. Perhaps this is related to the higher diversity of food and proximity to cover, which enables species to form associations on a temporary basis to enhance vigilance, but split to evade attack or seek optimal food. Vigilance against camouflaged lurking predators is more likely to be needed in structured habitats, where predators may be quite close before being detected. By contrast, in the open sea, predators always appear at the extreme of the visual range.

Mixed shoals of cyprinids have been investigated under semi-natural conditions by Allan (1985) who has demonstrated that many of the foraging benefits of larger groups also accrue when species are mixed. Diet and habitat shifts occur in mixed-species shoals, which appear to minimise competition while retaining the advantage of greater vigilance and search power. Cyprinid species sorting under threat (see Figure 12.13) reflects more effective shoal manoeuvres with conspecifics and perhaps an attempt to minimise conspicuousness (Allen and Pitcher 1985). Piscivores seem to pick out conspicuous individuals for attack, perhaps to overcome the confusion effect. This is exactly contrary to the prediction of apostatic selection theory, in which individuals which stand out are protected and therefore selected for. The theory dealing with circumstances in which conspicuousness does or does not pay therefore bears more careful examination (Allen and Pitcher 1985).

Krebs (1973) suggested that one of the benefits of mixed-species bird flocking was the greater diversity of food items which became available, not only through effects like flushing, but also by social observation for foraging specialisations of others. Several studies (e.g. Rubenstein, Barnett, Ridgely and Klopfer 1977) have supported this idea by demonstrating a broader diet for individuals in mixed flocks, and equivalent work on fish shoals would be timely.

The most rigorous investigation of mixed-species shoaling to date has been performed by diving on coral reefs by Nancy Wolf (Wolf 1983, 1985,

Figure 12.13: One Measure of Species-sorting in a Mixed-species Cyprinid Shoal under Predator threat (a Model Pike was Used). Vertical scale is the probability of a neighbour being of the same species. Solid circles (c) are control values when the pike was absent; open circles (p) when the threat was present. Shaded line (r) indicates the random expectation for neighbour identity (Allan and Pitcher 1985)

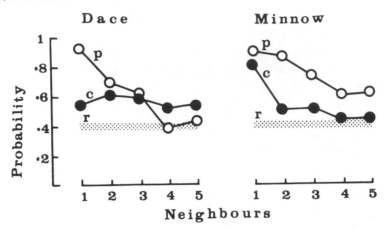

and in prep.). Wolf shows how the costs and benefits of shoaling vary continuously for different species, times and locations in shoals travelling around the reef. The diet and behaviour of three species (*Sparus, Sparisoma* and *Acanthurus* species) were recorded with group size and composition. In general, the pattern was that foraging and vigilance benefits increased with overall group size, and other factors, such as conspicuousness, varied with group composition. There were, not surprisingly, detailed differences between species. For example, *Sparisoma* was a cover-seeker under threat, whereas only small individuals of *Acanthurus* shoaled. Wolf's work raises the possibility of a revealing investigation of conflicts of interest between the species in her mixed-species shoals.

Wolf also confirmed the demonstration by Robertson, Sweatman, Fletcher and Cleland (1976) that mixed-species shoals allowing larger group size swamped the defences of territorial damselfish on reefs, giving shoalers access to otherwise unavailable resources.

As the species in mixed groups become more different, increasingly subtle relationships are possible. Benefits of larger group size accrue to juvenile French grunts (*Haemulon* sp.) and two species of mysid shrimp on coral reefs (McFarland and Kotchian 1982). The associations last for only 5 days while the fish grow, but are renewed every 2 weeks or so as fresh cohorts of postlarval grunts recruit from the plankton. Conspicuousness here may be reduced through interspecific mimicry, since the two organisms look remarkably similar. There is a clear differential distribution of benefits, as the young fish prey on their mysid companions as they grow.

McKaye and Oliver (1980) describe an interesting case of mixed-species shoaling in Lake Malawi where parental nesting catfish (*Bagrus meridionalis*) defend shoals of their own young mixed with those of two species of cichlid. Cichlid parents appear to release their young at the periphery of the catfish nest. Cichlid young are kept on the periphery of the defended shoal, however, an example of Hamilton's dangerous-margins effect.

In summary, much more work is required on the details of the costs and benefits of mixed-species shoaling. Where conflicts of interest occur, their resolution is likely to lead to profound insight into the behavioural ecology of shoaling, in a similar fashion to recent work on flocking birds.

Optimal Shoal Size

In most of the preceding discussion, benefits increase with shoal numbers, but this effect occurs most rapidly at small shoal sizes. Some costs, such as intraspecific competition, increase with shoal size as well, and some benefits, such as faster food location, reduce at very large shoal sizes. Cost and benefit curves which intersect at some group size may be envisaged. A simple-minded analysis suggests that there will be an optimal group size where the net benefit is at a maximum (see Bertram 1978).

In the best of possible worlds, we would tend to find groups of this optimal size in the wild. Using different hypothetical models, Sibly (1983), Pulliam and Carcaco (1984) and Clark and Mangel (1984) have shown independently that such an optimum is not stable, since individuals tend to join groups because they do better by joining a group than remaining on their own. Group sizes in the wild should therefore be much larger than the 'social' optimum predicted by such averaging. It does not, however, appear that anyone has rigorously tested this prediction, although Clark and Mangel (1984) cite some suggestive evidence.

A much simpler explanation of the non-achievement of group size giving maximum net benefit is based on a direct analogy with the well-known phenomenon of the 'tragedy of the commons' (Hardin 1968), termed 'common property rent dissipation' by mathematical economists (Clark 1976). This process leads unregulated 'open-access' resources such as fisheries inexorably to a point where no profit is made, termed the 'zero net revenue' outcome (Pitcher and Hart 1982). If no restrictions, such as licences, are placed upon access, new boats join a profitable fishery on the expectation of individual gains like those already working the resource. As successive new boats join, everyone's share of the benefit diminishes, until eventually the zero profit point is reached. (In some cases the process can overshoot, producing a loss for all.) If there is no alternative resource, the process is automatic and is impossible to halt. Joining of new boats is analogous to fish deciding to join a shoal, and the fishery yields are analogous to the net benefit of being a group member. Huge marine shoals may represent such 'common property' losses, zero benefit shoaling being favoured over dispersal as a bet hedged against predator encounters. (Since writing this review, I note that Clark and Mangel (1984) draw the same analogy.)

This hypothesis predicts that we should observe groups in which the average net benefit is zero rather than a maximum. The evolution of behaviour which kept a shoal at a certain size by deterring or evicting new-comers would render this analysis invalid in a way analogous to the issue of licences restricting access to a fishery.

Koslow (1981) speculated that the size of large pelagic shoals might be regulated by behaviour optimising plankton feeding in relation to plankton patch size. This is also unstable for the same reasons as above, unless such altruistic group-size regulating behaviours occur.

Anderson (1981) provided evidence that a surprising number of shoals off Southern California were of 15 m diameter, and constructed a mathe-matical model which fitted the observed distribution of shoal sizes. The model was based on two assumptions: (a) that fish joined the group irrespective of shoal numbers, and (b) that fish left the group in proportion to the numbers present. Both of these assumptions seem unrealistic in the light of the evidence we now have about shoal-size-related behaviours, which is reviewed in this Chapter.

Do fish exclude newcomers once the social optimum group size is reached? I speculate that some awareness of the current optimal group size may have evolved, but, clearly, some very carefully controlled experiments need to be devised in order to test this hypothesis.

Summary

Polarised and synchronised schooling behaviour is regarded as a category of the social grouping termed shoaling. The long search for structure in schools has not contributed much insight of the functions and evolution of shoaling. Homogeneity in fish shoals and schools has been over-emphasised. Detailed experimental examination of behaviour during foraging, and under predator threat, reveals tensions which reflect an indi-vidual's continual reappraisal of the costs and benefits of joining, staying with, or leaving the social group. The elective group size (EGS) provides a simple but effective measure of shoaling tendency.

There is little evidence supporting a hydrodynamic function in fish shoals, but the accuracy of homing and migration may be enhanced.

The chapter reviews critically ideas about how shoaling fish may counter predator attack through avoidance, dilution, abatement, evasion, con-fusion, detection, mitigation, inspection, inhibition and prediction.

Predator avoidance and attack dilution are insufficient singly as general advantages to account for shoaling, but the two effects combined provide an ESS favouring shoal behaviour, here termed the 'attack abatement effect'. A cognitive dilution benefit may also accrue in shoals. Shoaling aids early predator detection, and predator confusion is generated both passively by grouping and actively through shoal manoeuvres. Group tactics also aid predator evasion. Skittering and slow passing in minnows reveal conflicts between shoal members. Shoal members take risks to perform predator inspection visits (PIVs), mitigating attack. Present work is examining possible alarm signals and exploitation as a consequence of information gained.

Simple optimal foraging theory (OFT) and other optimality models such as the ideal free distribution (IFD) currently fail to explain foraging in shoals. In larger shoals fish find food faster, spend more time feeding despite predator threat, are less timid, more vigilant, sample the habitat more effectively and transfer information about feeding sites more quickly (passive information transfer, PIT). Forage area copying (FAC) is an important component of these shoaling advantages, but the behaviour is contingent upon an individual's feeding rate. In shoals, information about the type and quality of a food source, number of companions, threat and alternative feeding sites is acquired and decisions are taken surprisingly rapidly. The behavioural plasticity conferred by shoaling may therefore be seen as one of its major general advantages.

Fish in mixed-species shoals attempt to maximise their gains from foraging and vigilance, but sort into species under attack to optimise escape manoeuvres and reduce their conspicuousness among aliens.

Competition for food increases with shoal numbers, but the optimal shoal size that maximises the net benefit for all is unlikely to be observed since it is unstable for the same reasons as the degradation of common property resources. In the absence of behaviour which actively limits entry, we may expect to observe shoals in which the average net benefit is zero. This means that the plasticity in the face of risk referred to above may have been crucial to the evolution of shoaling. However, recently demonstrated kin recognition, genetic shoal affinity and the evolution of ostensibly altruistic shoal phenomena like 'fright substance' could mean that the search for behaviours which limit shoal size to the 'social optimum' may be worth while.

Acknowledgements

Much of my own work cited in this chapter was kindly supported by a grant from the Natural Environment Research Council (UK) and would not have been possible without the help and inspiration of Dr Anne Magurran. Professor J.M. Cullen kindly allowed me to discuss some of his experiments that are in preparation. I would like to thank John R. Allan, Dr Paul Hart, Dr Gene Helfman, Dr Anne Magurran, George Turner, the Gemini and the referee for their helpful comments upon the manuscript. Opinions and any mistakes remain my own responsibility, however.

References

Abrams, M.V. and Colgan, P. (1985) 'Risk of Predation, Hydrodynamic Efficiency and their Influence on School Structure', *Environmental Biology of Fishes*, in press

Allan, J.R. (1985) 'Studies of the Behavioural Ecology of Fish in the Semi-natural Conditions of a Fluvarium', *Proceedings of the 4th British Freshwater Fisheries Conference*

Allan, J.R. and Pitcher, T.J. (1985) 'Species Segregation during Predator Evasion in Cyprinid Fish Shoals' (submitted)

Anderson, J.J. (1981) 'A Stochastic Model for the Size of Fish Schools', *US Fishery Bulletin, 79*, 315-23

Andorfer, K. (1980) 'The Shoal Behaviour of *Leucaspius delineatus* (Heckel) in Relation to Ambient Space and the Presence of a Pike, *Esox lucius*', *Oecologia, 47*, 137-40

Barnard, C.J. (1983) *Animal Behaviour: Ecology and Evolution*, Croom Helm, London, 339 pp

Barnard, C.J. (1984) (ed.) *Producers and Scroungers*, Croom Helm, London, 303 pp

Barnard, C.J. and Sibly, R. (1981) 'Producers and Scroungers: a General Model and its Application to Captive Flocks of House Sparrows', *Animal Behaviour, 29*, 543-50

Barnard, C.J., Thompson, D.B.A. and Stephens, H. (1982) 'Time Budgets, Feeding Efficiency and Flock Dynamics in Mixed Species Flocks of Lapwings, Golden plovers and Gulls', *Behavior, 80*, 44-69

Belyayev, N. and Zuyev, G.V. (1969) 'Hydrodynamic Hypothesis of Schooling in Fishes', *Journal of Ichthyology, 9*, 578-84

Bertram, B.C.R. (1978) 'Living in Groups: Predators and Prey' in J.R. Krebs and N.B. Davies (eds), *Behavioural Ecology*, 1st edn, Blackwell, Oxford. Chapter 3, pp. 64-96

Blake, R.W. (1983) *Fish Locomotion*, Cambridge University Press, Cambridge

Blaxter, J.H.S., Gray, J.A.B. and Denton, E.J. (1981) 'Sound and Startle Responses in Herring Shoals', *Journal of the Marine Biological Association, UK, 61*, 851-69

Breder, C.M. Jr (1951) 'Studies on the Structure of Fish Shoals', *Bulletin of the American Museum of Natural History, 98*, 1-27

Breder, C.M. Jr (1954) 'Equations Descriptive of Fish Schools and Other Animal Aggregations', *Ecology, 35*, 361-70

Breder, C.M. Jr (1959) 'Studies on Social Grouping in Fishes', *Bulletin of the American Museum of Natural History, 117*, 393-482

Breder, C.M. Jr (1965) 'Vortices and Fish Schools', *Zoologica, 50*, 97-114

Breder, C.M. Jr (1976) 'Fish Schools as Operational Structures', *US Fishery Bulletin, 74*, 471-502

Broadbent, D.E. (1965) 'Information Processing in the Nervous System', *Science, 150*, 457-62

Brock, V.E. and Riffenburgh, R.H. (1960) 'Fish Schooling: a Possible Factor in Reducing Predation', *Journal du Conseil 25*, 307-17

Brockmann, H.J. and Barnard, C.J. (1979) 'Kleptoparasitism in Birds', *Animal Behaviour, 27*, 487-514

Caraco, T. (1980) 'Stochastic Dynamics of Avian Foraging Flocks', *American Naturalist, 115*, 262-75

Cerri, R.D. (1983) 'The Effect of Light Intensity on Predator and Prey Behaviour in Cyprinid Fish: Factors that Influence Prey Risk', *Animal Behaviour, 31*, 736-42

Charnov, E.L. (1976) 'Optimal Foraging: the Marginal Value Theorem', *Theoretical Population Biology, 9*, 129-36

Charnov, E.L. and Krebs, J.R. (1975) 'The Evolution of Alarm Calls: Altruism or Manipulation?', *American Naturalist, 109*, 107-11

Cowie, R.J. and Krebs, J.R. (1979) 'Optimal Foraging in Patchy Environments', in R.M. Anderson, B.D. Turner and R.L. Taylor, (eds), *Population Dynamics*, Blackwell, Oxford, pp. 183-205

Clark, C. (1976) *Bioeconomics*, Wiley, New York

Clark, C. and Mangel, M. (1984) 'Foraging and Flocking Strategies: Information in an Uncertain Environment', *American Naturalist, 123*, 626-41

Cullen, J.M., Shaw, E. and Baldwin, H. (1965) 'Methods for Measuring the Three-dimensional Structure of Fish Schools', *Animal Behaviour, 13*, 534-43

Cushing, D.H. and Harden-Jones, F.R. (1968) 'Why Do Fish School?', *Nature, London, 218*, 918-20

Dawkins, R. and Krebs, J.R. (1979) 'Arms Races between and within Species', *Proceedings of the Royal Society (London) B, 205*, 489-511

de Groot, P. (1980) 'A Study of the Acquisition of Information Concerning Resources by Individuals in Small Groups of Red-billed Weaver Birds', unpublished PhD thesis, University of Bristol, cited in J.R. Krebs and N.B. Davies (1981) *Introduction to Behavioural Ecology*, Blackwell, Oxford

Eaton, R.C., Bombardieri, R.A. and Meyer, D.L. (1977) 'Teleost Startle Responses', *Journal of Experimental Biology, 66*, 65-81

Ferguson, M.M. and Noakes, D.L.G. (1981) 'Social Grouping and Genetic Variation in the Common Shiner *Notropis cornutus*', *Environmental Biology of Fishes, 6*, 367-60

Foster, W.A. and Treherne, H.E. (1981) 'Evidence for the Dilution Effect in the Selfish Herd from Fish Predation on a Marine Insect', *Nature, London, 293*, 466-7

Fretwell, S.D. and Lucas, H.L. (1970) 'On Territorial Behaviour and Other Factors Influencing Habitat Distribution in Birds', *Acta Biotheoretica, 19*, 16-36

Giles, N. and Huntingford, F.A. (1984) 'Predation Risk and Interpopulation Variation in Anti-predator Behaviour in the Three-spined Stickleback', *Animal Behaviour, 32*, 264-75

Giraldeau, L-A. (1984) 'Group Foraging: the Skill-pool Effect of Frequency-dependent Learning', *American Naturalist, 124*, 72-9

Godin, J-G.J. and Keenleyside, M.H.A. (1984) 'Foraging on Patchily Distributed Prey by a Cichlid Fish: a Test of the Ideal Free Distribution Theory', *Animal Behaviour, 32*, 120-31

Goodey, W. and Liley, N.R. (1985) 'Grouping Fails to Influence the Escape Behaviour of the Guppy (*Poecilia reticulata*)' *Animal Behaviour, 33*, 1032-3

Grover, J.T. and Olla, B. (1983) 'The Role of the Rhinoceros Auklet *Cerorhinca monocerata* in Mixed Species Feeding Assemblages of Seabirds in the Strait of Juan de Fuca, Washington, *Auk, 100*, 979-82

Guthrie, D.M. (1980) *Neuroethology: an Introduction*, Blackwell, Oxford, 221 pp

Hamilton, W.D. (1971) 'Geometry for the Selfish Herd', *Journal of Theoretical Biology, 31*, 295-311

Hardin, G. (1968) 'The Tragedy of the Commons', *Science, 162*, 1243-8

Hart, P.J.B. and Connellan, B. (1984) 'The Cost of Prey Capture, Growth Rate, and Ration Size in Pike as Functions of Prey Weight', *Journal of Fish Biology, 25*, 279-91

Hasler, A.D. (1983) 'Synthetic Chemicals and Pheromones in Homing Salmon', in J.C. Rankin, T.J. Pitcher and R.T. Duggan (eds), *Control Processes in Fish Physiology*, Croom Helm, London, Chapter 6, pp. 103-16

Healey, M.C. and Prieston, R. (1973) 'The Interrelationships among Individuals in a Fish School', *Fisheries Research Board of Canada Technical Report, 389*, 1-15

Helfman, G.S. (1981) 'The Advantage to Fishes of Hovering in Shade', *Copeia, 1981*, 392-400

Helfman, G.S. (1984) 'School Fidelity in Fishes: the Yellow Perch Pattern', *Animal Behaviour, 32*, 663-72

Helfman, G.S. and Schultz, E.T. (1984) 'Social Transmission of Behavioural Traditions in a Coral Reef Fish', *Animal Behaviour, 32*, 379-84

Hemmings, C.C. (1966) 'Olfaction and Vision in Schooling', *Journal of Experimental Biology, 45*, 449-64

Hinde, R.A. and Stevenson-Hinde, J. (1973) *Constraints on Learning*, Academic Press, London, 488 pp

Hobson, E.S. (1968) 'Predatory Behaviour of Some Shore Fishes in the Gulf of California', *US Bureau of Sport Fisheries and Wildlife Research Report, 73*, 1-92

Howell, D.J. (1979) 'Flock Foraging in Nectar-eating Bats: Advantages to the Bats and to the Host Plants', *American Naturalist, 114*, 23-49

Hunter, J.R. (1966) 'Procedure for the Analysis of Schooling Behaviour', *Journal of the Fisheries Research Board of Canada, 23*, 547-62

Inagaki, T., Sakamoto, W., Aoki, I. and Kuroki, T. (1976) 'Studies on the Schooling Behaviour of Fish. III — Mutual Relationships between Speed and Form in Schooling', *Bulletin of the Japanese Society for Scientific Fisheries, 42*, 629-35

Itazawa, Y., Matsumoto, T. and Kanda, T. (1978) 'Group Effects on Physiological and Ecological Phenomena in Fish. I — Group Effect on the Oxygen Consumption of the Rainbow Trout and the Medaka', *Bulletin of the Japanese Society for Scientific Fisheries, 44*, 965-9

Jakobssen, S. and Jarvik, T. (1978) 'Antipredator Behaviour of Salmon Smolts', *Zool. Revy, 38*, 57-70

John, K.R. (1964) 'Illumination and Vision and the Schooling Behaviour of *Astyanax mexicalis*', *Journal of the Fisheries Research Board of Canada*, 21, 1453-73

Kanda, T. and Itazawa, Y. (1978) 'Group Effect on the Growth of Medaka', *Bulletin of the Japanese Society for Scientific Fisheries*, 44, 1197-1200

Keenleyside, M.H.A. (1955) 'Some Aspects of the Schooling Behaviour of Fish', *Behaviour*, 8, 83-248

Kennedy, G.J.A. and Pitcher, T.J. (1975) 'Experiments on Homing in Shoals of the European Minnow, *Phoxinus phoxinus* (L.)', *Transactions of the American Fisheries Society*, 104, 452-5

Kenward, R.E. (1978) 'Hawks and Doves: Attack Success and Selection in Goshawk Flights at Wood Pigeons', *Journal of Animal Ecology*, 47, 449-60

Koslow, J.A. (1981) 'Feeding Selectivity in Schools of Northern Anchovy in the Southern Californian Bight', *US Fishery Bulletin*, 79, 131-42

Krebs, J.R. (1973) 'Social Learning and the Significance of Mixed-species flocks of Chickadees (*Parus* spp.)', *Canadian Journal of Zoology*, 51, 1275-88

Krebs, J.R. (1974) 'Colonial Nesting and Social Feeding as Strategies for Exploiting food Resources in the Great Blue Heron, *Ardea herodius*', *Behaviour*, 51, 99-134

Krebs, J.R. (1978) 'Optimal Foraging: Decision Rules for Predators', in J.R. Krebs and N.B. Davies (eds), *Behavioural Ecology* 1st edn, Blackwell, Oxford, pp. 23-63

Krebs, J.R. and Barnard, C.J. (1980) 'Comments on the Functions of Flocking in Birds', in *Acta XVIIIth Congress of International Ornithology*, Berlin, 1978, pp. 795-9

Krebs, J.R. and Davies, N.B. (1978) (eds) *Behavioural Ecology*, 1st edn, Blackwell, Oxford, 494 pp.

Krebs, J.R. and Davies, N.B. (1981) *Introduction to Behavioural Ecology*, Blackwell, Oxford, 292 pp.

Krebs, J.R. and Davies, N.B. (eds) (1984) *Behavioural Ecology*, 2nd edn, Blackwell, Oxford, 493 pp.

Krebs, J.R. and Dawkins, R. (1984) 'Animal Signals: Mindreading and Manipulation', in *Behavioural Ecology*, 2nd edn, J.R. Krebs and R. Dawkins (eds), Blackwell, Oxford, pp. 380-402

Krebs, J.R., MacRoberts, M. and Cullen, J.M. (1972) 'Flocking and Feeding in the Great Tit *Parus major*: an Experimental Study', *Ibis*, 114, 507-30

Krebs, J.R., Kacelnik, A. and Taylor, P. (1978) 'Test of Optimal Sampling by Foraging Great Tits', *Nature, London*, 275, 539-42

Lauder, G.V. (1983) 'Food Capture', in P.W. Webb and D. Weihs (eds), *Fish Biomechanics*, Praeger, New York, Chapter 9, pp. 280-311

Larkin, P.A. and Walton, A. (1969) 'Fish School Size and Migration', *Journal of the Fisheries Research Board of Canada*, 26, 1372-4

Lazarus, J. (1979) 'The Early Warning Function of Flocking in Birds: an Experimental Study with Captive *Quelea*', *Animal Behaviour*, 27, 855-65

Lendrem, D. (1983) 'Predation Risk and Vigilance in the Blue Tit', *Behavioral Ecology and Sociobiology*, 13, 9-13

Lester, N.P. (1981) 'Feeding Decisions in Goldfish: Testing a Model of Time Allocation', unpublished PhD thesis, University of Sussex

Lester, N.P. (1984) 'The Feed-Feed Decision: How Goldfish Solve the Patch Depletion Problem', *Behavior*, 89, 175-99

Loekle, D.M., Madison, D.M. and Christian, J.J. (1982) 'Time Dependency and Kin Recognition of Cannibalistic Behaviour among Poecilid Fishes', *Behavioral and Neurological Biology*, 35, 315-18

McFarland, W.N. and Kotchian, N.M. (1982) 'Interaction between Schools of Fish and Mysids', *Behavioral Ecology and Sociobiology*, 11, 71-76

McKaye, K.R. and Oliver, M.K. (1980) 'Geometry of a Selfish School: Defence of Cichlid Young by a Bagrid Catfish in Lake Malawi, Africa', *Animal Behaviour*, 28, 1278-90

McNicol, R.G. and Noakes, D.L.G. (1984) 'Environmental Influences on Territoriality of Juvenile Brook Char *Salvelinus fontinalis* in a Stream Environment', *Environmental Biology of Fishes*, 10, 29-42

Magurran, A.E. (1984) 'Gregarious Goldfish', *New Scientist*, 9th August, 32-33

Magurran, A.E. and Girling, S. (1985) 'Predator Model Recognition and Response Habituation in Shoaling Minnows', *Animal Behaviour*, in press

Magurran, A.E. and Pitcher, T.J. (1983) 'Foraging, Timidity and Shoal Size in Minnows and

Goldfish', *Behavioral Ecology and Sociobiology, 12,* 142-152
Magurran, A.E. and Pitcher, T.J. (1986) 'Provenance, Shoal Size and the Organisation of Predator Evasion Behaviours in Minnow Shoals' (submitted)
Magurran, A.E., Oulton, W. and Pitcher, T.J. (1985) 'Vigilant Behaviour and Shoal Size in Minnows', *Zeitschrift für Tierpsychologie, 67,* 167-78
Milinski, M. (1979a) 'An Evolutionarily Stable Feeding Strategy in Sticklebacks', *Zeitschrift für Tierpsychologie, 51,* 36-40
Milinski, M. (1979b) 'Can an Experienced Predator Overcome the Confusion of Swarming Prey More Easily?', *Animal Behaviour, 27,* 1122-6
Milinski, M. (1984a) 'A Predator's Costs of Overcoming the Confusion Effect of Swarming Prey', *Animal Behaviour, 32,* 1157-62
Milinski, M. (1984b) 'Competitive Resource Sharing: an Experimental Test of a Learning Rule for ESSs', *Animal Behaviour, 32,* 233-42
Milinski, M. (1985) 'Risk of Predation Taken by Parasitised Sticklebacks under Competition for Food', *Behaviour,* in press
Motta, P.J. (1983) 'Response by Potential Prey to Coral Reef Fish Predators', *Animal Behaviour, 31,* 1257-8
Neill, S.R. St. J. and Cullen, J.M. (1974) 'Experiments on Whether Schooling by their Prey Affects the Hunting Behaviour of Cephalopod and Fish Predators', *Journal of Zoology (London), 172,* 549-69
Nursall, J.R. (1973) 'Some Behavioural Interactions of Spottail Shiners (*Notropis hudsonius*), yellow perch (*Perca flavescens*) and northern pike (*Esox lucius*)', *Journal of the Fisheries Research Board of Canada, 30,* 1161-78
Olsen, F.C.W. (1964) 'The Survival Value of Fish Schooling', *Journal du Conseil, 29,* 115-16
Orians, G.H. (1981) 'Foraging Behaviour and the Evolution of Discriminating Abilities', in A.C. Kamil and T.D. Sargent (eds), *Foraging Behaviour,* Garland Press, New York, pp. 389-405
Parker, F.R. (1973) 'Reduced Metabolic Rate in Fishes as a Result of Induced Schooling', *Transactions of the American Fisheries Society, 102,* 125-31
Parker, G.A. (1978) 'Searching for Mates', in J.R. Krebs and N.B. Davies (eds), *Behavioural Ecology,* 1st edn, Blackwell, Oxford, pp. 214-44
Parker, G.A. (1984) 'Evolutionarily Stable Strategies', in J.R. Krebs and N.B. Davies (eds), *Behavioural Ecology,* 2nd edn, Blackwell, Oxford, pp. 30-61
Partridge, B.L. (1982a) 'Structure and Function of Fish Schools', *Scientific American, 245,* 114-23
Partridge, B.L. (1982b) 'Rigid Definitions of Schooling Behaviour are Inadequate', *Animal Behaviour, 30* 298-9
Partridge, B.L. and Pitcher, T.J. (1979) 'Evidence against a Hydrodynamic Function of Fish Schools', *Nature, London, 279,* 418-19
Partridge, B.L. and Pitcher, T.J. (1980) 'The Sensory Basis of Fish Schools: Relative Roles of Lateral Line and Vision', *Journal of Comparative Physiology A, 135,* 315-25
Partridge, B.L., Pitcher, T.J., Cullen, J.M. and Wilson, J.P.F. (1980) 'The Three-dimensional Structure of Fish Schools', *Behavioral Ecology and Sociobiology, 61,* 277-88
Partridge, B.L., Johansson, J. and Kalish, J. (1983) 'The Structure of Schools of Giant Bluefin Tuna in Cape Cod Bay', *Environmental Biology of Fishes, 9,* 253-62
Pitcher, T.J. (1973) 'The Three-dimensional Structure of Schools in the Minnow, *Phoxinus phoxinus* (L.)', *Animal Behaviour, 21,* 673-86
Pitcher, T.J. (1975) 'A Periscopic Method for Determining the Three-dimensional Positions of Fish in Schools', *Journal of the Fisheries Research Board of Canada, 32,* 1533-8
Pitcher, T.J. (1979a) 'The Role of Schooling in Fish Capture', *International Commission for the Exploration of the Sea, CM 1979/B:5,* 1-12
Pitcher, T.J. (1979b) 'Sensory Information and the Organisation of Behaviour in a Shoaling Cyprinid', *Animal Behaviour, 27,* 126-49
Pitcher, T.J. (1980) 'Some Ecological Consequences of Fish School Volumes', *Freshwater Biology, 10,* 539-44
Pitcher, T.J. (1983) 'Heuristic Definitions of Shoaling Behaviour', *Animal Behaviour, 31,* 611-13
Pitcher, T.J. and Hart, P.J.B. (1982) *Fisheries Ecology,* Croom Helm, London, 415 pp.
Pitcher, T.J. and Lawrence, J.E.T. (1984) 'A Simple Stereo Television System with Application to the Measurement of Coordinates of Fish in Schools', *Behavior Research*

Methods, Instruments and Computers, 16, 495-501

Pitcher, T.J. and Magurran, A.E. (1983) 'Shoal Size, Patch Profitability and Information Exchange in Foraging Goldfish', *Animal Behaviour, 31*, 546-55

Pitcher, T.J. and Partridge, B.L. (1979) 'Fish School Density and Volume', *Marine Biology, 54*, 383-94

Pitcher, T.J. and Wyche, C.J. (1983) 'Predator-avoidance Behaviours of Sand-eel Schools: Why Schools Seldom Split? in D.L.G. Noakes, B.G. Lindquist, G.S. Helfman and J.A. Ward (eds), *Predators and Prey in Fishes*, Junk, The Hague, pp. 193-204

Pitcher, T.J., Partridge, B.L., Wardle, C.S. (1976) 'A Blind Fish can School', *Science, 194*, 963-5

Pitcher, T.J., Magurran, A.E. and Winfield, I. (1982) 'Fish in Larger Shoals Find Food Faster', *Behavioral Ecology and Sociobiology, 10*, 149-51

Pitcher, T.J., Wyche, C.J. and Magurran, A.E. (1982) 'Evidence for Position Preferences in Schooling Mackerel', *Animal Behaviour, 30*, 932-4

Pitcher, T.J., Magurran, A.E. and Allan, J.R. (1983) 'Shifts of Behaviour with Shoal Size in Cyprinids', *Proceedings of the British Freshwater Fisheries Conference, 3*, 220-8

Pitcher, T.J., Magurran, A.E. and Edwards, I. (1985a) 'Schooling Mackerel and Herring Choose Neighbours of Similar Size', *Marine Biology, 86*, 319-22

Pitcher, T.J., Magurran, A.E. and Allan, J.R., (1985b) 'Size Segregative Behaviour in Minnow Shoals', *Journal of Fish Biology* (submitted).

Pitcher, T.J., Green, D.A. and Magurran, A.E. (1986) 'Dicing with Death: Predator Inspection Behaviour in Minnow Shoals', *Journal of Fish Biology* (in press)

Poole, T.B. and Dunstone, N. (1975) 'Underwater Predatory Behaviour of the American Mink *Mustela vison*', *Journal of Zoology (London), 178*, 395-412

Potts, G.W. (1970) 'The Schooling Ethology of *Lutjianus monostigma* in the Shallow reef Environment of Aldabra', *Journal of Zoology (London), 161*, 223-35

Potts, W.K. (1984) 'The chorus line hypothesis of manoeuvre coordination in avian flocks', *Nature, London, 309*, 344-5

Powell, G.V.N. (1974) 'Experimental Analysis of the Social Value of Flocking by Starlings (*Sturnus Vulgaris*) in Relation to Predation and Foraging', *Animal Behaviour, 23*, 504-8

Pulliam, H.R. and Caraco, T. (1984) 'Living in Groups: Is There an Optimal Group Size?' in J.R. Krebs and N.B. Davies (eds) *Behavioural Ecology*, 2nd edn, Blackwell, Oxford, pp. 122-42

Quinn, T.P. and Busack, C.A. (1985) 'Chemosensory Recognition of Siblings in Juvenile Coho Salmon *Oncorhyncus kisutch*', *Animal Behaviour, 33*, 51-6

Quinn, T.P. and Fresh, K. (1984) 'Homing and Straying in Chinook Salmon *O. tschawytscha* from Cowlitz River Hatchery, Washington', *Canadian Journal of Fisheries and Aquatic Sciences, 41*, 1078-82

Radakov, D.V. (1973) *Schooling in the Ecology of Fish*, Israel Programme for Scientific Translations, Wiley, New York, 173 pp.

Reist, J.D. (1983) 'Behavioral Variation in Pelvic Phenotypes of Brook Stickleback *Culaea inconstans* in Response to Predation by Northern Pike *Esox Lucius*', *Environmental Biology of Fishes, 8*, 255-68

Robertson, D.R., Sweatman, H.P.A., Fletcher, G.A. and Cleland, M.G. (1976) 'Schooling as a Means of Circumventing the Territoriality of Competitors', *Ecology, 57*, 1208-20

Rubenstein, D.I., Barnett, R.J., Ridgely, R.S. and Klopfer, P.H. (1977) 'Adaptive Advantages of Mixed Species Feeding Flocks among Seed-eating Finches in Costa Rica', *Ibis, 119*, 10-21

Rüppell, G. and Gösswein, E. (1972) 'Die Shwärme von *Leucaspius delineatus* (Cyprinidae, Teleostei) be Gofahr im Hellen und im Dunkeln' *Zeitschrift für vergh. Physiologie, 76*, 333-40

Savino, J.F. and Stein, R.A. (1982) 'Predator-Prey Interaction between Largemouth Bass and Bluegills as Influenced by Simulated Submersed Vegetation', *Transactions of the American Fisheries Society, 11*, 255-66

Seghers, B. (1974) 'Schooling by the Guppy: an Evolutionary Response to Predation', *Evolution, 28*, 488-9

Seghers, B. (1981) 'Facultative Schooling Behaviour in the Spottail Shiner (*Notropis hudsonius*): possible costs and benefits', *Environmental Biology of Fishes, 6*, 21-4

Shaw, E. (1962) 'The Schooling of Fishes', *Scientific American, 206*, 128-38

Shaw, E. (1965) 'The Optomotor Response and the Schooling of Fishes', *International*

Commission for North Atlantic Fisheries, Special Publication 6, 753-79
Shaw, E. (1970) 'Schooling Fishes: Critique and Review' in L. Aronson, E. Tobach, D.S.
    Lehrmann and J.S. Rosenblatt (eds), *Development and Evolution of Behaviour*, W.H.
    Freeman, San Francisco, pp. 452-80
Shaw, E. (1978) 'Schooling Fishes', *American Scientist*, 66, 166-75
Shaw, E. and Tucker, A. (1965) 'The Optomotor Reaction of Schooling Carangid Fishes',
    Animal Behaviour, 13, 330-66
Shettleworth, S.J. (1984) 'Learning and Behavioural Ecology', in J.R. Krebs and N.B. Davies
    (eds), *Behavioural Ecology*, 2nd edn, Blackwell, Oxford, pp. 170-94
Sibly, R.M. (1983) 'Optimal Group Size is Unstable', *Animal Behaviour*, 31, 947-8
Sih, A. (1980) 'Optimal Foraging: Can Foragers Balance Two Conflicting Demands?,
    Science, 210, 1041-3
Street, N.G. and Hart, P.J.B. (1985) 'The Influence of Group Size upon Patch Location by a
    Non-visually Foraging Predator — the Stone Loach (*Noemacheilus barbatalus*)', *Journal of
    Fish Biology* (in press)
Street, N.G., Magurran, A.E. and Pitcher, T.J. (1984) 'The Effects of Increasing Shoal Size
    on Handling Time in Goldfish *Carassius auratus*', *Journal of Fish Biology*, 25, 561-6
Sutherland, W.S. (1983) 'Aggregation and the "Ideal Free Distribution"', *Journal of Animal
    Ecology*, 52, 821-8
Taylor, R.J. (1984) *Predation*, Chapman & Hall, London, 166 pp.
Tinbergen, L. (1981) 'Foraging Decisions in Starlings', *Ardea*, 69, 1-67
Treherne, J.E. and Foster, W.A. (1981) 'Group Transmission of Predator Avoidance in a
    Marine Insect: the Trafalgar Effect', *Animal Behaviour*, 29, 911-17
Treisman, M. (1975) 'Predation and the Evolution of Gregariousness. I. Models of
    Concealment and Evasion', *Animal Behaviour*, 23, 779-800
Tremblay, D. and FitzGerald, G.J. (1979) 'Social Organisation as an Antipredator Strategy in
    Fish', *Naturaliste Canadienne*, 105, 411-13
Turner, G. and Pitcher, T.J. (1985) 'Attack Dilution, Search and Abatement', submitted
Uematsu, T. and Takamuri, J. (1976) 'Social Facilitation in Feeding Behaviour of the
    Himedaka *Oryzias latipes*. I. Continuous Observation during a Short Period', *Japanese
    Journal of Ecology*, 26, 135-40
Ward, P. and Zahavi, A. (1973) 'The Importance of Certain Assemblages of Birds as
    Information Centres for Food Finding', *Ibis*, 115, 517-34
Webb, P.W. (1976) 'The Effect of Size on the Fast-start Performance of Rainbow Trout and
    a Consideration of Piscivorous Predator-Prey Interactions', *Journal of Experimental
    Biology*, 65, 157-77
Webb, P.W. (1980) 'Does Schooling Reduce Fast Start Response Latencies in Teleosts?'
    Comparative Biochemistry and Physiology, 65A, 231-34
Weihs, D. (1973) 'Hydromechanics and Fish Schooling', *Nature, London*, 241, 290-1
Weihs, D. (1975) 'Some Hydrodynamical Aspects of Fish Schooling', in T.Y. Wu, C.J.
    Broklaw and C. Brennan (eds), *'Symposium on Swimming and Flying in Nature*', Plenum
    Press, New York, pp. 703-18
Weihs, D. and P.W. Webb (1983) 'Optimisation of Locomotion', in P.W. Webb and D.
    Weihs (eds) *Fish Biomechanics*, Praeger Press, New York, pp. 339-71
Welty, J.L.C. (1934) 'Experiments on Group Behaviour of Fishes', *Physiological Zoology*,
    B7, 85-128
Wijffels, W., Thines, G. Dijkgraaf, S. and Verheijen, F.J. (1967) 'Apprentissage d'un
    Labyrinthe Simple par des Teleosteans Isoles ou Groupes de l'Espece *Barbus tincto*',
    Archives Néerlandaises de Zoologie, 3, 376-402
Williams, G.C. (1964) 'Measurements of Consociation among Fishes and Comments on the
    Evolution of Schooling', *Publications of the Museum of Michigan State University*, 2,
    351-83
Wilson, E.O. (1975) *Sociobiology: the Modern Synthesis*, Harvard University Press,
    Cambridge, Mass. 697pp.
Wolf, N.G. (1983) 'The Behavioral Ecology of Herbivorous Fishes in Mixed Species Groups',
    unpublished PhD thesis, Cornell University, Ithaca, New York
Wolf, N.G. (1985) 'Odd Fish Abandon Mixed-species Groups when Threatened', *Behavioral
    Ecology and Sociobiology*, in press
Yates, G.T. (1983) 'Hydrodynamics of Body and Caudal Fin Propulsion', in P.W. Webb and
    D. Weihs (eds), *Fish Biomechanics*, Praeger Press, New York,pp. 177-213

13 INDIVIDUAL DIFFERENCES IN FISH BEHAVIOUR

Anne E. Magurran

No one supposes that all individuals of the same species are cast in the very same mould. Darwin (1859)

But they do! In ethology individual differences in behaviour have traditionally been ignored, treated as white noise (Ringler 1983) or, as Arak (1984) points out, considered to be maladaptive deviations from optimal strategies. Many classical ethology textbooks proceed with the implicit assumption that intraspecific differences in behaviour are less important and less interesting than differences between species. Fish behaviour, where most students begin by looking at the stereotyped response of the breeding male stickleback (*Gasterosteus aculeatus*) to things coloured red (Tinbergen 1951), is no exception. Closer inspection, however, reveals pronounced individual differences in the rate at which male sticklebacks attack red dummies (Rowland 1982).

Despite first impressions, rigid behaviour patterns are not the rule. This chapter will show that there can be considerable variation between individual fish in a whole host of behaviours, including the methods they adopt to find food, or to ensure that their genes are represented in future generations. There are even individual differences in the behaviour of members of fish schools (Pitcher, Wyche and Magurran 1982; Helfman 1984; and see Chapter 12 by Pitcher, this volume), long considered the most uniform and egalitarian of piscine social structures (Shaw 1970).

Why are individual differences so important? Darwin (1859) was the first to recognise that variation between individuals is the raw material for natural selection. Although Darwin was primarily concerned with morphological variation, his arguments apply equally to physiological and behavioural variation; indeed, behaviour, morphology and physiology are often linked and may have profound influences on one another. Contemporary evolution biology, which stresses the importance of the individual (or gene) rather than the group or species as the unit of selection (Dawkins 1976), has reawakened an interest in individual differences. Game theory, with its associated concept of the evolutionarily stable strategy (Maynard Smith 1982; Parker 1984) has provided us with an explanation as to how two or more different types of behaviour can be advantageous.

The aim of this chapter is to investigate the nature and significance of individual differences in teleost behaviour. It will begin with a brief review of explanations of variable behaviour, followed by an examination of five areas where individual differences are particularly well documented:

foraging behaviour, predator avoidance, mating strategies, habitat use and dominance hierarchies.

The Nature of Individual Differences

There are many ways in which the behaviour of individual animals can vary, and many reasons why such variation occurs. Individual differences in behaviour may, for instance, reflect sex or body size or be a response to food availability, predation pressure, parasitism, or competition from conspecifics. These differences can be genetically determined, or the result of development, environment or experience. Differences can be found between populations, between individuals within a population, or even within an individual.

The propensity of individuals to differ in their behaviour varies between species. Inevitably, differences are smaller in species where individuals spend much of their time in polarised and synchronised schools (see Chapter 12 by Pitcher, this volume). As shoaling tendency decreases, differences become more evident. For instance, Magurran and Pitcher (1983) noted that individual differences are more pronounced in goldfish (*Carassius auratus*) than in minnows (*Phoxinus phoxinus*), the more strongly shoaling species.

Three mechanisms have been proposed to account for individual differences in behaviour (Krebs and Davies 1981; Partridge and Green 1985). These are a variable environment, phenotypic differences and the behaviour of other individuals.

(1) A Variable Environment.   Behavioural flexibility is obviously appropriate in an environment which varies in space or time. For example, a fish's foraging behaviour may depend on the patchiness of the habitat (see Chapter 8 by Hart, this volume) or on the degree of predation it is exposed to (see page 334).

(2) Phenotypic Differences between Individuals.   An animal's phenotype may constrain its behaviour. This can happen in one of two ways. First, different sizes, sexes or morphs may have different requirements in terms of food, cover, etc. For instance, a Mexican cichlid (the Cuatro Cienegas cichlid *Cichlasoma minckleyi*) occurs as two distinct morphs with very different dentition, and, as a result, very different diets (Kornfield, Smith, Gagnon and Taylor 1982; Kornfield and Taylor 1983). Secondly, if an animal is competitively inferior, it may be forced to modify its behaviour in order to 'make the best of a bad job' (Dawkins 1979). This situation, where the less successful individuals have access to a smaller portion of the resource, contrasts with the last category, where behaviour depends on what others are doing.

(3) Best Behaviour Depends on the Behaviour of Others.   Individuals can employ different strategies to achieve the same ends. It appears, for example, that territory holders and sneak males in the bluegill sunfish (*Lepomis macrochirus*) enjoy, on average, approximately equal mating success (Gross 1982, and see page 352) Despite the fact that short-term benefits may be unequal, different behaviour patterns can be selected for if they bring equivalent lifetime success (Dunbar 1982). Since the payoff for a given behaviour pattern will be affected by what other individuals are doing (for example, it is not worth while being a bluegill sneak if too many other fish are doing the same thing), the proportions of the various strategies in a population will be maintained through frequency-dependent selection (Partridge and Green 1985).

These three mechanisms do not necessarily act independently since an individual's phenotype may affect its use of a heterogenous environment, as well as its response to the behaviour of other individuals.

Although some individual differences in behaviour are genetically based (Oliverio 1983), it is often difficult to disentangle the effects of genes, environment (Partridge 1983; Huntingford 1984) and development (Bateson 1983; and see Chapter 3 by Huntingford, this volume). Consequently, there are relatively few clear demonstrations that variable behaviour has been inherited. This is especially true in the case of complex social behaviour (Huntingford 1984). As Slater (1983) points out, the degree of variability of a behaviour is no guide to whether it is environmentally or genetically determined. In several cases, for example, the presence of distinct stereotyped behaviour patterns has erroneously been used as evidence of the genetic basis of behaviour (Dunbar 1982).

Genes may be responsible for individual differences in behaviour. Individual differences in behaviour, especially those differences that exist between populations, may in turn enhance genetic diversity. Ferguson and Mason (1981) have shown that there are at least three reproductively isolated, sympatric stocks of brown trout (*Salmo trutta*) in Lough Melvin, a small lake (16 km²) in the north west of Ireland. Although this reproductive isolation was initially brought about by ice movements during the last glaciation, it is maintained through behaviour. In the final analysis, of course, individual differences in behaviour which are sufficiently great to stop interbreeding between populations may be important in bringing about speciation (Bush 1975).

Foraging

Individual fish can vary considerably in their choice of food and method of capturing and handling prey (Ringler 1983). Curio (1976) has termed this 'predator versatility', and proposed two foraging strategies. Individuals, he suggests, may specialise in a few tactics and/or prey types, or may alternatively be generalists which adopt most if not all tactics available to the species. Individual differences will be found in species where all individuals are specialists, or where there is a combination of specialists and generalists. Specialisation may be advantageous as the work of Werner, Mittelbach and Hall (1981) on the bluegill sunfish (*Lepomis macrochirus*) shows. These fish can be divided into specialists which forage on either plankton or benthos, and generalists which feed in both habitats. Specialists are more successful and obtain more prey than generalists. In situations where the food supply is unpredictable, however, the generalist strategy is more appropriate (Dill 1983).

On an even finer scale, variability in foraging tactics can occur within individual fish. Short-term changes in foraging tactics will often be related to motivational state (see Chapter 2 by Colgan, this volume), especially hunger levels (Ringler 1983). Hungry sticklebacks, for instance, are less selective in their choice of prey (Beukema 1968). Learning and experience are important in bringing about longer-term changes (Dill 1983; Giraldeau 1984), and experienced fish search more and with greater efficiency (Beukema 1968; Atema, Holland and Ikehara 1980; Potts 1980), attack more speedily (Godin 1978) and are more successful in capturing prey (Webb and Skadsen 1980; Vinyard 1982). Changes in the abundance and type of prey will also cause a fish to modify its foraging behaviour.

Nineteen studies demonstrating individual differences in the foraging behaviour of fish have been documented by Ringler (1983). For example, Bryan and Larkin (1972) investigated the feeding tactics of brook trout (*Salvelinus fontinalis*), cutthroat trout (*Salmo clarki*) and rainbow trout (*Salmo gairdneri*) in a stream in British Columbia. Known individuals were tagged, released and recaught at intervals of up to one year. Analysis of the stomach contents of marked fish showed that individuals ate different types of food, that degree of food specialisation varied between individuals, and that specialisations persisted for periods of up to six months. There was no relationship between the composition of the diet and size or growth rate of the fish, the weight of food eaten or the site of recapture. Some individuals moved around the stream but nevertheless maintained their specialisation, demonstrating that diet choice was not simply a reflection of fish remaining in a small home range and eating the food available there. Allan (1981) found that wild brook trout also showed individual differences in diurnal feeding rhythm.

Bryan and Larkin (1972) confirmed their field observations by conducting experiments on hatchery-reared rainbow trout. In these trials the absolute and relative abundance of available prey items was kept constant. A significant interaction between individuals and food type in an analysis of variance confirmed that food specialisation was occurring. They also noted that individuals had slightly different searching behaviour, and concluded that this, along with capturing technique and feeding rhythm, was important in bringing about the observed specialisations.

It is likely that any feeding specialisation such as that shown by Bryan and Larkin's trout will be positively reinforced since skill at finding and capturing prey will increase with experience. Once an animal has begun to specialise on a particular type of food, it may also be energetically disadvantageous to switch to another specialisation. Partridge and Green (1985) review work showing that gut morphology can be related to diet and that the efficiency of digestion may be reduced if food is varied. Windell (1978) provides a discussion of the changes in gut morphology and digestion efficiency that can take place in teleosts.

The feeding specialisations of trout in Bryan and Larkin's work were not dependent on physiological or morphological differences in the fish concerned. In many cases, however, variation in feeding tactics occurs as a direct result of physical differences between individuals. One of the most important of these is size. Large fish often have an advantage over their smaller conspecifics (Werner 1984). In a recent experiment, shoals of 20 large (mean length 60 mm) and 20 small (mean length 40 mm) minnows (*Phoxinus phoxinus*) foraged together on an artificial food patch consisting of 90 gravel-filled pots (Pitcher, Magurran and Allan in prep). Minnows in possession of a food-filled pot were able to defend it successfully against similar sized or smaller fish, but were evicted by larger fish (Figure 13.1).

Figure 13.1: Feeding Interactions of Large and Small Minnows. Twenty large and twenty small minnows foraged together on an artificial food patch comprised of 90 gravel-filled pots, each pot big enough to accommodate only one fish. Nine pots held food. Interactions between a minnow in possession of a pot (the owner) challenged for occupancy by another fish (the intruder) were scored. The results, which summarise data from eight 15-min trials and show the probability of a challenged owner remaining in possession, indicate that size of fish and ownership of a pot were important in determining the outcome of contests. It is also interesting to note that minnows invested most effort in defending those pots that held food. With the exception of small owners, which were always evicted by large intruders, there was no significant difference in the number of times minnows won or lost possession of a no-food pot

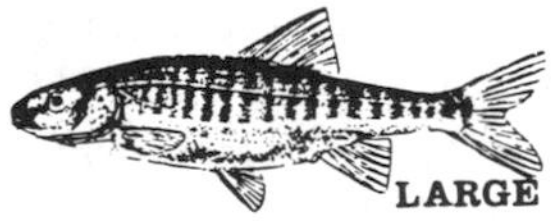

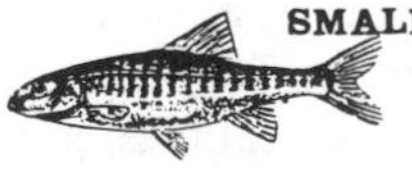

	OWNER			
	FOOD		NO FOOD	
	LARGE	SMALL	LARGE	SMALL
INTRUDER LARGE	Owner Wins ☆☆☆	Owner Loses ☆☆☆	Outcome Uncertain Ns	Owner Loses ☆☆☆
SMALL	Owner Wins ☆☆☆	Owner Wins ☆☆☆	Outcome Uncertain Ns	Outcome Uncertain Ns

☆☆☆ p<0.001

Figure 13.2: Neighbour Size in Minnow Shoals. The neighbour size of large and small minnows was recorded every 5s for 5min (*n*─60) on 36 occasions (*n*─36: 4 minnows per size class × 9 days). This figure shows a frequency histogram (for both size-classes) of the proportion of nearest neighbours (out of the 60 records) that were the same size as the reference fish. The median, and interquartile range, of the proportion of same-sized nearest neighbours is also shown

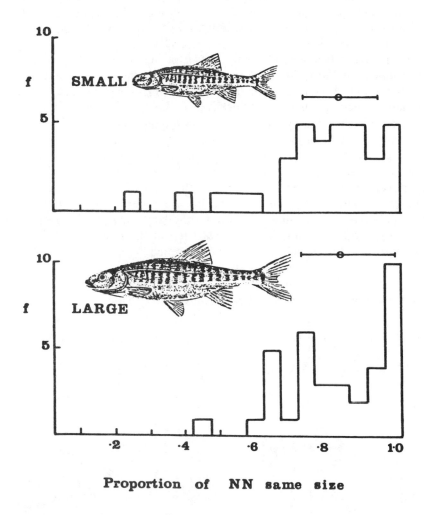

Proportion of NN same size

When these 20 large and 20 small minnows were allowed to shoal freely in a large arena tank (2 m × 1.6 m × 1 m deep), it was found that fish tended to associate with individuals of similar size; small minnows had small minnows as nearest neighbours and large minnows had large nearest neighbours (Figure 13.2). It is likely that one function of this size sorting is to reduce competition between large and small fish.

In addition to superior strength, large fish may also enjoy improved powers of food detection. Breck and Gitter (1983), working on bluegill sunfish, were able to show that an increase in fish size is accompanied by an increase in lens diameter and lens acuity, which in turn are associated with an increase in reactive distance, the distance at which fish respond to prey. Since larger bluegills are also able to swim faster while searching for food (Mittelbach 1981), they encounter more prey than smaller fish (Figure 13.3). These benefits are enhanced by an exponential decrease in handling time (above a critical ratio of prey length/fish length) with increasing fish size.

A further advantage of being big is a decreased vulnerability to predation (Dill 1983). Foraging fish must balance the conflicting demands of finding food (see Chapter 8 by Hart, this volume) and avoiding predators (see Chapter 9 by Milinski, this volume). Werner, Gilliam, Hall and Mittelbach (1983) investigated the foraging behaviour of three size classes of bluegill sunfish in the wild. In the presence of a natural predator, the largemouth bass (*Micropterus salmoides*), the two larger size classes of bluegill, which were not susceptible to predation, chose the most profitable open-water areas to feed in. By contrast the small bluegills, which were in danger of being eaten, spent more time in cover where foraging return rates were only one-third of those enjoyed by the larger fish. In the absence of predators, both large and small bluegills fed in the most profitable areas (Werner and Mittelbach 1981). The presence of a predator therefore brings about differences in foraging behaviour that reduce intraspecific competition and keep small fish small and vulnerable to predation longer.

In situations where some individuals are better at locating food it could

Figure 13.3: Rate of Prey Encounter and Bluegill Size. Three size classes of bluegills (20, 60 and 100 mm) foraged in vegetation for damselfly naiads. Prey density ranged from 10 to 45 prey m^{-3}, and prey length from 5 to 12 mm. This graph plots the rate of prey encounter as a function of prey density, prey length and bluegill size. (Redrawn from Mittelbach 1981.)

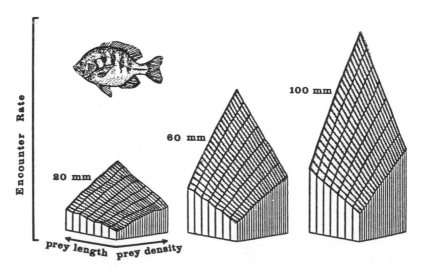

be advantageous for the less-successful feeders to cease hunting independently and forage instead in places where they have seen other fish find food. Barnard and Sibly (1981) have termed these two categories of foragers 'copiers' and 'searchers', and this method of foraging 'area copying'. There is good evidence to suggest that fish do area copy. In goldfish (*Carassius auratus*) searching individuals are attracted to fish that have already found food and are foraging with a characteristic 'head-down' posture (Pitcher *et al.* 1982; Magurran 1984 and see Pitcher, Chapter 12, this volume). Helfman (1984) found that some yellow perch (*Perca flavescens*) tended to follow and observe individuals which were actively foraging in the benthos. It is unlikely, however, that fish consistently adopt searcher and copier strategies, as is the case in birds (Barnard 1984).

Individual differences become especially important when limited food acts to increase levels of intraspecific competition. Inevitably body size is one factor influencing the success of individuals (Wilson 1975), though as Keen (1982) showed for juvenile brown bullhead (*Ictalurus nebulosus*) larger size may need to be coupled with higher levels of aggression before a fish can capture a disproportionate share of the resource. Rubenstein (1981), in a series of experiments on the everglades pygmy sunfish (*Elassoma evergladei*), found that individual differences in growth rate were magnified under competitive conditions. In Rubenstein's work the sunfish were raised in densities of one, four, eight and sixteen fish per laboratory tank or field cage. Food supply to the whole tank or cage was kept constant to ensure more vigorous competition in the larger group sizes. Rubenstein showed that as fish density increased from four to sixteen fish, the growth rate of all fish fell. However, the growth of the best competitors decreased much more slowly than the growth of the average fish and was not adversely affected by the higher fish density. The coefficient of variation (COV) of growth rate increased from 35 per cent for solitary fish and 45 per cent for fish in a group of four, to 209 per cent for the largest group. Average fecundity of both males and females was also reduced in larger groups and, as with growth rate, individual differences in fecundity were exaggerated. Reduced growth rates of subordinate individuals have also been found in bluegill sunfish (Drager and Chizar 1982).

Rubenstein provides a clear demonstration of the consequences of inequalities in competitive ability. Milinski's (1982, 1984a-c) work on sticklebacks (*G. aculeatus*) shows precisely how the foraging behaviour of good and poor competitors differs, and indicates the methods that poor competitors adopt in an attempt to compensate for these inequalities. Milinski (1982) fed daphnia (*Daphnia magna*) to pairs of sticklebacks and used the proportion of daphnia that each fish caught as a measure of its relative competitive ability. Each pair of sticklebacks was then presented with daphnia belonging to two size classes. Since handling time was the same for both prey sizes, the larger daphnia were more profitable. In each pair of sticklebacks tested, poorer competitors caught significantly fewer large daphnia. Surprisingly, when allowed to forage alone, these poorer competitors still continued to direct over half of their attacks towards small daphnia; good competitors, by contrast, preferred to attack large daphnia. Milinski argues that lower success rate in capturing large daphnia makes it advantageous for the less-successful sticklebacks to incorporate more small daphnia in their diet.

Evidence for a link between individual competitive ability and the profitability of prey consumed is also provided by Coates' (1980) work on the humbug damselfish (*Dascyllus aruanus*). Coates found that there was a clear relationship between the competitive rank of an individual within the social group and the size of prey eaten. In contast to Milinski's sticklebacks, however, less-competitive damselfish switched to eating larger prey in the absence of the more-successful individuals.

Direct competition for the most profitable prey is not the only problem that poor competitors meet. It is usual for food to be patchily distributed in both space and time (for example marine plankton (Steele 1976) and freshwater invertebrates (Allan 1975; Brooker and Morris 1980)). Individuals must therefore decide how long to stay on one patch before moving on to another (see Chapter 8 by Hart, this volume). Milinski (1984a) investigated the foraging behaviour of sticklebacks in a patchy environ-

Figure 13.4: Competitive Rank and Foraging Success. Each individual in a group of six sticklebacks was assigned a competitive rank on the basis of number of daphnia captured during a trial. Nine trials were conducted in all. This figure shows the mean number of daphnia consumed by the nine most successful fish, the nine second-most successful fish, and so on to the nine least successful fish. (Redrawn from Milinski 1984a.)

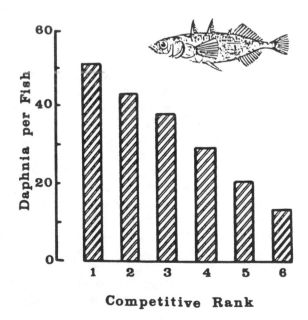

ment. In this experiment shoals of six sticklebacks were allowed to forage on two patches of cryptic daphnia. Daphnia were pipetted into these patches at a known rate. Since the patches were nearly 0.5 m apart, fish moving from one to the other incurred a cost in travel time. The best patch was twice as profitable and fish allocated themselves between patches in a ratio of 2 : 1, apparently in accordance with the predictions of the ideal free distribution (Fretwell and Lucas 1970). As Figure 13.4 shows, however, foraging returns were not equal and there was a positive and significant relationship between competitive rank and number of daphnia captured. Milinski found that the poorer competitors took longer at the beginning of a trial to ascertain the profitabilities of the patches, and switched patches more frequently (see Figure 13.5).

In situations where prey are conspicuous, switching can benefit poor competitors. In a similar further experiment using non-cryptic daphnia,

Figure 13.5: Foraging Success and Switching by Good and Poor Competitors. The mean number of daphnia consumed by good and poor competitors and the mean number of switches between patches by good and poor competitors. (Redrawn from Milinski 1984a.)

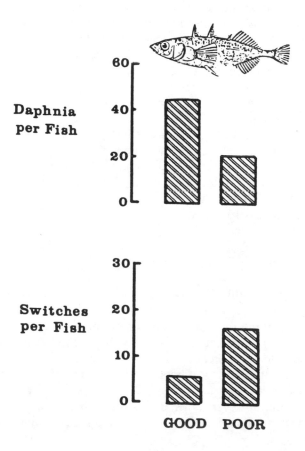

Milinski (1984a) found that 'switchers' (or poor competitors) often returned to the centre of the tank and waited there to pounce on any daphnia escaping from the ends of the tanks where the better competitors, or 'stayers' remained. Godin and Keenleyside (1984) observed similar switching and staying strategies in a cichlid fish (*Aequidens curviceps*) foraging on patchy prey. (See also Chapter 12 by Pitcher, this volume.)

Parasite infestation reduces an individual's competitive ability. Crowden and Broom (1980) showed that the reactive distance of dace (*Leuciscus leuciscus*) to *Gammarus* was greater for fish heavily infected with eye flukes (*Diplostomom* sp.). Milinski (1984b) found that sticklebacks parasitised with *Glugea anomala* or *Schistocephalus solidus* attacked small daphnia more often than unparasitised fish did. The percentage of attacks directed at small prey was highest in sticklebacks infected with both parasites. By shifting their diet away from the larger, more profitable daphnia, the parasitised sticklebacks avoided direct confrontation and maximised their food intake. Giles (1983) found that parasitised sticklebacks started foraging sooner after exposure to a frightening overhead stimulus than did unparasitised fish. Although Giles suggests that this behaviour is a response to the metabolic demands that the parasite places on the host, it is also possible that parasitised, and less competitive, sticklebacks are more willing to risk predation in order to feed in the absence of competition. For instance, Milinski (see Chapter 9, this volume) has shown that parasitised sticklebacks forage closer to a predator than unparasitised sticklebacks.

These examples illustrate how it can be misleading to frame optimal foraging predictions in terms of the average animal; the appropriate behaviour for each individual will be different depending on its competitive ability, with, in many cases less-successful individuals being forced to make the best of a bad job. They also show how an individual's foraging behaviour is a product of the interactions between its phenotype, the behaviour of other fish and the heterogeneity (in food availability and predator risk) of the environment.

Avoiding Predators

Since the most frequently described response of fish threatened by a predator is to clump together in a school, the members of which all appear to behave in exactly the same way (see, for example, Keenleyside 1979), anti-predator behaviour might not seem a very fruitful area in which to search for individual differences. Joining a school is unquestionably advantageous, and the larger a school, the less successful most predators will be in attacking it (Neill and Cullen 1974; Chapter 12 by Pitcher, this volume). There are, however, other tactics that fish can employ to evade predators, and variation in behaviour may be found even within the special case of schooling. Schooling tendency can vary within individuals as the costs and benefits of belonging to a school change. It may, for instance, pay an individual to remain solitary. Savino and Stein (1982) have shown that bluegill sunfish under predator threat are less likely to school in dense vegetation. In this case the advantage of being able to hide alone in thick cover outweighs the benefits of belonging to a more detectable school. If a school is small, fish may choose to break ranks. Minnows in groups of ten or less are more likely to abandon the school and seek cover when attacked by a pike (*Esox lucius*) (Magurran and Pitcher 1986).

One bonus of belonging to a shoal of fish is that individuals are able to spend less time being vigilant for predators. Magurran, Oulton and Pitcher (1985) found that minnows in large shoals detected a stalking pike model sooner than did minnows in small shoals (see also Chapter 9 by Milinski, this volume). This result helps explain the finding that the behaviour of individual minnows and goldfish is contingent on shoal size (Magurran and Pitcher 1983). In small shoals, fish of both species are timid, do little foraging and spend time hiding in weed beds; as shoal size, and corporate vigilance increase, the same individuals become less timid and devote progressively more time to foraging.

Hamilton (1971) pointed out that individuals positioning themselves close to other members of a group are less at risk of being captured by a randomly striking predator. Experiments on daphnia swarms have confirmed that stragglers and animals at the edge are especially vulnerable (Milinski 1977). It has not yet been ascertained whether fish compete for the relatively safer positions at the centre of a school, but if this were the case, the benefits of schooling would not be enjoyed equally by all its members. Evidence for position preferences in mackerel (*Scomber scombrus*) schools has been provided by Pitcher *et al.* (1982), and a mixed-species 'selfish school' consisting of catfish (*Bagrus meridionalis*) and peripherally positioned cichlids (*Haplochromis* spp.) has been described by McKaye and Oliver (1980).

Schooling tendency can also vary between populations. Seghers (1974), working on guppies (*Poecilia reticulata*) in Trinidad, found that fish from populations sympatric with piscivorous predators are more likely to school than those from predator-free populations. Since fish raised through three generations in a controlled, predator-free environment retained their schooling tendency, Seghers concluded that the propensity of fish from the different populations to school is genetically based. Guppies from predated populations also have better developed avoidance behaviour, including increased reaction distance, quicker transmission of alarm responses and differences in habitat usage (Liley and Seghers 1975). Reznick and Endler (1982) have demonstrated that Trinidad guppies experiencing high natural levels of predation increase their investment in reproduction relative to other populations. This increased investment is brought about by devoting more of their body weight to reproductive effort, commencing reproduction at a smaller size, having a shorter interbrood interval and producing larger numbers of (but smaller) offspring.

Magurran and Pitcher (1986) have shown that minnows from a provenance sympatric with pike in southern England form larger shoals and change shoals less frequently than North Wales minnows from a population which have lived in the absence of pike for thousands of years. Minnows from both provenances were very similar in their execution of synchronised anti-predator manoeuvres (such as fountain and flash expansion: see Pitcher, Chapter 12, this volume), but those living with pike in the wild recovered normal behaviour more quickly and started foraging sooner after exposure to pike.

Other anti-predator behaviours are also related to the natural level of predation and the evolutionary history of the population. Giles and Huntingford (1984) found that sticklebacks from high-risk populations showed greatest fright response to fish (pike) and bird (heron: *Ardea cineria*) predators. A spectrum of behaviours was implicated in the differences in anti-predator responses between populations. There was also individual variation in anti-predator behaviour within populations, and in general, adult female sticklebacks had better developed fright responses than adult males.

Individual differences in phenotype have important consequences for anti-predator behaviour. Reist (1983) looked at the behaviour of Canadian brook sticklebacks (*Culaea inconstans*). These sticklebacks differ with respect to the pelvic skeleton and its associated spines, and vary from forms where it is complete through a range of intermediates to forms where it is totally absent. Pike (*Esox lucius*) prefer the large, spineless sticklebacks, and Reist observed that these fish are most likely to spend time in cover to avoid initial detection, and to retreat if discovered. Reist scored a number of behaviours and classified them as being either 'adaptive', that is, reducing risk of detection and increasing chance of evasion, or 'non-adaptive', defined as those behaviours that do not reduce the fishes' chance of being caught. Sticklebacks without spines are more likely to show adaptive behaviour, and individual differences in behaviour are least prominent in these vulnerable fish. Some populations of *G. aculeatus* also contain spine-deficient morphs but, unlike the Canadian brook sticklebacks studied by Reist, these spineless sticklebacks have no specialised anti-predator behaviour to compensate for their increased vulnerability (Giles and Huntingford 1984).

Reist noted differences in the anti-predator behaviour of sticklebacks (*Culaea inconstans*) belonging to different size classes. Other within-individual differences in anti-predator behaviour may occur as a response to changes in hunger: Milinski (1984c) showed that hungry sticklebacks (*G. aculeatus*) are less likely to detect a predator; or breeding condition: Huntingford (1976) observed that male sticklebacks guarding a nest are more reluctant to leave and seek cover when threatened by a pike. Giles

(1984) has shown that experience is not necessary in the development of the fright response in sticklebacks, though work by Dill (1974) suggests that anti-predator behaviour may be modified through experience (see also Chapter 3 by Huntingford, this volume).

Much of the variation in anti-predator behaviour can be explained by environmental heterogeneity, especially in terms of predator risk, and phenotypic differences which render some individuals particularly susceptible to predation. The behaviour of other individuals also contributes to the extent that shoaling confers a certain amount of protection on shoal members. It is also possible, but not proven, that some individuals selfishly manipulate the shoaling behaviour of others.

Mating Behaviour

Recent reviews (Rubenstein 1980; Dunbar 1982; Arak 1984) testify to the diversity of species where intraspecific variation in mating behaviour occurs. Fish are particularly interesting since their alternative mating strategies can involve separate developmental pathways, distinct morphologies and different behaviour patterns. Differences in male behaviour are most widespread and occur when male reproductive success is dependent on aggression and male-male competition (Dominey 1981). One classic, but surprisingly neglected, case involves the Atlantic salmon (*Salmo salar*). It has been known for at least 150 years that male salmon parr mature precociously, and work by Orton, Jones and King (1938) and Jones and King (1950) has proved conclusively that these fish are sexually active and successful breeders. (See also Chapter 10 by Turner, this volume). Precocious males are found across the geographic range of salmon (Jones and Orton 1940) although the proportion of precociously maturing males varies between stocks (Schaffer and Elson 1975). In Newfoundland rivers, for instance, the incidence of precociousness ranges from 12 per cent to 100 per cent with a mean of 73 per cent (Dalley, Andrews and Green 1983). Approximately 75 per cent of male parr in British rivers are sexually mature (Jones 1959). Salmon parr must compete for females with adult males that have been to sea and returned home to the spawning ground. Jones (1959) found that females were reluctant to spawn in the presence of the precocious males. The strategy of these fish, which resemble immature fish in appearance, is to take up position under the bellies of the spawning adult male and female, to remain as inconspicuous as possible and to fertilise eggs as they are laid in the gravel. A dominance hierarchy exists within the male parr, with the dominant fish achieving the majority of fertilisations. Little is known about the factors that control the development of sexual precociousness in these males. Schaffer and Elson (1975) provide evidence that age of first spawning may be heritable, and Dalley, Andrews and Green (1983) show that faster growing males become precocious. It is not known how their reproductive success and subsequent migration patterns compare with those of the other male salmon.

Male salmon parr can be classified as sneaky breeders (Arak 1984), a strategy that is widespread in fish (see, for example, Fernald and Hirata 1977; Kodric-Brown 1977; Robertson and Hoffman 1977; Wirtz 1978). In the typical case the sneak (or scrounger: Barnard 1984; or submissive male: Dominey 1981) reduces the costs of mating, such as attracting females, defending territories and being more visible to predators, by stealing fertilisations from dominant (or producer: Barnard 1984) fish. These lower costs must be balanced against reduced attractiveness to females (Brown 1981) and lower reproductive success. An excellent example is provided by Constanz (1975)· who worked on the gila topminnow (*Poeciliopsis occidentalis*). These fish belong to a family in which male growth ceases at sexual maturity. Since position in the dominance hierarchy is related to size, small fish are condemned to remain subordinate. Large dominant males form territories and court females; small males remain at a distance from the females, seeking an opportunity to enter the territory and perform sneak copulations. The differences between the males and the costs and benefits of the two strategies are summarised in Table 13.1. If the large males are removed, small males will, within a matter of minutes, take on the dominant colour pattern and start defending territories and courting females. Although it is likely that a male's propensity to become a subdominant sneak is at least in part genetically based, these topminnows are not locked into one behaviour pattern. Bluegill sunfish, by contrast, are committed early on in their development to become either a 'parental male' or a sneak (Gross 1982). Parental males, which only gain sexual maturity at seven years of age, build nests, attract females and care for the eggs after fertilisation. Sneaks become reproductively active after two years and commence breeding attempts by dashing into the nest and releasing sperm as the female spawns. At a later stage of their development, they take on adult female size, colour and behaviour and swim between the spawning female and parental male. Dominey (1981) has put forward two reasons to explain why these female mimics, or 'satellites' (Dunbar 1982) are tolerated in nests. Spawning females are attracted to areas where other females are aggregating. By having female mimics around him, the parental male might give real females the impression that his nest is a centre of spawning activity. Alternatively, parental males that allocate too much time to evicting female mimics may decrease the success of their encounters with functional females.

Approximately 20 per cent of bluegill sunfish males become sneaks and are responsible for fertilising 20 per cent of the eggs (Gross 1982). An evolutionarily stable equilibrium of sneaky males and parental males will be maintained through frequency-dependent payoffs. At low densities sneaky males have many opportunities to fertilise eggs, but as their proportion in the population increases, the reproductive success of each sneak will

Table 13.1: Costs and Benefits of Territorial and Sneak Mating Strategies in *Poeciliopsis occidentalis* (after Constanz 1975)

	LARGE TERRITORIAL MALE			SMALL SNEAK MALE		
		COST	BENEFIT		COST	BENEFIT
(1) Body size	LARGE	Delayed sexual maturity	Higher competitive ability	SMALL	Lower competitive ability	Quicker maturation
(2) Body colour	JET BLACK	More visible to predators. Encourages aggression in territorial neighbours	Encourages submissives to flee	LIGHT COLOURED	Greater energy expenditure required in attempting to mate with females	Not so visible to predators. Trespassing into territories easier
(3) Agonistic competition and territoriality	AGGRESSIVE; HOLDS TERRITORY	Time and energy expended in territory defence	Territory permits relatively undisturbed courtship	SUBMISSIVE; NO TERRITORY	Movement inhibited. Less knowledge geography of territories	Time saved in territory defence potentially available for mating attempts
(4) Courtship and copulation	COURTS FEMALE; LONGER COPULATION	—	Female easier target. Mating less likely to be disturbed	DOES NOT COURT FEMALE; SHORTER COPULATION	Female evasive target. Mating activities readily disrupted	—
(5) Gonopodial length	SHORT	Possibly less accuracy in contacting moving female	Less investment in gonopodium	LONG	Greater investment in gonopodium	Greater accuracy for moving target female

diminish (Arak 1984). It seems likely that the frequencies of the two phenotypes in the adult population in conjunction with nutritional status at a critical stage of development (W.J. Dominey and P. Bateson personal communication) determine which reproductive strategy a male adopts.

As Rubenstein (1980) has pointed out, two mechanisms can produce sneaky males. First, frequency-dependent effects may determine the proportion of sneaks. In this case the average lifetime reproductive success of parental males and sneaks will be equal, as appears to be happening with the bluegill sunfish. Secondly, some males could adopt the sneaky strategy in order to make the best of a bad job. Here males may become sneaks while they are small and competitively inferior before switching to become conventional males. For instance Wirtz (1978) found that one-year-old black-faced blennies (*Tripterygion triperonotus*) are sneaks whereas two-year-old fish hold territories and court females. Alternatively, as with Constanz's (1975) topminnows, males may remain sneaks unless the superior competition of the territory holders is removed. 'Best of a bad job' sneaks enjoy lower mating success than conventional males in the short term (Dunbar 1982), but since they invest less (for instance, they do not have the energetic costs or risk of injury involved in defending a territory), their lifetime reproductive fitness may be greater than is at first apparent. Unfortunately, the difficulties involved in measuring lifetime reproductive fitness mean that there is little conclusive evidence on the relative success of sneaky and conventional males.

A more drastic solution to the problems of competing with large dominant males for females is to change sex (Krebs and Davies 1981; see Shapiro 1979 for a review of sex changes in fish). One species (of many) where this occurs is the blue-headed wrasse (*Thalassoma bifasciatum*). Large males, which are brightly coloured, corner the market in fertilising females and make it virtually impossible for small males to achieve a mating. Fish therefore commence their reproductive life as females, only switching to maleness when they are large enough to make it worth while. This switch is socially controlled since the removal of the large males is sufficient to initiate a sex change in the largest females (Warner, Robertson and Leigh 1975). A few primary males (those that stay male through their life) do exist but are only found in large populations where there is a greater chance of achieving sneak matings. The alternative change, from male to female, is less common and likely to occur only when there is a premium on female size. The anemone fish (*Amphiprion akallopisos*) provides one example. These fish live in pairs on sea anemones. Since their reproductive success depends on the ability of the female to produce eggs, it is advantageous for the biggest fish to be female (Fricke 1979).

The majority of examples of variable strategies discussed derive from competitive inequalities between males. Rather than compete directly with dominant males, the competitively inferior males can adopt a different mating strategy and thereby maximise their reproductive success.

Habitat Use

A patchy environment combined with differences in competitive ability make individual variation in habitat use inevitable. One way in which individual fish differ is in the size and quality of their home range, that is, the restricted area in which a fish spends most of its time (Kennedy 1981). Fish and Savitz (1983) studied largemouth bass, yellow perch (*Perca flavescens*), bluegill sunfish and pumpkinseed sunfish (*Lepomis gibbosus*) in an Illinois lake, and found variation in home range size between individuals in all four species. For instance, home range size varied from 0.18 ha to 2.07 ha for the largemouth bass. In the bluegill sunfish a significant negative correlation between fish size and home range size was recorded, with some individuals found predominantly in weedy areas and some in sandy areas. Fish and Savitz suggest that social dominance is responsible for this finding, with the larger, more successful bluegills occupying the smaller, more profitable home ranges.

The tendency of fish to occupy a home-range also varies between individuals. Kennedy (1981) found that a majority (approximately 85 per cent) of minnows (*Phoxinus phoxinus*) tested would accept a home range in aquarium tanks, and return to it if displaced. The remainder of minnows could be divided into 'colonisers', which spent most of their time in a strange area, and 'mobile fish', which moved a lot but showed no preference for any one area. Since all fish in Kennedy's experiment were similar-sized, mature minnows, the individual differences in mobility and use of home range could not be explained by size of fish.

Fish size can, however, be important in determining an individual's use of the habitat (Werner 1984). Laughlin and Werner (1980) found differences in the distribution of small ($<$ 39 mm), medium (39-75 mm) and large ($>$ 75 mm) size classes of fish in both the pumpkinseed (*Lepomis gibbosus*) and northern longear (*L. megalotis peltastes*) sunfishes in Michigan lakes. Large longears, for example, were found almost exclusively close to the shore whereas small longears were confined to deep water, with medium-sized fish split between the two habitats. Seasonal variation in habitat use, reflecting changes in vegetation cover, was found within the small and medium size classes of longears (see Figure 13.6), and a difference in the distribution of the two sexes in both species was also recorded.

The sunfish within-individual differences in habitat use are relatively long term, taking place seasonally or over a lifetime. More rapid within-individual variation in habitat use has also been recorded. For instance, Gore (1983) found frequent changes in the spacing behaviour of individual butterflyfish (*Chaetodon capistratus*) on shallow-water reefs in the West Indies. These fish could be classified as 'residents' which remained in and fed on particular areas of reef, 'transients', which were mobile and not associated with any one area, and 'randoms', which were intermediate in

Figure 13.6: Seasonal Variation in Longear Sunfish Abundance. Average proportional abundance of three size classes of longear sunfish at different positions and distances from the shore in a Michigan lake during May, July and August. The positions are 1: in shallow water 1 m from shore; 3: in shallow water 3 m from shore; L: on the lip of a slope to deeper water; S: on the slope leading to deeper water. (Redrawn from Laughlin and Werner 1980.)

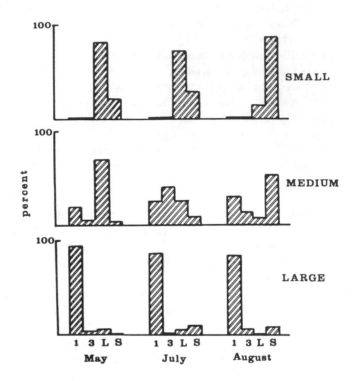

behaviour and represented fish changing from one class of habitat user to the other. Although residents suffered more harassment from neighbouring fish, they gained feeding advantages by remaining in one place. Fish changed classes frequently and there was no difference in the number of times each individual was observed as a resident or transient. Gore suggests that it only pays fish to be residents at times when they want to feed.

Territories, which differ from home ranges in that they are actively defended by the occupier, are particularly likely to be found when portions of desirable habitat are in short supply. Male sticklebacks, for instance, must defend a territory and build a nest in order to breed (Wootton 1984). In *G. aculeatus* territory size affects success in both courtship (males with large territories are most successful at attracting females and persuading them to enter the nest), and egg rearing. Male sticklebacks unable to defend a territory must resort to the lottery of sneaky mating if they are to stand any chance of reproducing.

Time invested in territory defence can modify a fish's foraging behaviour. Green, Martel and Martin (1984) found that territorial male cunners (*Tautogolabrus adspersus*), non-territorial male cunners and female cunners all had different diets despite the fact that they shared the same habitat in which the complete range of food items was widely available. Analysis of stomach contents showed that territorial males chose food which was easily handled and allowed them to devote maximum time to defending their territory, whereas females selected items with the highest energy content. Non-territorial males showed an intermediate diet choice since they could either go for high-energy food and behave as sneaks, or eat food that was quickly handled and spend more time seeking to usurp territory from its holder.

Variation in habitat use is often the direct result of a heterogeneous environment with competitive ability determining an individual's access to limited resources.

Dominance Hierarchies

Many of the examples of individual differences in behaviour so far discussed in the chapter have been the result of social dominance. Thus, the competitive ability of an individual determines its rank in a dominance hierarchy, with high-ranking individuals having preferential access to food, mates or whatever commodity is in demand (Morse 1980). Although subordinate individuals are at a competitive disadvantage, they can still gain benefits such as food location or predator avoidance by belonging to a social group, and will be in a position to work their way up the hierarchy should the chance arise (Colgan 1983). High rank has its own costs, of course, since superior status needs to be defended (Nelissen 1985). Territoriality can be considered as a form of dominance, though some ethologists prefer to restrict the term 'dominance hierarchy' to situations in which there is no clear reference point in space (Morse 1980).

Dominance hierarchies can either be 'statistical' (high-ranking individuals usually win contests) or 'absolute' (high-ranking individuals always win), 'linear' or 'non-linear' (Morse 1980; Appleby 1985), and are found in a wide range of animals, including fish (Gauthreaux 1978). If a hierarchy is to persist over a period of time, individuals must be able to recognise each other or, at the very least, respond to cues correlated with dominance. There is evidence of individual recognition in teleost dominance hierarchies (Colgan 1983), occurring, for instance, in guppies (*Poecilia reticulata*: Gorlick 1976), platies (*Xiphophorus maculatus*: Braddock 1945), damselfish (*Dascyllus aruanus*: Fricke and Holzberg 1974) and glandulocaudine fish (Nelson 1964). Recognition can be based on smell, as is the case in the yellow bullhead (*Ictalurus natalis*: Todd, Atema and Bardach 1967), or on behaviour and morphology (for example, *Xiphophorus* spp.: Zayan 1974, 1975). Barnard and Burk (1979, 1981) have suggested that dominance hierarchies may have an important role in the evolution of individual recognition.

In many species of fish, dominant individuals differ in colour or markings from those of inferior status. One of the best-known examples is that of the midas cichlid (*Cichlasoma citrinellum*), found in lakes in Nicaragua (Barlow 1983). In this species, which is polychromatic, the majority of adult fish are a grey, cryptic morph, known as 'normal' and the remainder, some 8 per cent are a conspicuous morph known as 'gold'. The gold morphs, which lack melanophores, range in colour from yellow to red, with orange the predominant hue. Although these brightly coloured fish are at a disadvantage as they are more visible to predators, being gold brings benefits since gold morphs are dominant over normal morphs of similar size (Barlow and Ballin 1976). Golds also suffer less fin damage in contests than normals (Barlow 1983), and, as a result of their domiance over normals, gold juveniles grow at a faster rate (Barlow 1973; Barlow and McKaye 1982). Surprisingly, gold morphs are not intrinsically more aggressive. Rather, they become dominant by virtue of the effect of their colour on opponents. This appears to be a result of the relative rarity of gold morphs (Barlow, Bauer and McKaye 1975; Barlow 1983), combined with the intimidating nature of the gold colour itself. Normal midas cichlids take on a red, orange or yellow coloration during breeding, and Barlow (1983) suggests that the effectiveness of the gold hue stems from its association with threat displays. The gold effect seems to be enhanced if the gold morph is in a competitive situation in which it is already at an advantage, for example if it is actively defending a territory (Barlow 1983).

The gold midas cichlids are dominant because of their colour. In most cases, however, the situation is reversed, with high-status individuals taking on special colouring as a result of their dominance. One such species is the Japanese freshwater serranid (*Coreoperca kawamebari*). When placed in a tank togeher, these fish engage in fights until, after a few days, a dominance hierarchy is established (Kohda and Watanabe 1982). The most dominant fish have conspicuous eyespots on their bodies but are otherwise uniform in colour. The eyespots become progressively less clear, and the body progressively more striped, as rank decreases. If new fish are added, the dominance hierarchy will be reordered to accommodate the competitive ability of the newcomers, with each fish taking on the colouring appropriate to its rank. Behaviour differences accompany these colour patterns; in encounters with other fish, high-ranking individuals are aggressive and low-ranking individuals retreat.

Gorlick (1976) has found that high-status male guppies indulge in more mating behaviour than males of low rank, and that in both male and female hierarchies dominant individuals gain substantial feeding benefits. Gorlick suggests that an important function of dominance hierarchies is to ensure that the fittest individuals have the best chance of surviving periods of hardship and of reproducing. Mating success is certainly linked to rank in many species (for example *Malanochromis auratus*, a cichlid from Lake Malawi: Nelissen 1984, 1985). Nelissen's work provides a cautionary note for ethologists interested in dominance hierarchies; he found that the hierarchy obtained by staging 'round-robin' contests between all possible pairs of individuals differed from the one that existed when all fish were allowed to live together in a group.

An explicit demonstration of the benefits that can accrue to subordinate fish is provided by Taborsky (1984). Taborsky investigated the behaviour of *Lamprologus brichardi*, a Lake Tanganyika cichlid. In this species, adult pairs defend territories whereas young of previous broods either remain with their parents as helpers, participating in brood-care and territorial defence, or join non-breeding family-independent aggregations. Due to their low rank in the family hierarchy, helpers grow at a slower rate than the non-helper fish in the aggregations. Compensating for this cost are two benefits. First, helpers contribute to the reproductive success of their parents, ensuring the production of more siblings. Secondly, membership of the territorial unit gives helpers protection against predation. Taborsky set up an experiment in which groups of helpers, with their parents, and equal-sized groups of controls (non-helper aggregation members), were exposed to one of their natural predators, *Lamprologus elongatus*. All fish had access to cover. Figure 13.7 shows that controls were caught before helpers, as well as in greater numbers.

Conclusions

Dominance hierarchies are an expression of individual differences in com-

Figure 13.7: Helpers and Protection from Predation. Order in which helpers and unprotected control fish of the same size were caught in nine experiments. A total of 55 fish were captured. (Redrawn from Taborsky 1984.)

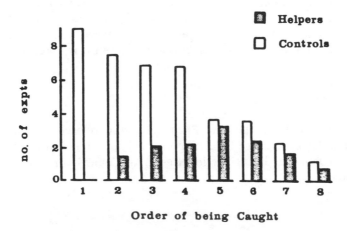

petitive ability. Indeed, many of the examples of variable behaviour discussed in this chapter have resulted from the ability of certain individuals to capture a disproportionate share of limited resources. In many cases the best way of making a living will depend on an individual's competitive ability. Rather than compete directly for scarce resources, subordinate individuals may find it advantageous to behave in different ways from the dominant fish. Competitive differences in conjunction with a variable environment, different individual requirements and a variety of tactics that can be used to achieve the same ends make individual differences in behaviour inevitable. It can therefore be incorrect to assume that optimal behaviour is the same for all individuals.

Variable behaviour also has important ecological consequences. Population stability is affected by individual differences in mortality and reproduction (Lomniki 1978, 1982), dominance (Gauthreaux 1978) and spatial distribution (Hassell and May 1985), and variation in habitat use or foraging tactics can influence community structure (Ringler 1983; Werner 1984).

Summary

Recent advances in evolutionary biology have highlighted the importance of individual differences in behaviour. By reviewing examples from foraging behaviour, predator avoidance, mating strategies, habitat use and dominance hierarchies, this chapter has shown that intraspecific variation in fish behaviour is widespread. Three mechanisms are generally proposed to account for variable behaviour: a patchy environment, phenotypic differences and the behaviour of other individuals. Each of these is important, for example, many differences in foraging behaviour arise from competitive inequalities whereas anti-predator behaviour often varies with predation risk, and the best reproductive strategy can depend on what other individuals are doing. Since the appropriate behaviour for an individual fish will depend on the nature of its phenotype in conjunction with the variability of the environment and the activities of others, it can be misleading to frame optimal-behaviour arguments in terms of the average individual.

Acknowledgements

I would like to thank Dr T.J. Pitcher and the referee for their useful comments, Dr L. Patridge for letting me have an unpublished manuscript, and the Science and Engineering Research Council for a Postdoctoral Fellowship.

References

Allan, J.D. (1975) 'The Distributional Ecology and Diversity of Benthic Insects in Cement Creek, Colorado', *Ecology, 56*, 1040-53

Allan, J.D. (1981) 'Determinants of Diet of Brook Trout (*Salvelinus fontinalis*) in a Mountain Stream', *Canadian Journal of Fisheries and Aquatic Sciences, 38*, 184-92

Appleby, M. (1985) 'Hawks, Doves and Chickens', *New Scientist*, 10 January 1985, pp. 16-18

Arak, A. (1984) 'Sneaky Breeders', in *Producers and Scroungers*, C.J. Barnard (ed.), Croom Helm, London, pp. 154-94

Atema, J., Holland, K. and Ikehara, W. (1980) 'Olfactory Responses of Yellowfin Tuna (*Thunnus albacares*) to Prey Odors: Chemical Search Image', *Journal of Chemical Ecology, 6*, 457-65

Barlow, G.W. (1973) 'Competition between Color Morphs of the Polychromatic Midas Cichlid *Cichlasoma citrinellum*', *Science 179*, 806-7

Barlow, G.W. (1983) 'The Benefits of Being Gold: Behavioral Consequences of Polychromatism in the Midas Cichlid *Cichlasoma citrinellum*', in *Predators and Prey in Fishes*, D.L.G. Noakes, D.G. Lindquist, G.S. Helfman and J.A. Ward (eds), Junk, The Hague, pp. 73-86

Barlow, G.W. and Ballin, P.J. (1976) 'Predicting and Assessing Dominance from Size and Coloration in the Polychromatic Midas Cichlid', *Animal Behaviour, 24*, 793-814

Barlow, G.W. and McKaye, K.R. (1982) 'A Comparison of Feeding, Spacing and Aggression in Color Morphs of the Midas Cichlid. II. After 24 Hours without Food', *Behavior, 80*, 127-42

Barlow, G.W., Bauer, D.H. and McKaye, K.R. (1975) 'A Comparison of Feeding, Spacing and Aggression in Color Morphs of the Midas Cichlid. I. Food Continuously Present', *Behavior, 54*, 72-96

Barnard, C.J. (1984) 'When Cheats May Prosper', in *Producers and Scroungers*, C.J. Barnard (ed.), Croom Helm, London, pp. 6-33

Barnard, C.J. and Burk, T. (1979) 'Dominance Hierarchies and the Evolution of "Individual Recognition"', *Journal of Theoretical Biology, 81*, 65-73

Barnard, C.J. and Burk, T. (1981) 'Individuals as Assessment Units — Reply to Breed and Bekoff', *Journal of Theoretical Biology, 88*, 595-7

Barnard, C.J. and Sibly, R.M. (1981) 'Producers and Scroungers: a General Model and its Application to Feeding Flocks of House Sparrows', *Animal Behaviour, 29*, 543-50

Bateson, P.P.G. (1983) 'Genes, Environment and the Development of Behaviour', in *Animal Behaviour 3: Genes, Development and Learning*, T.R. Halliday and P.J.B. Slater (eds), Blackwell, Oxford, pp. 52-81

Beukema, J.J. (1968) 'Predation by the Three-spined Stickleback (*Gasterosteus aculeatus* L.): the Influence of Hunger and Experience', *Behavior, 31*, 1-126

Braddock, J.C. (1945) 'Some Aspects of the Dominance-Subordination Relationship in the Fish *Platypoecilus maculatus*', *Physiological Zoology, 18*, 176-95

Breck, J.E. and Gitter, M.J. (1983) 'Effect of Fish Size on the Reactive Distance of Bluegill (*Lepomis macrochirus*) Sunfish', *Canadian Journal of Fisheries and Aquatic Sciences, 40*, 162-7

Brooker, M.P. and Morris, D.L. (1980) 'A Survey of the Macroinvertebrate Riffle Fauna of the River Wye', *Freshwater Biology, 10*, 437-58

Brown, L. (1981) 'Patterns of Female Choice in Mottled Sculpins (Cottidae, Teleostei)', *Animal Behaviour, 29*, 375-82

Bryan, J.E. and Larkin, P.A. (1972) 'Food Specialization by Individual Trout', *Journal of the Fisheries Research Board of Canada, 29*, 1615-24

Bush, G.L. (1975) 'Modes of Animal Speciation', *Annual Review of Ecology and Systematics, 6*, 339-64

Coates, D. (1980) 'Prey-size Intake in Humbug Damselfish, *Dascyllus aruanus* (Pisces, Pomacentridae) Living within Social Groups', *Journal of Animal Ecology, 49*, 335-40

Colgan, P. (1983) *Comparative Social Recognition*, Wiley, New York

Constanz, G.D. (1975) 'Behavioral Ecology of Mating in the Male Gila Topminnow, *Poeciliopsis occidentalis* (Cyprinodontiformes: Poeciliidae)', *Ecology, 56*, 966-73

Crowden, A.E. and Broom, D.M. (1980) 'Effects of the Eyefluke, *Diplostomum spathaceum*, on the Behaviour of Dace (*Leuciscus leuciscus*)', *Animal Behaviour, 28*, 287-94

Curio, E. (1976) *The Ethology of Predation*, Springer-Verlag, New York

Dalley, D.L., Andrews, C.W. and Green, R.H. (1983) 'Precocious Male Atlantic Salmon Parr (*Salmo salar*) in Insular Newfoundland', *Canadian Journal of Fisheries and Aquatic Sciences, 40*, 647-52

Darwin, C. (1859) *The Origin of Species* (1968 edition), Penguin, Harmondsworth

Dawkins, R. (1976) *The Selfish Gene*, Oxford University Press, Oxford

Dawkins, R. (1979) 'Good Strategy or Evolutionarily Stable Stategy', in *Sociobiology: Beyond Nature/Nurture?* G.W. Barlow and J. Silverberg (eds), Westview Press, Boulder, Colorado, pp. 331-70

Dill, L.M. (1974) 'The Escape Response of the Zebra Danio (*Brachydanio rerio*), II. The Effect of Experience', *Animal Behaviour, 22*, 723-30

Dill, L.M. (1983) 'Adaptive Flexibility in the Foraging Behavior of Fishes', *Canadian Journal of Fisheries and Aquatic Sciences, 40*, 398-408

Dominey, W.J. (1981) 'Maintenance of Female Mimicry as a Reproductive Strategy in Bluegill Sunfish (*Lepomis macrochirus*)', in *Ecology and Ethology of Fishes*, D.I.G. Noakes and J.A. Ward (eds), Junk, The Hague, pp. 59-64

Drager, B. and Chizar, D. (1982) 'Growth Rate of Bluegill Sunfish (*Lepomis macrochirus*) Maintained in Groups and in Isolation', *Bulletin of the Psychonomic Society, 20*, 284-6

Dunbar, R.I.M. (1982) 'Intraspecific Variations in Mating Strategy', in *Perspectives in Ethology*, vol. 5, P.P.G. Bateson and P. Klopfer (eds), Plenum Press, New York, pp. 385-431

Ferguson, A. and Mason, F.M. (1981) 'Allozyme Evidence for Reproductively Isolated Sympatric Populations of Brown Trout *Salmo Trutta* L. in Lough Melvin, Ireland', *Journal of Fish Biology, 18*, 629-42

Fernald, R.D. and Hirata, N.R. (1977) 'Field Study of *Haplochromis burtoni*: Quantitative Behavioural Observations', *Animal Behaviour, 25*, 964-75

Fish, P.A. and Savitz, J. (1983) 'Variations in Home Ranges of Largemouth Bass, Yellow Perch, Bluegills, and Pumpkinseeds in an Illinois Lake', *Transactions of the American Fisheries Society, 112*, 147-53

Fretwell, S.D. and Lucas, H.L. (1970) 'On Theoretical Behaviour and Other Forms Influencing Habitat Distribution in Birds. I. Theoretical Development', *Acta Biotheoretica, 19*, 16-36

Fricke, H.W. (1979) 'Mating System, Resource Defence and Sex Changes in the Anemone Fish, *Amphiprion akallopisos*', *Zeitschrift für Tierpsychologie, 50*, 313-26

Fricke, H.W. and Holzberg, S. (1974) 'Social Units and Hermaphroditism in a Pomacentrid Fish', *Naturwissenschaften, 60*, 367-8

Gauthreaux, S.A. (1978) 'The Ecological Significance of Behavioral Dominance', in *Perspectives in Ethology* vol. 3, P.P.G. Bateson and P.H. Klopfer (eds), Plenum Press, New York, pp. 17-54

Giles, N. (1983) 'Behavioural Effects of the Parasite *Schistocephalus solidus* (Cestoda) on an Intermediate Host, the Three-spined Stickleback, *Gasterosteus aculeatus* L.', *Animal Behaviour, 31*, 1192-4

Giles, N. (1984) 'Development of the Overhead Fright Response in Wild and Predator-naive Three-spined Sticklebacks, *Gasterosteus aculeatus* L.', *Animal Behaviour, 32*, 276-9

Giles, N. and Huntingford, F.A. (1984) 'Predation Risk and Interpopulation Variation in Anti-predator Behaviour in the Three-spined Stickleback, *Gasterosteus aculeatus* L.', *Animal Behaviour, 32*, 264-75

Giraldeau, L.-A. (1984) 'Group Foraging: the Skill Pool Effect and Frequency-dependent Learning', *American Naturalist, 124*, 72-9

Godin, J.-G. J. (1978) 'Behavior of Juvenile Pink Salmon (*Oncorhynchus gorbuscha* Walbaum) Toward Novel Prey: Influence of Ontogeny and Experience', *Environmental Biology of Fishes, 3*, 261-6

Godin, J.-G. J. and Keenleyside, M.H.A. (1984) 'Foraging on Patchily Distributed Prey by a Cichlid Fish (Teleostei, Cichlidae): a Test of the Ideal Free Distribution Theory', *Animal Behaviour, 32*, 120-31

Gore, M.A. (1983) 'The Effect of a Flexible Spacing System on the Social Organization of a Coral Reef Fish, *Chaetodon capistratus*', *Behavior, 85*, 118-45

Gorlick, D.L. (1976) 'Dominance Hierarchies and Factors Affecting Dominance in the Guppy *Poecilia reticulata* (Peters)', *Animal Behaviour, 24*, 336-46

Green, J.M., Martel, G. and Martin, D.W. (1984) 'Comparisons of the Feeding Activity and Diets of Male and Female Cunners *Tautoglabrus adspersus* (Pisces: Labridae)', *Marine Biology*, *84*, 7-11

Gross, M.R. (1982) 'Sneakers, Satellites and Parentals: Polymorphic Mating in North American Sunfishes', *Zeitschrift für Tierpsychologie*, *60*, 1-26

Hamilton, W.D. (1971) 'Geometry for the Selfish Herd', *Journal of Theoretical Biology*, *31*, 295-311

Hassell, M.P. and May, R.M. (1985) 'From Individual Behaviour to Population Dynamics', in *Behavioural Ecology Symposium, British Ecological Society*, R. Sibly and R. Smith (eds), Blackwell, Oxford, pp. 3-32

Helfman, G. (1984) 'School Fidelity in Fishes: the Yellow Perch Pattern', *Animal Behaviour*, *32*, 663-72

Huntingford, F.A. (1976) 'A Comparison of the Reaction of Sticklebacks in Different Reproductive Conditions towards Conspecifics and Predators', *Animal Behaviour*, *24*, 694-7

Huntingford, F.A. (1984) *The Study of Animal Behaviour*, Chapman & Hall, London

Jones, J.W. (1959) *The Salmon*, Collins, London

Jones, J.W. and King, G.M. (1950) 'Further Experimental Observation of the Spawning Behaviour of Atlantic Salmon', *Proceedings of the Zoological Society of London*, *120*, 317-23

Jones, J.W. and Orton, J.H. (1940) 'The Paedogenetic Male Cycle in *Salmo salar* L.', *Proceedings of the Royal Society of London (B)*, *128*, 485-99

Keen, W.H. (1982) 'Behavioral Interactions and Body Size Differences in Competition for Food among Juvenile Brown Bullhead', *Canadian Journal of Fisheries and Aquatic Sciences*, *39*, 316-20

Keenleyside, M.H.A. (1979) *Diversity and Adaptation in Fish Behaviour*, Springer-Verlag, New York

Kennedy, G.J.A. (1981) 'Individual Variation in Homing Tendency in the European Minnow, *Phoxinus phoxinus* (L.)', *Animal Behaviour*, *29*, 621-5

Kodric-Brown, A. (1977) 'Reproductive Success and the Evolution of Breeding Territories in Pupfish (*Cyprinidon*)', *Evolution*, *31*, 750-66

Kohda, Y. and Wantanabe, M. (1982) 'Relationship of Color Pattern to Dominance Order in a Freshwater Serranid Fish, *Coreoperca kawamebari*', *Zoological Magazine*, *91*, 140-5

Kornfield, I., Smith, D.C., Gagnon, P.S. and Taylor, J.N. (1982) 'The Cichlid Fish of Cuatro Cienegas, Mexico: Direct Evidence of Conspecificity among Distinct Trophic Morphs', *Evolution*, *36*, 658-64

Kornfield, I. and Taylor, J.N. (1983) A New Species of Polymorphic Fish, *Cichlasoma minckleyi*, from Cuarto Ciénegas, Mexico (Teleostei: Cichlidae). *Proceedings of the Biological Society of Washington*, *96*, 253-69

Krebs, J.R. and Davies, N.B. (1981) *An Introduction to Behavioural Ecology*, Blackwell, Oxford

Laughlin, D.R. and Werner, E.E. (1980) 'Resource Partitioning in Two Coexisting Sunfish: Pumpkinseed (*Lepomis gibbosus*) and Northern Longear Sunfish (*Lepomis megalotis peltastes*)', *Canadian Journal of Fisheries and Aquatic Sciences*, *37*, 1411-20

Liley, N.R. and Seghers, B.H. (1975) 'Factors Affecting the Morphology and Behaviour of Guppies in Trinidad', in *Function and Evolution in Behaviour*, G. Baerends, C. Beer and A. Manning (eds), Clarendon Press, Oxford, pp. 92-118

Lomniki, A. (1978) 'Individual Differences between Animals and Natural Regulation of their Numbers', *Journal of Animal Ecology*, *47*, 461-75

Lomniki, A. (1982) 'Individual Heterogeneity and Population Regulation', in *Current Problems in Sociobiology* Kings College Sociobiology Group (eds), Cambridge University Press, Cambridge, pp. 153-67

McKaye, K.R. and Oliver, M.K. (1980) 'Geometry of a Selfish School: Defence of Cichlid Young by Bagrid Catfish in Lake Malawi, Africa', *Animal Behaviour*, *28*, 1287-90

Magurran, A.E. (1984) 'Gregarious Goldfish', *New Scientist*, 9 August, 32-3

Magurran, A.E. and Pitcher, T.J. (1983) 'Foraging, Timidity and Shoal Size in Minnows and Goldfish', *Behavioural Ecology and Sociobiology*, *12*, 147-52

Magurran, A.E. and Pitcher, T.J. (1986) 'Provenance, Shoal Size and the Organisation of Predator Evasion Behaviours in Minnow shoals' (submitted)

Magurran, A.E., Oulton, W.J. and Pitcher, T.J. (1985) 'Vigilant Behaviour and Shoal Size in

Minnows', *Zeitschrift für Tierpsychologie, 67*, 167-78

Maynard Smith, J. (1982) *Evolution and the Theory of Games*, Cambridge University Press, Cambridge

Milinski, M. (1977) 'Do All Members of a Swarm Suffer the Same Predation?', *Zeitschrift für Tierpsychologie, 43*, 373-88

Milinski, M. (1982) 'Optimal Foraging: the Influence of Intraspecific Competition on Diet Selection', *Behavioral Ecology and Sociobiology, 11*, 109-15

Milinski, M. (1984a) 'Competitive Resource Sharing: an Experimental Test of a Learning Rule for ESS's', *Animal Behaviour, 32*, 233-42

Milinski, M. (1984b) 'Parasites Determine a Predator's Optimal Feeding Strategy', *Behavioral Ecology and Sociobiology, 15*, 35-7

Milinski, M. (1984c) 'A Predator's Costs of Overcoming the Confusion-effect of Swarming Prey', *Animal Behaviour, 32*, 1157-62

Mittelbach, G.G. (1981) 'Foraging Efficiency and Body Size: a Study of Optimal Diet and Habitat Use by Bluegills', *Ecology, 62*, 1370-86

Morse, D.H. (1980) *Behavioral Mechanisms in Ecology*, Harvard University Press, Cambridge, Mass.

Neill, S.R. St J. and Cullen, J.M. (1974) 'Experiments on Whether Schooling by their Prey Affects the Hunting Behaviour of Cephalopods and Fish Predators', *Journal of Zoology (London), 172*, 549-69

Nelissen, M.H.J. (1984) 'Social Organisation in *Melanochromis auratus* (Cichlidae): the Dominance Hierarchy', *Newsletter of the International Association of Fish Ethologists, 7*, 3-4

Nelissen, M.H.J. (1985) 'Structure of the Dominance Hierarchy and Dominance Determining 'Group Factors' in *Melanochromis auratus* (Pisces, Cichlidae), *Behavior*

Nelson, K. (1964) 'Behavior and Morphology in the Glandulocaudine Fishes (Ostariophysi, Characidae)', *University of California Publications in Zoology, 75*, 59-152

Oliverio, A. (1983) 'Genes and Behaviour: an Evolutionary Perspective', in *Advances in the Study of Behaviour* vol. 13, J.S. Rosenblatt, R.A. Hinde, C. Beer and M.C. Busnel (eds), Academic Press, New York, pp. 191-217

Orton, J.H. Jones, J.W. and King, G.M. (1938) 'The Male Sexual Stage in Salmon Parr (*Salmo salar* L. Juv.)', *Proceedings of the Royal Society of London (B), 125*, 103-14

Parker, G.A. (1984) 'Evolutionarily Stable Strategies', in *Behavioural Ecology*, 2nd edn, J.R. Krebs, and N.B. Davies (eds), Blackwell, Oxford, pp. 30-61

Partridge, L. (1983) 'Genetics and Behaviour', in *Animal Behaviour 3: Genes, Development and Learning*, T.R. Halliday and P.J.B. Slater (eds), Blackwell, Oxford, pp. 11-51

Partridge, L. and Green, P. (1985) 'Intraspecific Feeding Specialisations and Population Dynamics', in *Behavioural Ecology Symposium, British Ecology Society*, R. Sibley and R. Smith (eds), Blackwell, Oxford, pp. 207-26

Pitcher, T.J., Wyche, C. and Magurran, A.E. (1982) 'Evidence for Position Preferences in Mackerel Schools', *Animal Behaviour, 30*, 932-4

Pitcher, T.J., Magurran, A.E. and Winfield, I.J. (1982) Fish in Larger Shoals Find Food Faster', *Behavioral Ecology and Sociobiology, 10*, 149-51

Potts, G.W. (1980) 'The Predatory Behaviour of *Caranx melamphygus* (Pisces) in the Channel Environment of Aldabra Atoll (Indian Ocean)', *Journal of Zoology (London), 192*, 323-50

Reist, J. (1983) 'Behavioral Variation in Pelvic Phenotypes of Brook Stickleback, *Culaea inconstans*, in Response to Predation by Northern Pike, *Esox lucius*', in *Predators and Prey in Fishes*, D.I.G. Noakes, D.G. Lindquist, G.S. Helfman and J.A. Ward (eds), Junk, The Hague, pp. 93-105

Reznick, D. and Endler, J.A. (1982) 'The Impact of Predation on Life History Evolution in Trinidadian Guppies (*Poecilia reticulata*)', *Evolution, 36*, 160-77

Ringler, N.H. (1983) 'Variation in Foraging Tactics of Fishes', in *Predators and Prey in Fishes*, D.L.G. Noakes, D.G. Lindquist, G.S. Helfman and J.A. Ward, (eds), Junk, The Hague, pp. 159-72

Robertson, D.R. and Hoffman, S.S. (1977) 'The Roles of Female Mate Choice and Predation in the Mating Systems of Some Tropical Labroid Fishes', *Zeitschrift für Tierpsychologie, 45*, 298-20

Rowland, W.J. (1982) 'The Effects of Male Nuptual Coloration on Stickleback Aggression: a Reexamination', *Behaviour, 80*, 118-26

Rubenstein, D.I. (1980) 'On the Evolution of Alternative Mating Strategies', in *Limits to Action: the Allocation of Individual Behavior*, J.E.R. Staddon (ed.), Academic Press, New York, pp. 65-100

Rubenstein, D.I. (1981) 'Individual Variation and Competition in the Everglades Pygmy Sunfish', *Journal of Animal Ecology, 50*, 337-50

Savino, J.F. and Stein, R.A. (1982) 'Predator-Prey Interactions between Largemouth Bass and Bluegills as Influenced by Simulated, Submerged Vegetation', *Transactions of the American Fisheries Society, 111*, 255-66

Schaffer, W.M. and Elson, P.F. (1975) 'The Adaptive Significance of Variation in Life History among Local Populations of Atlantic Salmon in North America', *Ecology, 56*, 577-90

Seghers, B. (1974) 'Schooling Behavior in the Guppy (*Poecilia reticulata*): an Evolutionary Response to Predation', *Evolution, 28*, 486-9

Shapiro, D.Y. (1979) 'Social Behavior, Group Structure and the Control of Sex Reversal in Hermaphroditic Fish', in *Advances in the Study of Behavior*, vol. 10, J.S. Rosenblatt, R.A. Hinde, C. Beer and M.-C. Busnel (eds), Academic Press, New York, pp. 43-102

Shaw, E. (1970) 'Schooling in Fishes: Critique and Review', in *Development and Evolution of Behavior*, L.R. Aronson, B. Tobach, D.S. Lehrman and J.S. Rosenblatt (eds), W.H. Freeman, San Fransisco, pp. 452-80

Slater, P.J.B. (1983) 'The Development of Individual Behaviour', in *Animal Behaviour 3: Genes, Development and Learning*, T.R. Halliday and P.J.B. Slater (eds), Blackwell, Oxford, pp. 82-113

Steele, J.H. (1976) 'Patchiness' in *The Ecology of the Seas*, D.H. Cushing and J.J. Walsh (eds), Blackwell, Oxford, pp. 98-115

Taborsky, M. (1984) 'Broodcare Helpers in the Cichlid Fish *Lamprologus birchardi*: their Costs and Benefits', *Animal Behaviour, 32*, 1236-52

Tinbergen, N. (1951) *The Study of Instinct*, Oxford University Press, London

Todd, J.H., Atema, J. and Bardach, J.E. (1967) 'Chemical Communication in Social Behavior of a Fish, the Yellow Bullhead (*Ictalurus natalis*)', *Science, 158*, 672-3

Vinyard, G.L. (1982) 'Variable Kinematics of Sacramento Perch (*Archoplites interruptus*) Capturing Evasive and Nonevasive Prey', *Canadian Journal of Fisheries and Aquatic Sciences, 39*, 208-11

Warner, R.R., Robertson, D.R. and Leigh, E.G. (1975) 'Sex Change and Sexual Selection', *Science, 190*, 934-44

Webb, P.W. and Skadsen, J.M. (1980) 'Strike Tactics of *Essox*', *Canadian Journal of Zoology, 58*, 1462-9

Werner, E.E. (1984) 'The Mechanisms of Species Interactions and Community Organization in Fish', in *Ecological Communities: Perceptual Issues and the Evidence*, D.R. Strong, D. Simberloff, L.G. Abele and A.B. Thistle (eds), Princeton University Press, Princeton, pp. 360-82

Werner, E.E. and Mittelbach, G.G. (1981) 'Optimal Foraging: Field Tests of Diet Choice and Habitat Switching', *American Zoologist, 21*, 813-29

Werner, E.E., Mittelbach, G.G. and Hall, D.J. (1981) 'The Role of Foraging Profitability and Experience in Habitat Use by the Bluegill Sunfish', *Ecology, 62*, 116-25

Werner, E.E., Gilliam, J.F., Hall, D.J. and Mittelbach, G.G. (1983) 'An Experimental Test of the Effects of Predation Risk on Habitat Use in Fish', *Ecology, 64*, 1540-8

Wilson, D.S. (1975) 'The Adequacy of Body Size as a Niche Difference', *American Naturalist, 109*, 769-84

Windell, J.T. (1978) 'Digestion and the Daily Ration of Fishes', in *Ecology of Freshwater Fish Production*, S.D. Gerking (ed.), Blackwell, Oxford, pp. 159-83

Wirtz, P. (1978) 'The Behaviour of the Mediterranean *Tripterygion*', *Zeitschrift für Tierpsychologie, 48*, 142-74

Wooton, R.J. (1984) *A Functional Biology of Sticklebacks*, Croom Helm, London

Zyan, R.C. (1974) 'Le Rôle de la Reconnaissance Individuelle dans la Stabilité des Relations Hierarchiques chez *Xiphophorus* (Pisces, Poeciliidae)', *Behavior, 49*, 268-312

Zayan, R.C. (1975) 'Defense du Territoire et Reconnaissance Individuelle chez *Xiphophorus* (Pisces, Poeciliidae)', *Behavior, 52*, 266-312

FISH BEHAVIOUR BY DAY, NIGHT AND TWILIGHT

Gene S. Helfman

The simple diel cycle of rising and setting of the sun imposes on the behaviour and activity of fishes a dramatic, overriding set of predictable constraints. As a direct result, many kinds of behaviour and the species that engage in them follow characteristic convergent patterns that transcend geographic and taxonomic boundaries. These patterns can be recognised in such fundamental activities as the times when fishes feed, breed, aggregate and rest, in the transitions between activities, in the kinds of things that fishes eat, and in the ways in which fishes feed and avoid being eaten. The objectives of this chapter are to review the available information concerning the influences of day, night and twilight on various classes of fish behaviour; to delimit general diel activity patterns that characterise fishes in different habitat types; and to explore the environmental, ecological, physiological and developmental factors that interact with the cycle of daylight and darkness in determining diel patterns of fish behaviour. Throughout the chapter, day and daytime refer to daylight hours; night and nighttime refer to periods of darkness; crepuscular refers to twilight periods of dusk and dawn (sunset and sunrise); and diel refers to the 24-hour cycle.

Scope of the Chapter

The emphasis in this chapter is on direct field observations of diel patterns in the wild, particularly as influenced by changing ambient illumination. Circadian rhythms of fish activity (24-hour patterns driven by endogenous factors with external cues often involving daylength and daylight) will not be considered. However, their importance should not be discounted (see Thorpe 1978). Other exogenous factors besides photoperiod also affect activity cycles in fishes. Seasonal progressions involve changing temperature, with 'normal' diel patterns disrupted during the cold-water period of late autumn, winter and early spring in temperate and polar regions (Emery 1978; Müller 1978). Tide also influences and sometimes overrides diel activity cycles (see Chapter 15 by Gibson, this volume). Certain environments lack daily light cues and contain fishes that are continually active or arrhythmic. The deep sea is one example, but little is known of the diel activities of bathypelagic fishes. The behaviour of cave-dwelling fishes is covered in Chapter 17 by Parzefall, this volume. Diel vertical migrations, particularly characteristic of temperate marine species in the Atlantic, have been thoroughly reviewed by Woodhead (1966) and Blaxter (1976).

Activity Periods in Fishes

Feeding

Feeding behaviour involves several categories of behaviour, but as used here generally refers to the major period of foraging, including movement between foraging areas and other sites. In the wild, most of a fish's day appears to be spent either pursuing food or avoiding predators; many fishes appear to separate the day into an active, food-gathering phase and a relatively inactive, resting phase that is intimately linked with predator avoidance.

The majority of fishes at tropical and temperate latitudes feed primarily during either the day or night, a smaller number foraging during the crepuscular or twilight periods of dusk and dawn (Emery 1978). In general, the timing of activity is a familial characteristic (Table 14.1), but exceptions to this generalisation are certainly common. For example, many species of nominally nocturnal fishes will feed during the day if food is available, as any fisherman knows. Such an exception reflects the well-known opportunism that characterises many fishes, particularly predatory forms (Larkin 1979); it also suggests that activity patterns in fishes may be strongly determined by the activity patterns of their prey.

The relative numbers of fishes active at different periods of the diel cycle in different assemblages are remarkably similar. A survey of such 'temporal ratios', defined as the percentage of species active by day, night or twilight in different fish assemblages, shows that about a half to two-thirds of the species in most assemblages are diurnal, a quarter to a third are nocturnal, and the remaining 10 per cent or so are primarily crepuscular (Helfman 1978). These temporal ratios are probably a result of the relationship between phylogenetic position, temporal habits and trophic patterns among teleost fishes, as noted by Hobson (1974) and Hobson and Chess (1976). More primitive or generalised fishes are typically large-mouthed, nocturnal or crepuscular predators, whereas more advanced species have often specialised towards diurnality and feeding on smaller animals or on plants. Based on the food web or pyramid that typifies most animal communities, one would expect herbivores to be most numerous, and specialised carnivores least numerous as individuals in a community.

Breeding

A notable exception to characteristic diel activity patterns involves the relative breakdown of normal patterns during spawning periods (e.g. Nash 1982). Many species that usually feed during the day and then seek shelter at or shortly after sunset may breed actively at sunset (Lobel 1978). Some strongly diurnal species will even spawn at night (e.g. the percid *Perca flavescens*, Helfman 1981). Such night spawning may be even more common than we suspect, given our inability to make detailed nocturnal observations. Normal rhythms of activity are also disturbed during the

Table 14.1: Diel Activity Periods, Defined as Major Feeding Period, of Better-known Groups and Families of Teleost Fishes. In many cases, information is available for only a fraction of the species in a group. See Helfman 1978 for references and details

All or most species studied diurnal	All or most species studied nocturnal	Both diurnal and nocturnal species	Several crepuscular species (although also active at other times)	Several species without well-defined activity periods
Acanthuridae	Anguillidae	Carangidae	Aulostomidae	Muraenidae
Balistidae	Apogonidae	Centrarchidae	Carangidae	Scombridae
Blenniidae	Clupeidae	Congridae	Elopidae	Scorpaenidae
Characidae	Diodontidae	Gadoidei	Fistulariidae	Serranidae
Cichlidae	Gymnotidae	Mullidae	Serranidae	
Cirrhitidae	Haemulidae	Pleuronectiformes		
Cyprinidae	Holocentridae	Salmonidae		
Embiotocidae	Lutjanidae	Serranidae		
Esocidae	Mormyridae	Sphyraenidae		
Gobiidae	Ophidiidae			
Labridae	Pempheridae			
Percidae	Sciaenidae			
Pomacentridae	Siluriformes			
Scaridae				
Siganidae				

reproductive season. 'Nocturnal' or 'diurnal' fish may be continuously active during the day and night when engaged in parental care (e.g. diurnal damselfishes, Emery 1973b; the nocturnal ictalurid, *Ictalurus nebulosus,* Helfman 1981).

A simple explanation for the apparent breakdown in diel patterns during breeding is not immediately apparent. Obviously, parental care cannot be limited to just one period of the day, particularly if the young are relatively mobile or if eggs occur in relatively exposed locales. But what of diurnal species that spawn in the evening? (see Chapter 10 by Turner, this volume.) A possible explanation is that adults release their gametes into the water column late in the day because both light levels and numbers of active planktivores are reduced at that time. Floating eggs, because of their relatively small size, may also be less subject to predation by zooplanktivores at night (Hobson and Chess 1978). Parents are thus maximising the likelihood of initial survival of their offspring, before the fertilised eggs are dispersed by water movements (see Johannes 1978). Breeding at dusk would appear to entail some additional hazards for the adults engaged in spawning, given the abundance and activity of predatory fishes at that time (see below). However, many fishes show a surprising disregard for their own safety during spawning periods, a phenomenon termed 'spawning stupor' (Johannes 1978). Twilight spawning could represent an instance where adult survival is jeopardised in favour of maximising survival of offspring.

Diel Patterns in Different Habitat Types

Our knowledge of activity patterns in fishes in the wild is strongly influenced by factors that affect our ability to conduct research, particularly water clarity and water temperature, not to mention proximity of study locales to centres of research. Not surprisingly, the best-known species occur on coral reefs, followed by temperate lakes and temperate marine assemblages. The following accounts summarise information on diel activity patterns in these different habitat types, focusing on the families of fishes active at different times, the activities in which they engage, and the behavioural, ecological and evolutionary factors that may have influenced the observed patterns. Emphasis will be placed on coral-reef habitats, and information from other systems will be summarised and compared. Geographically, attention is focused on tropical Pacific and Caribbean faunas, temperate lakes of North America, and California kelp-bed fishes. Diel patterns have received less study than other ecological traits in tropical fresh waters; see Lowe-McConnell (1975) for a review of general day/ night differences in tropical lakes and rivers.

Coral-reef Fishes

Daytime

Coral reefs during daytime are characterised by a great diversity of forms and activity among the fishes. Diurnal fishes on coral reefs include many boldly 'poster-coloured' forms such as butterflyfishes (Chaetodontidae), angelfishes (Pomacanthidae), damselfishes (Pomacentridae), wrasses (Labridae), parrotfishes (Scaridae), surgeonfishes (Acanthuridae) and triggerfishes (Balistidae). The functions of these colour patterns have been variously attributed to inter- and intraspecific territoriality, species recognition, aposematic properties, crypsis, or combinations of the above (e.g. Hamilton and Peterman 1971; Kelly and Hourigan 1983). Regardless, these striking colour patterns probably relate to the bright light/clear water conditions that prevail on coral reefs during the day and the apparent importance of vision and visual displays to these fishes (Levine, Lobel and MacNichol 1980; see Chapter 4 by Guthrie, this volume).

Foraging by diurnal reef fishes is carried out by both solitary and grouped individuals. Many of the group feeders are herbivores. In fact, practically all herbivorous fishes on coral reefs are diurnal. Solitary herbivores are generally stationary territory holders, whereas shoaling species are usually roving and non-territorial. In some species, the distinction between shoaling and solitariness is an ontogenetic one, with younger fishes forming shoals and older fishes holding territories that serve as both feeding and breeding sites (Ogden and Buckman 1973). Shoals among herbivorous fishes probably serve as predator protection, but may also help overcome the territorial defence of algal patches by herbivorous damselfishes and parrotfishes (Barlow 1974; Ehrlich 1975; Robertson, Sweatman, Fletcher and Cleland 1976; see Chapter 12 by Pitcher, this volume).

Some of the most abundant diurnal fishes on reefs are zooplanktivorous, the smaller individuals and species of which generally feed as parts of large, relatively stationary aggregations. This group includes many damselfishes and streamlined groupers (e.g. *Anthias* spp.), snappers (Lutjanidae) and wrasses. The most abundant diurnal zooplanktivores occur over drop-off regions such as outer reef faces or in lagoon areas where at least moderate tidal currents occur (Hobson and Chess 1978). The food of these fishes is generally quite small (< 1 mm), consisting largely of calanoid copepods, cladocerans and larval invertebrates transported by currents and originating upstream from the feeding locale. Group formation in these fishes probably serves both for predator protection and in maintaining social and sexual hierarchies. Both functions appear to be intimately related to vision and high levels of ambient illumination.

These distantly related zooplanktivores have remarkably convergent morphologies and behaviours, apparently influenced by predator avoidance interacting with the distance above the refuge sites over which they feed, the strength of water currents and the common need to take tiny prey

(Davis and Birdsong 1973; Hobson and Chess 1978). The most stream-lined forms typically forage higher above the bottom and in stronger currents. Zooplanktivores, as well as other diurnal reef fishes, have also evolved visual capabilities that permit both high resolution of small prey items (e.g. small zooplankters) as well as sensitive detection of motion of prey or predators (Munz and McFarland 1973). These visual tasks are again dependent on the high light levels characteristic of daytime on coral reefs. Because of the 'many watchful eyes' nature of such aggregations, detection of a predator and flight by one prey individual generally results in the entire group diving rapidly to the protective refuge of the reef below (see Chapter 12 by Pitcher, this volume).

Other feeding guilds of coral-reef fishes active by day include relatively solitary species that feed on either mobile or well-defined sessile inverte-brates, such as corals, hydrozoans, sponges and sea urchins (see Hobson 1974, 1975). Some of the fishes have striking colorations or shapes, including butterflyfishes, angelfishes, triggerfishes and pufferfishes (Tetraodontidae). These fishes may rely on anatomical characteristics such as deep bodies, stout spines, tough skins, toxins and relatively large size, which allow them to feed in exposed locations in broad daylight while over-coming the defences evolved by their prey. Bright coloration in these fishes may serve as optical reinforcement of relative inedibility to visually hunting predators.

Fishes in sand and seagrass areas during daylight hours are relatively scarce and difficult to distinguish from the background of white sand, green grass or algae. These small fishes (pipefishes, Syngnathidae; wrasses; parrot-fishes; pike blennies, Chaenopsidae; dragonets, Callionymidae; gobies, Gobiidae) generally feed on small invertebrates. Larger fishes in the grass beds either have exceptional escape abilities, such as the sand-diving razor-fish (*Hemipteronotus*, Labridae), or have specialised abilities for probing in the sand in pursuit of buried prey (e.g. goatfishes, Mullidae), or are able to blow sand away from prey with expelled jets of water (e.g. boxfish, Ostraciontidae).

The activities of cleanerfishes, species that feed on parasites and necrotic tissue of other fishes, appear to be limited to daylight. These cleaners are usually wrasses and gobies or juvenile angelfishes and butterflyfishes. They often defend relatively permanent 'cleaning stations' to which host fishes are attracted (Limbaugh 1961; Gorlick, Atkins and Losey 1978). Visual cues, including distinctive striped guild coloration and a zigzag/jerky swimming pattern, are used by some cleaners to attract host fishes; hosts in turn often use colour changes and unusual body postures when soliciting cleaning. Since the taxonomic groups and life-history stages of the cleaners are predominantly diurnal, and since visual displays play an important role in stimulating cleaning interactions, it appears likely that cleaning is a diurnal phenomenon. However, cleaning could occur at night (possibly by invertebrate cleaners) and not be noticed because of the practical difficulties associated with making nocturnal observations of behaviour under water.

During the day, nocturnal fishes rest on or in reef structures. Some medium- and large-sized nocturnal fishes occur singly in holes or under ledges (e.g. drums, Sciaenidae; bigeyes, Priacanthidae; porcupinefishes, Diodontidae). Most nocturnal fishes form daytime resting aggregations (Starck and Davis 1966; Hobson 1975). Many small species (squirrelfishes, Holo-centridae; cardinalfishes, Apogonidae; sweepers, Pempheridae) enter holes and small caverns. Others form resting aggregations under ledges, at the mouths of caves, or over coral heads. This latter category includes some of the most abundant fishes on reefs, such as anchovies (Engraulidae), herrings (Clupeidae), silversides (Atherinidae), Indo-Pacific apogonids, and grunts (Haemulidae).

Predators that feed during the day tend to be cryptic, benthic species (e.g. lizardfishes, Synodontidae; scorpionfishes, Scorpaenidae; flatfishes, Bothidae) or relatively stationary, water-column hoverers (e.g. groupers, Serranidae; barracuda, Sphyraenidae). Some active predators, such as the jacks (Carangidae) and mackerels (Scombridae), may also form travelling, synchronised schools (Potts 1980) and make periodic, medium-speed passes at prey, perhaps testing their responsiveness (Smith 1978). Only a few predators, such as trumpetfishes (Aulostomidae), can be frequently observed stalking prey in broad daylight. They employ mimetic coloration and behaviour (Kaufman 1976; Aronson 1983), often matching the bold hues of the prey themselves, which presumably exploits colour vision in the prey, a strategy limited to daytime conditions.

Given the acute vision of diurnal prey fishes, the likelihood of detection of a predator involved in a bold, frontal attack is high. As a consequence, most diurnal predators rely on infrequent errors in judgement on the part of the prey for success (Hobson 1979): a prey fish that moves too close to a hovering or concealed predator, or that leaves its aggregation, or that habituates to the presence of a predator and is distracted by other events, is a candidate for an opportunistic attack. In sum, because of differential visual capabilities, the protection afforded by shoaling, and in some cases defensive anatomical specialisations, prey fishes have a behavioural and physiological advantage during the bright light of daytime.

Night-time

Information on fish activities at night are often inferred from static obser-vations of distribution or analyses of stomach contents (e.g. Starck and Davis 1966; Vivien 1973; Hobson 1974; Gladfelter and Johnson 1983). None the less, a fairly clear picture of the locales and activities of the more abundant species on coral reefs has emerged.

Many diurnal species have analogues among the night-active fishes. Such 'replacement sets' include all functional groups except the herbivores and cleanerfishes. Probably the most conspicuous of such replacement sets consists of the many silvery-red/brown, large-mouthed, large-eyed zoo-planktivores that hover above the reef (squirrelfishes, bigeyes, cardinal-fishes, sweepers). Their prey is usually larger than that encountered by diurnal planktivores and often originates more locally, consisting of invertebrates (polychaete worms, amphipods, isopods) that spend daytime secluded in the reef or buried in the sand and emerge at dusk (Hobson and Chess 1978). Typically, weak current areas are more densely populated with planktivores at night than during the day. The fishes involved are relatively deep bodied and robust, in contrast to their more streamlined diurnal counterparts. This lack of streamlining may reflect reduced predation at night, but is probably also linked to the weak currents that typify the foraging areas of the fishes.

Fish abundance often increases over sandy areas at night. For example, Caribbean seagrass beds at night abound with squirrelfishes, cardinalfishes and grunts that have migrated away from the reef during twilight and which feed on invertebrates that are now relatively exposed, moving over the sand and vegetation (Robblee and Zieman 1984). Preliminary distributional information indicates the possibility of nocturnal feeding territories, at least among grunts (McFarland and Hillis 1982).

Piscivores and invertebrate feeders also frequent the reef at night. As Hobson (1974) has pointed out, nocturnal reef fishes tend to be relatively primitive and generalised carnivores with large mouths (squirrelfishes, scorpionfishes, groupers, snappers), in contrast to the relatively advanced, specialised teleosts that feed on invertebrates during the day. These nocturnal predators apparently rely on good night vision to find moving, relatively large crustaceans or resting fishes. Olfaction and tactile sensation are important in food detection for the moray and snake eels (Muraenidae, Ophichthidae) that move about the reef and grass beds at night. Some parrotfishes sleep in mucous cocoons, which presumably seal in both olfactory and tactile cues that might be used by foraging eels (Winn and Bardach 1959) or predatory molluscs.

One striking difference between day and night is the dearth of group-associated activities in the dark. Nocturnal fishes that had formed distinctive resting aggregations by day occur either singly or in much looser aggregations while foraging at night. Shoaling depends partly on visual signals, but olfaction and water turbulence are also important. Therefore this reduction in shoaling could reflect reduced predation and hence less need for the predator-deterrent attributes of aggregating. (This may be coupled with a stronger response towards avoidance of competition than is possible during the light. Ed.) Diurnal fishes that have sought refuge are seldom found in groups.

An additional difference concerns the coloration of nocturnally active fishes as well as resting diurnal fishes. The bold colours of daytime in general give way to solid washes of brown, red and silver, often in combination. Nocturnal fishes that were relatively colourful by day take on a paler appearance, losing their bold stripes, bars and other marks. Diurnal fishes tend to assume a muted coloration, either uniformly pale or somewhat blotchy (see Emery 1973a; Hobson 1975). The absence of colourful patterns is probably related to poor colour vision via retinal cones at night (see Chapter 4 by Guthrie, this volume). The absence of bold contrasting marks in many species suggests that contrast visual signals are no longer useful in shoal maintenance, or are discouraged for reasons of predator avoidance. (The detection of outline silhouettes against a light background is probably more important for these large-eyed nocturnal fishes. Ed.) They may also be abandoned in favour of other sensory modes for social communication, such as acoustic cues (i.e. nocturnal foragers such as squirrelfishes, drums and grunts). (See Chapter 5 by Hawkins, this volume.)

Twilight

Two twilight, or crepuscular, periods occur during each diel cycle: at dusk and dawn. Evening twilight lasts from sunset until the sun is at a certain angle below the horizon (18° below for astronomical twilight). At tropical latitudes, twilight generally lasts between 70 and 85 min, depending on the time of year. Taken together, dusk and dawn amount to about 5 per cent of the 24 hour diel cycle, but ecologically, twilight may play a role in the lives of fishes out of proportion to the actual time involved. Two keys to understanding this role are first, that twilight is a period of environmental, behavioural and ecological transition, and, secondly, that predators exploit the transitional nature of these periods.

Environmental transitions primarily involve changes in light. Illumination levels at the surface fall from approximately 100 lx at sunset to about 0.01 lx 30 min later. During this period, diurnal fishes 'change over' from their daytime activity mode to their night-time inactivity mode, whereas nocturnal fishes initiate activity. During twilight, predators seem to be maximally active and successful. Twilight activities of coral-reef fishes have been studied in the Gulf of California, Hawaii, Australia and the Caribbean (see Helfman 1981). Despite substantial faunal differences, events at the different locales are surprisingly similar, suggesting similar selective pressures and behavioural convergence in twilight activities. The most complete studies are those by Hobson (1968, 1972); his descriptions and terminology depict activities at most locales and are now summarised.

Hobson found a sequence of five overlapping events at evening twilight that were essentially repeated in reverse at dawn. The events are (1) vertical and horizontal migrations of diurnal fishes; (2) cover-seeking of diurnal fishes; (3) evacuation of the water column; (4) emergence of nocturnal fishes; and (5) vertical and horizontal migrations of nocturnal fishes.

(1) Migrations of Diurnal Fishes. During the hour preceding sunset,

zooplanktivorous fishes descend to locations immediately above the reef. Smaller individuals and species move earlier. Many fishes, such as surgeon-fishes and parrotfishes, also move horizontally from feeding areas to resting areas along predictable paths (Hobson 1973; Dubin and Baker 1982).

(2) Cover-seeking of Diurnal Fishes.   Beginning shortly before sunset and continuing over the next 15-20 min (i.e. overlapping with the next period), diurnal fishes seek shelter for the night. This event defines 'changeover' in most studies (Helfman 1981). Smaller fishes in particular enter holes or cracks in the reef and are lost from sight. Larger species, such as butterflyfishes, parrotfishes, surgeonfishes and rabbitfishes, often rest in slight depressions or at the bases of overhanging coral heads. Some resting-site fidelity occurs, an individual occupying the same site on successive evenings (Hobson 1972; Ehrlich, Talbot, Russell and Anderson 1977; Robertson and Sheldon 1979). Agonistic interactions observed at this time suggest that sites constitute resting territories.

A distinctive sequence in cover-seeking appears to exist, with non-overlapping times of changeover characteristic of each group of species. For example, wrasses usually change over first, followed approximately by zooplanktivorous damselfishes, butterflyfishes, larger damselfishes, surgeonfishes and parrotfishes. The seeming predictability of this sequence led Domm and Domm (1973) to postulate that it existed to minimise con-fusion resulting from fights over limited resting sites. Such confusion could be exploited by predators, but this hypothesis has not been adequately tested. One problem is that the reported 'species' sequence may be an arti-fact of having watched the same individual on different evenings and then extrapolated to species. Individual fish may have relatively precise, repeated times at which they seek shelter, but the range of changeover times may in fact overlap considerably among species.

(3) Evacuation of the Water Column: the Quiet Period.   At about 10 to 15 min after sunset, the degree of activity above the reef rapidly decreases. The number of small and medium-sized fishes that have been hovering over the reef quickly diminishes. For approximately 15 to 20 min, the water column is essentially devoid of small fishes, and movement above the reef has come to a standstill. It is an eerie period for a human observer, which Hobson (1972) has aptly named the quiet period, the time when neither diurnal nor nocturnal fishes are truly active.

The quiet period is, however, a time of major activity for larger reef predators, such as groupers, jacks and snappers (Hobson 1968; Major 1977; see also Protasov 1970). These fishes typically lurk just over the bottom, striking at fishes above that remain in the water column. Such actions give predators a considerable strategic advantage. With the sun below the horizon, light levels are relatively dim. Dark-coloured predators moving close to the darkened reef face are particularly difficult to see from above. However, prey in the water column are silhouetted against the back-lighting of the evening sky. These predators typically strike upwards at their prey, which have apparent difficulty detecting the predator's initial movement off the bottom. Hobson (1968, 1972) gained direct evidence of use of this tactic by recording the times at which predators broke the water surface during strikes. He found that the number of such attacks increased dramatically during the quiet period. It seems likely that the quiet period of inactivity by small diurnal and nocturnal fishes is an avoidance response to the potential of being eaten by such predators that are maximally active at this time.

(4) Emergence of the Nocturnal Fishes.   Although some nocturnal fishes are visible, most remain in hiding through the quiet period. The quiet period ends as most nocturnal forms initiate activity about half an hour after sunset, near the limits of comfortable vision for humans. Some species move up into the water column, whereas others undertake extensive horizontal migrations between resting and feeding areas.

(5) Migrations of Nocturnal Fishes.   Species most often involved in intra-reef movements are squirrelfishes, copper sweepers and grunts (Ogden and Ehrlich 1977; Gladfelter 1979; McFarland, Ogden and Lythgoe 1979). The twilight migrations of juvenile grunts in the Caribbean are the most intensively studied of any twilight behaviours and exemplify the behavioural and evolutionary forces that have influenced such migrations.

Most grunts (Haemulidae) spend daylight hours in resting aggregations over reefs. At night they feed on invertebrates in sand or seagrass areas. At dusk, they migrate from daytime to night-time locales and the schools disband in the grass beds. At dawn, shoals of the same individuals migrate back to the same coral head they occupied the previous day, using the same route as the previous evening. Resting site locations, migration times and migration routes are all remarkably precise and predictable. Differences between shoals can often be explained on the basis of age differences of group members.

The dusk migration itself has several components that occur between 5 and 20 min after sunset (i.e. overlapping the quiet period). After a period of noticeable restlessness, the travelling school converges on a staging area, usually at the base of the coral head nearest the grass bed. Next, individuals make short excursions ($<$ 1m) along the migration route, but quickly return to the group (= 'ambivalence'). Finally, at about 20 min after sunset, the school streams off the reef and into the grass bed, where individuals progressively leave the shoal and commence foraging. Times of migration can differ by less than a minute on successive evenings if cloud cover is similar. The route used on different nights also differs by only a few centimetres. It is not unusual for fish to travel several hundred metres during migration. McFarland *et al.* (1979) concluded that the timing of migrations was driven by changing light.

The above description characterises the migrations of juvenile grunts 40 to 120 mm long and two months to two years old. The remarkable precision and constancy between nights in these juveniles also has a longer-term basis: resting-site locales and migration routes are social traditions, remaining relatively constant over periods of up to three years, even though shoal members seldom exceed two years in age.

Helfman, Meyer and McFarland (1982) addressed the question of whether precise knowledge of migration times and routes can change during development, and if such knowledge is open to modification due to experience. They found that larger juveniles were more precise, that migrations occurred later, and that variability in times decreased with increasing age. Possible influences on these changes included (1) development of the eye, which allowed larger fish to migrate at lower light levels (= later times); (2) social influences, whereby young fish joining a shoal might learn migration routes by following larger fish; and (3) crepuscular predators, which selected out young juveniles that migrated at the wrong time or over the wrong terrain.

The question of possible learning of routes was further examined by Helfman and Schultz (1984), who transplanted small juveniles between sites and found that transplanted fish could in fact learn a new migration route by following resident fish at the new site. These findings also indicated that fishes can acquire knowledge about social traditions via learning, i.e. can engage in cultural behaviour.

Determinants of Twilight Behaviour in Coral-reef Fishes

The changing nature of light during twilight, and the increased activity of predators at that time, suggest a direct link between predation and vision during crepuscular periods. Munz and McFarland (1973), and McFarland and Munz (1976) found strong correlates between the structure of reef-fish eyes, the timing of twilight activities, and the behaviour of predators (Figure 14.1; see also Lythgoe 1979). Diurnal fishes typically have eyes with many small cone cells in their retinae that function optimally at high light levels; this maximises resolution and motion detection at the expense of night-time vision (see Chapter 4 by Guthrie, this volume). Nocturnal fishes typically contain rod-dominated eyes with fewer but larger cones. These eyes maximise light capture, but sacrifice resolution and motion detection and are overloaded by the intense light of daytime. Surprisingly, both diurnal and nocturnal fishes have rod visual pigments that are most sensitive to light at around 590 nm, which is a better match to prevailing wavelengths during twilight than to night-time conditions. Apparently both diurnal and nocturnal forms are sacrificing some nocturnal abilities in favour of better vision during twilight. The selective force driving this twilight match is apparently crepuscular predation. Twilight-active piscivores possess intermediate eyes with an intermediate number of large cones. The predators' intermediate number of large cones. The predators' eyes function poorly eyes function poorly relative to potential prey capabilities during daylight and darkness, but may function better than either a diurnal or nocturnal eye during twilight. The quiet period corresponds closely to the time when predators have both a relative visual and behavioural advantage over prey. The result is that few small fishes can afford to be in the water column during the quiet period, when the rate of change of light is most rapid and illumination is at levels at which both diurnal and nocturnal eyes function poorly. This may also explain why attacks during daylight are limited to behavioural errors by prey, and why there are relatively few nocturnal piscivores.

Diel Patterns in Other Habitat Types

Extensive observational studies of day/night patterns of fish behaviour have not been numerous in habitats other than coral reefs. A brief

Figure 14.1: Comparison of Evening Twilight Changeover between Diurnal and Nocturnal Fishes in Different Habitat Types. Data are the ranges of mean times at which diurnal fishes cease moving or seek shelter and at which nocturnal fishes initiate movement. The quiet period, when neither diurnal nor nocturnal fishes on coral reefs are active, occurs when light levels are changing most rapidly and when dark adaptation of the fish eye is taking place. Note particularly the greater overlap of diurnal and nocturnal species in temperate regions. Number of species is shown to the right of each line. Times for temperate lake fishes are standardised to a 100-min astronomical twilight length to account for seasonal changes in twilight lengths; absolute times relative to sunset would generally be 5 min later. SS = sunset. Data from Hobson (1972), Munz and McFarland (1973), Helfman (1981) and Hobson *et al.* (1981)

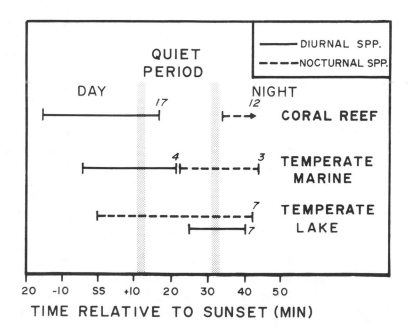

summary of the literature on fish activities in temperate lakes and kelp beds is given below, emphasizing assemblage level characteristics, and comparing diel patterns among the habitat types.

Temperate Lakes

Knowledge of day/night patterns among temperate lake fish assemblages is primarily from observational studies in North America (Emery 1973a; Helfman 1981). As with coral reefs, temperate lake assemblages consist primarily of diurnal and nocturnal fishes, although the distinction between the two blurs in comparison with coral-reef fishes (see Helfman 1981 for review). Phylogenetic patterns of diel behaviour are also less distinct. Families of temperate lake fishes are also more likely to contain both diurnal and nocturnal species.

Diurnal guilds in lakes include zooplanktivores (minnows, Cyprinidae; sunfishes, Centrarchidae; perches, Percidae), which often form large feeding aggregations. Herbivorous fishes are relatively lacking in North American lakes, although they are more common in European lentic environments (Presj, 1984). Feeders on benthic or plant-associated invertebrates are abundant, including members of the minnow, sunfish and perch families as well as mudminnows (Umbridae), suckers (Catostomidae) and topminnows (Cyprinodontidae). Diurnal piscivores include pike and pickerel (Esocidae) and larger black basses (Centrarchidae); the latter are also active at twilight and during the night. Cleaning behaviour exists but is not apparently as highly evolved as in tropical marine fishes. The dominant cleaner in North American lakes is the bluegill sunfish (*Lepomis macrochirus*); much of its cleaning activity is performed during twilight (Helfman 1981). Nocturnal fishes generally rest during the day, either singly on the bottom among vegetation or other structure (e.g. ictalurid catfishes and the percid *Stizostedion*), or in daytime resting shoals associated with various kinds of structure.

Nocturnal fishes in temperate lakes are abundant, replacing most of the diurnal trophic categories. Zooplanktivores swim more in open-water limnetic regions than above littoral regions (Hall *et al.* 1979; Bohl 1980). Abundant zooplanktivores occur in the whitefish family (Coregonidae) and also among the minnows (Cyprinidae). Fishes that feed on invertebrates forage mostly near the bottom and also take small, resting fishes at night. This group includes eels (Anguillidae), trout (Salmonidae), catfishes, sculpins (Cottidae), sunfishes, and drums (Sciaenidae). The catfish, *Ictalurus nebulosus*, also feeds on filamentous algae, which makes it a nocturnal herbivore, another contrast with coral-reef patterns. Nocturnal piscivores include bottom-feeding forms such as bowfin (Amiidae) and burbot (Gadidae), as well as those that also swim in the water column (e.g. salmonids; temperate basses, Percichthyidae; sunfishes; percids). In contrast to coral-reef fishes which seek shelter at night, diurnal lake fishes tend to rest in relatively exposed locales on barren or sparsely vegetated bottoms, or in clearings among vegetation.

As with coral-reef fishes, twilight in temperate lakes is a period of transition when diurnal fishes cease activity and nocturnal fishes initiate feeding. In fact, twilight events in temperate lakes bear a number of marked similarities to coral reefs. Helfman (1981) found a series of six general activity patterns among the fishes that were repeated in reverse order at dawn: increased movement of diurnal fishes, group break-up, cessation of feeding, first stop or slowdown, initiation of activity by nocturnal fishes, and cessation of activity of diurnal fishes. Increased movement involved some migrations from feeding to resting areas as well as descent from the water column of smaller planktivores. Shoals disbanded in both habitat types and diurnal fishes then ceased feeding. Cessation of activity probably corresponds to cover-seeking in coral-reef species, the main difference being that few lake fishes sought shelter for the evening.

Size and age differences in timing were also found in both systems. Larger individuals were often active later in the evening than smaller individuals. In some species, diurnal juveniles changed to nocturnally foraging adults (see also Helfman 1978). Young lake fishes had a greater affinity for structure while resting, a greater likelihood of defending resting sites, and faster shoal formation at dawn. Such ontogenetic differences may relate to greater predation pressure on smaller, younger fish (Hobson 1972).

Other similarities between the two systems include: (1) dawn activities occurring at lower light levels than corresponding dusk activities; (2) upsurge in non-feeding activities (e.g. agonistic behaviour) at dusk; (3) ecological replacement sets, shifting to older individuals and larger-mouthed species at dusk; (4) a shift from group to individual activities in the evening; (5) a breakdown in characteristic twilight activities during the breeding season; and (6) an upsurge in predation at twilight.

These similarities imply that similar selective pressures have operated with respect to behaviour at twilight. However, several differences between the assemblages indicate potentially important differences in the determinants of diel patterns. Differences in the lake fishes included: (1) less resting site constancy and structural affinity in diurnal lake fishes; (2) fewer twilight migrating groups; (3) increased cleaning activity at dusk; (4) increased intraspecific and ontogenetic variability in timing; (5) later times and less light for most twilight activities of diurnal species at dusk; (6) more prolonged changeover; (7) greater apparent between-species overlap in changeover time; and (8) lack of a quiet period.

The last four categories indicate less precision and predictability of twilight events in the temperate lake fauna. Helfman (1981) postulated that differences in the lake fishes might result from the combined influence of lower and variable species diversity, reduced crepuscular predation, variable and often poor underwater visibility, variable and longer twilight length, and long-term and short-term climatic instability. Variability in all these parameters would select for fishes with an ability to change their behaviour as a function of immediate conditions, i.e. that were behaviourally plastic. Helfman predicted that the twilight activities of temperate lake fishes would converge towards those of coral-reef fishes in locales where species diversity, water clarity, twilight length and crepuscular predation levels more closely resembled those found in the coral-reef environment. A recent study by Hanych, Ross, Magnien and Suggars (1983) in a Minnesota lake found twilight migratory patterns of one minnow species that differed from those reported by Helfman (1981). The authors suggested that the differences could be explained as a plastic response by the species to disparate levels of predation in the two lakes.

Temperate Marine Fishes

Studies by Ebeling and Bray (1976) in south–central California and by Hobson and Chess (1976) and Hobson, Chess and McFarland (1981) in southern California have explored diel patterns in kelp-bed-associated assemblages and compared them with tropical regions. The faunas studied contained 25 to 27 species, 14 in common, with affinities to tropical or temperate groups. The southern California assemblage contained more 'tropical derivative' species and the south–central California assemblage contained more temperate derivatives. This zoogeographic/phylogenetic component may reveal possible interactions between ecological and systematic factors in determining activity patterns.

Both faunas contained diurnal, nocturnal and crepuscular species, although some species were difficult to characterise with respect to temporal patterns. As with coral-reef assemblages, diurnal species included abundant shoaling zooplanktivores (Atherinidae, Serranidae, Embiotocidae, Pomacentridae, Labridae, Scorpaenidae); foragers on small, hidden invertebrates (Serranidae, Embiotocidae, Clinidae, Gobiidae) and cleanerfishes (Labridae, Embiotocidae). As in temperate lakes, herbivores were relatively rare (two kyphosid species). Many nocturnal species in southern California formed daytime resting aggregations, but these were absent in the south–central California assemblage. When active, nocturnal species (Scorpaenidae, Haemulidae, Sciaenidae, Embiotocidae) fed primarily on zooplankton or other invertebrates that were larger than the food of their diurnal counterparts. Cleanerfish, and probably herbivores, were inactive at night; one herbivore (*Girella nigricans*, Kyphosidae) may shift to feeding on invertebrates after sunset. At night, some diurnal fishes rested in holes or sought other shelter, whereas many species rested in comparatively exposed locales. Piscivores (Serranidae, Scorpaenidae, Hexagrammidae) were primarily nocturnal or crepuscular.

As elsewhere, twilight was a period of transition. In southern California, diurnal fishes migrated from foraging areas to resting areas, some along established routes but at less predictable times than coral-reef fishes. Ebeling and Bray (1976) felt that such migrations were uncommon at their site. As in both coral reefs and temperate lakes, twilight movements were more common in nocturnal fishes; some migrations were extensive (e.g. the embiotocid, *Hyperprosopon argenteum,* moved 1.6 km: Ebeling and Bray 1976; the haemulid, *Xenistius californiensis,* moved > 400 m: Hobson *et al.* 1981). Two ontogenetic patterns also found elsewhere were shifts from diurnal to nocturnal foraging with growth in some species, and a tendency for small zooplanktivores to remain relatively close to cover. Resting-site fidelity for diurnal fishes appeared slight, perhaps more like the temperate lake than the tropical reef situation. Interestingly, two primarily diurnal species (*Brachyistius frenatus,* Embiotocidae, and *Chromis punctipinnis,* Pomacentridae) aggregated at night, a pattern not observed in either coral reefs or temperate lakes.

Diel patterns at these two warm temperate locales share obvious similarities with other habitat types, suggesting that certain patterns can be expected in almost any fish assemblage. However, some obvious differences occurred between the two sites, as did disagreement on interpretation of results. Hobson *et al.* (1981) emphasised the similarities between kelp-bed and coral-reef fishes with respect to diel replacement sets, daytime resting aggregations and twilight migrations. Ebeling and Bray (1976) felt that these phenomena were noticeably lacking at their site, particularly when nocturnal zooplanktivores were concerned. They stated that 'after the dusk period of intensified activity, the notably lackluster night life gives the kelp forest an aura of desolation' (page 714), and emphasised that turbid water conditions might discourage nocturnal zooplanktivory.

Some of the disagreement may relate to zoogeographic and phylogenetic differences: six tropical derivative species occurred at the southern California site only, four of which migrated at twilight or fed in the water column at night (see Hobson *et al.* 1981, Table 4). Ebeling and Bray (1976) interpreted differences in cover-seeking on the basis of phylogenetic affinities: tropical derivative species sought shelter at night, whereas temperate derivatives rested in relatively exposed locations, perhaps because 'tropical derivatives compete more successfully against the primarily temperate species for shelter' (page 714). Hobson *et al.* (1981) countered that the same temperate species rest in exposed locales at higher latitudes where tropical derivative species are lacking. They also pointed out that on coral reefs larger species tended to rest in relatively exposed locales, suggesting that the exposed resting fishes in south–central California were relatively large.

The various authors generally agree on the distinctiveness of diel patterns. The overall assemblage patterns of activity in warm temperate marine fishes appear to be less clearly defined than on coral reefs, more like the temperate lake pattern. Ebeling and Bray (1976: 715) concluded that 'the general program of diel activity in the kelp forest appears to be comparatively unstructured'. Hobson *et al.* (1981: 22) found 'little evidence of the detailed community transition patterns that typically characterise these phenomena on tropical reefs'. This lack of distinctiveness results from several factors, including: (1) relatively high variance in times of changeover; (2) species that feed both by day and by night; (3) diurnal species that çease activity relatively late, and nocturnal species that initiate activity relatively early, producing an indistinct or nonexistent quiet period (Figure 14.1); and (4) possible incompleteness of replacement sets. The overall assemblage pattern strongly parallels the pattern described above for temperate lake fishes.

The causes underlying attributes of indistinctiveness during changeover remain unresolved. Again, zoogeographic and phylogenetic considerations may partially explain this pattern. Hobson *et al.* (1981) pointed out that the most distinctive twilight activities characterised representatives of basically tropical families. Another possible determinant is a comparative lack of crepuscular predation in warm temperate faunas. Predators appear to be a primary selective force promoting precision among coral-reef fishes (see above); reduced crepuscular predation in kelp-bed assemblages could lead to increased variability in twilight activities. A third explanation is that the longer twilight of temperate regions results in a relaxation of visual con-straints on twilight behaviours. Crisply distinct behaviours in coral-reef fishes could be a response to short twilight, when light levels decrease at a rate equal to or faster than the rate at which the fish eye dark-adapts. The longer twilight lengths at temperate latitudes may exceed the dark-adaptation rate, and thereby remove the necessity for completing twilight activities in a minimum of time with a minimum of variability (see Helfman 1981).

Support for the idea that duration of twilight behaviours is directly related to length of twilight comes from recent observations of predator-prey interactions in Alaska at latitude 56° N (E.S. Hobson, in press). Hobson found that sand lances (*Ammodytes hexapterus*) experi-enced intensified predation by four species of benthic piscivores when the prey sought night-time shelter in sediments through much of a 2.5 h twilight changeover period. The period of danger during twilight apparently increases in relation to the length of twilight, at least where predators are abundant.

Two additional factors may influence diel patterns in temperate as well as tropical assemblages. Hobson *et al.* (1981) studied the retinal pigments of the fishes at their locale and compared their findings with expectations based on Munz and McFarland's 'twilight hypothesis' (Munz and McFarland 1973; McFarland and Munz 1976; see above). This hypothesis predicts that, because crepuscular periods are particularly dangerous, visual adaptations will arise to counter crepuscular predation at the cost of nocturnal function. As with coral-reef fishes, the spectral sensitivity of rod pigments in kelp-bed fishes matched twilight visual conditions better than nocturnal conditions. However, Hobson *et al.* found that pigment charac-teristics were an even better match to the light generated by bioluminescent organisms, which typically luminesce when disturbed by movement in the water column. They felt that natural selection would strongly favour sensitivity to bioluminescence because such flashes of light in the dark would often signify the movement of both predators and prey. It would be difficult to separate the selective influence of light quality at twilight from that related to bioluminescence, since a pigment with an absorption maximum around 500 nm, as found in kelp bed fishes, would be maximally sensitive to both.

The final confounding factor concerns the invertebrates active at night in temperate kelp beds. C.A. Stepien (personal communication) has found that cirolanid isopods, species listed as prey of some nocturnal kelp-bed fishes by Hobson and Chess (1976), may in fact also be predators on diurnal fishes that rest in the kelp beds. Caged fishes placed in the kelp beds at night are attacked and often killed by these invertebrates. The disputed topic of different resting locales of different species may be influenced by the density of these killer isopods and the escape manoeuvres a fish must employ to avoid them.

Summary

Most fish species forage primarily during the day or night, with a few species primarily active during crepuscular periods of dawn and dusk. Several similarities in these diel patterns of activity occur in very different habitat types. In coral reef, temperate lake, and temperate marine kelp-bed assemblages, diurnal fishes include herbivores, zooplanktivores, invertebrate feeders, piscivores and cleaners 'Replacement sets' of these various functional groups are active at night, except for herbivores and cleaners. Twilight is a time of transition between daytime and night-time modes, when diurnal fishes cease activity and nocturnal fishes initiate feeding. It is also a period of major activity and success of piscivores. Predator/prey interactions at twilight have had an apparently strong influence on the behaviour and particularly the visual physiology of both diurnal and nocturnal fishes. Predators are relatively rare in some temperate assemblages, which may explain a lower degree of precision in the twilight activities of temperate species. Differences in water clarity, species diversity, phylogenetic constraints and twilight length may also have influenced behavioural differences in temperate versus tropical habitats.

Acknowledgement

Dr E.S. Hobson kindly commented on the draft manuscript of this chapter.

References

Aronson, R.B. (1983) 'Foraging Behavior of the West Atlantic Trumpetfish, *Aulostomus maculatus*: Use of Large, Herbivorous Reef Fishes as Camouflage', *Bulletin of Marine Science, 33*, 166-71

Barlow, G.W. (1974) 'Extraspecific Imposition of Social Grouping among Surgeonfishes (Pisces: Acanthuridae)', *Journal of Zoology, London, 174*, 333-40

Blaxter, J.H.S. (1976) 'The Role of Light in the Vertical Migration of Fish — a Review', in G.C. Evans, R. Bainbridge and O. Rackham (eds), *Light as an Ecological Factor: II*, Blackwell, London, Chapter 8

Bohl, E. (1980) 'Diel Pattern of Pelagic Distribution and Feeding in Planktivorous Fish', *Oecologia (Berlin), 44*, 368-75

Davis, W.P. and Birdsong R. (1973) 'Coral Reef Fishes which Forage in the Water Column', *Helgolander wissenschaftliche Meeresuntersuchungen, 24*, 292-306

Domm, S.B. and Domm, A.J. (1973) 'The Sequence of Appearance at Dawn and Disappearance at Dusk of Some Coral Reef Fishes', *Pacific Science, 27*, 128-35

Dubin, R.E. and Baker, J.E. (1982) 'Two Types of Cover-seeking at Sunset by the Princess Parrotfish, *Scarus taeniopterus*, at Barbados, West Indies', *Bulletin of Marine Science, 32*, 572-83

Ebeling, A.W. and R.N. Bray (1976) 'Day versus Night Activity of Reef Fishes in a Kelp Forest off Santa Barbara, California', *United States Fishery Bulletin, 74*, 703-17

Ehrlich, P.R. (1975) 'The Population Biology of Coral Reef Fishes', *Annual Review of Ecology and Systematics, 6*, 211-47

Ehrlich, P.R., Talbot, F.H., Russell, B.C. and Anderson, G.R.V. (1977) 'The Behavior of Chaetodontid Fishes with Special Reference to Lorenz' 'Poster Colouration' Hypothesis', *Journal of Zoology, London, 138*, 213-28

Emery, A.R. (1973a) 'Preliminary Comparisons of Day and Night Habits of Freshwater Fish in Ontario Lakes', *Journal of the Fisheries Research Board of Canada, 30*, 761-74

Emery, A.R. (1973b) 'Comparative Ecology and Functional Osteology of Fourteen Species of Damselfish (Pisces: Pomacentridae) at Alligator Reef, Florida Keys', *Bulletin of Marine Science, 23*, 649-770

Emery, A.R. (1978) 'The Basis of Fish Community Structure: Marine and Freshwater Comparisons', *Environmental Biology of Fishes, 3*, 33-47

Gladfelter, W.B. (1979) 'Twilight Migrations and Foraging Activities of the Copper Sweeper, *Pempheris schomburgki* (Teleostei, Pempheridae)', *Marine Biology, 51*, 109-19

Gladfelter, W.B. and Johnson, W.S. (1983) 'Feeding Niche Separation in a Guild of Tropical Reef Fishes (Holocentridae)', *Ecology, 64*, 552-63

Gorlick, D.L., Atkins, P.D. and Losey, G.S. (1978) 'Cleaning Stations as Water Holes, Garbage Dumps, and Sites for the Evolution of Reciprocal Altruism?', *American Naturalist, 112*, 341-53

Hall, D.J., Werner, E.E., Gilliam, J.F., Mittelbach, G.G., Howard, D., Doner, C.G., Dickerman, J.A. and Stewart, A.J. (1979) 'Diel Foraging Behavior and Prey Selection in the Golden Shiner (*Notemigonus chrysoleucas*)', *Journal of the Fisheries Research Board of Canada, 36*, 1029-39

Hamilton, W.J. and R.M. Peterman (1971) 'Countershading in the Colourful Reef Fish *Chaetodon lunula*: Concealment, Communication or Both', *Animal Behaviour, 19*, 357-64

Hanych, D.A., Ross, M.R., Magnien, R.E. and Suggars, A.L. (1983) 'Nocturnal Inshore Movement of the Mimic Shiner (*Notropis volucellus*): a Possible Predator Avoidance Behavior', *Canadian Journal of Fisheries and Aquatic Sciences, 40*, 888-94

Helfman, G.S. (1978) 'Patterns of Community Structure in Fishes: Summary and Overview', *Environmental Biology of Fishes, 3*, 129-48

Helfman, G.S. (1981) 'Twilight Activities and Temporal Structure in a Freshwater Fish Community', *Canadian Journal of Fisheries and Aquatic Sciences, 38*, 1405-20

Helfman, G.S. and Schultz, E.T. (1984) 'Social Transmission of Behavioural Traditions in a Coral Reef Fish', *Animal Behaviour, 32*, 379-84

Helfman, G.S., Meyer, J.L. and McFarland, W.N. (1982) 'The Ontogeny of Twilight Migration Patterns in Grunts (Pisces: Haemulidae)', *Animal Behaviour, 30*, 317-26

Hobson, E.S. (1968) 'Predatory Behavior of Some Shore Fishes in the Gulf of California',

United States Bureau of Sport Fisheries and Wildlife, Research Report, 73, 1-92

Hobson, E.S. (1972) 'Activity of Hawaiian Reef Fishes during Evening and Morning
  Transitions between Daylight and Darkness', *United States Fishery Bulletin, 70,* 715-40

Hobson, E.S. (1973) 'Diel Feeding Migrations in Tropical Reef Fishes', *Helgolander
  wissenschaftliche Meeresuntersuchungen, 24,* 361-70

Hobson, E.S. (1974) 'Feeding Relationships of Teleostean Fishes on Coral Reefs in Kona,
  Hawaii', *United States Fishery Bulletin, 72,* 915-1031

Hobson, E.S. (1975) 'Feeding Patterns among Tropical Reef Fishes', *American Scientist, 63,*
  382-92

Hobson, E.S. (1979) 'Interactions between Piscivorous Fishes and their Prey', in H. Clepper
  (ed.), *Predator-Prey Systems in Fisheries Management,* Sport Fishing Institute,
  Washington, DC, pp. 231-42

Hobson, E.S. (in press) 'Predation on the Pacific Sand Lance, *Ammodytes hexapterus*
  (Pisces: Ammodytidae), during the transition between day and night in southeastern
  Alaska, *Copeia*

Hobson, E.S. and J.R. Chess (1976) 'Trophic Interactions among Fishes and Zooplankters
  near Shore at Santa Catalina Island, California', *United States Fishery Bulletin, 74,*
  567-98

Hobson, E.S. and J.R. Chess (1978) 'Trophic Relationships among Fishes and Plankton in
  the Lagoon at Enewetak Atoll, Marshall Islands', *United States Fishery Bulletin, 76,*
  133-53

Hobson, E.S., Chess, J.R. and McFarland, W.N. (1981) 'Crepuscular and Nocturnal
  Activities of Californian Nearshore Fishes, with Consideration of their Scotopic Visual
  Pigments and the Photic Environment', *United States Fishery Bulletin, 79,* 1-30

Johannes, R.E. (1978) 'Reproductive Strategies of Coastal Marine Fishes in the Tropics',
  Environmental Biology of Fishes, 3, 65-84

Kaufman, L. (1976) 'Feeding Behavior and Functional Coloration of the Atlantic
  Trumpetfish, *Aulostomus maculatus*', *Copeia, 1976,* 377-8

Kelly, C.D. and Hourigan, T.F. (1983) 'The Function of Conspicuous Coloration in
  Chaetodontid Fishes: a New Hypothesis', *Animal Behaviour, 31,* 615-17

Larkin, P.A. (1979) 'Predator-Prey Relations in Fishes: an Overview of the Theory' in H.
  Clepper (ed.), *Predator-Prey Systems in Fisheries Management,* Sport Fishing Institute,
  Washington, DC, pp. 13-22

Levine, J.S., Lobel, P.S. and MacNichol, E.F. Jr. (1980) 'Visual Communication in Fishes', in
  M.A. Ali (ed.), *Environmental Physiology of Fishes,* Plenum, New York, pp. 447-75

Limbaugh, C. (1961) 'Cleaning Symbiosis', *Scientific American, 205,* 42-9

Lobel, P.S. (1978) 'Diel, Lunar, and Seasonal Periodicity in the Reproductive Behavior of the
  Pomacanthid Fish, *Centropyge potteri,* and Some Other Reef Fishes in Hawaii', *Pacific
  Science, 32,* 193-207

Lowe-McConnell, R.H. (1975) *Fish Communities in Tropical Freshwaters: their Distribution,
  Ecology and Evolution,* Longman, London

Lythgoe, J.N. (1979) *The Ecology of Vision,* Clarendon Press, Oxford

McFarland, W.N. and Hillis, Z.-M. (1982) 'Observations on Agonistic Behavior between
  Members of Juvenile French and White grunts — family Haemulidae', *Bulletin of Marine
  Science, 32,* 255-68

McFarland, W.N. and Munz, F.W. (1976) 'The Visible Spectrum during Twilight and its
  Implications to Vision', in G.C. Evans, R. Bainbridge and O. Rackham (eds), *Light as an
  Ecological Factor: II,* Blackwell, Oxford, pp. 249-70

McFarland, W.N., Ogden, J.C. and Lythgoe, J.N. (1979) 'The Influence of Light on the
  Twilight Migrations of Grunts', *Environmental Biology of Fishes, 4,* 9-22

Major, P.F. (1977) 'Predator-Prey Interactions in Schooling Fishes during Periods of
  Twilight: a Study of the Silverside *Pranesus insularum* in Hawaii', *United States Fishery
  Bulletin, 75,* 415-26

Müller, K. (1978) 'The Flexibility of Circadian Rhythms Shown by Comparative Studies at
  Different Latitudes', in J.E. Thorpe (ed.), *Rhythmic Activity of Fishes,* Academic Press,
  London, pp. 91-104

Munz, F.W. and McFarland, W.N. (1973) 'The Significance of Spectral Position in the
  Rhodopsins of Tropical Marine Fishes', *Vision Research, 13,* 1829-74

Nash, R.D.M. (1982) 'The Diel Behaviour of Small Demersal Fish on Soft Sediments on the
  West Coast of Scotland Using a Variety of Techniques: with Special Reference to

Lesuerigobius friesii (Pisces: Gobiidae)', *Marine Ecology, 3*, 161-78

Ogden, J.C. and Buckman N.S. (1973) 'Movements, Foraging Groups, and Diurnal Migrations of the Striped Parrotfish *Scarus croicensis* Bloch (Scaridae)', *Ecology, 54*, 589-96

Ogden, J.C. and Ehrlich, P.R. (1977) 'The Behavior of Heterotypic Resting Schools of Juvenile Grunts (Pomadasyidae)', *Marine Biology, 42*, 273-80

Potts, G.W. (1980) 'The Predatory Behaviour of *Caranx melampygus* (Pisces) in the Channel Environment of Aldabra Atoll (Indian Ocean)', *Journal of Zoology, London, 192*, 323-50

Presj, A. (1984) 'Herbivory by Temperate Freshwater Fishes and its Consequences', *Environmental Biology of Fishes, 10*, 281-96

Protasov, V.R. (1966) *Vision and Near Orientation of Fish*, M. Raveh (transl.), Israel Program for Scientific Translation, IPST No. 5738, United States Department of the Interior, Washington, DC

Robblee, M.B. and Zieman, J.C. (1984) 'Diel Variation in the Fish Fauna of a Tropical Seagrass Feeding Ground', *Bulletin of Marine Science, 34*, 335-45

Robertson, D.R. and Sheldon, J.M. (1979) 'Competitive Interactions and the Availability of Sleeping Sites for a Diurnal Coral Reef Fish', *Journal of Experimental Marine Biology and Ecology, 40*, 285-98

Robertson, D.R., Sweatman, H.P.A., Fletcher, E.A. and Cleland, M.G. (1976) 'Schooling as a Mechanism for Circumventing the Territoriality of Competitors', *Ecology, 57*, 1208-20

Smith, C.L. (1978) 'Coral Reef Fish Communities: a Compromise View', *Environmental Biology of Fishes, 3*, 109-28

Starck, W.A., II and Davis, W.P. (1966) 'Night Habits of Fishes of Alligator Reef, Florida', *Ichthyologia, 38*, 313-56

Thorpe, J.E. (1978) *Rhythmic Activity of Fishes*, Academic Press, London

Vivien, M.L. (1973) 'Contribution a la Connaissance de l'Ethologie Alimentaire de l'Ichtyofaune de Platier Interne des Recifs Coralliens de Tulear (Madagascar)', *Tethys, Supplement 5*, 221-308

Winn, H.E. and Bardach, J.E. (1959) 'Differential Food Selection by Moray Eels and a Possible Role of the Mucous Envelope of Parrotfishes in Reduction of Predation', *Ecology, 40*, 296-8

Woodhead, P.M.J. (1966) 'The Behaviour of Fish in Relation to Light in the Sea', *Oceanography and Marine Biology Annual Review, 4*, 337-403

15 INTERTIDAL TELEOSTS: LIFE IN A FLUCTUATING ENVIRONMENT

R.N. Gibson

Teleost fishes are numerous and diverse and they can be found in a wide range of habitats. Some of these habitats, like caves (see Chapter 17 by Parzefall, this volume) and the deepest parts of the sea, change relatively slowly in their physico-chemical characteristics. Others are subject to marked diurnal and seasonal variation, (see Chapter 14 by Helfman, this volume). In the narrow area between the tidemarks, the situation is further complicated by the ebb and flow of the tides. For fish, unprotected by the thick exoskeleton of many littoral invertebrates, the seashore is a difficult region to occupy, although the rewards of colonisation may be high because there are likely to be few competitors. Naturally, there are also disadvantages, such as the risk of predation by both terrestrial and aquatic predators, regular and possibly rapid changes in environmental conditions and, perhaps most drastic of all for fish, the almost complete removal of their normal locomotory medium for hours at a time.

Fish that utilise the intertidal zone have overcome these problems in two ways. The first and simplest is to avoid the zone at low tide and make use of its resources, usually food, only when it is submerged. The strategy of migration in and out with the tide is employed by large numbers of fish on both rocky and sandy shores throughout the world. It is particularly favoured by juveniles, who may find, in addition to food, refuge from larger predators in the shallow water. Movement into and out of the intertidal zone over a seasonal time scale can also be seen in some species whose distribution is mainly subtidal. In these forms the juveniles are found intertidally for the first few months of their life but gradually move into deeper water as they grow. This shift in distribution with size may be a response to the relatively greater risk of predation for small fishes in deeper water. It also results in these species only being present on the shore at certain times of the year. On sandy shores these seasonal visitors, of which juvenile flatfish (Pleuronectiformes) are good examples, migrate up and down the beach with each tide (Tyler 1971; Gibson 1973; Kuipers 1973) and frequently make up a large proportion of the fish population of such 'nursery areas'.

The second strategy is to remain in the intertidal zone at low tide. This strategy requires the presence of some form of shelter to alleviate the dangers of exposure, of which predation and desiccation are the most severe. Consequently most species of fish which remain between the tidemarks do so mainly on rocky shores where shelter is provided by space beneath boulders or algae, crevices in rocks and the temporary refuges of rock pools. An important exception to this generalisation is the presence of the amphibious mudskippers (Gobiidae) on tropical shores. They may be present in very large numbers on mudflats where they construct their own shelter in the form of burrows in the mud.

It is possible, therefore, to make a basic distinction between the two types of fishes that can be found intertidally: there are the 'visitors' or 'transients', which are only present for part of the year or the tidal cycle, and the 'residents', which spend most, if not all, of their life there. This chapter describes the behavioural adaptations of resident fishes to life on the sea shore and begins by describing the nature of the environmental changes the fishes are likely to experience and the structure of the fishes themselves. Detailed reviews of the biology of intertidal fishes are given by Gibson (1969, 1982).

Morphological and Physiological Adaptations to Intertidal Life

At high tide, conditions in the intertidal zone will reflect those of the sea and will be relatively uniform as regards such factors as temperature, salinity, pH and oxygen content of the water. The ebbing tide exposes the shore to air, however, and the environment changes suddenly from an aquatic to a virtually terrestrial one. The location of a fish after the tide has receded may therefore be extemely critical. If it is in a rock pool, then ambient conditions may change little over the few hours the pool is isolated from the sea. On the other hand, depending on the air temperature, time of day and prevailing weather conditions, the water in the pool may be heated or cooled, its salinity reduced or increased. At night, its pH and oxygen content may be drastically lowered by respiring algae (Truchot and Duhamel-Jouve 1980; Morris and Taylor 1983). Under such conditions of reduced oxygen, some species possess the ability to move out of pools and respire aerially (Davenport and Woolmington 1981). When out of water the fish can take advantage of the greater concentration of oxygen in air, but at the same time it becomes exposed to the dangers of desiccation. The duration of exposure to such hazards will depend on the vertical position of an individual on the shore, with those at higher levels being subjected to more severe conditions for longer periods than those at lower levels. Consequently resident species are physiologically very versatile. They are eurythermal and euryhaline and often possess the ability to respire equally well in air or water (see references in Gibson 1969, 1982). Many species can survive out of water for several hours and tolerate a loss of water equivalent to 20 per cent or more of their body weight (Horn & Riegle 1981). In some cases the ability to survive out of water has led to an almost terrestrial existence as exemplified by the mudskippers, but also seen in some tropical blennies (Blenniidae) and clinids (Clinidae).

This semi-terrestrial mode of life is accompanied by numerous morphological, physiological and behavioural adaptations, many of which relate to maintaining water and temperature balance. Mudskippers, for example, regulate their body temperature by actively selecting those sites in their habitat that provide optimum thermal conditions (Tytler and Vaughan 1983). The number of secondary gill lamellae is reduced to prevent both gill collapse and evaporation. The skin and bucco-pharyngeal epithelium are often heavily vascularised to promote aerial respiration and the loss of carbon dioxide, the fish never allowing these surfaces to dry. Mudskippers and some blennies frequently roll on their sides on moist substrata or return to the water to wet their skin and renew the water held in their mouth and gill cavity. Other less amphibious species simply remain in moist conditions or just above the water line, where they are continually wetted by waves or spray.

It is interesting that aerial respiration and amphibious behaviour seem to have evolved in marine and freshwater fishes for different reasons (Graham 1976). The great majority of marine air breathers live in an environment where the water is normoxic for most of the time. Their requirement for an air-breathing capability is mainly to survive exposure at low tide and perhaps only secondarily to avoid hypoxic conditions that may arise occasionally in pools. Truly amphibious forms have been able to exploit a habitat where there are few competitors, and there is some evidence (Graham 1973; Zander 1983) that by so doing they also avoid aquatic predators. Air breathing in freshwater fishes, on the other hand, assists survival in habitats where the water frequently becomes hypoxic and may even dry up for long periods. Few freshwater species are amphibious and this may be because the environment around fresh waters has already been fully exploited by the Amphibia themselves. Further discussion of air breathing in marine and freshwater fishes can be found in Graham (1976) and Johansen (1970) respectively.

Turbulence is a further factor of major significance in the lives of intertidal fishes. In contrast to most other factors which are likely to be stressful at low tide, the turbulence caused by wave action will affect fishes only when the intertidal zone is submerged. Only on the calmest day in very sheltered areas is turbulence likely to be absent. On most shores some degree of water movement is always present, and it may vary from a gentle surge to the full force of breaking waves. Life in such conditions imposes certain constraints on morphology and behaviour if the fish are not to be swept away from their habitat or subject to damage by being dashed against rocks. Consequently it is possible to recognise considerable similarity in the body forms of those families that are especially frequent as intertidal residents.

The most important families that have colonised the intertidal zone are the blennies and their relatives the gunnels (Pholidae), pricklebacks (Stichaeidae) and clinids, the gobies (Gobiidae), the clingfishes (Gobiesocidae) and sculpins (Cottidae). The main features that these

Figure 15.1: Drawings of Selected Intertidal Fishes Together with Diagrammatic Cross-sections of the Body at its Deepest Point. A, *Istiblennius lineatus* (Blenniidae, Indian Ocean). B, *Oligocottus maculosus* (Cottidae, north–eastern Pacific). C, *Gobius paganellus* (Gobiidae, western Europe and Mediterranean). D, *Lepadogaster lepadogaster* (Gobiesocidae, south–western Europe and Mediterranean). E, *Pholis gunnellus* (Pholidae, northern Atlantic and Arctic). (Original.)

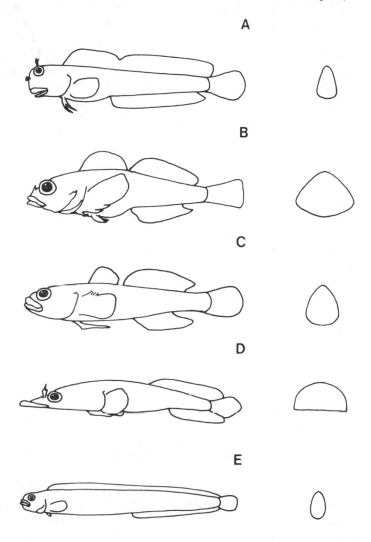

groups have in common (Figure 15.1) are small size (rarely more than 15-20 cm), negative buoyancy resulting from the absence or reduction of the swimbladder, compressed or depressed body form, and some modifications of the fins which act as attachment devices. In the gobies and clingfishes, the pelvic fins are united to form suction pads, whereas in the blennies the rays of the pectoral, pelvic and anal fins are often curved at their distal ends, allowing them to cling to rough surfaces. All these features combined enable the fish to keep close to the bottom where water velocity is lowest, to resist sudden surges and to take advantage of small apertures for shelter if conditions become too harsh. Thigmotaxy, the distinctive behavioural tendency to keep in contact with solid objects, also reduces the amount of time the fish spend away from shelter and ensures that when not active they come to rest in holes, crevices or beneath stones. Coupled with this 'cryptobenthic' mode of life is a style of locomotion in which the fish is propelled as much by its large pectoral fins as by its tail. Movement is therefore in short hops and darts along the bottom, and the fish never undertake sustained bouts of swimming in open water.

Those species which are active out of water have to contend with the problem of the much lower density of air compared with that of water. Air provides no buoyancy, and greater muscular strength is required to move equivalent distances on land. Mudskippers overcome this problem by using their strong pectoral fins to propel themselves forward, and their method of progression is analogous to that of a man on crutches; the pectoral fins are the 'crutches' and the pelvic fins the 'legs'. This is a slow means of loco-motion in which the tail plays little part. If danger threatens, however, the fish escape by skipping rapidly over the ground, propelled forwards and upwards by sudden vigorous thrusts of their tail and pelvic fins (Harris 1960). Amphibious blennies such as the rockskipper *Alticops* (Zander 1967) and the clinid *Mnierpes* (Graham 1970) also move on land using alternate strokes of their long tails, and are capable of jumping many times their own body length when disturbed.

Habitat Selection

One of the keys to survival for mobile foraging species in the intertidal zone is the ability to find and occupy a favourable position at low tide. Many behaviour patterns of intertidal fishes can be interpreted as adaptations for protecting either the individual or its progeny from adverse conditions when the tide is out.

Perhaps the most general way of ensuring that a favourable position is obtained is to select a level on the shore where conditions are likely to correspond to the physiological capabilities which a fish possesses. Species capable of resisting desiccation or high temperatures, for example, will be able to survive at higher levels than others which lack this ability. Differences in physiological performance coupled with differences in behaviour are thus responsible for the differential distribution or zonation of species which have been described in many parts of the world. The two examples of zonation patterns illustrated in Figure 15.2 are taken from regions with a marked tidal range, but even in the Mediterranean, where fluctuations in water level due to tides are small, field observations have

Figure 15.2: Zonation Patterns of Some Common Intertidal Fishes from British Columbia (A, from data in Green 1971a) and the Atlantic coast of France (B, redrawn from Gibson 1972). In (A) the total numbers of fish from a series of tide pools are plotted. In (B) the scales represent the mean number of fish per pool over the range examined. In both cases species which occupy the upper shore levels can be distinguished from those mainly inhabiting the lower levels. The species abbreviations are: A.l., *Artedius lateralis*; B.p., *Blennius pholis*; C.e., *Clinocottus embryum*; C.ga., *Coryphoblennius galerita*; C.gl., *Clinocottus globiceps*; G.c., *Gobius cobitis*; G.p., *Gobius paganellus*; N.l., *Nerophis lumbriciformis*; O.m., *Oligocottus maculosus*; O.s., *Oligocottus snyderi*

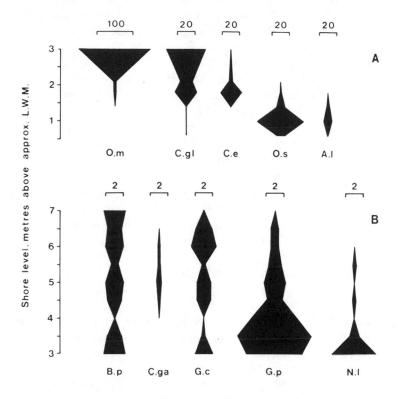

shown that blennies occupy distinct zones and microhabitats within one or two metres' depth (Abel 1962; Zander 1972, 1980).

For the most part, the behavioural mechanisms underlying these zonation patterns are unknown, but there is some experimental evidence which suggests that larvae are able to select particular substrata on which to settle and so might be responsible for the initial choice of habitat. This choice is then refined by the habitat selection behaviour of the meta-morphosed juveniles and adults. Table 15.1 presents the results of a series of experiments by Marliave (1977) in which the planktonic larvae of six species were presented with different substrata and their settling prefer-ences were investigated. There was a marked tendency for the larvae of five of the species to settle on one substratum in preference to others. The exception, *Artedius lateralis*, simply settled on substrata in proportion to their surface area so that most settled on the bottom of the experimental tank. This lack of preference probably reflects the wide variety of habitats occupied by the adult. For the other species, the chosen substratum corresponded to a large extent to that of the adult habitat. Where they did not, as in the case of *Xiphister atropurpureus* and *Gobiesox maeandricus*, the discrepancies could be explained by factors related to body size. *Xiphister atropurpureus* prefers interstitial spaces slightly greater than the cross-section of its body, so that its requirements change with growth, i.e. from gravel to the larger spaces between pebbles and rocks. The attachment disc of very small *Gobiesox maeandricus* is not fully developed and may only be suitable for adhering to smooth surfaces like plant leaves. As the fish grows, the disc develops papillae which enable it to attach to the rougher surfaces of rocks, the adult habitat. Such settling behaviour of the larvae could be responsible for the initial and fairly coarse choice of habitat. It is not yet clear whether the final zonation patterns of adults are the result of precise initial settlement or of random settlement followed either by active choice of the exact zone to be occupied, or by passive elimination of individuals from unfavourable zones, or all three mechanisms.

Much more information is available on the behaviour involved in maintaining position in a habitat once selected. Numerous studies (Gibson 1982) have shown that cover of some form is of overwhelming importance, and in a general sense this serves to keep the fish in areas where shelter is likely to be available over the low-tide period. Requirements for a specific type of cover have also been demonstrated, and by choosing to shelter in an alga, for example, which itself grows at a particular level on the shore, the fish ensures that an appropriate zone is selected.

A demonstration of how different behavioural responses to the same stimulus can lead to a separation of species on the shore is provided by Nakamura's experiments with two sculpins (Nakamura 1976). The common tidepool sculpin *Oligocottus maculosus* and its relative *O. snyderi* can both be found in tide pools on the Pacific coast of North America. The former inhabits the upper part of the shore whereas *O. snyderi* tends to be limited to vegetated pools at lower levels and is also found subtidally (Figure 15.2). When given a choice of three habitat types in aquaria (eelgrass, rocks or open sand), *O. maculosus* showed a strong preference for rocks, and *O. snyderi* greatly preferred the eelgrass (Figure 15.3A). The results were the same whether the species were tested separately or together, showing that the differences in distribution were not caused by aggressive interactions or competition for a particular habitat type. Furthermore, when tested in an experimental depth gradient with a simulated tidal cycle and artificial 'pools', *O. maculosus* was consistently found in the shallower depths whereas *O. snyderi* showed a random distribution with depth (Figure 15.3B). The effect of temperature on distribution was also tested with *O. snyderi*, the more stenothermal of the two species. Raising the temperature in the 'pools' at 'low tide' altered the 'tidepool'

Table 15.1: Substratum Preferences of Settling Fish Larvae. The symbols in the table represent preferred substratum (X); substratum not preferred (0); not investigated (—); particle size in millimetres is given where appropriate. Data from Marliave (1977)

Species and family	Mud	Sand (< 1)	Gravel (2-12)	Substratum Pebbles (20-40)	Rocks (60-100)	Tank bottom	Plants	Adult habitat
Artedius lateralis (Cottidae)	—	0	0	0	—	X	—	All types of habitat, sandy, rocky and mixed beaches, tidepools
Leptocottus armatus (Cottidae)	—	X	—	0	0	0	0	Sandy and muddy grounds
Bothragonus swani (Agonidae)	—	0	X	0	0	0	—	Boulder and gravel beaches, kelp beds, tidepools with rock and gravel
Xiphister atropurpureus (Stichaeidae)	—	0	X	0	0	0	—	Boulder beaches, under rocks with pebble, rock and sand substrata
Pholis laeta (Pholidae)	0	0	0	—	0	0	X	Among intertidal algae and rocks, under boulders and in eelgrass beds
Gobiesox maeandricus (Gobiesocidae)	—	0	0	0	0	0	X	On and under rocks

distribution to lower levels on subsequent 'tides'. These results, together with experiments on the temperature tolerances of the two species, help to explain their observed distribution. The more eurythermal *O. maculosus*, with its preference for shallow water and generalised rock habitats, occupies the upper shore. The stenothermal *O. snyderi*, in contrast, avoids the upper pools because of their unstable temperature regime, even though suitable vegetation may be present, and is consequently restricted to lower levels. Sculpins may be unusual in that they seem to show no intra- or interspecific aggressive behaviour. In other studies of habitat selection such interactions have been observed. The gobies *Gobiosoma robustum* and *G. bosci*, for example, which live in eelgrass and among oysters, alter their habitat preference in the presence of the other species (Hoese 1966).

Comparable results were found with the Californian blennies *Hypsoblennius jenkinsi* and *H. gilberti*. The latter lives intertidally and subtidally on rocky shores and wanders over a large home range. In contrast, *H. jenkinsi* is found only subtidally and inhabits the holes of boring molluscs, gastropod tubes or mussel beds, where it is sedentary and highly territorial. When tested alone in aquaria, *H. jenkinsi* was found frequently among the calcareous tubes of the gastropod *Serpulorbis*. However, *H. gilberti* usually occupied spaces at the bases of the tubes or open sand. When the two species were mixed, these preferences were greatly reinforced (Figure 15.4). The change in distribution when the two species were together is partly explicable by the greater territorial tendencies and aggressive superiority of *H. jenkinsi*. It rarely moved from the immediate area of its 'home' tube, could readily displace *H. gilberti* from its home position, and

Figure 15.3: Habitat Preferences of *Oligocottus maculosus* (Black Columns) and *O. snyderi* (Open Columns) in Aquaria. A, Distribution of the two species when presented with a choice of eelgrass (G), rocks (R) and open sand (S). B, Distribution with respect to water depth. (From data in Nakamura 1976.)

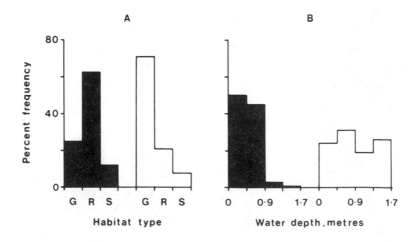

won over 90 per cent of aggressive encounters between the two species (Stephens, Johnson, Key and McCosker 1970).

Habitat Maintenance

Once established in a favourable zone it is clearly an advantage to maintain position there and the simplest way of doing that is not to move very far. If movement has to be over relatively large distances, to obtain sufficient food for example, then it is advantageous to be able to return to the place formerly occupied. Restricted movement is a common feature of intertidal fishes and many studies have demonstrated that some individuals can be found in the same pool at low tide on numerous consecutive occasions. Figure 15.5 illustrates the results of two such studies from widely separated areas. *Clinocottus analis* clearly tended to stay longer in pools than *Acanthocottus bubalis*. The difference in the patterns of residence of the two species illustrated does not necessarily represent a real interspecific difference because several other factors can affect how long a fish inhabits a particular pool. Probably the most important factor regulating residence time is the amount of disturbance to which the pool is subjected. Disturbance is usually caused by wave action which, when severe, can totally alter the suitability of a pool, mostly by removing or altering the

Figure 15.4: Habitat Preferences of *Hypsoblennius jenkinsi* (A) and *H. gilberti* (B) when Tested in Aquaria Alone (Black Columns) and Together (Open Columns). The histograms plot the percentage of observations in which the fish were seen among the tubes of *Serpulorbis* (T), at the base of the tubes (B), or on the open sand (S). (Redrawn from Stephens *et al.* 1970.)

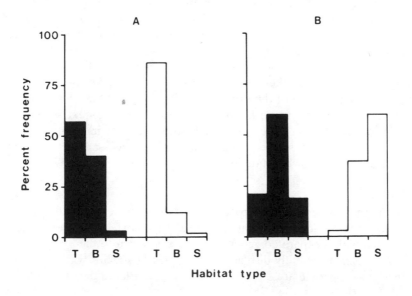

Figure 15.5: Frequency Distribution of the Time that Individual Fish Stay in Tide Pools. Open columns, *Clinocottus analis* in California (from data in Richkus 1978); stippled columns *Acanthocottus bubalis* in Britain. (Redrawn from Gibson 1967.)

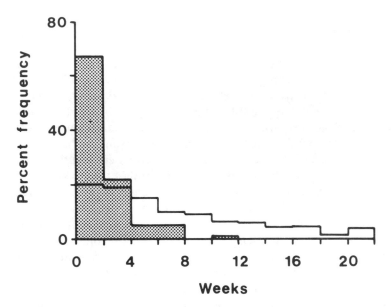

configuration of cover provided by rocks and boulders.

There is, however, a biological factor which affects residence time or fidelity to a particular pool and that is a fish's age. Numerous observations suggest the young fish are more mobile and do not stay in the same pool as long as older ones. The reason for the greater mobility of young fish seems to be that they are at the colonising phase of their life history. When pools are depopulated, either naturally or experimentally, it is usually the young stages that are the first to repopulate the vacant areas. It also seems likely that, for one species at least (*Oligocottus maculosus*: Craik 1981), the young stages during their initial period of extensive movement are acquiring a knowledge of their immediate locality, and that this knowledge is retained for future use in homing as discussed in the next section. The general pattern that emerges from such studies is that individuals tend to move over a rather restricted range which is not necessarily centred upon one particular pool or low-tide location (Figure 15.6).

Homing

It seems that these restricted patterns of movement cannot be accounted for solely by poor locomotory abilities because many species from different families in widely separated geographical areas return to their original location when experimentally displaced. Examples of species which will home to their pool of origin are known in the Cottidae, Blenniidae, Gobiidae and

Figure 15.6: Diagrams Showing the Extent of Movement of Fish between Tide Pools. (A) *Clinocottus analis* over 20 weeks: the thicker the arrows the greater the number of fish which moved between pools. The thinnest line represents two fish (simplified from Richkus 1978). (B) *Blennius pholis* over 15 months. (Redrawn from Gibson 1967.) Note the different scales in the two diagrams

A

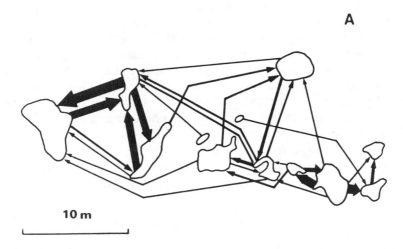

10 m

B

30 m

10

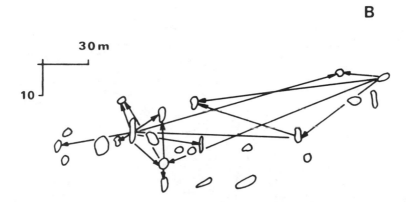

Kyphosidae, from places as far apart as Europe, the Caribbean and the Pacific coast of North America. Such experimental displacement probably mimics the fishes' high-water excursions and provides evidence that they are able to recognise and learn topographical details of their immediate environment and use them in navigating back to their home site. The possession of a topographic memory seems to be a common feature of blennies and gobies, and a detailed knowledge of the environment can be rapidly acquired.

Convincing evidence for both the rapid learning and long-term retention of the knowledge of an area comes from experiments on the small tropical goby *Bathygobius soporator* (Aronson 1951, 1971). This fish can be found either on sandy beaches or in rock pools. When stimulated, fish in pools at low tide will jump out of their pool into another close by with remarkable accuracy and with very few failures (Figure 15.7). The function of the jumping behaviour remains unknown but it provides a convenient means of assessing the learning and orientational capabilities of the fish. Those accustomed to their surroundings will jump readily from one pool to the next, but others transferred to the pool from elsewhere will not unless allowed to familiarise themselves with the area first. Experiments in artificial pools showed that familiarisation is obtained by swimming over the pool surroundings at high tide. Only one tide is necessary for this familiarisation process and the topographic knowledge so acquired can be retained for at least 40 days.

A knowledge of local geography, pressumably acquired visually, does not, however, completely account for homing behaviour because some fish

Figure 15.7: Results of an Experiment Demonstrating the Topographic Knowledge of *Bathygobius soporator*. A fish was disturbed in pool A and after continual stimulation swam through pools B to G jumping over intervening ledges until it reached the open sea at H. (Modified from Aronson 1951.)

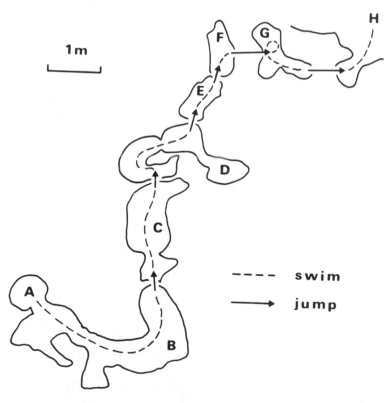

can still home even when displaced over distances greater than they would normally be expected to travel. Experiments with *Oligocottus maculosus* have shown that vision is not essential for this fish to home because blinded individuals can return to their home pool when displaced, although not as successfully as normal ones. In this species olfaction also apparently plays a large role in homing because destruction of the olfactory organs greatly impairs the ability to return to the place formerly occupied (Khoo 1974). These results provide a comparison over short distances with the long-range olfactory homing mechanisms of salmonids (Brannon 1981), and with short-range homing by cyprinids in the laboratory (Kennedy and Pitcher 1975).

The function of homing and fidelity to a particular area seems to be to enable a fish to utilise and return to a location which provides a suitable refuge over the critical low tide period. (See also Chapter 14 by Helfman, this volume.) Tide pools are the most obvious refuges, and all homing studies have been conducted with tide-pool species, mainly, perhaps, because the pools act as natural reservoirs or collecting basins for the fish as well as convenient sampling points for the investigator. The fish are not passively collected in pools by the outgoing tide, however, as the homing studies show, and homing behaviour ensures that the fish return only to those pools which do provide a favourable environment at low tide. To some extent this interpretation of the function of homing by tide-pool fishes is borne out by the few observations that have been made on the species which also occupy habitats outside pools. Such fish, the gunnels and pricklebacks for example, do not seem to show a fidelity to one place for any length of time, and have a much wider range of habitat preference. Hence the need to return to one particular location is not as great.

Territoriality

One other form of restricted movement which ensures that a favourable location is maintained is territoriality, in which a specific area, usually much smaller than a home range, is actively defended against others of the same and sometimes different species (Keenleyside 1979). The possession of a territory guarantees to the holder exclusive rights to feeding, shelter and a spawning site in the defended area. Although food and reproduction are of obvious importance, shelter as a protection against turbulence and predation is equally vital for these small fishes. Territoriality, dominance hierarchies and the aggressive behaviour that accompanies them are particularly common among gobies, blennies, clingfishes and mudskippers, although the degree to which it is developd varies between species and sometimes even between the sexes of one species. Some species chase others away from their immediate vicinity wherever they happen to be; others are rigidly territorial and defend a well-defined area. Several studies have demonstrated experimentally that dominant fish, or those possessing territories, have better access to food and shelter than subordinates. In the striped blenny, *Chasmodes bosquianus*, this priority of access to shelter greatly decreases the risk of predation (Phillips and Swears 1979).

Remarkable examples of territoriality are known among the mud-skippers of tropical and subtropical shores. In Madagascar, *Periophthalmus sobrinus* can be found on the banks of tidal channels among mangrove swamps. At high tide the fish occupy a weakly defended territory at the top of the bank where their main burrow is situated (Figure 15.8). As the tide recedes, the fish move down the banks of the channel, some defending their areas of descent, until they reach the floor of the channel which is mostly left empty by the outgoing tide. Here they have other strongly defended territories in which they feed until the tide returns and they move back up to their burrows (Brillet 1975). In Kuwait, another species, *Boleophthalmus boddaerti* lives on shallow tidal flats in large numbers and demarcates its territory by building mud walls along the boundaries. The walls are formed from mouthfuls of mud which the fish carry from their burrows and deposit at the edge of their territory. In some densely popu-lated areas the boundaries abut one another on all sides and form a striking polygonal mosaic (Clayton and Vaughan 1982).

Reproduction

Although many intertidal species defend territories throughout the year, territoriality is much more common during the reproductive season (see Chapter 10 by Turner, this volume). In virtually all resident species that have been studied, the male selects and defends a territory, where spawn-ing takes place. Spawning females of blennies, gobies, clingfishes and sculpins usually lay eggs in batches, which are deposited individually in a single layer and adhere to the substratum by means of adhesive filaments or special attachment areas. The gunnels and pricklebacks also lay eggs in batches, but these adhere to one another and are not attached to the sub-stratum itself. This habit of laying demersal eggs is common to many shallow-water fishes and may serve as a mechanism for retaining the offspring in or close to their optimum environment. It does mean, however, that the static egg masses of species that spawn in this way are vulnerable to both benthic and pelagic predators, to damage by turbulence and to all the other physical risks that intertidal life imposes on the adults. Consequently the eggs are usually laid in sheltered places where these risks are minimised. Suitable sites are in holes, crevices and on the undersides of stones. It has also been suggested (Wirtz 1978) that such sites offer the added advantage in turbulent water of increasing the probability of sperm reaching and fertilising the ova.

Such cryptic spawning sites, although reducing stress from physical factors, pose several other problems not met by free-spawning pelagic fishes. The males, who generally seem to be responsible for selecting the

Figure 15.8: Territories of the Mudskipper *Periophthalmus sobrinus* in a Tidal Channel of a Madagascar Mangrove Swamp. (A) Cross-section of the channel showing the position of zones referred to in (B). (B) Plan of four territories. Each fish, 1 to 4, had a resting territory (a) on the bank containing its burrow and a feeding territory (c) on the floor of the channel. Fishes 1 and 2 also defended a narrow 'corridor' (b) connecting their resting and feeding territories. (Modified from Brillet 1975.)

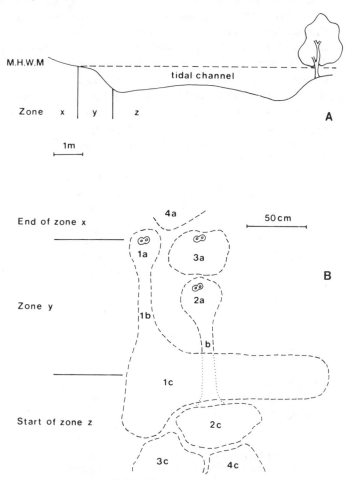

spawning site, must attract females to the chosen location before spawning can begin. Successful rapid attraction of a mate may be particularly important for intertidal fishes because their reproductive activities are limited by the tidal cycle as well as the daily fluctuations in light intensity. This limitation arises partly because the time available for mating may be curtailed over the low-tide period for those individuals which are left out of water, but also because the majority of species are diurnal. Consequently, courtship can only take place during daylight because it relies so heavily on visual communication between the partners. In most groups this initial

attractive phase of reproductive behaviour consists of some form of movement off the bottom, which advertises the male's position and is in marked
contrast to the normal bottom-orientated methods of locomotion. The
mudskippers, for example, leap into the air off the mud (Brillet 1969), and
some blennies swim vertically upwards and may 'loop the loop' (Phillips
1977; Wirtz 1978). In other blennies these vertical movements are reduced
to lifting the anterior portion of the body or rearing up. The head in particular is an important signalling organ during the later phases of courtship in
hole-dwelling blennies, and frequently bears dramatic colour patterns of
contrasting light and dark areas (Figure 15.9). The striking appearance of
the head is often further enhanced by the presence of crests and tentacles.
In addition to these visual signals, recent evidence suggests that some
blennies may also use pheromones secreted from special glands on the anal
fin rays to attract females (Laumen, Pern and Blüm 1974; Zander 1975;
see also Chapter 6 by Hara, this volume).

Once one or more females have been attracted and induced to spawn in
the nest site, the male guards the eggs until they hatch. (See Chapter 11 by
Sargent and Gross, this volume.) Guarding provides the obvious benefit of
deterring predation, but the parental activities of the male also include
active care of the eggs themselves. This care involves removing dead eggs,
keeping developing eggs clean and free from detritus, and providing an
adequate supply of oxygen by fanning water over the egg mass. Fanning is
particularly important when the eggs are laid in confined spaces where
water circulation may be restricted.

Synchronisation of Behaviour with the Tides

The fluctuating nature of the environment in the intertidal zone imposes on
its inhabitants a continual rhythm of change. Certain phases of this rhythm
may be more favourable than others for the performance of particular

Figure 15.9: Diagrammatic Representation of Head Markings of Male Blennies in the Breeding
Season. (A) *Blennius canevae.* (B) *Entomacrodus vermiculatus.* (C) *Lipophrys velifer.* All redrawn
from Zander (1975), Abel (1973) and Wirtz (1980) respectively

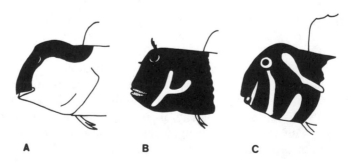

A   B   C

activities. Feeding, for example, is probably best undertaken at high water when movement is easier and a greater variety of food items may be accessible. Some form of synchronisation with the changes in the environment (which to a large extent are predictable) is therefore likely to make more efficient use of the limited periods when conditions are favourable. Synchronisation can be achieved by a direct response to change; flooding of a pool or sudden immersion by the rising tide, for example, could be used to signal the start of that tide's activities, but many species show evidence of the presence of an inherent timing mechanism, 'a biological clock', which is phased with, and operates independently of, current environmental conditions. In the laboratory the presence of this clock manifests itself as a locomotory rhythm in which periods of rest alternate with periods of high activity (Figure 15.10). Such rhythms, which seem to be a common feature of mobile intertidal animals (Palmer 1974), can be further modulated by the light/dark cycle or by interaction with an inherent circadian rhythm. In intertidal fish, whose rhythmic behaviour is much less well known than that of intertidal invertebrates, examples of both types of modulation are known (Gibson 1978). The pattern of activity a fish shows in the laboratory, where most extraneous stimuli are deliberately excluded from experiments, is not necessarily the pattern it is likely to exhibit in nature. Several studies (Green 1971b; Gibson 1975; Taborsky and Limberger 1980) have shown discrepancies between field and laboratory activity patterns. The pattern shown in the field is likely to be a combination of direct responses to concurrent stimuli underlain by a basic

Figure 15.10: An Example of an Endogenous Tidal Rhythm. The activity pattern of a single *Blennius pholis* recorded in darkness in the laboratory. The vertical dotted lines indicate the predicted times of high tide on the shore where the fish was captured (original)

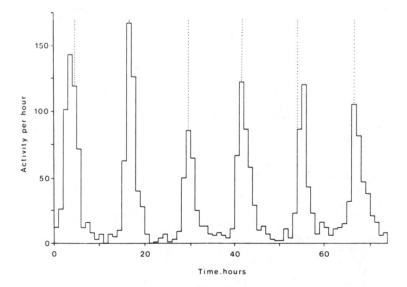

rhythm which regulates the level of behaviour, whatever the behaviour may be. When conditions are exceptionally severe so that no activity is possible, during storms for example, the rhythm may then serve as a true clock so that when conditions return to normal the animal can perform the activity appropriate to the state of the tide.

Summary

Many teleost fish utilise the intertidal zone either as a permanent habitat or as a feeding ground when it is submerged. Those that live there permanently show many morphological, physiological and behavioural adaptations which enable them to survive and reproduce in a habitat subject to regular change. They are small, negatively buoyant and frequently possess fin modifications which allow them to cling to the bottom. To survive the critical low-tide period they are eurythermal and euryhaline; they can also respire efficiently out of water and tolerate considerable water loss. Some have these abilities developed to such a high degree that they lead a semi-terrestrial life. The great majority lay demersal eggs which are attached to the substratum and guarded by the male until they hatch. Behavioural patterns that ensure they acquire and maintain suitable low-tide refuges include thigmotaxy and the ability to select zones and microhabitats on the shore which are compatible with their physiological capabilities. They can learn and remember details of their environment, limit their movements to a restricted area within the selected zone, and return (home) to such areas if displaced. Many species possess an inherent 'biological clock' which is phased with the tides and probably enables them to synchronise their activities with the tidal cycle.

References

Abel, E.F. (1962) 'Freiwasserbeobachtungen an Fischen in Golf von Neapel als Beitrag zur Kenntnis ihrer Ökologie und ihres Verhaltens', *Internationale Revue der Gesamten Hydrobiologie und Hydrographie, 47*, 219-90

Abel, E.F. (1973) 'Zur Öko-Ethologie des amphibisch lebenden Fisches *Alticus saliens* (Forster) und von *Entomacrodus vermiculatus* (Val.) (Blennioidea, Salariidae) unter besonderer Berucksichtigung des Fortpflanzungsverhaltens', *Sitzungsberichte der Österreichischen Akademie der Wissenschaften. Abteil I, 181*, 137-53

Aronson, L.R. (1951) 'Orientation and Jumping Behaviour in the Gobiid Fish *Bathygobius soporator*', *American Museum Novitates, No. 1486*, 22 pp.

Aronson, L.R. (1971) 'Further Studies on Orientation and Jumping Behaviour in the Gobiid Fish *Bathygobius soporator*', *Annals of the New York Academy of Sciences, 1-8*, 378-92

Brannon, E.L. (1981) 'Orientation Mechanisms of Homing Salmonids', in E.L. Brannon and E.L. Sale (eds), *Proceedings of the Salmon and Trout Migratory Behaviour Symposium*, University of Washington, pp. 219-27

Brillet, C. (1969) 'Première Description de la Parade Nuptiale des Poissons Amphibies Periophthalmidae', *Compte rendu hebdomadaire des seances de l'Academie des Sciences, 269D*, 1889-92

Brillet, C. (1975) 'Relations entre Territoire et Comportement Aggressif chez
  Periophthalmus sobrinus Eggert (Pisces, Periophthalmidae) au Laboratoire et en Milieu
  Natural', *Zeitschrift für Tierpsychologie, 39*, 283-331
Clayton, D.A. and Vaughan, T.C. (1982) 'Pentagonal Territories of the Mudskipper
  Boleophthalmus boddarti (Pisces: Gobiidae)', *Copeia, 1982*, 232-4
Craik, G.J.S. (1981) 'The Effects of Age and Length on Homing Performance in the
  Intertidal Cottid, *Oligocottus maculosus* Girard', *Canadian Journal of Zoology, 59*,
  598-604
Davenport, J. and Woolmington, A.D. (1981) 'Behavioural Responses of Some Rocky Shore
  Fish Exposed to Adverse Environmental Conditions', *Marine Behaviour and Physiology,
  8*, 1-12
Gibson, R.N. (1967) 'Studies on the Movements of Littoral Fish', *Journal of Animal
  Ecology, 36*, 215-34
Gibson, R.N. (1969) 'The Biology and Behaviour of Littoral Fish', *Oceanography and
  Marine Biology Annual Review, 7*, 367-410
Gibson, R.N. (1972) 'The Vertical Distribution and Feeding Relationships of Intertidal Fish
  on the Atlantic Coast of France', *Journal of Animal Ecology, 41*, 189-207
Gibson, R.N. (1973) 'The Intertidal Movements and Distribution of Young Fish on a Sandy
  Beach with Special Reference to the Plaice (*Pleuronectes platessa* L.)', *Journal of
  Experimental Marine Biology and Ecology, 12*, 79-102
Gibson, R.N. (1975) 'A Comparison of the Field and Laboratory Activity Patterns of
  Juvenile Plaice', in H. Barnes (ed.), *Proceedings of the 9th European Marine Biology
  Symposium*, Aberdeen University Press, Aberdeen, pp. 13-28
Gibson, R.N. (1978) 'Lunar and Tidal Rhythms in Fish', in J.E. Thorpe (ed.), *Rhythmic
  Activity in Fishes*, Academic Press, London, pp. 201-13
Gibson, R.N. (1982) 'Recent Studies on the Biology of Intertidal Fishes', *Oceanography and
  Marine Biology Annual Review, 20*, 363-414
Graham, J.B. (1970) 'Preliminary Studies on the Biology of the Amphibious Clinid *Mnierpes
  macrocephalus*', *Marine Biology, 5*, 136-40
Graham, J.B. (1973) 'Terrestrial Life of the Amphibious Fish *Mnierpes macrocephalus*',
  Marine Biology, 23, 83-91
Graham, J.B. (1976) 'Respiratory Adaptations of Marine Air-breathing Fishes', in G.M.
  Hughes (ed.), *Respiration of Amphibious Vertebrates*, Academic Press, London, pp.
  165-87
Green, J.M. (1971a) 'Local Distribution of *Oligocottus maculosus* Girard and other Tidepool
  Cottids on the West Coast of Vancouver Island, British Columbia', *Canadian Journal of
  Zoology, 49*, 1111-28
Green, J.M. (1971b) 'Field and Laboratory Activity Patterns of the Tidepool Cottid
  Oligocottus maculosus Girard', *Canadian Journal of Zoology, 49*, 255-64
Harris, V.A. (1960) 'On the Locomotion of the Mudskipper *Periophthalmus koelreuteri*
  (Pallas) Gobiidae', *Proceedings of the Zoological Society of London, 134*, 107-35
Hoese, H.D. (1966) 'Habitat Segregation in Aquaria between Two Sympatric Species of
  Gobiosoma', *Publications of the Institute of Marine Science, University of Texas, 11*, 7-11
Horn, M.H. and Riegle, K.C. (1981) 'Evaporative Water Loss and Intertidal Vertical
  Distribution in Relation to Body Size and Morphology of Stichaeoid Fishes from
  California', *Journal of Experimental Marine Biology and Ecology, 50*, 273-88
Johansen, K. (1970) 'Air Breathing in Fishes', in W.S. Hoar and D.J. Randall (eds), *Fish
  Physiology*, vol. 4, Academic Press, London, 361-411
Keenleyside, M.H. (1979) *Diversity and Adaptation in Fish Behaviour*, Springer-Verlag,
  Berlin, 208 pp.
Kennedy, G.J.A. and Pitcher, T.J. (1975) 'Experiments on Homing in Shoals of the
  European Minnow, *Phoxinus phoxinus* (L.)', *Transactions of the American Fisheries
  Society, 104*, 454-7
Khoo, H.W. (1974) 'Sensory Basis of Homing in the Intertidal Fish *Oligocottus maculosus*
  Girard', *Canadian Journal of Zoology, 52*, 1023-9
Kuipers, B. (1973) 'On the Tidal Migration of Young Plaice (*Pleuronectes platessa* L.) in the
  Wadden Sea', *Netherlands Journal of Sea Research, 6*, 376-88
Laumen, J., Pern, U. and Blüm, V. (1974) 'Investigations on the Function and Hormonal
  Regulation of the Anal Appendices in *Blennius pavo* (Risso)', *Journal of Experimental
  Zoology, 190*, 47-56

Marliave, J.B. (1977) 'Substratum Preferences of Settling Larvae of Marine Fishes Reared in the Laboratory', *Journal of Experimental Marine Biology and Ecology, 27*, 47-60

Morris, S. and Taylor, A.C. (1983) 'Diurnal and Seasonal Variations in Physico-chemical Conditions within Intertidal Rock Pools', *Estuarine Coastal and Shelf Science, 17*, 339-55

Nakamura, R. (1976) 'Experimental Assessment of Factors Influencing Microhabitat Selection by the Two Tidepool Fishes *Oligocottus maculosus* and *O. snyderi*', *Marine Biology, 37*, 97-104

Palmer, J.D. (1974) *Biological Clocks in Marine Organisms*, Wiley, London

Phillips, R.R. (1977) 'Behavioural Field Study of the Hawaiian Rockskipper *Istiblennius zebra* (Teleostei, Blenniidae) I. Ethogram', *Zeitschrift für Tierpsychologie, 32*, 1-22

Phillips, R.R. and Swears, S.B. (1979) 'Social Hierarchy, Shelter Use, and Avoidance of Predatory Toadfish (*Opsanus tau*) by the Striped Blenny (*Chasmodes bosquianus*)', *Animal Behaviour, 27*, 1113-21

Richkus, W.A. (1978) 'A Quantitative Study of Intertidepool Movement of the Wooly Sculpin *Clinocottus analis*', *Marine Biology, 49*, 277-84

Stephens, J.S. Jr, Johnson, R.K., Key, G.S. and McCosker, J.E. (1970) 'The Comparative Ecology of Three Sympatric Species of the Genus *Hypsoblennius* Gill (Teleostomi, Blenniidae)', *Ecological Monographs, 40*, 213-33

Taborsky, M. and Limberger, D. (1980) 'The Activity Rhythm of *Blennius sanguinolentus* Pallas, an Adaptation to its Food Source', *Marine Ecology, Pubblicazioni della Stazione Zoologica di Napoli, 1*, 143-53

Truchot, J.P. and Duhamel-Jouve, A. (1980) 'Oxygen and Carbon Dioxide in the Marine Environment: Diurnal and Tidal Changes in Rock Pools', *Respiration Physiology, 39*, 241-54

Tyler, A.V. (1971) 'Surges of Winter Flounder, *Pseudopleuronectes americanus*, into the Intertidal Zone', *Journal of the Fisheries Research Board of Canada, 28*, 1727-32

Tytler, P. and Vaughan, T. (1983) 'Thermal Ecology of the Mudskipper *Periophthalmus koelreuteri* (Pallas) and *Boleophthalmus boddarti* (Pallas) of Kuwait Bay', *Journal of Fish Biology, 23*, 327-37

Wirtz, P. (1978) 'The Behaviour of the Mediterranean *Tripterygion* Species (Pisces, Blennioidei)', *Zeitschrift für Tierpsychologie, 48*, 142-74

Wirtz, P. (1980) 'A Revision of the Eastern-Atlantic Tripterygiidae (Pisces, Blennioidei) and Notes on some West African Blennioid Fish', *Cybium, 3rd series, 1980*, 83-101

Zander, C.D. (1967) 'Beitrage zur Ökologie und Biologie littoralbewohnender Salariidae und Gobiidae (Pisces) aus dem Roten Meer', *'Meteor' Forschungsergebnisse, D(2)*, 69-84

Zander, C.D. (1972) 'Beiträge zur Ökologie und Biologie von Blenniidae (Pisces) des Mittelmeeres', *Helgoländer wissenschaftliche Meeresuntersuchungen, 23*, 193-231

Zander, C.D. (1975) 'Secondary Sex Characteristics of Blennioid Fishes (Perciformes)'. *Pubblicazioni della Stazione Zoologica di Napoli, 39 (Supplement)*, 717-27

Zander, C.D. (1980) 'Morphological and Ecological Investigations on Sympatric *Lipophrys* Species (Blenniidae, Pisces)', *Helgoländer wissenschaftliche Meeresuntersuchungen, 34*, 91-110

Zander, C.D. (1983) 'Terrestrial Sojourns of Two Mediterranean Blennioid Fish (Pisces, Blennioidei, Blenniidae)', *Senckenbergiana Maritima, 15*, 19-26

16 BEHAVIOURAL ECOLOGY OF STICKLEBACKS

G.J. FitzGerald and R.J. Wootton

Four problems currently dominate the study of animal ecology. First, electrophoretic studies have shown that natural populations are characterised by high levels of genetic variation, but there is controversy about the processes which maintain this variation (Lewontin 1974). Secondly, natural populations show greater or lesser fluctuations in abundance; some of these fluctuations are caused by abiotic factors but some reflect the effects of biotic interactions such as predation and parasitism. The relative importance of abiotic and biotic factors and the effect of the population's own density on its changes in abundance have to be estimated. Age-specific birth and death rates categorise the life-history pattern of a species. This pattern is assumed to have adaptive significance, but the third problem is to develop a theory of life histories that will identify that significance and yield predictions of how the life-history pattern is likely to change in the face of changes in the environment. The fourth problem is to determine the factors that control the number of species that can coexist in a given area, particularly to assess the relative role of deterministic and stochastic processes in the pattern of colonisations and extinctions in a community. Because the behaviour of an animal mediates its interaction with the environment, any solutions to these problems will demand a knowledge of the behavioural responses of the animal to its abiotic and biotic environment (see also Pitcher, Kennedy and Wirjoatmodjo 1979). The sticklebacks (Gasterosteidae) provide unusually favourable material for revealing the importance of the study of behaviour to the analysis of ecological problems. These small fish have a wide geographical distribution. They frequently adapt well to laboratory conditions and they have a sufficiently short lifespan that all stages in their ontogeny can be studied experimentally. Their behaviour not only provides a behavioural ecologist with the raw material for research; it also yields a continuing aesthetic pleasure.

The Gasterosteidae form a natural (monophyletic) family containing five genera (Table 16.1) whose evolutionary relationship to other teleosts is obscure. Their geographical distribution is entirely in the Northern Hemisphere between about 35 and 74°N. In many areas two species are sympatric, and in north-eastern America four or five species may be sympatric. All have similar life histories characterised by a short lifespan and elaborate reproductive behaviour, although in areas of sympatry, the species can show significant differences in their life-history traits (Table 16.2). During the breeding season in late spring and summer, the reproductively active male defends a territory within which it builds a nest. Once eggs are laid in the nest, they are guarded and ventilated by the male. In most species, the

409

female can spawn several times during a breeding season. Some species have populations that are anadromous, moving from coastal waters into fresh or brackish water in late spring to breed, with the young and any surviving adults migrating seaward in autumn (Wootton 1976, 1984).

Sticklebacks also exhibit considerable intra- and interpopulation variation in easily measured morphological characteristics. Breeding experiments show that for many of the characteristics, the variability reflects an underlying genetic variation. A well-studied example is the lateral-plate polymorphism of the three-spined stickleback, *Gasterosteus aculeatus*. Sticklebacks lack typical teleost scales, but the flanks of the body may be arotected by bony lateral plates (Wootton 1976). In *G. aculeatus* three distinct plate morphs are recognised. The completely plated morph (*trachurus*) has a full row of plates that runs from just behind the head on to a keeled caudal peduncle; the partially plated morph (*semiarmatus*) has an anterior row and a caudal row separated by a gap; the low-plated morph (*leiurus* has only an anterior row and an unkeeled caudal peduncle. Within

Table 16.1: The Sticklebacks (Gasterosteidae)

Species	Common names	Geographical distribution
Apeltos quadracus	Four-spined	North America
Culaea inconstans	Five-spined, brook	North America
Gasterosteus aculeatus	Three-spined	North America and Eurasia
Gasterosteus wheatlandi	Black-spotted	North America
Pungitius pungitius	Nine-spined, ten-spined	North America and Eurasia
Pungitius platygaster	Ukrainian	Black and Aral Sea
Spinachia spinachia	Sea, fifteen-spine	Europe

Table 16.2: Selected Life-history Traits of Four Sympatric Sticklebacks from the Isle Verte Area of the St. Lawrence Estuary, Quebec. (Modified from Craig and FitzGerald 1982.) Means ± SD given, details of sample sizes, ranges, etc. given in Craig and FitzGerald (1982)

Trait	*Gasterosteus aculeatus*	*Gasterosteus wheatlandi*	*Pungitius pungitius*	*Apeltes quadracus*
Standard length (mm)	64 ± 4.0	33 ± 2.9	44 ± 3.7	41 ± 6.9
Wet weight (g)	3.958 ± 0.917	0.596 ± 0.189	0.933 ± 0.250	1.042 ± 0.466
Egg number (clutch size)	366 ± 175	80.1 ± 20.3	76.1 ± 22.4	36.1 ± 11.9
Egg diameter (mm)	1.39 ±0.06	1.25 ± 0.10	1.14 ± 0.08	1.40 ± 0.06
Longevity (years)	2+	1+	1+	2+
Breeding season	May and June	May and June	May and June	May to end of July
Post-hatching care of fry	Yes	Yes	Yes	No

each morph, there is significant variation in the number of plates, both within and between populations. Plate morph and life history are correlated. Anadromous populations of *G. aculeatus* are monomorphic for, or usually contain a high proportion of, the completely plated morph. Resident populations in fresh water are commonly monomorphic for the low-plated morph, though in parts of the geographical range, resident freshwater populations monomorphic for the completely plated morph are common (Wootton 1976, 1984; Bell 1978, 1983). Some river systems contain anadromous and resident populations which coexist during the summer period with little or no gene flow between them, thus acting as good biological species (Hagen 1967).

The behavioural processes which may allow the coexistence of such closely related but genetically distinct populations and species include the selection of a habitat, foraging, defence against predation and reproductive behaviour.

Habitat Selection

The habitat of an animal is the complex of physical and biotic factors that determines or describes the place where an animal lives (Partridge 1978). The chosen habitat depends not only on the response of the animal to abiotic factors but also to the biotic factors of food supply, intra- and interspecific competition, predation, parasitism and disease. The presence of other animals may prevent an animal from living in its optimal habitat, defined as that habitat in which it can maximise its lifetime production of offspring. Geographical and historical factors may also prevent dispersal into a potentially optimal habitat, though in the sticklebacks that have a circumpolar range, *G. aculeatus* and *Pungitius pungitius*, the absence from an area is unlikely to be due to a failure to disperse there.

Both *G. aculeatus* and *P. pungitius* occur in streams, ponds, small and large lakes, estuaries, tidal marshes, littoral pools and coastal waters. Recent studies on *G. aculeatus* suggest that the species is split into many genetic units which are reproductively isolated from each other (Bell 1983). Within a genetic unit, the habitat preferences may be much more restricted than is suggested by the range of habitats in which the species is found. This may also be true for *P. pungitius*, but the species has attracted less research. *Apeltes quadracus* and *G. wheatlandi* are confined to sheltered coastal waters, bays, lagoons and a few freshwater sites on the north-eastern seaboard of North America. *Culaea inconstans* is restricted to inland waters of North America east of the Rockies and north from the Great Lakes. *Spinachia spinachia* is found only in coastal waters of western Europe. Although *S. spinachia* lives in low salinities in the Baltic Sea, it is the only stickleback with no freshwater populations (Wootton 1976).

Habitat Cues

Sticklebacks show clear behavioural responses to gradients in temperature and salinity and also respond to low oxygen concentrations. They can show selective responses to vegetation, and such responses become particularly important during the reproductive season. Aggressive interactions between species of sticklebacks and with other species may be important when the reproductively active males are strongly territorial.

(1) Salinity.   Sticklebacks are euryhaline. The least tolerant of high salinities, *C. inconstans*, can live in alkaline prairie lakes tolerating salinities up to about 20 parts per thousand (ppt) (sea water is about 35 ppt). The other species will tolerate salinities up to and greater than sea water. Within a species, there is significant interpopulation variation in tolerance, with the anadromous populations typically tolerating higher salinities than resident freshwater populations (Wootton 1976).

When tested in a salinity gradient, fish from an anadromous population of *G. aculeatus* showed changes in their salinity preferences which correlated with their migratory pattern. In late winter and spring, the fish showed a preference for fresh water. This is when the fish normally move into water of low salinity to breed. By late summer, the preference was for saline water. Young-of-the year, though born in fresh water, developed a preference for saline water two months after hatching (Baggerman 1957).

Responses to a salinity gradient may partly explain the habitat segregation among sympatric species within a restricted geographic area. Worgan and FitzGerald (1981a) found that in Rivière des Vases, a tidal river on the south shore of the St. Lawrence Estuary near Isle Verte, Quebec *G. aculeatus* and *P. pungitius* occurred throughout a 2.5 km stretch of the river including the upstream freshwater stretches. *A. quadracus* and especially *G. wheatlandi* were found only in the downstream brackish water. *A. quadracus* did not occur in nearby tidal pools, although it could tolerate the salinities of the pools. *Gasterosteus aculeatus, G. wheatlandi* and *P. pungitius* were most abundant in the more saline pools closest to the Estuary. Both *G. aculeatus* and *P. pungitius*, but not *G. wheatlandi*, also occurred in freshwater pools about 300 m from the Estuary. A laboratory study showed that the salinity preferences of the four species corresponded with their distribution in the salinity gradient (Audet, FitzGerald and Guderlay 1985).

The ability to select a salinity along a gradient may also be crucial for reproductive success. The optimum salinities for egg hatching in anadromous and resident populations of *G. aculeatus* in Belgium were both restricted and relatively distinct. At 19°C, the maximum hatching success for the resident population occurred at 0.33 ppt, with a second optimum at 10 ppt, whereas the optimum salinities for the anadromous population were at 0.04 ppt and 3.3 ppt (Heuts 1947).

(2) Temperature. In ectothermic animals, temperature is likely to be a major factor limiting their distribution. Although the upper or lower lethal temperature are rarely experienced, the effect of temperature on metabolism will have consequences for feeding, growth and reproduction (Elliott 1981).

Using conditioning techniques, Bull (1957) found that *S. spinachia* could detect temperature increments as small as 0.05°C, a sensitivity that indicates that the stickleback could choose precisely the temperature at which it lives. When *G. aculeatus* collected from tidal pools in Nova Scotia were tested in a vertical temperature gradient, their preferred temperature depended on the acclimation temperature and the water salinity. The value at which the preferred and the acclimation temperatures were the same was 16°C in fresh water but 18°C in sea water (Garside, Heinze and Barbour 1977). In contrast, *G. aculeatus* collected in the Oslofjorden, Norway, had a preferred temperature between 4 and 9°C, irrespective of the acclimation temperature (Roed 1979).

Typically, *C. inconstans* reproduces at temperatures between 15 and 19°C, with reproduction inhibited at temperatures much above 19°C. In spring, the fish move from deep, cold waters into warmer, shallow waters to spawn. In a horizontal temperature gradient, prespawning *C. inconstans* selected a range of temperatures between 14.9 and 20.2°C, but a considerably broader range, 8.9-25.6°C, after spawning (MacLean and Gee 1971).

In tidal pools at Isle Verte, St. Lawrence Estuary, *G. aculeatus*, *G. wheatlandi* and *P. pungitius* ceased breeding when the water temperature rose above 30°C (FitzGerald 1983). Part of the *P. pungitius* population remained in the pools throughout July and August, whereas the adults of the *Gasterosteus* species returned to the Estuary. Perhaps the high (> 30°C) July temperatures were better tolerated by *P. pungitius* than *Gasterosteus*. *G. aculeatus* seemed to be more tolerant of sudden temperature changes than the other two species (A. Gaudreault, personal communication).

The optimum salinity for egg hatching in *G. aculeatus* is a function of temperature (Heuts 1947), which emphasises the importance of the ability of the fish to select appropriate temperature and salinity regimes in which to reproduce.

(3) Dissolved Oxygen. Sticklebacks in fresh water may encounter low oxygen conditions in summer in productive lakes and in winter when the lake surface becomes covered by a blanket of ice and snow. There is no evidence that sticklebacks can detect a gradient in oxygen concentration, but in deoxygenated water they become agitated and the resulting movements may take them into more oxygenated water (Jones 1964). *C. inconstans* and *P. pungitius* make use of localised regions of high oxygen concentration such as the surface film, at cracks in ice cover or the gas

bubbles at the ice-water interface (Gee, Tallman and Smart 1978; Klinger, Magnuson and Gallepp 1982). Collections of *G. aculeatus* and *P. pungitius* from fresh waters in eastern England suggested that *P. pungitius* was more likely to be found in habitats where oxygen concentrations fell to low levels (Lewis, Walkey and Dartnall 1972). In tide pools, *G. aculeatus*, *G. wheatlandi* and *P. pungitius* can tolerate oxygen levels as low as 1 ppm (G.J. FitzGerald, unpublished observations).

Sticklebacks may nest in habitats such as tidal saltmarsh pools with pronounced diel variations in abiotic factors (FitzGerald 1983). The reproductive behaviour of male *G. aculeatus* in a saltmarsh pool was studied over 24 h periods to determine if their behaviour varied with diel changes in dissolved oxygen and temperature (Reebs, Whoriskey and FitzGerald 1984). Although dissolved oxygen concentrations in the tide pools decreased during the night the proportion of time the male spent ventilating the nest by fanning did not differ significantly between night and day. Nocturnal fanning bouts were significantly longer but less numerous than daytime bouts. The decrease in dissolved oxygen and water temperature in the night did not cause an increase in ventilatory behaviour. *G. aculeatus* has evolved behavioural as well as physiological adaptations to cope with unstable environmental conditions, and it would be useful to study the strategies of the other sticklebacks in response to such environmental challenges.

(4) Vegetation and Substrate.   With the exception of *Gasterosteus* and some populations of *P. pungitius*, male sticklebacks build their nests in vegetation (Wootton 1976), so the nature of the vegetation is likely to play an important role in habitat choice. Fish from a freshwater population of *A. quadracus* showed a clear preference for *Elodea* over *Potamogeton* when presented with a choice in a laboratory tank. Both plant species occurred in the stream in which the fish lived (Baker 1971). *A. quadracus* collected from a tidal pool tended to avoid nesting in *Fucus* that was covered with filamentous algae (Courteney and Keenleyside 1983).

Such preferences for types of vegetation or other features of the habitat may be important in allowing the coexistence of closely related but genetically distinct populations or species in relatively small areas.

In Paxton Lake, British Columbia, two distinct phenotypes, labelled 'benthics' and 'limnetics', of *G. aculeatus* coexist within a 17 ha area, but seem to be adapted to different parts of the lake (Larson 1976). The spatial and trophic segregation of the two phenotypes was at least partly due to differences in habitat choice. Both phenotypes bred near the shoreline between April and early July, with nests of the 'limnetics' in open regions and those of 'benthics' in cover. In the summer, the 'limnetics' were usually in shallower water than large 'benthics' and were associated with beds of *Chara* or branches of dead, submerged trees. Small 'benthics' had similar depth and substrate preferences but also occupied open mud areas, whereas larger 'benthics' ($>$ 50 mm) were in deeper water over mud close to cover. In increased cover, 'limnetics' tended to aggregate but 'benthics' dispersed. Both in the field and in laboratory tests, the 'limnetics' associated with floating cover, whereas the 'benthics' associated with cover on the bottom. 'Benthics' were more aggressive than 'limnetics', and this behavioural difference also contributed to the segregation of the two phenotypes. (Two distinct phenotypes in Lake Enos, British Columbia have also recently been described by McPhail and co-workers,' see McPhail, 1984.)

A resident population of the low-plated morph coexists during the summer with an anadromous population of the completely plated morph of *G. aculeatus* in the Little Campbell River, British Columbia. Field observations and choice experiments in the laboratory revealed distinct habitat preferences. The low-plated morph tended to nest in an open location on a mud substrate in still water among plants of *Oenanthe*. The completely plated morph nested on sand, within or just downstream of stands of *Elodea*, usually in water with a measurable flow. These preferences were correlated with considerable spatial segregation of the two morphs within the same small river (Hagen 1967).

Differences in habitat choice based on preferred nesting sites also occur among sympatric species. In tide pools where *G. aculeatus, G. wheatlandi* and *P. pungitius* coexist during the breeding season, *P. pungitius* nests more often in the algae than in open areas, and *G. aculeatus* and *G. wheatlandi* nest more often in open areas (FitzGerald 1983).

Behavioural Interactions

Little is known of the use of space by non-breeding sticklebacks, particularly in the winter (Blouw and Hagen 1981). Anadromous *G. aculeatus* probably shoal (*sensu* Pitcher 1983), but MacLean (1980) found that in a resident population in a small lake in British Columbia some fish maintained feeding territories outside the breeding season. Goyens and Sevenster (1976) found that juveniles from a resident, freshwater population of *G. aculeatus* showed well-developed intraspecific aggression, but juveniles from an anadromous population showed virtually no aggression. Juvenile aggression may serve to disperse the young fish in freshwater populations, but in anadromous populations the young migrate seawards in large shoals.

Interspecific aggression may play a key role in the habitat segregation of sympatric species. *G. aculeatus* could displace both *A. quadracus* and *G. wheatlandi* from their nest sites in laboratory aquaria (Rowland 1983a,b). However, in the pools at Isle Verte, territorial *G. wheatlandi* were more numerous than *G. aculeatus*, particularly early in the breeding season (FitzGerald 1983), so whether such interspecific interactions have the same effect in natural environments remains to be investigated.

Interspecific interactions between sticklebacks and other species may also be important. Gaudreault and FitzGerald (1983) reported intense interspecific aggression between *P. pungitius* and brook trout, (*Salvelinus fontinalis*) fry in the Matamek River, Quebec. Territorial *P. pungitius* attacked fry which were holding feeding territories in the same areas of the river.

In the Chehalis River system in Washington State, the males of an endemic, freshwater esocid fish, *Novumbra hubbsi*, take up breeding territories in the summer. There is probably intense interspecific competition for breeding territories between *N. hubbsi* and *G. aculeatus*, and this may have been the major causal factor in the evolution of the unusual breeding colours of the males in the populations of *G. aculeatus* that are sympatric with *N. hubbsi*. Such males become black during the breeding season, rather than developing the typical breeding colours of red throat and blue-green back (McPhail 1969). Territorial male *N. hubbsi* also adopt a black or dark-brown breeding coloration, and experimental evidence suggested that 'black' stickleback males suffered fewer territorial intrusions by *N. hubbsi* than typically coloured males (Hagen and Moodie 1979; Hagen, Moodie and Moodie 1980).

Significance of Habitat Selection

The ability to select preferred habitats leads to selective segregation between sympatric species that are potential competitors. Such segregation minimises overt competition between the species in the form of exploitation of the same prey species, interspecific territoriality and fighting. The use of a restricted habitat may also allow further adaptive specialisation to and efficient exploitation of that habitat, reinforcing the distinctiveness between sympatric taxa. Any hybrids between the taxa are likely to be at a selective disadvantage in the preferred habitats of the parental taxa which will reinforce their reproductive isolation (Hagen 1967).

Foraging Behaviour

Although the diet of sticklebacks is catholic, their small mouth limits the range of potential prey. Despite the wide geographical distribution and habitat diversity of the sticklebacks, the same types of prey form the major part of the diet. The commonest prey include copepods, cladocerans, ostracods, amphipods, isopods, chironomid larvae and pupae, and the nymphs or larvae of other aquatic insects such as the Ephemeroptera. Algae and other plant material are also eaten (Wootton 1976). (General reviews of foraging behaviour are given in Chapters 8 and 9, by Hart and by Milinski, this volume.)

Detection of Prey

Sticklebacks are primarily visual predators (Wootton 1976, 1984). Prey characteristics which facilitate their detection by a stickleback are size, movement and colour relative to the background.

(1) Size. The distance at which sticklebacks can detect potential prey depends on the size of the prey and on the turbidity of the water. *G. aculeatus* detected 10-mm-long *Asellus* 440 mm away, a tubificid worm at 150 mm and a 2.8-mm-long daphnid at 120 mm. In turbid water, the *Asellus* was only detected 260 mm from the stickleback (Beukema 1968; Moore and Moore 1976; Ohguchi 1981). This dependence on vision means that sticklebacks are essentially diurnal feeders. During the night, feeding either stops or is greatly reduced (Allen and Wootton 1984).

Both field and laboratory observations showed that *S. spinachia* had a preferred prey size which was a function of the size of the stickleback, with the larger fish taking larger mysids. Kislalioglu and Gibson (1976a) found that the cost of handling a mysid, defined as the time taken to manipulate and swallow the prey divided by the prey weight, was a function of the prey and stickleback size (see Chapter 8 by Hart, this volume). There was a strong correlation between the average prey size taken by fish in the wild and the prey size which minimised this cost. Of the *G. aculeatus* in Paxton Lake (see above) the 'benthics' reached a greater size, and at a given size had a larger mouth than the 'limnetics'. Correlated with this difference, there was a difference in diet, with the 'limnetics' tending to feed on zoo-plankton but the 'benthics' feeding on the larger-sized benthic prey, especially amphipods and chironomid larvae. In laboratory tests, 'benthics' took significantly larger amphipods than those consumed by 'limnetics' of the same body length (Larson 1976).

(2) Movement. *Spinachia spinachia* directed more attacks at mysids that were moving than at still ones, and the difference was greatest when the mysid's speed was about 3 mm s^{-1} (Kislalioglu and Gibson 1976b).

(3) Colour and Contrast. Tests in which *G. aculeatus* attacked *Daphnia* that were either red or pale yellow showed that the initial attacks were directed against *Daphnia* of the colour which did not match the yellow or red background (Ohguchi 1981). *G. aculeatus* searching for the white larvae of *Drosophila* showed a sharp drop in the detection rate and an increase in the reaction distance when patches of white cloth were placed on the normally dark bottom of the tank (Beukema 1968). In some situations oddly coloured prey may be taken. *G. aculeatus* tended to attack *Daphnia* that differed in colour from others in the same swarm, irrespective of the colour of the background (Ohguchi 1981).

(4) Palatability. Once the prey has been detected, its palatability will determine whether it is eaten or rejected. Observations on *G. aculeatus* feeding on *Tubifex* showed that if the worm was eaten, the fish tended to

restrict its search for more worms to the area where it had eaten, but if the worm was rejected, the fish moved out of the area (Thomas 1977). When *G. aculeatus* that were used to feeding on *Tubifex* were given highly palatable enchytraeid worms as well as *Tubifex*, the risk of the *Tubifex* being eaten declined sharply once the sticklebacks learned to take the enchytraeids. Even in subsequent feeding sessions in which only *Tubifex* were present, their risk of being eaten was reduced (Beukema 1968).

(5) Apostatic Selection.   The ability of sticklebacks to learn the characteristics of the prey can lead them to feeding selectively on the most common prey form, that is, apostatic selection protecting the rarer morph (Clarke 1962). Experiments in which *G. aculeatus* preyed upon mixtures of pale and dark *Asellus* showed that the most common form was preyed upon at a rate in excess of its relative abundance in the population of *Asellus* (Maskell, Parkin and Verspoor 1977). So, although in some circumstances sticklebacks preferentially prey on odd and conspicuous prey, in other circumstances they preferentially attack the commonest type of prey (Visser 1981). (See also Chapter 12 by Pitcher, this volume.)

Effect of Prey Density

Experiments in which *G. aculeatus* were allowed to attack swarms of *Daphnia* suggested that although the fish could maximise the rate of capture by attacking the densest parts of a swarm, dense swarms also had a 'confusion' effect on the fish. When attacking a dense swarm, a fish would have to pay close attention to the prey at the cost of paying less attention to potential danger. The region of the swarm which the fish attacked was interpreted as being that density at which the costs of being hungry and being at risk were minimised, a density which changed as the fish satiated (Milinski 1977a,b; Milinski and Heller 1978; Heller and Milinski 1979; and see also Chapter 9 by Milinski, this volume).

Prey density can be important in determining which prey sizes are profitable for the stickleback. Two theories attempt to predict prey-size selection by planktivorous fishes. The 'apparent-size' hypothesis assumes that the fish always takes the prey that appears largest, irrespective of its true size. Optimal diet theory predicts that the fish will take the most profitable prey sizes (the prey that yield the highest energy income per unit time), though the theory does not specify the causal mechanisms by which the selection is made (see Chapter 8 by Hart, this volume). When *G. aculeatus* preyed on a mixture of large and small *Daphnia*, the 'apparent size' hypothesis correctly predicted the size selection at low densities of prey, but not at high densities where use of an 'apparent-size' rule would not have maximised the rate of food consumption. Optimal diet theory successfully predicted that at high densities, the fish continued to take both small and large *Daphnia*, probably because even at the high densities the time taken to search for the large prey was never sufficiently short for the rate of food consumption to be maximised by ignoring small *Daphnia* (Gibson 1980). Optimal foraging theory was also reasonably successful at predicting the diet composition when *G. aculeatus* were presented with *Daphnia* and nymphs of the ephemeropteran *Cleon* at various densities (Visser 1982).

Effect of Stickleback Density

Sticklebacks frequently forage in loose shoals (see Chapter 12 by Pitcher, this volume), which tend to become looser as the fish become hungry. If food is discovered by one fish in shoal, the others will rush towards it and start searching in the same area (Keenleyside 1955). During the breeding season, gangs of females and non-reproductive males will raid the nests of males, eating the eggs (Whoriskey and FitzGerald, in press). This social facilitation of foraging may allow the sticklebacks to exploit locally abundant supplies of food such as eggs.

If the distribution of food is patchy, optimal foraging theory predicts that a fish foraging on its own should select the patch with the highest density of prey so that it can maximise its rate of food intake. But when the fish is in a shoal, its rate of feeding in a patch will be inversely proportional to the number of fish feeding in the same patch. In this situation, the decision of an individual fish on where to forage depends on the decisions made by the other fish. If the fish distribute themselves between food patches in the ratio of the patch profitabilities each fish should have the same rate of feeding irrespective of which patch it is feeding in. The fish would be in an ideal free distribution. (For a discussion of this distribution see Milinski 1984, and Chapters 8 and 12 by Hart and by Pitcher, this volume). Fish that deviated from this distribution would have a reduced rate of feeding.

When *Daphnia* were supplied to groups of *G. aculeatus* so that two food patches were formed at the end of an aquarium, the fish distributed themselves between the patches in a ratio which reflected the profitabilities of the patches. As the *Daphnia* were supplied at rates which caused one patch to be five times more profitable than the other, the fish distributed themselves between the patches in a ratio that approximated 5:1. When the profitabilities of the two patches were reversed, the distribution of the fish also changed to reflect the reversal. Thus the sticklebacks seemed to adopt the ideal free distribution. However, a detailed analysis of the behaviour of the sticklebacks showed that two of the assumptions of that distribution were not met. All fish did not have the same rate of feeding and the fish were not of equal competitive ability. The observed distribution and behaviour of the fish could be explained by assuming that the fish were following a rule-of-thumb learning rule whose effect was to produce a distribution close to that expected in an ideal free distribution (Milinski 1984). (See discussion in Chapters 8 and 9 by Hart and Milinski, this volume.)

A potential disadvantage of group foraging is that two or more fish can find themselves competing for the same prey. Milinski (1982) found that when *G. aculeatus* were feeding on *Daphnia*, the good competitors took a higher proportion of large *Daphnia* than less-successful fish. When the fish were subsequently isolated, the poor competitors failed to include a higher proportion of large *Daphnia* in their diet. Milinski suggested that the poor competitor may learn which type of prey it can successfully attack when in competition and then retains that preference even when competitors are not present; however, other explanations are not excluded.

Intertaxon Competition for Food

Although the similarities between the diets of the sticklebacks suggest that in sympatry, intertaxa competition for food is possible, there is little field evidence for such competition (Wootton 1976). In tidal pools in the St. Lawrence Estuary, *G. aculeatus*, *G. wheatlandi* and *P. pungitius* showed significant overlap in their diet, which largely consisted of copepods, chironomid larvae and pupae, corixid adults, gammarids and stickleback eggs. Early in the reproductive period in May and early June, the diet of *G. wheatlandi*, the smallest of the species (Table 16.2), included a higher proportion of copepods, and the mean prey size was significantly smaller than in *G. aculeatus*. In July, after the breeding season, all three species were feeding mostly on gammarids and chironomid pupae, and the mean prey sizes were relatively similar. Experiments in which one or two species were held in enclosures within the tidal pools provided no evidence that the composition of the diet of one species changed significantly in the presence of another (Walsh and FitzGerald 1984). It is unlikely that food is a limiting resource for the sticklebacks in these tidal pools, for when high densities of sticklebacks were kept in enclosures for two months, no significant impact on the macrobenthos was detected (Ward and FitzGerald 1983). Analysis of the stomach contents of fish collected in different years has revealed few empty stomachs (Worgan and FitzGerald 1981b; FitzGerald 1983; Ward and FitzGerald 1983).

Nevertheless, differences in diet may be one of the mechanisms that allow the coexistence of reproductively isolated populations of *G. aculeatus* in rivers or lakes. In the Little Campbell River in July, the anadromous population was feeding on chironomids and floating or swimming organisms whereas the resident population was feeding mostly on bottom-dwelling prey. The anadromous fish had a significantly higher number of gill rakers compared with the residents, which also suggests the former are better adapted for feeding pelagically (Hagen 1967). In Paxton Lake there were significant differences between the diets of the 'limnetics' and 'benthics', which were correlated with differences in morphology and habitat (Larson 1976).

Conclusion

Experimental studies on *G. aculeatus* have shown that both growth rate and the total fecundity of the females are a function of the rate of food consumption (Wootton 1984). Some aspects of the reproductive behaviour of the males are also sensitive to food availability. Thus the reproductive success of a population is dependent on the food supply experienced both before and during the breeding season. This implies that a stickleback population will 'track' changes in its food supply closely, so that changes during the breeding season will be quickly reflected in changes in reproductive output.

Defence against Predation

Predation probably has been and is a major selective factor in the evolution of the sticklebacks. Adult sticklebacks are preyed on by piscivorous birds, mammals, reptiles and large carnivorous fish. Eggs, larvae and juveniles, because of their small size, have a wider range of enemies. During the breeding season, sticklebacks themselves are significant predators of their eggs and larvae. Defence against this range of predators has both structural and behavioural components, with the relative importance of these varying both between species and between populations within a species (Wootton 1976, 1984).

Structural Defence against Predation

The structural adaptations take two forms. The first is the cryptic coloration which characterises sticklebacks (with the exception of males in breeding colours), and reduces the chance that the stickleback will be detected by the predator. The effectiveness of this crypsis is augmented by the tendency of sticklebacks to remain still, moving only in short, slow bursts by sculling with their large pectoral fins.

In some populations of *G. aculeatus* there is evidence that heavy predation by trout has led to the evolution of males that fail to develop the typical breeding coloration. In Lake Wapato (Washington State), only about 15 per cent of the males had the typical red throat. Rainbow trout (*Salmo gairdneri*) were major predators of the sticklebacks in this lake. Choice tests showed that the courting females preferred males with red throats to other males, but trout were more likely to attack the red males (Semler 1971; Moodie 1972b).

The second structural adaptation consists of the skeletal defences, especially the dorsal and pelvic spines and their associated skeletal elements, but also the lateral plates. Experiments have shown that predator fish take species that lack spines in preference to sticklebacks, but that if the sticklebacks are de-spined, the predators take them as readily as species that lack spines (Hoogland, Morris and Tinbergen 1957; Reist 1980a). An important feature of the spines is that they can be locked when erect so that a predator cannot depress them but may break them if it can exert sufficient force. The anterior lateral plates, when present, may act to stabilise the erect spines and so reduce tissue damage if the spines are displaced by the action of the predator (Reimchen 1983). For *G. aculeatus*, a positive correlation between spine length and the presence of fish predators has been found both in North America (Hagen and Gilbertson 1972) and Europe (Gross 1977). In the low-plated morph, the number of anterior lateral plates also tends to be higher when predator fish are present (Reimchen 1983).

Behavioural Defence against Predation

Behavioural defence has two components: behaviour that reduces the chances of the stickleback being detected, and behaviour that reduces the chance of capture after detection. Of three species that have been studied in some detail, *G. aculeatus* seems to rely less on cryptic behaviour and more on its morphological and escape capacities to evade predation. *P. pungitius* and *C. inconstans* both tend to be associated with thick vegetation and may avoid open areas. Correlated with this behavioural difference, the spines and lateral plates in *P. pungitius* and *C. inconstans* are smaller and less formidable than those of *G. aculeatus*. Experiments showed that small fish predators, perch (*Perca fluviatilis*) and pike (*Esox lucius*) took *P. pungitius* in preference to *G. aculeatus* (Hoogland *et al.* 1957). The behaviour of *G. aculeatus* may expose it to greater risks. In the St. Lawrence Estuary, the diet of black-crowned night herons (*Nycticorax nycticorax*) included *G. aculeatus* but not the sympatric *G. wheatlandi*, *P. pungitius* and *A. quadracus*, (FitzGerald and Dutil 1981). In areas of sympatry, where there may be competition for nest sites or food, selective predation on an ecologically dominant or behaviourally dominant species may diminish interspecific competition (Rowland 1983a,b).

Sticklebacks in a loose shoal respond to a disturbance by clumping more closely together and swimming away in a polarised school (Kenleyside 1955, 1979; Pitcher 1983). This may generate a 'confusion' effect in the predator (see Chapter 12 by Pitcher, this volume).

A solitary stickleback encountering a potential fish predator shows a distinct repertoire of behaviour. The stickleback stops and fixes the predator with its binocular vision. *G. aculeatus* tends to react to the predator at a shorter distance than *P. pungitius* and is more likely to approach. Two extreme types of escape behaviour can be shown. The stickleback may flee, jumping away in an unpredictable direction (protean display), or it may freeze, becoming completely motionless except for slight opercular movements. *P. pungitius* and *C. inconstans* are more likely to conceal themselves in vegetation than *G. aculeatus*. The spines are usually erected during the encounter, and even if the stickleback is captured, the erect spines may be sufficient to prevent the predator swallowing the stickleback. The spines are more effective against small predators and those predators such as perch and trout that tend to manipulate the stickleback in the mouth. Sticklebacks can form a high proportion of the diet of large predators, which indicates the relative ineffectiveness of the spines against such predators (Wootton 1976).

Sticklebacks that have had prior experience of a predator will react at a greater distance than naive sticklebacks. The length of time that the stickleback is in the presence of its father can also have an effect. *G. aculeatus* whose father had been removed at hatching were more easily frightened in their first encounter with a fish predator than sticklebacks whose father had cared for them as fry in the first few days after hatching. This effect tended to disappear after the stickleback had experienced a predator (Benzie 1965).

In male *G. aculeatus*, there is evidence of a positive correlation between the level of aggressive behaviour shown in the breeding season and the 'boldness' the male will show in the presence of a predator or when placed in an unfamiliar environment. This correlation is found both within populations and between populations. In Scottish populations that were judged to differ in the extent that they were exposed to fish and bird predation, the mean levels of aggression and 'boldness' were higher in the populations that experienced the lowest predatory pressures. Though relatively high levels of aggression and conspicuousness may be advantageous in competing for and defending a territory against conspecifics and attracting a mate, it may be disadvantageous if it leads to easier detection by a visually hunting predator. The correlation between aggression and 'boldness' may provide an economical way of adjusting the levels of intra- and interspecific aggression in relation to predatory pressures (Huntingford 1976, 1977, 1982).

Giles and Huntingford (1984) also found significant differences in the response of sticklebacks from several Scottish populations to both a live pike and a model heron. There was some correlation between the levels of avian and fish predation that the populations experienced and the level of response to the two types of predator. The response to the model of the avian predator developed in young fish independently of specific experience of the predator (Giles 1984a).

'Boldness' towards potential predators also increases when the males have eggs in the nest. Pressley (1981) presented models of a predator, the prickly sculpin (*Cottus asper*), to male *G. aculeatus* which were at different stages in the reproductive cycle and had different numbers of eggs in their nests. Males with fewer or younger eggs in their nests were more likely to desert than males with more or older eggs. In Lake Wapato, males with eggs or fry were less likely to desert when a predatory rainbow trout was manoeuvred near them than males with empty nests (Kynard 1978). This increase in 'boldness' with increasing reproductive investment may be adaptive because males that lose their eggs early in a parental cycle will retain the nest and revert to courting females, but later in the parental cycle, the male destroys the nest and may have to re-establish a territory before it can resume courtship behaviour.

The relative balance between behavioural and morphological defences against predation can vary intraspecifically. There are populations of *G. aculeatus, P. pungitius* and *C. inconstans* that show a reduction or complete loss of the pelvic spines and pelvic skeleton, and sometimes a reduction in the number of dorsal spines. Such populations are not restricted to habitats that are free of predators. Experiments in which phenotypes of *C. inconstans* that possessed or lacked pelvic spines were exposed to predation by small pike or by invertebrate predators suggested that, though the pike preferred to take fish lacking pelvic spines, the behaviour of the spineless phenotype made it less likely to be captured. The lack of spines and, in *G. aculeatus* lack of anterior lateral plates may improve the streamlining and manoeuvrability of the spineless or plateless phenotypes (Reist 1980a,b, 1983).

Conclusion

There is now good evidence that predation has acted as a major selective influence in the evolution of the sticklebacks. It may play a role both in allowing sympatric species to coexist and in maintaining morphological and behavioural variability within a species. Its role in determining the abundance of stickleback populations cannot yet be evaluated because of a lack of adequate quantitative data.

Behavioural Ecology of Reproduction

The reproductive biology of the sticklebacks is characterised by the division of behavioural roles between the males and females. The males construct and defend a nest in which the eggs are concealed from predators. Ventilatory behaviour by the male ensures that the eggs are provided with a flow of oxygenated water and kept clear of detritus. Courtship takes place within a male's territory and provides an opportunity for male and female to exert a choice of mate. Apart from the courtship the female is free to forage for food and so increase her fecundity during the breeding season (Wootton 1976). Females may also be more cautious than males in the presence of potential predators because they do not have to defend a nest and its contents from conspecifics (Giles 1984b). (For a general review of mating strategies in teleosts, see Chapter 10 by Turner, this volume.)

Mate Choice

Courtship provides a mechanism by which reproductive isolation between closely related taxa can be maintained without the wastage of gametes. In the Little Campbell River, although the resident low-plated morph and the anadromous completely plated morph of *G. aculeatus* have different habitat preferences (see above), the heterogeneity of the environment does bring them into contact so that some hybridisation occurs. However, mate-choice tests showed that there was significant positive assortative mating. Both sexes showed a significant preference for mating with a partner of the same morph. During the courtship, there were differences in the behaviour of the anadromous and resident males which may provide a basis for this positive assortative mating. Significantly, the initial response of a male to a gravid female differed between the two morphs, which suggests that the differentiation between morphs can be made early in courtship (Hay and McPhail 1975; McPhail and Hay 1983).

Courtship also contributes to the reproductive isolation between sympatric species. Differences in the male breeding colour, form of court-ship dance to the female and the way in which the male leads her to the nest are all cues that can be used to prevent crosses between *P. pungitius*, *G. wheatlandi* and *G. aculeatus*. Attempts at interspecific crosses usually break down early in courtship, frequently with the female failing to respond to the approach of a heterospecific male (Wootton 1976).

Territorial, Nesting and Parental Behaviour

The courtship takes place within a territory established by the male. A territory is essential for the reproductive success of a male. Although terri-torial males may steal fertilisations in the nests of other males, there are no reports of non-territorial males achieving this (Wootton 1976).

The choice of a nest site is also crucial for a male's reproductive success. Sargent and Gebler (1980) allowed male *G. aculeatus* to compete for a limited number of flower pots as nest sites in circular wading pools. Males nesting inside pots spawned earlier and more often, had a higher mean and lower variance for hatching success and suffered fewer stolen fertilisations, nest raids and territorial encounters than males nesting outside pots. In natural populations of *G. aculeatus*, concealed nests were more likely to contain fry than nests in open areas, probably because the concealed nests suffered less intraspecific egg predation (Moodie 1972a; Kynard 1978).

Both experimental and field studies on *G. aculeatus* have suggested that phenotypes of the low-plated morph with different numbers of lateral plates nest in different microhabitats because of variations in their compe-titive abilities. Some plate phenotypes are more likely to nest in the pre-ferred sites which are in or near cover and in relatively deep water (Moodie 1972a; Kynard 1978, 1979). Plate phenotypes may also differ in aggressiveness (Huntingford 1981).

Suitable nest sites may be in short supply in certain habitats. In tidal salt-marsh pools along parts of the eastern seaboard of North America, three and sometimes four species may be trying to build nests simultaneously. FitzGerald (1983) found that during the month of May reproductively mature *P. pungitius* did not reproduce in the pools, presumably because of a lack of algae, whereas *G. aculeatus* and *G. wheatlandi* nested in open areas. When algae appeared in the pools in early June, four or five male *P.*

pungitius per square metre were able to establish territories and spawn.

The density of males and food supply can also affect the proportion of males that can establish territories and build nests. Experiments with *G. aculeatus* suggest that if the area available for territories is restricted, the proportion of males that become territorial and successfully build a nest is inversely related to their density (van den Assem 1967; Stanley 1983). At low food levels the proportion is also reduced. At high densities and food levels, some males that are physiologically capable of building nests fail to do so, although if isolated they will then build. At low food levels, some males become physiologically unable to build nests even in isolation (Stanley 1983).

In a relatively homogeneous environment, breeding success can be correlated with territory size. Laboratory studies showed that *G. aculeatus* males with large territories were more likely to be successful in courtship, suffered fewer egg losses and had a better chance of hatching their clutches than males with small territories (van den Assem 1967; Black 1971). Thus both nest location and territory size contribute to reproductive success, though the former may be more important (Sargent and Gebler 1980; Sargent 1982). No evidence of a relationship between male quality (e.g. size) and nesting or reproductive success independent of nest location was found by Sargent (1982).

The importance of nest location was also illustrated in Kynard's (1978) study of the population of *G. aculeatus* in Lake Wapato. Males nesting in rocky areas had fewer eggs in their nest than males from vegetated areas. The reproductively active males tended to clump in the preferred habitats, so that some areas which superficially looked suitable for nesting sticklebacks were underutilised. A similar clumping of *G. aculeatus* and *G. wheatlandi* nests was observed in tidal pools, which again suggests that nest location is more important than territory size (FitzGerald 1983). There was a negative correlation between internest distance and the number of eggs in the nest of *G. aculeatus* in the tidal pools, whereas in Lake Wapato there was a positive correlation (Kynard 1978). In the tidal pools, some males that were not observed in aggressive interactions with other males nevertheless had high numbers of eggs in their nests (F.G. Whoriskey, personal communication).

A fuller understanding of the relationships between male phenotype, territory size, nest location and reproductive success will require further experimental and field analysis.

There is preliminary evidence for both intra- and interspecific competition in the tidal pools, but the relative intensity of the intraspecific competition seems to differ between *G. wheatlandi* and *G. aculeatus*. When the densities were manipulated in pools that contained only one of the two species, there was no consistent relationship between the number of eggs per square metre and fish density over a range of densities from 4 to 32 fish m^{-2} (sex ratio $1:1$) in both *G. wheatlandi* and in *G. aculeatus* (Whoriskey

1985). The proportion of the egg biomass that was consumed by stickle-backs increased significantly with fish density in pools that contained only *G. aculeatus* but not in pools containing *G. wheatlandi* (Whoriskey and FitzGerald, in press). A similar experimental analysis of interspecific competition has shown that, *G. aculeatus* does attack and destroy *G. wheatlandi* nests, but there is no evidence of the smaller *G. wheatlandi* destroying *G. aculeatus* nests (FitzGerald and Whoriskey, in press).

A feature of the reproductive biology of sticklebacks is the phenomenon of nest raiding and egg cannibalism. In *G. aculeatus* both breeding and non-breeding males and females will raid nests. Sometimes a reproductively active male will carry stolen eggs back to his nest, but most frequently the stolen eggs are eaten (Wootton 1976; Kynard 1978). Parental males do raid, but since the eggs in the stomachs of parental males are usually at a different stage of development from those in the nest (Kynard 1978), the speculation by Rohwer (1978) that males consume some eggs from their own nest to keep themselves in good condition for further reproductive cycles seems unfounded. Ridley and Rechten (1981) found that females preferred to spawn in nests that already contained eggs, which may be an advantage because there is evidence that defence of the nest is stronger when the number of eggs is greater (Pressley 1981). This may also reflect choice of competent males (see Chapter 10 by Turner, this volume). Under artificially high densities for a single species (32 fish m^{-2}), nest raiding by females can lead to destruction of all the nests in *G. aculeatus* pools but not in *G. wheatlandi* pools (Whoriskey and FitzGerald, in press). The ecological significance of nest raiding and cannibalism still has to be clarified.

The parental success of a male will also depend on his ventilatory behaviour (fanning in all the species except *A. quadracus*, see Wootton 1976). Experimental studies of *G. aculeatus* suggest that the average length of the fanning bouts and an even temporal distribution of fanning bouts and interfanning intervals may contribute to hatching success (Wootton 1976; Sargent and Gebler 1980). Thus the optimal allocation of available time between nest ventilation, nest defence and foraging during the parental phase of a male poses a significant problem for analysis.

Kynard (1978) drew attention to the high production of young by the sticklebacks in Lake Wapato during the early part of the breeding season, when males were rearing a high percentage of their eggs (76.8 per cent) through to fry. On average each male was producing about 400 free-swimming fry. The potential for numerical increase by *G. aculeatus* was illustrated when 4000 were put in Marion Lake, a small forest lake in British Columbia. The following year, the population was estimated at 120 000 (McPhail, cited in Krebs 1978).

Summary

Many of the major themes of behavioural ecology are illustrated by research on the sticklebacks. Significant progress in the understanding of habitat selection, foraging behaviour, defence against predation and reproductive behaviour has resulted from these studies. The suitability of sticklebacks for both laboratory and field studies helps to ensure a close relationship between theory, experiment and observation. For many years, studies on the behaviour of sticklebacks concentrated on the problem of causation. Recently the ecological and evolutionary significance of their behaviour has been appreciated. They provide such suitable material that their study is likely to yield as much significant information for behavioural ecology and sociobiology as the original studies did for the young discipline of ethology.

Acknowledgements

We thank H. Guderley and F.G. Whoriskey for their valuable comments on the manuscript. R.J.W. thanks the British Council for an Academic Travel Grant which made possible a trip to Université Laval (Quebec).

References

Allen, J.R.M. and Wootton, R.J. (1984) 'Temporal Patterns in Diet and Rate of Food Consumption of the Three-spined Stickleback (*Gasterosteus aculeatus*) in Llyn Frongoch, an Upland Welsh Lake', *Freshwater Biology, 14*, 335-46

Audet, C., FitzGerald, G.J. and Guderley, H. (1985) 'Salinity Preferences of Four Sympatric Sticklebacks (Pisces: Gasterosteidae) during their Reproductive Season', submitted

Baggerman, B. (1957) 'An Experimental Study on the Timing of Breeding and Migration in the Three-spined Stickleback', *Archives neerlandaises de Zoologie, 12*, 105-317

Baker, M.C. (1971) 'Habitat Selection in Fourspine Sticklebacks (*Apeltes quadracus*)', *American Midland Naturalist, 85*, 239-42

Bell, M.A. (1976) 'Evolution of Phenotypic Diversity in *Gasterosteus aculeatus* Superspecies on the Pacific Coast of North America', *Systematic Zoology, 25*, 211-27

Bell, M.A. (1983) 'Evolutionary Phenetics and Genetics: Threespine Stickleback, *Gasterosteus aculeatus*, and Related Species', in B.J. Turner (ed.), *Evolutionary Genetics of Fishes*, Plenum Press, New York

Benzie, V.L. (1965) 'Some Aspects of the Anti-predator Responses of Two Species of Stickleback', Unpublished D.Phil. thesis, University of Oxford

Beukema, J.J. (1968) 'Predation by the Three-spined Stickleback (*Gasterosteus aculeatus*): the Influence of Hunger and Experience', *Behaviour, 31*, 1-126

Black, R. (1971) 'Hatching Success in the Three-spined Stickleback (*Gasterosteus aculeatus*) in Relation to Changes in Behaviour during the Parental Phase', *Animal Behaviour, 19*, 532-41

Blouw, D.M. and Hagen, D.W. (1981) 'Ecology of the Fourspine Slickleback, *Apeltes quadracus*, with Respect to a Polymorphism for Dorsal Spine Number', *Canadian Journal of Zoology, 591*, 1677-92

Bull, H.O. (1957) 'Behavior: Conditioned Responses', in M.E. Brown (ed.), *The Physiology of Fishes*, Academic Press, London, pp. 211-28

Clarke, B. (1962) 'Balanced Polymorphism and the Diversity of Sympatric Species', *Publications of the Systematics Association, 41*, 47-70

Courteney, S.C. and Keenleyside, M.H.A. (1983) 'Nest Site Selection by the Fourspine Stickleback, *Apeltes quadracus*', *Canadian Journal of Zoology, 61*, 1443-7

Craig, D. and FitzGerald, G.J. (1982) 'Reproductive Tactics of Four Sympatric Sticklebacks (Gasterosteidae)', *Environmental Biology of Fishes, 7*, 369-75

Elliott, J.M. (1981) 'Some Aspects of Thermal Stress on Freshwater Teleosts', in A.D. Pickering (ed.), *Stress and Fish*, Academic Press, London, pp. 209-45

FitzGerald, G.J. (1983) 'The Reproductive Ecology and Behavior of Three Sympatric Sticklebacks (Gasterosteidae) in a Saltmarsh', *Biology of Behaviour, 8*, 67-79

FitzGerald, G.J. and Dutil, J-D. (1981) 'Evidence for Differential Predation on an Estuarine Stickleback Community', *Canadian Journal of Zoology, 59*, 2394-5

FitzGerald, G.J. and Whoriskey, F.G. (in press) 'The Effects of Interspecific Interactions upon Male Reproductive Success in Two Sympatric Sticklebacks *Gasterosteus aculeatus* and *G. wheatlandi*', *Behaviour*

Garside, E.T., Heinze, D.G. and Barbour, S.E. (1977) 'Thermal Preference in Relation to Salinity in the Threespine Stickleback, *Gasterosteus aculeatus* L., with an Interpretation of its Significance', *Canadian Journal of Zoology, 53*, 590-4

Gaudreault, A. and FitzGerald, G.J. (1983) 'The Interactions of Brook Trout Fry (*Salvelinus fontinalis*) and Nine-spine Sticklebacks (*Pungitius pungitius*) in the Matamek River, Quebec', *Woods Hole Oceanographic Institute Technical Report, WH 01-83-371*, 65-80

Gee, J.H., Tallman, R.F. and Smart, H.J. (1978) 'Reactions of Some Great Plains Fishes to Progressive Hypoxia', *Canadian Journal of Zoology, 56*, 1962-6

Gibson, R.M. (1980) 'Optimal Prey-size Selection by Three-spined Sticklebacks (*Gasterosteus aculeatus*): a Test of the Apparent Size Hypothesis', *Zeitschrift für Tierpsychologie, 52*, 291-307

Giles, N. (1984a) 'Development of the Overhead Fright Response in Wild and Predator-naive Three-spined Sticklebacks, *Gasterosteus aculeatus* L.', *Animal Behaviour, 32*, 276-9

Giles, N. (1984b) 'Implications of Parental Care for the Anti-predator Behaviour of Adult Male and Female Three-spined Sticklebacks, *Gasterosteus aculeatus* L.', in G.W. Potts and R.J. Wootton (eds), *Fish Reproduction: Strategies and Tactics*, Academic Press, London, pp. 275-89

Giles, N. and Huntingford, F.A. (1984) 'Predation Risk and Interpopulation Variation in Anti-predator Behaviour in the Three-spined Stickleback, *Gasterosteus aculeatus* L.', *Animal Behaviour, 32*, 264-75

Goyens, J. and Sevenster, P. (1976) 'Influence du Facteur Hereditaire et de la Densité de Population sur l'Ontogénèse de l'Aggressivité chez l'Epinoche (*Gasterosteus aculeatus* L.)', *Netherlands Journal of Zoology, 26*, 427-31

Gross, H.P. (1977) 'Adaptive Trends of Environmentally Sensitive Traits in 3-spined Stickleback, *Gasterosteus aculeatus* L.', *Zeitschrift für Zoologische Systematik und Evolutionsforschung, 15*, 252-77

Hagen, D.W. (1967) 'Isolating Mechanisms in Three-spine Stickelbacks (Gasterosteidae)', *Journal of the Fisheries Research Board of Canada, 24*, 1637-92

Hagen, D.W. and Gilbertson, L.G. (1972) 'Geographic Variation and Environmental Selection in *Gasterosteus aculeatus* L. in the Pacific Northwest, America', *Evolution, 26*, 32-51

Hagen, D.W. and Moodie, G.E.E. (1979) 'Polymorphism for Breeding Colours in *Gasterosteus aculeatus* I. Their Genetics and Geographical Distribution', *Evolution, 331*, 641-8

Hagen, D.W., Moodie, G.E.E. and Moodie, P.F. (1980) 'Polymorphism for Breeding Colours in *Gasterosteus aculeatus* II. Reproductive Success as a Result of Convergence for Threat Display', *Evolution, 34*, 1050-9

Hay, D.E. and McPhail, J.D. (1975) 'Mate Selection in Three-spine Sticklebacks (*Gasterosteus*)', *Canadian Journal of Zoology, 53*, 441-50

Heller, R. and Milinski, M. (1979) 'Optimal Foraging of Sticklebacks on Swarming Prey', *Animal Behaviour, 27*, 1127-41

Heuts, M. (1974) 'Experimental Studies on Adaptive Evolution in *Gasterosteus aculeatus* L.', *Evolution, 1*, 89-102

Hoogland, R.D., Morris, D. and Tinbergen, N. (1957) 'The Spines of Sticklebacks

(*Gasterosteus* and *Pygosteus*) as a Means of Defence against Predators (*Perca* and *Esox*)', *Behaviour*, *10*, 205-37

Huntingford, F.A. (1976) 'The Relationship between Anti-predator Behaviour and Aggression among Conspecifics in the Three-spined Stickleback, *Gasterosteus aculeatus*', *Animal Behaviour*, *24*, 245-60

Huntingford, F.A. (1977) 'Inter- and Intraspecific Aggression in Male Sticklebacks', *Copeia*, *1977*, 158-9

Huntingford, F.A. (1981) 'Further Evidence for an Association between Lateral Scute Number and Aggressiveness in the Threespine Stickleback, *Gasterosteus aculeatus*', *Copeia*, *1981*, 717-19

Huntingford, F.A. (1982) 'Do Inter- and Intraspecific Aggression Vary in Relation to Predation?', *Animal Behaviour*, *30*, 909-16

Jones, J.R.E. (1964), *Fish and River Pollution*, Butterworths, London

Keenleyside, M.H.A. (1955) 'Some Aspects of the Schooling Behaviour of Fish', *Behaviour*, *8*, 183-248

Keenleyside, M.H.A. (1979) *Diversity and Adaptation in Fish Behaviour*, Springer-Verlag, Berlin

Kislalioglu, M. and Gibson, R.N. (1976a) 'Prey "Handling Time" and its Importance in Food Selection by the 15-spined Stickleback, *Spinachia spinachia* (L.)', *Journal of Experimental Marine Biology and Ecology*, *25*, 151-8

Kislalioglu, M. and Gibson, R.N. (1976b) 'Some Factors Governing Prey Selection by the 15-spined Stickleback, *Spinachia spinachia* (L.)', *Journal of Experimental Marine Biology and Ecology*, *25*, 159-70

Klinger, S.A., Magnuson, J.J. and Gallepp, G.W. (1982) 'Survival Mechanisms of the Central Mudminnow (*Umbra limi*), Fathead Minnow (*Pimephales promelas*) and Brook Stickleback (*Culaea inconstans*) for Low Oxygen in Winter', *Environmental Biology of Fishes*, *7*, 113-20

Krebs, C.J. (1978) *Ecology: the Experimental Analysis of Distribution and Abundance* (2nd edn), Harper & Row, New York

Kynard, B.E. (1978) 'Breeding Behaviour of a Lacustrine Population of Threespine Sticklebacks (*Gasterosteus aculeatus* L.)', *Behaviour*, *67*, 178-207

Kynard, B.E. (1979) 'Nest Habitat Preference of Low Plate Number Morphs in Threespine Sticklebacks (*Gasterosteus aculeatus*)', *Copeia*, *1979*, 525-8

Larson, G.L. (1976) 'Social Behaviour and Feeding Ability of Two Phenotypes of *Gasterosteus aculeatus* in Relation to their Spatial and Trophic Segregation in a Temperate Lake', *Canadian Journal of Zoology*, *54*, 107-21

Lewis, D.B., Walkey, M. and Dartnall, H.J.G. (1972) 'Some Effects of Low Oxygen Tensions on the Distribution of the Three-spined Stickleback *Gasterosteus aculeatus* L. and the Nine-spined Stickleback *Pungitius pungitius* (L.)', *Journal of Fish Biology*, *4*, 103-8

Lewontin, R.C. (1974) *The Genetic Basis of Evolutionary Change*, Columbia University Press, New York

MacLean, J. (1980) 'Ecological Genetics of Threespine Sticklebacks in Heisholt Lake', *Canadian Journal of Zoology*, *58*, 2026-39

MacLean, J.A. and Gee, J.H. (1971) 'Effects of Temperature on Movements of Prespawning Brook Sticklebacks, *Culaea inconstans*, in the Roseau River, Manitoba', *Journal of the Fisheries Research Board of Canada*, *28*, 919-23

McPhail, J.D. (1969) 'Predation and the Evolution of a Stickleback (*Gasterosteus*)', *Journal of the Fisheries Research Board of Canada*, *26*, 3183-208

McPhail, J.D. (1984) 'Ecology and Evolution of Sympatric Sticklebacks (*Gasterosteus*): Morphological and Genetic Evidence for a Species Pair in Enos Lake, British Columbia', *Canadian Journal of Zoology*, *50*, 1402-8

McPhail, J.D. and Hay, D.E. (1983) 'Differences in Male Courtship in Freshwater and Marine Sticklebacks (*Gasterosteus aculeatus*)', Canadian Journal of Zoology, *61*, 292-7

Maskell, M., Parkin, D.T. and Verspoor, E. (1977) 'Apostatic Selection by Sticklebacks upon a Dimorphic Prey', *Hereditary*, *39*, 83-9

Milinski, M. (1977a) 'Experiments on the Selection by Predators on the Spatial Oddity of their Prey', *Zeitschrift für Tierpsychologie*, *43*, 311-25

Milinski, M. (1977b) 'Do All Members of a Swarm Suffer the Same Predation?', *Zeitschrift für Tierpsychologie*, *45*, 373-88

Milinski, M. (1982) 'Optimal Foraging: the Influence of Intraspecific Competition on Diet

Selection', *Behavioural Ecology and Sociobiology, 11,* 109-15

Milinski, M. (1984) 'Competitive Resource Sharing: an Experimental Test of a Learning Rule for ESSs', *Animal Behaviour, 321,* 233-42

Milinski, M. and Heller, R. (1978) 'Influence of a Predator on the Optimal Foraging Behaviour of Sticklebacks (*Gasterosteus aculeatus* L.)', *Nature London, 275,* 642-4

Moodie, G.E.E. (1972a) 'Morphology, Life History and Ecology of an Unusual Stickleback (*Gasterosteus aculeatus*) in the Queen Charlotte Islands, Canada', *Canadian Journal of Zoology, 50,* 721-32

Moodie, G.E.E. (1972b) 'Predation, Natural Selection and Adaptation in an Unusual Stickleback', *Heredity, 28,* 155-67

Moore, J.W. and Moore, I.A. (1976) 'The Basis of Food Selection in Some Estuarine Fishes', *Journal of Fish Biology, 9,* 375-90

Ohguchi, O. (1981) 'Prey Density and Selection against Oddity by Three-spined Sticklebacks', *Advances in Ethology, 23,* 1-79

Partridge, L. (1978) 'Habitat Selection', in J.R. Krebs and N.B. Davies (eds), *Behavioural Ecology,* Blackwell, Oxford, pp. 351-76

Pitcher, T.J. (1983) 'Heuristic Definitions of Fish Shoaling Behaviour', *Animal Behaviour, 31,* 611-13

Pitcher, T.J., Kennedy, G.J.A. and Wirjoatmodjo, S. (1979) 'Links between the Behaviour and Ecology of Fishes', *Proceedings of 1st British Freshwater Conference,* pp. 162-75

Pressley, P.H. (1981) 'Parental Effort and the Evolution of Nest Guarding Tactics in the Threespine Stickleback *Gasterosteus aculeatus*', *Evolution, 35,* 282-95

Reebs, S.G., Whoriskey, F.G. and FitzGerald, G.J. (1984) 'Diel Patterns of Fanning Activity, Egg Respiration, and the Nocturnal Behaviour of Male Threespine Sticklebacks, *Gasterosteus aculeatus* L. (form *trachurus*)', *Canadian Journal of Zoology, 62,* 320-41

Reimchen, T.E. (1983) 'Structural Relationships between Spines and Lateral Plates in Threespine Stickleback (*Gasterosteus aculeatus*)', *Evolution, 37,* 931-46

Reist, J.D. (1980a) 'Selective Predation upon Pelvic Phenotypes of Brook Stickleback, *Culaea inconstans,* by northern pike, *Esox lucius*', *Canadian Journal of Zoology, 58,* 1245-52

Reist, J.D. (1980b) 'Predation upon Pelvic Phenotypes of Brook Stickleback, *Culaea inconstans,* by Selected Invertebrates', *Canadian Journal of Zoology, 58,* 1253-8

Reist, J.D. (1983) 'Behavioural Variation in Pelvic Phenotypes of Brook Stickleback, *Culaea inconstans,* in Response to Predation by Northern Pike, *Esox lucius*', *Environmental Biology of Fishes, 8,* 255-67

Ridley, M. and Rechten, C. (1981) 'Female Sticklebacks Prefer to Spawn with Males whose Nests Contain Eggs', *Behaviour, 76,* 152-61

Rocd, K.H. (1979) 'The Temperature Preference of the Three-spined Stickleback, *Gasterosteus aculeatus* L. (Pisces), Collected at Different Seasons', *Sarsia, 64,* 137-41

Rohwer, S. (1978) 'Parent Cannibalism of Offspring and Egg Raiding as a Courtship Strategy', *American Naturalist, 112,* 429-40

Rowland, M.J. (1983a) 'Interspecific Aggression and Dominance in *Gasterosteus*', *Environmental Biology of Fishes, 81,* 269-77

Rowland, W.J. (1983b) 'Interspecific Aggression in Sticklebacks — *Gasterosteus aculeatus* Displaces *Apeltes quadracus*', *Copeia, 1983,* 541-4

Sargent, R.C. (1982) 'Territory Quality, Male Quality, Courtship Intrusions, and Female Choice in the Threespine Stickleback *Gasterosteus aculeatus*', *Animal Behaviour, 30,* 364-74

Sargent, R.C. and Gebler, J.B. (1980) 'Effects of Nest Site Concealment on Hatching Success, Reproductive Success, and Paternal Behaviour of the Threespine Stickleback, *Gasterosteus aculeatus*', *Behavioural Ecology and Sociobiology, 7,* 137-42

Semler, D.E. (1971) 'Some Aspects of Adaptation in a Polymorphism for Breeding Colours in the Threespine Stickleback (*Gasterosteus aculeatus*)', *Journal of Zoology, London, 165,* 291-302

Stanley, B.V. (1983) 'Effect of Food Supply on Reproductive Behaviour of Male *Gasterosteus aculeatus*', Unpublished PhD thesis, University of Wales

Thomas, G. (1977) 'The Influence of Eating and Rejecting Prey Items upon Feeding and Food Searching Behaviour in *Gasterosteus aculeatus* L.', *Animal Behaviour, 25,* 52-66

van den Assem, J. (1967) 'Territory in the Three-spined Stickleback, *Gasterosteus aculeatus* L. An Experimental study in Intra-specific Competition', *Behaviour Supplement, 16,* 1-164

Visser, M. (1981) 'Prediction of Switching and Counter Switching Based on Optimal Foraging', *Zeitschrift für Tierpsychologie, 55*, 129-38

Visser, M. (1982) 'Prey Selection by the Three-spined Stickleback (*Gasterosteus aculeatus* L.)', *Oecologia, 55*, 395-402

Walsh, G. and FitzGerald, G.J. (1984) 'Resource Utilization and Coexistance of Three Species of Sticklebacks (Gasterosteidae) in Tidal Salt Marsh Pools', *Journal of Fish Biology, 25*, 405-20

Ward, G. and FitzGerald, G.J. (1983) 'Fish Predation on the Macrobenthos of Tidal Salt Marsh Pools', *Canadian Journal of Zoology, 61*, 1358-61

Whoriskey, F.G. (1985) 'Le Rôle de Facteurs Choisis, Biotiques et Abrotiques, dans la Structuration d'une Communante d'Epinoches', Unpublished D.Phil. thesis, Laval University

Whoriskey, F.G. and FitzGerald, G.J. (in press) 'Sex, Cannibalism and Sticklebacks', *Behavioural Ecology and Sociobiology*

Wootton, R.J. (1976) *The Biology of the Sticklebacks*, Academic Press, London

Wootton, R.J. (1984) *The Functional Biology of Sticklebacks*, Croom Helm, London

Worgan, J.P. and FitzGerald, G.J. (1981a) 'Habitat Segregation in a Salt Marsh among Adult Sticklebacks (Gasterosteidae)', *Environmental Biology of Fishes, 6*, 105-9

Worgan, J.P. and FitzGerald, G.J. (1981b) 'Diel Activity and Diet of Three Sympatric Sticklebacks in Tidal Salt Marsh Pools', *Canadian Journal of Zoology, 59*, 2375-9

17 BEHAVIOURAL ECOLOGY OF CAVE-DWELLING FISHES

Jakob Parzefall

The ecological conditions in caves are characterised by two main factors: nearly all caves have complete darkness and more or less constant temperature. The animals found in this habitat form a heterogeneous assembly. Some animals use caves only occasionally to avoid unfavourable conditions outside. Others, such as bats, enter caves regularly to rest during the day and in winter. But there are also many species which live permanently in caves. Omitting all the different classifications (Vandel 1965) of cave-living animals, we can call these true cave-dwellers 'troglobionts'. Their striking morphological differences in comparison with their epigean relatives concern the reduction of the eye dark pigmentation. These reduction phenomena can be observed in many groups of animals. The degree of reduction in different species studied seems to be connected with the phylogenetical age of cave colonisation (Wilkens 1982).

In the teleostean fish, members of about 12 families have colonised caves successfully (Table 17.1). In seven of these families, epigean members are known to be nocturnal and so it is not surprising that those species with a preference for activity in darkness enter caves. But among the ancestors of cave fish we also note Characidae, Poeciliidae and Cyprinidae, which include many species with clear diurnal activity.

Table 17.1: Families with Obligate Cave-dwelling Fishes. F = freshwater, M = marine (from Thines 1955, 1969)

| | Number of: | | | | | | General activity: | |
| | Known populations | | Species | | Epigean relatives | | | |
Family	F	M	F	M	F	M	Diurnal	Nocturnal
Amblyopsidae	—	—	4[1]	—	+	—	—	+[4]
Brotulidae (Ophidiidae)	—	—	4	—	—	+[2]	—	+[2]
Characidae	29[3]	—	1	—	+	—	+	—
Clariidae	—	—	5	—	+	—	—	+
Cyprinidae	—	—	5	—	+	—	+	—
Gobiidae	—	—	6	2	—	+	+	—
Ictaluridae	—	—	4	—	+	—	—	+
Mastacembalidae	—	—	1	—	+	—	—	+
Pimelodidae	—	—	3	—	+	—	—	+
Poeciliidae	1[5]	—	—	—	+	—	+	—
Trichomycteridae	—	—	5	—	+	—	—	+
Synbranchidae	—	—	6	—	+	—	—	+

Additional data: [1]Cooper and Kuehne (1974); [2] Riedl (1966); [3] Mitchell, Russel and Elliot (1977); [4] Poulson (1963); [5] Gordon and Rosen (1962)

What enables certain families of fish to survive and to reproduce in caves? We must look to see whether all these families show behavioural preadaptation for a life in darkness. Comparing the cave-living fishes with their epigean relatives it is interesting to study the possible behavioural adaptations to the cave habitat. But this is difficult because the behaviour of cave-dwelling fishes has not been very well examined. This chapter therefore concentrates mainly on the three best-studied families: the Amblyopsidae, the Characidae and the Poeciliidae.

Comparison of Behaviour Patterns in Cave-dwelling Fishes and their Epigean Relatives

Potential cave-dwellers must have the sense organs and the behaviour necessary to find food and to reproduce in caves. Such animals may be said to be preadapted to cave life. In this chapter we will first examine feeding and reproductive behaviour and then look at other behaviour patterns which have been studied.

Feeding Behaviour

Suitable food sources and quantity vary from cave to cave. Since there are no green plant producers in the dark, cave fish depend upon food brought in from the outside by animals or by floods. Compared with above-ground habitats, most caves do not have an abundance of food.

An exception is the sulphur cave where live-bearing toothcarp *Poecilia mexicana* (formerly described as *P. sphenops* (Schultz and Miller 1971) exist at high population density (Parzefall 1974, 1979) and which has an exceptional abundance of food (Table 17.1). These fishes mainly feed on white material of different algae and bacteria (J. Parzefall, unpublished) which develops in connection with hydrogen-sulphide springs and covers all the rocks in this cave. They obtain this in the same way as their epigean conspecifics, which scrape off green algae from rocks. The rich food production in the sulphur cave leads to a higher population density in the cave form of *Poecilia mexicana* than in epigean populations. In another part of the cave with a bat colony, the fish also feed on guano or invertebrates living on the guano (Gordon and Rosen 1962; Peters, Peters, Parzefall and Wilkens 1973). In this cave system, abundance of food means that no changes in food detection and feeding are required.

In various cave populations of *Astyanax mexicanus* the population density is generally higher in the omnivorous fish of epigean habitats (Table 17.2) but here feeding behaviour is different in cave fish. In the various cave habitats studied, all animals swim without contact in slow zigzag movements dispersed over the pool. The reaction to clay balls and food balls of 5 mm diameter has been tested in Pachon, Micos and Chica caves, in test areas of 1m^2. In Pachon and Micos cave fish, a falling clay ball induced higher swimming activity within a distance of 0.80 m and animals started searching for food near the bottom or on the water surface. However, there was only a slight increase in fish in the test area here (Figure 17.1A,B). On the other hand, in the Chica cave no reaction was observed (Figure 17.1C) and this population showed no reaction to food balls either. In the two other populations food balls produced a significant increase of individuals in the test area. The fish bit off pieces with a head-shaking movement or tried to carry the food ball away, followed by others. After the food offered in the Pachon cave was stopped, most fish continued searching in the test area for some hours. In epigean habitats the eyed fish react to a similar protocol with a rapid increase in numbers during the feeding period and a rapid decrease without food in the test area (Figure 17.2).

The epigean forms normally school near the riverbank down to a depth of 2 m. In a small pool of the Rio Naranjos, which was separated by flat passages of the river, territoriality has been observed. Schools with smaller fish stay near the water surface. The larger ones school in deeper water. The fish follow by rapid swimming movements the smaller particles that arrive, and test them by direct contact. They also try to feed on larger immobile prey. All smaller animals such as insects which fall into the water or which swim unprotected (e.g. young fishes) are eaten within seconds. The smaller fish near the water surface catch the food first. When it sinks into deeper water, the larger *Astyanax* schooling there catch it. In test areas of 1 m water depth, the sinking food seldom reaches the bottom. In general, the epigean fish get their food visually in open water, whereas cave fish are guided to food by water movements and chemical signals from the food (Parzefall 1983).

Table 17.2: Population Density and Food Sources of Cave-dwelling Fishes and their Epigean Relatives

Species	Density (fish per m²)	Food sources
Poeciliidae:		
Poecilia mexicana[1]		
epigean fish	2-50	Algae
cave fish	100-200	Bat guano, sulphur bacteria and algae
Characidae:		
Astyanax mexicanus[2]		
epigean fish	15-200	Omnivorous
cave fish	5-15	Bat guano
Amblyopsidae:[3]		
Chologaster cornuta, C. agassizi	0.01	Invertebrates
cave habitat	0.005	Invertebrates
epigean habitat	0.01	Invertebrates
Typhlichthys subterraneus	0.03	Invertebrates
Amblyopsis spelea	0.05	Invertebrates
Amblyopsis rosae	0.15	Invertebrates, young conspecifics

[1] Parzefall 1979; [2] 1983; [3] Poulson 1963, 1969

Figure 17.1: Density of Cave-dwelling *Astyanax mexicanus* Counted Every 5 min in a Test Area. Each arrow marks the counting after the offer of a clay ball (empty circle), or a food ball (circled F) of 5mm diameter (from Parzefall 1983)

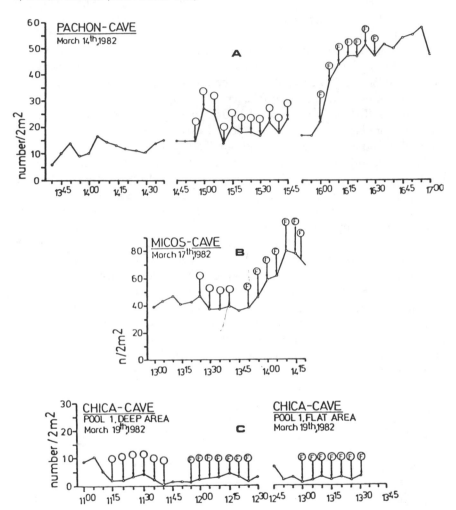

Laboratory studies in darkness with the aid of infra-red videorecording showed that there was a clear difference in the method of picking up food from the bottom (Schemmel 1980). The cave fish fed at an angle of about 45° subtended from the ground whereas the epigean form stood vertically on its head. This difference in behaviour depends on genetic factors (Figure 17.3). Environmental factors cannot explain the difference because the possibility of visual orientation is excluded in darkness and the bottom is flat. Genetic analysis shows that a trifactorial but essentially Mendelian inheritance explains the data from backcrosses (see Chapter 1 by Noakes,

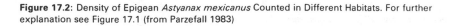

Figure 17.2: Density of Epigean *Astyanax mexicanus* Counted in Different Habitats. For further explanation see Figure 17.1 (from Parzefall 1983)

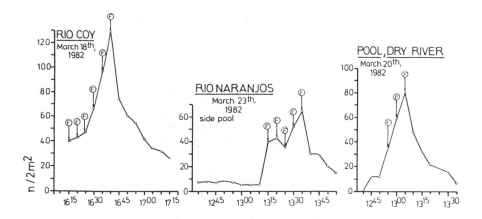

this volume). This behavioural difference is inherited independently of eye size and number of melanophores. In addition the taste buds, restricted to the mouth region in epigean fish, are found over the lower jaw and cover ventral areas of the head in the cave fish. Schemmel (1980) states that such an evolutionary improvement of the gustatory equipment needs a lower angle of the body position during feeding, and both these traits could be achieved by small genetic steps during a process of integrative interaction.

The Amblyopsidae comprise six species in four genera. *Chologaster cornuta* is nocturnal and lives in swamps in the Atlantic coastal plain of the United States; *Chologaster agassizi* is found in springs and caves, more commonly in springs, and the remaining four species (see Table 17.2) are limited to caves. The fish population density in caves is extremely low (Poulson 1969). Their food, consisting of aquatic invertebrates like isopods, amphipods and copepods, is also very scarce. As a consequence cave fish species must swim a great deal to get enough food. Comparing this behaviour in the partial troglodyte *Chologaster agassizi* and in the hypogean *Typhlichthys subterraneus*, it was shown that swimming behaviour is more efficient in *Typhlichthys*. In addition the maximal prey detection distance is greater in the cave species: *Daphnia* was detected by *C. agassizi* within 10mm and in *T. subterraneus* within 30 to 40mm (Poulson 1963). Food-finding ability at low prey densities in the dark in *Amblyopsis spelea* is much better than in *C. agassizi*. When one *Daphnia* was introduced into a 100-litre aquarium in the dark, *A. spelea* found a prey hours before *C. agassizi* did. In contrast *C. agassizi* ate all ten *Daphnia* introduced in a 5-litre aquarium before *A. spelea* had eaten half of them (Poulson and White 1969). These results at high prey densities probably result from a low maximum food intake in the cave fish. The

Figure 17.3: Feeding Behaviour in *Astyanax mexicanus*. Frequency distributions of the angle of feeding in darkness in a cave population, an epigean one, and in the hybrids (redrawn from Schemmel 1980)

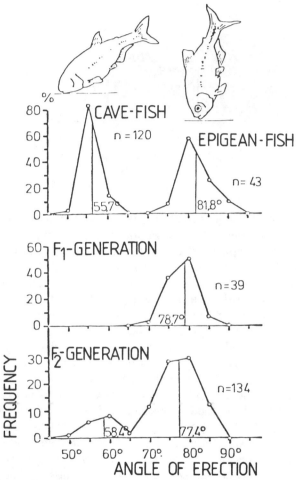

behavioural changes are combined with adaptive changes in the system of free neuromasts (see Chapter 7 by Bleckmann, this volume) and brain anatomy. As a result of these changes, obstacle avoidance and spatial memory are also enhanced in the cave-limited species (Poulson 1963). The *A. spelea* studied in the Upper Twin Cave (Mohr and Poulson 1966) is distributed irregularly over the cave area. The biggest fish are in deeper water, especially along underwater ledges. Each swims and feeds regularly in a limited area, moving upstream along a ledge for about 20 m and floating downstream to its starting point. Upstream-moving fish search under the rocks for prey. The time spent under a rock was directly related to the number of isopods to be found there. Defence of food territories has not been observed although the food is so scarce that adult fish get barely enough to stay alive.

The behaviour of *Speoplatyrhinus poulsoni* (Cooper and Kuehne 1974), an endemic species of a cave in north-western Alabama, has not been studied.

Sexual Behaviour

Having found enough food to reach maturity, the next problem to be solved by cave fish is finding a sexual partner in darkness. Subsequently they need behaviour patterns which provide effective fertilisation in the absence of any visual orientation.

For male *Poecilia mexicana*, which actively seek mates, it is easy to find the female because of the very high population density in the cave. The male checks conspecifics by nipping at the genital region. Females are recognised by a species-specific chemical substance (Zeiske 1971) which is produced continuously by mature females. In addition to this species-specific substance the female also produces a chemical substance in the genital region and becomes attractive for males, but only for several days after the birth of young. Coming in contact with such a female, the nipping behaviour of the male becomes faster and he tries to copulate. The attractive female can only be recognised through direct nipping contact since the chemical signals cannot be transferred through water (Parzefall 1973). (See also review by Hara, Chapter 5, this volume).

Comparative laboratory and field studies have shown that the epigean populations of this species have similar sexual behaviour: in a school of adult *P. mexicana*, the females are checked continuously with nipping by dominant males. On contact with an attractive female, the colour of the male fish darkens. The male then tries to separate the attractive female from the school to defend her against other males, and to copulate. This behaviour gives the bigger dominant males a reproductive advantage. The smallest males counteract this one-sided selective advantage by ambushing the attractive females and attempting to copulate without any aggressive interaction against the defending dominant male. For this reason a small male waits with an inconspicuous body coloration in typical head-down position near an attractive female for the possibility of quick copulation. The best moment for that is during an aggressive interaction of the dominant male with another male (Parzefall 1969, 1979). (See Chapter 13 by Magurran, this volume.) In contrast to epigean fish, cave females are not defended by males, and small males do not perform ambushing behaviour. The consequences of these behavioural differences for the population structure in cave males will be discussed below.

Comparative studies with the closely related species *Poecilia velifera* and *Poecilia latipinna* show that they possess a visual display in addition to the chemical signals. All the Poeciliidae studied, with the exception of *P.*

mexicana, P. sphenops and *P. vivipara*, perform similar species-specific swimming movements with erect fins in front of or around the female (Parzefall 1969). The absence of such a visual display in *P. mexicana* is clearly associated with successful reproduction in a lightless environment. The prior existence of the chemical communication system does not require any further special adaptations of the reproductive behaviour in order to operate in darkness.

In the characid fish *Astyanax mexicanus*, with its several blind cave populations which have been studied in the laboratory, we have a similar situation (Wilkens 1972). A female ready for reproduction produces a chemical substance in the genital region and remains swimming in a small area. After the first contact between a male and an attractive female, the male maintains contact by touching the genital region of the female. Whenever the male loses contact with the female, he searches for her very actively. The male seems only to be able to identify the female through nipping contact in the genital region. After some contacts the male tries to swim into a position parallel with the female. In this position the animals turn their ventral side very rapidly to the water surface and then release the genital products. The female seems to be able to produce the chemical substance without being stimulated by males. This substance stimulates only males and is perceived by them with the olfactory sense. There is also weak evidence that females sometimes mark the substrate by short contacts with the genital region. Sexual activities were released experimentally by putting the fish in aquaria with fresh water. Sometimes, after the first spawning of one pair, some other fish started sexual activities. Comparing an epigean population with hypogean ones, Wilkens (1972) could neither show differences concerning sexual behaviour, nor find any optical display in the epigean population.

In natural habitats sexual behaviour has not yet been observed (Parzefall 1983), probably because most field studies have been conducted during the dry season. In all the populations examined, young animals were absent. It is therefore likely that reproductive activity in *A. mexicanus* takes place during the rainy season, perhaps in connection with an increase of food supply and rising water levels.

Such an annual reproductive cycle is known in the Amblyopsidae (Poulson 1963). The swampfish *Chologaster cornuta* leaves the cypress swamp habitats in early April when it spawns. It comes into open streams for the rich food along edges of submerged weed banks. *Chologaster agassizi*, the spring cave fish, seems to spawn in caves in February when water levels are at the year's maximum. A yearly cycle exists also in the three cave-dwelling species. Ova reach mature size and breeding occurs during high water from February through April. *Amblyopsis spelea* shows an especially well-defined cycle. In this species the females carry the eggs in their gill cavities until hatching and carry the young until they lose their yolk sacs, a total period of 4-5 months. Thus the young appear in late summer and early autumn (Poulson 1963). In the Upper Twin Cave, Indiana, USA, the larger fish of *A. spelea* in the early winter appear to be spread out along the stream passage in pairs. Male and female seem to establish a territory. But neither Poulson (1963) nor Bechler (1983) have observed agonistic behaviour in the field. During a seven-year study (Poulson 1963), the reproduction rate in this cave was low. In the population, which never had less than 81 fish nor more than 130, on average about five females a year produced 40 to 60 large, heavily yolked eggs each. Other females did not breed at all. After a year of better food supply and higher reproduction rate by more egg-producing females, the young fish were reduced by cannibalism from adults.

Comparing the lifespans of the different amblyopsid species, we can observe that, with increasing evolutionary time in caves, the time for hatching and the attainment of sexual maturity are increased (Figure 17.4), and the mean number of ova becomes smaller and the volume of ova larger. This corresponds to the classic pattern of K-selected species (Culver 1982).

Agonistic Behaviour

Agonistic behaviour includes aggressive and submissive patterns. In some cave-dwelling fish, agonistic behaviour remains unchanged, but in others a striking reduction of this complex behaviour has been observed.

Most of the cave-dwelling fish studied seem to have agonistic behaviour. In the Clariidae, one of the five cave-dwelling species *Uegitglanis zammaranoi*, an anophthalmic phreatic fish from Somalia, shows an aggressive behaviour which is clearly based on a dominance relationship.

Figure 17.4: Lifespans of Species of Amblyopsid Fish. Black bar is time to hatching, striped bar is time from hatching to first reproduction, open bar is reproductive lifespan (data from Poulson 1963 and Culver 1982)

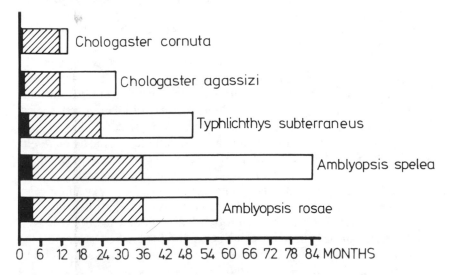

When two specimens are placed together in an aquarium, they start fighting with gasping movements of the mouth, yawning, chasing, butting with the head or mouth, biting and mouth-locking. The dominant fish tends to occupy the lower part of the aquarium. The subordinate fish exhibits rubbing of the bottom, a vertical position or rapid flight reactions. The intensity of fights increases when a dominant fish is introduced into the aquarium of a subordinate one (Ercolini, Berti and Cianfanelli 1981). Some of these aggressive patterns have been observed in the natural habitat where the specimens studied have been collected (Berti and Ercolini 1979). Though this blind species has no possibility for visual communication, a complicated intraspecific aggressive behaviour still exists.

For some other cave-living fish we have only a few data, and detailed studies are missing. Studied in aquaria, the blind cyprinid *Caecobarbus geertsi*, from caves in the lower Congo, attacks conspecifics by blows in the middle of the body. When two fish alternate these attacks, they perform circling movements (Thines 1969). The blind intertidal *Thyplogobius californiensis*, which lives in pairs in burrows built by the ghost shrimp *Callianassa affinis* attacks intruding conspecifics. The aggressive behaviour is released by chemical signals (MacGinitie 1939). The synbranchid eel *Furmastix infernalis* studied in the Hoctun cave (Yucatan, Mexico) and in aquaria seems to have individual territories; biting and tail-beating against intruders have been observed in aquaria (J. Parzefall unpublished). During a short observation period of the blind species *Typhliasina pearsei* (Brotulidae) in the Cueva del Pochote and in aquaria, head-shaking movements between animals at a distance of 10-20 cm have been registered which release flight reactions, but no other aggressive patterns have been found (J. Parzefall unpublished; Schemmel 1977).

In more detailed comparative studies the reduction of agonistic behaviour in caves has been demonstrated in three examples. Bechler (1983) has compared this behaviour in five amblyopsid species. Six aggressive acts and two submissive acts have been exhibited (Table 17.3). Males and females of the four species that behaved agonistically did not appear different. The fifth species, the swampfish *Chologaster cornuta*, did not show any agonistic behaviour. In the highly adapted cave species *Amblyopsis rosae* only tail-beating and the submissive acts are still existent. The detailed analysis of the behavioural diversity demonstrates that *Chologaster agassizi* and *Amblyopsis spelea*, the least cave-adapted subterranean species, engage in relatively intense, complex agonistic bouts. In contrast, the more highly cave-adapted species *Typhlichthys subterraneus* and *Amblyopsis rosae* engage in simpler, less intense bouts, which were considerably shorter in length. The most frequent act by any of the species studied was tail-beating. Reduced selective pressure with increasing adaptation to cave life is the most probable explanation for the observed reduction in agonistic behaviour. Concluding the field data, Bechler (1983) states that food is a primary selective force in amblyopsids. In adaptation to this factor, metabolic rate and fecundity decrease in the cave, longevity is increased and swimming efficiency is improved. These adaptations confer the advantage of energy conservation on the more highly evolved subterranean species. It is suggested that this conservation of energy serves to reduce selection pressures produced by a scarce food supply, and allows for a reduction in overt agonistic behaviour.

Another case of reduction of aggressive behaviour is found in cave-living populations of the characid *Astyanax mexicanus*. Epigean fish usually live in schools, and territoriality only exists as a special case in small pools. Except for weak ramming attacks in the epigean fish, no aggression has been observed in field studies during the dry season conducted in various natural habitats in Mexico (Parzefall 1983). Laboratory studies have demonstrated that, with increasing size of the aquaria, the frequency of ramming attacks decreases and the territorial fish tend to school (Figure 17.5). Ramming and biting in some epigean populations can lead to the death of smaller animals after several attacks. This is especially the case in small aquaria where they cannot escape. There is no difference in aggression between male and female. Hungry fish become much more aggressive (Burchards, Dölle and Parzefall 1985). In the blind cave populations, however, fights are very rare and no animal has ever been seen to die because of attack.

To obtain quantitative data, three different populations have been tested in visible light and in the infra-red (Figure 17.6). The epigean fish originates from Teapao river. The Pachon population is completely blind, and the Micos population reared in light is variable in eye size, indicating that this population is a phylogenetically young one (Wilkens 1976). The Pachon population, which is believed to be a phylogenetically old cave form, shows a loss of aggressive behaviour. The cave form from the Micos cave performs aggressive patterns in light and darkness. However, there is

Table 17.3: Aggressive Repertoire for Each Amblyopsid Species. '+' indicates act observed in a species; '-' indicates act not observed (from Bechler 1983)

	Chologaster cornuta	*Chologaster agassizi*	*Typhlichthys subterraneus*	*Amblyopsis spelea*	*Amblyopsis rosae*
Aggressive acts					
Tail-beat	−	+	+	+	+
Head-butt	−	+	+	+	−
Attack	−	+	+	+	−
Bite	−	+	+	+	−
Chase	−	+	+	+	−
Jaw-lock	−	+	−	+	−
Submissive acts					
Freeze	−	+	+	+	+
Escape	−	+	+	+	+

Figure 17.5: Frequency of Two Aggressive Patterns and the Tendency for Territoriality and Schooling of Epigean *Astyanax mexicanus* in Aquaria of Different Size (from Burchards *et al.* 1985)

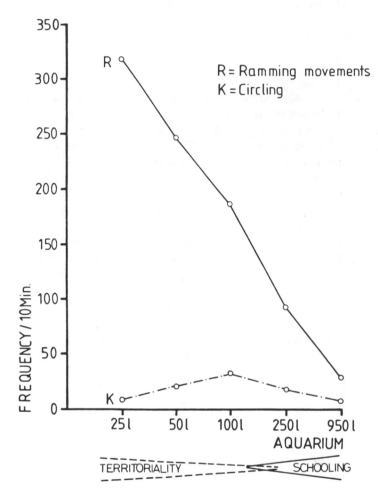

less aggression in darkness. These results demonstrate first that absence of visual communication only reduces the aggression to a certain degree, and secondly that this behaviour seems to undergo a reduction process in the cave biotope. In a subsequent study, ramming and ramming attempts are compared in two epigean populations, one blind cave population and the two *F*1 generations (Figure 17.7). In darkness, all fish tested exhibit the same level of aggression. In visible light, there is no difference between the two epigean populations either. But there is a significant difference in the *F*1 generations. The average in *F*1 generations in a test group with a good optomotor response (minimum separable 15′) cannot be distinguished from a group with a lower optomotor response (Burchards in prep.). From this data we can conclude that a genetically based reduction of the aggressive behaviour exists in cave-dwelling *Astyanax mexicanus*. We do not know enough about the function of the aggressive behaviour in epigean *Astyanax*. For the cave-dwelling populations we do not know enough about density and food sources to explain this reduction phenomenon.

In *Poecilia mexicana*, the third example of reduction in aggressive behaviour, the cave population still has eyes. Aggressive behaviour gradually reduces and the eye diameter decreases from the entrance to the end of the cave (Parzefall 1974). But there is no topographic, physical, chemical or biotic factor (including light) to be found that can explain this gradient (Peters *et al.* 1973). The comparative studies in aggressive behaviour have been concentrated on the cave form collected near the end of the cave (chamber XIII after Gordon and Rosen 1962). In the epigean fish the aggressive behaviour has the function of establishing a size-dependent rank order in the males within the schools of adults. As already described in the section on sexual behaviour, the males separate especially attractive females from the school and defend them vehemently against rivals. In the cave, where the fish live in much higher density (Table 17.2),

Figure 17.6: Frequency of Ramming Attacks during a Light-Dark (LD) Cycle in Two Cave Populations and an Epigean One of *Astyanax mexicanus*. In the specimen belonging to the epigean population the one on the right ramming his opponent. R: ramming; RV: ramming attempt; R_1, R_2: test groups with LD 3:3; KR: test group with LD 12:12 (data from Burchards *et al.* 1985)

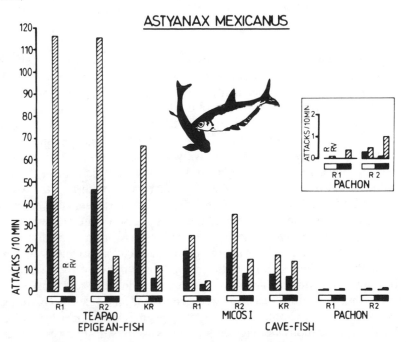

Figure 17.7: Ramming and Ramming Attempts of Two Epigean, One Cave-fish Population and the two *F*1 Generations of *Astyanax mexicanus*. Each fish tested against an epigean opponent in light (L) and darkness (D). DT: optomotor response (minimum separable of 15′ of arc) (from Burchards in prep.)

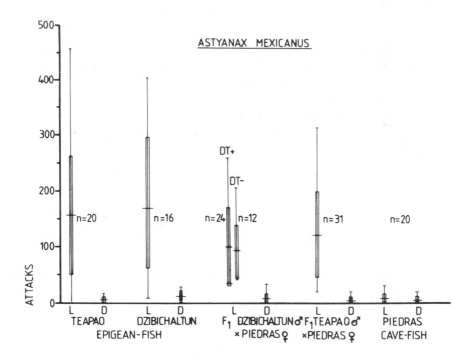

no aggressive behaviour was to be observed. After testing cave specimens that show a good optical orientation with males of their own or of the epigean population in visible light, the reduction of aggression remains unchanged (Figure 17.8). In all parts of the test, the mean for the number of aggressive acts is about zero. In the mixed tests the majority of the cave males answered the initial attacks of the epigean males with sexual behaviour. Only a few reacted defensively for a short time. No cave male showed an attack reaction. Tests in darkness with epigean males, carried out with the aid of infra-red, showed one aggressive pattern, namely tail-beating. Aggressive behaviour in *P. mexicana* therefore seems to be based mainly on visual communication. For this reason, the reduction of aggressiveness in the cave could be the result of less optical orientation. If this is the case, the tail-beating should not be reduced as it can be perceivd non-visually. But there is no significant difference between the number of tail-beatings and other aggressive movements in the cave fish. This reduction must therefore be considered genetic for all aggressive patterns.

To confirm the hypothesis cave and epigean populations were cross-bred and all generations were tested for aggressiveness. In Figures 17.9 and

Figure 17.8: Aggressive Patterns in Males of *Poecilia mexicana*. XIII, cave population; O, epigean population; N, standard test of a male against an opponent of the same population; M, male tested against an opponent of the other population. The total variability, mean, standard error and standard deviation are shown (from Parzefall 1979)

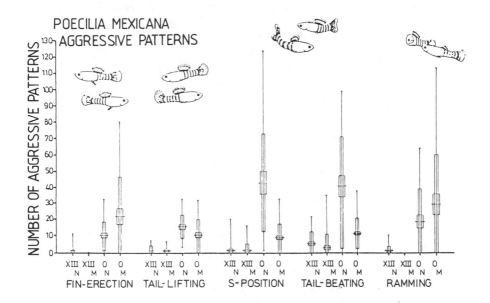

17.10 the frequency distribution for fin erection and S-position are presented. The parental generations are clearly separated. The $F1$ and $F2$ crosses show a more or less intermediate value for the average, as shown by the arrow. The expected separation in the backcrosses is only present in fin erection. Without detailed examination it can be stated that these data confirm the hypothesis.

Follow-up tests of the $F2$ generation failed to find a correlation between eye diameter and the aggressiveness of the males. However, positive correlation despite free recombination exists in the $F2$ generation for several aggressive patterns found in the epigean form. The reduction of all aggressive patterns as compared with surface dwellers in the cave population is about equal (Parzefall 1979).

From these findings it was concluded that a closely linked polygenic system exhibiting additive gene interaction controls the inheritance of aggressive behaviour in *P. mexicana*.

As a result of the reduction in aggression, we should expect a change in the average size of males, because there is no longer any reproductive advantage in being large. Plotting the total length of different samples from outside to the very back of the cave, we can indeed see a significant diminution in the size of males in the cave (Figure 17.11).

There are two hypotheses explaining the reduced aggression in *P.*

Figure 17.9: Aggressive Fin Erection of *Poecilia mexicana*. Frequency distribution in the epigean population (PO), the cave population (PXIII) and hybrid generations. The mean is shown by the arrow (from Parzefall 1979)

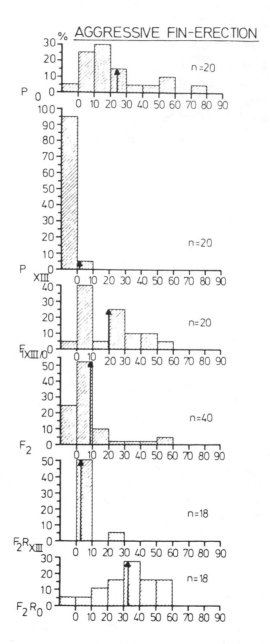

Figure 17.10: S-Positioning of *Poecilia mexicana*. For explanation see Figure 17.9 (from Parzefall 1979)

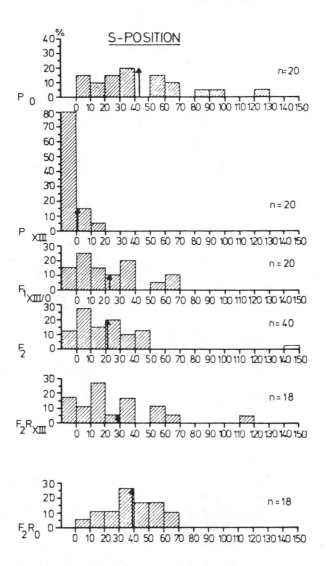

mexicana in this cave. Aggressive behaviour patterns cannot be seen in darkness. For this reason, stabilising selection acting on aggressive behaviour is no longer effective. Over a long period, this leads to a degeneration of the aggression. But one can argue that tail-beating is still performed and could be detected (by the lateral lines) in darkness, and so could serve as a basis for necessary aggression in the cave. The fact that tail-beating shows the same degree of reduction as the other aggressive

Figure 17.11: Total Length of Males and Females in Different Samples of *Poecilia mexicana* Collected in the Natural Habitat (from Parzefall 1979)

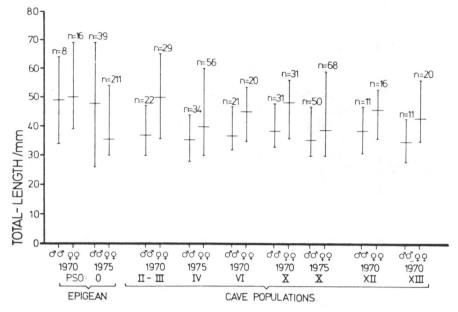

patterns, and the much higher degree of reduction in aggressive behaviour in comparison with the slightly reduced eye, suggest that these changes could be better explained by a selection pressure acting against aggression directly. Selection in general can cause changes more rapidly than the absence of the stabilising selection. The latter generally causes the eye reduction in caves.

Schooling Behaviour

For two of the species already presented, *Astyanax mexicanus* and *Poecilia mexicana*, schooling has been reported as common organisation in the epigean habitat. However, in the cave populations this behaviour has not been observed (Parzefall 1979, 1983). In the following cave-fish species studied by Berti and Thines (1980) and Jankowska and Thines (1982), schooling behaviour was also absent: *Caecobarbus geertsi* (Cyprinidae), *Barbobsis devecchii* (Cyprinidae) and *Uegitglanis zammaranoi* (Clariidae). All these cave-dwelling fish show a random spatial distribution in their habitats and in aquaria. In *P. mexicana* and in *A. mexicanus* the fish swim without contact, dispersed throughout the cave pool. This is more striking in *Astyanax* with its low population density. The absence of schooling has also been found in one epigean population of *Astyanax mexicanus* which lives near a cave entrance and shows an affinity to the cave (Romero 1983).

From different studies on schooling behaviour (Partridge and Pitcher

1980) the important role of visual orientation is well known. Individual saithe (*Pollachius virens*) that had been blinded with opaque eye covers were able to join schools of 25 normal saithe probably using their lateral line organs (see also Chapter 7 by Bleckmann, this volume). But this result does not mean that blind saithe would school in the wild (Pitcher, Partridge and Wardle 1976), and cave fish do not seem to operate in this manner.

The cave form of *P. mexicana* with its functional eyes has not been tested in visible light for schooling. But epigean fishes of *A. mexicanus* observed in darkness with infrared cease schooling. Therefore the loss of visual orientation could be the direct reason for the absence of schooling in the cave. The other possibility is that the genetical basis for this behaviour has changed after the long period of cave life without any stabilising selection for schooling.

To answer these questions, the tendency for schooling in *A. mexicanus* has been tested in *F*1 and *F*2 hybrids between the blind Piedras cave population and the epigean one of the Teapao river. For these experiments, only hybrids that have demonstrated a good optomotor response (minimum separable of 15′ of arc) are used. Despite good visual orientation, these hybrids show a weaker tendency for schooling (Figure 17.12) than epigean fish, and an increased variability. Similar experiments have been conducted with animals of the phylogenetically young cave fish from Micos cave after selective breeding for functional eyes. In these tests the tendency for schooling was weaker than in the *F*2 cross shown in Figure 17.12 (Senkel 1983). From the data presented one can conclude that in cave-dwelling *A. mexicanus* there seems to exist a genetically based reduction of schooling behaviour.

Alarm Reaction

The alarm or fright reaction appears to be confined to the Ostariophysi (Pfeiffer 1977). The 'Schreckstoff' is an alarm pheromone released from club cells in the epidermis when damaged (see Chapter 6 by Hara, this volume). The alarmed conspecifics may react by cover-seeking, closer crowding, rapid swimming or immobility. Up to now in cave-dwelling fishes, only *Astyanax mexicanus* and *Caecobarbus geertsi* have been studied. In the *A. mexicanus* population of Chica cave tested by Pfeiffer (1963) and Thines and Legrain (1973), no alarm reaction was found against the own-skin extract or the alarm substance of the epigean *A. mexicanus*. However, it was demonstrated that the cave population does have the alarm substance since the epigean fish reacted well to extracts from the skin of the Chica fish. It is very probable that the cave population can perceive the alarm substance with its well-developed olfactory system, since it searches for food immediately when skin extract is poured into the aquarium. Thines and Legrain (1973) describe a similar reaction in *Caecobarbus geertsi*: the alarm substance causes an alimentary exploration oriented towards the bottom.

Figure 17.12: The Tendency to Follow a School of *Astyanax mexicanus* in min/10 min. Tests of single specimens of the epigean Teapao population, the *F*1 and *F*2 hybrids with the blind Piedras population. In the hybrids, only specimens with a good optomotor response (minimum separable of 15′ of arc) have been used (redrawn from Senkel 1983)

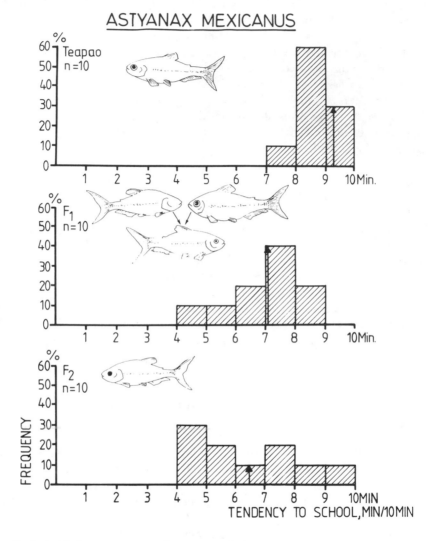

In hybrids between an epigean population and the Pachon cave population of *A. mexicanus* the inheritance of the alarm reaction has been studied (Pfeiffer 1966). All *F*1 hybrids tested responded to the alarm substance. The segregation rate in the *F*2 and *F*2R generations leads to the interpretation that two dominant factors are responsible for the fright reaction. But the results with the backcross to the cave fish do not fit completely with this hypothesis.

Pfeiffer (1963) tried to explain the loss of the fright reaction by the absence of predators in the cave habitat. Thines and Legrain (1973) follow this explanation for *Caecobarbus geertsi.*

In fact in all cave habitats of *Astyanax* studied, no predator has been detected (Mitchell *et al.* 1977; Parzefall 1983). However, cannibalism probably occurs. It has been observed in the Micos cave that the bigger fish seem to show predatory behaviour against smaller ones (Parzefall 1983). This is in accordance with studies by Wilkens and Burns (1972) which have noted the presence of small *Astyanax* in the stomach of the Micos fish.

Circadian Clock

The circadian clock in animals has to be synchronised by external stimuli which are called *Zeitgeber* (forcing signals), principally light and temperature (Bünning 1973). These stimuli are normally absent in caves. Epigean animals tested in darkness show an endogenous free-running daily rhythm in their locomotory activity. Therefore the question arises as to whether, in cave-dwellers that have already been living in darkness and more or less constant temperature for a long time, such an endogenous circadian rhythm still exists.

In a group of blind *Astyanax mexicanus* of unknown origin, Erckens and Weber (1976) frequently found one to four postoscillations after the animals had been transferred from light-dark cycles (LD) into the darkness. After the postoscillations had damped out, the animals seemed to become aperiodically active. In subsequent studies with the Pachon cave fish, weak circadian periodicities have been detected in constant darkness after a light-dark cycle of 12h:h only. Using cycles of other lengths, this phenomenon was lacking. These periodicities were detected by the mathematical procedure of complex demodulation. In the eyed epigean *A. mexicanus,* circadian activity cycles were quite obvious in darkness (Erckens 1981; Erckens and Martin 1982a,b). Thines and Weyers (1978) have described an aperiodic locomotory behaviour in darkness in the blind Pachon fish.

In the Amblyopsidae three species have been tested (Poulson and Jegla 1969). The springfish *Chologaster agassizi,* which is functionally blind and cannot focus on an object, possesses a circadian clock. This species is nocturnal in epigean habitats and goes underground in winter and breeds there. In contrast, the obligate cave-dwellers *Typhlichthys subterraneus* and *Amblyopsis rosae* show no circadian components in their behaviour in darkness.

These two examples of cave-living fishes reflect the general situation in all obligate cave-dwelling animals. More extremely specialised cavernicolous animals exhibit no or doubtful locomotor periodicities, or periodicities of very poor precision. As the change between activity and rest reflects the most drastic changes in metabolism, Lamprecht and Weber (1982) conclude that a periodic organisation of resting and activity metabolism is not necessary, perhaps even unfavourable, inside caves.

Regressive Evolution and Behaviour

The discussion about regressive evolution mainly centres on the reduction of eyes and pigmentation. But regressive evolution is by no means unique to caves. There are also examples in parasites or animals that become sessile (Dzwillo 1984).

What makes regressive evolution in cave animals especially interesting is the possibility that selection may play a minor role compared with the accumulation of neutral mutations and genetic drift. The question of whether selection or neutral mutation plays the dominant role cannot be answered on the basis of our current knowledge (Culver 1982). Kosswig (1948; 1963) has demonstrated that in animals entering caves various patterns (e.g. eyes and pigmentation) first show a high variability. His hypothesis was that the absence of stabilising selection allows an accumulation of selectively neutral mutations, and that variability decreases during cave life within a population because of mutation pressure. In phylogenetically old cave animals this leads to genetic homozygosity and diminished variability for the reduced patterns. Other authors favour selective explanations involving selection for increased metabolic economy (Poulson 1963) or indirect effects of pleiotropy (Barr 1968). There is only a general consensus that differentiation of cave populations, especially with regard to regressive evolution, cannot occur when there is gene exchange with surface populations.

The regressive evolution in the different behavioural traits presented cannot be explained by only one of the above-mentioned hypotheses. The neutral-mutation hypothesis seems to me to be the most plausible explanation for the reduction of the circadian clock, and the lack of schooling behaviour and alarm reaction. To test this hypothesis, more comparative data on cave-dwellers of different phylogenetic age are necessary. If it is valid, it should be possible to demonstrate higher variability for these behaviour patterns in phylogenetically younger cave populations in parallel with the degree of eye reduction. Circadian activity, schooling and alarm reaction are relatively independent of motivational changes and are therefore good examples for testing. One cannot, however, neglect the opposing arguments that these characters of little or no selective value in caves change as a pleiotropic byproduct of selection for adaptively advantageous characters (Barr 1968), or, on the other hand, that the reduction of these characters is an economic advantage: increased metabolic economy conserves energy for cave-living organisms (Poulson 1963). So the material compensation for behavioural traits that are unnecessary in caves can be adaptive.

On the other hand, the reduction of the agonistic behaviour seems more likely to be caused by selection. The great reduction of agonistic behaviour in comparison to the eye, which is only slightly reduced, favours the explanation of a selection pressure against aggression in *Poecilia mexicana*.

The aggressive males seem to have no more reproductive advantage because of the risk of losing contact with the female in darkness during an aggressive encounter against an opponent. This risk is increased by the fact that only tail-beating, which requires a short distance to the opponent, is effective in darkness. The most successful strategy in darkness should be a quick copulation after a short nipping contact. Such a behaviour already exists among small males in epigean habitats. The absence of bigger males in the cave population supports this hypothesis (Parzefall 1979).

The abundance of food in the *Poecilia* cave allows us to neglect energy conservation as another reason for the reduction of aggression. However, the study of Bechler (1983) on the Amblyopsidae has pointed to food as a primary selective force. But several facts argue against the hypothesis that the reduction of agonistic behaviour serves specifically to conserve energy. So this reduction can be explained as a consequence of the decreased metabolic rate and fecundity, which allows a reduction in overt agonistic behaviour with continuing adaptation to cave life.

For *Astyanax mexicanus* more data from different cave populations are necessary in order to understand the reduction of the agonistic behaviour. In epigean fish, ramming an opponent occurs even in darkness and increases in hungry animals studied in light (Brust-Burchards 1980; Dölle 1981). So *A. mexicanus* should be able to defend food sources in the cave. But in the Pachon cave the fish fed together without any aggression, and in Chica cave there was no reaction to food offered (Parzefall 1983). Up to now we can only speculate that the low density of *A. mexicanus* in the caves, in connection with enough food in the form of bat guano, does not require agonistic behaviour in this habitat.

The changes of head-standing behaviour in *A. mexicanus* are surely adaptive. The evolutionary improvement of the gustatory equipment presupposes changes in the head-standing movement (Schemmel 1980). If the food is mainly distributed on the bottom of the habitat, it seems to be a useful tactic to search at an angle of about 45° in slow zigzag movements (Parzefall 1983).

Conclusion

In the examples of different behaviour traits presented, chemical communication is the most important mode of communication in caves. Acoustic and tactile stimuli also exist, but here again few exact data are available.

It is striking that, despite regressive evolution for agonistic behaviour, schooling, fright reaction and circadian activity, no example exists that shows the gradual reduction of an optical display among cave-dwelling animals. For this reason the absence of visual displays is probably a necessary assumption for the successful colonisation of a lightless habitat.

In every case studied, the type of communication which is realised in the cave form exists already in their epigean relatives.

In some of the caves, food is extremely scarce. Cave-dwelling fish respond to the severe food limitation by a decreased metabolic rate and fecundity. Many individuals of the cave-limited species in the Amblyopsidae fail to reproduce.

Summary

In the teleost fish, members of about 12 families have colonised caves successfully. Mainly three of these families have been studied for possible behavioural adaptations to the cave habitat. Potential cave-dwellers seem to need a preadaptation for cave life in their sexual behaviour which is mainly based on chemical communication. The most striking phenomenon is reduction of different behavioural traits ranging from circadian loco-motory activity, fright reaction, schooling and aggressive behaviour to parts of the feeding behaviour.

These behavioural differences between cave-dwellers and their epigean relatives allow the use of these animals for studies on evolutionary genetics in behaviour because of fertile cross-breeding. On the basis of the present data it is very likely in cave animals that, with some important exceptions, such as aggressive behaviour, selection may play a minor role compared with the accumulation of neutral mutations. For these reasons field and laboratory studies on cave-dwelling fish seem to hold great promise.

References

Barr, T.C. (1968) 'Cave Ecology and the Evolution of Troglobites', *Evolutionary Biology, 2,* 35-102

Bechler, D.L. (1983) 'The Evolution of Agonistic Behavior in Amblyopsid Fishes', *Behavioral Ecology and Sociobiology, 12,* 35-42

Berti, R. and Ercolini, A. (1979) 'Aggressive Behaviour in the Anophthalmic Phreatic Fish *Uegitglanis zammaranoi* Gianferrari (Clariidae, Siluriformes)', *Monitore Zoologico Italiano, 13,* 197

Berti, R. and Thines, G. (1980) 'Influence of Chemical Signals on the Topographic Orientation of the Cave Fish *Caecobarbus geertsi* Boulenger (Pisces, Cyprinidae)', *Experientia, 36,* 1384-5

Brust-Burchards, H. (1980) 'Das Aggressionsverhalten von Fischen. Eine vergleichende Betrachtung unter besonderer Berücksichtigung von *Astyanax mexicanus*', Unveröff, Staatsexamensarbeit der Universität Hamburg

Bünning, E. (1973) *The Physiological Clock, 3,* Auflage Springer, Berlin

Burchards, H. (in prep.) 'Das Aggressionsverhalten von Bastarden oberirdisch und unterirdisch lebender Populationen des Salmlers *Astyanax mexicanus* (Pisces)'

Burchards, H., Dölle, A. and Parzefall, J. 'The Aggressive Behaviour of an Epigean Population of *Astyanax mexicanus* (Characidae, Pisces) and Some Observations of Three Subterranean Populations', *Behavioural Processes,* in press

Cooper, J.E. and Kuehne, R.A. (1974) '*Speoplatyrhinus poulsoni,* a New Genus and Species of Subterranean Fish from Alabama', *Copeia, 1974,* 486-93

Culver, D.C. (1982) *Cave Life. Evolution and Ecology.* Harvard University Press, Cambridge

Dölle, A. (1981) 'Über Ablauf und Funktion des Aggressionsverhaltens von *Astyanax mexicanus* (Characidae, Pisces) unter Berücksichtigung zweier Höhlenpopulationen', Unveröff, Diplomarbeit, Universität Hamburg

Dzwillo, M. (1984) 'Regressive Evolution in der Phylogenese des Tierreiches', *Fortschritte in der Zoologischen Systematik und Evolutionsforschung, 3,* 115-26

Erckens, W. (1981) 'The Activity Controlling Time-system in Epigean and Hypogean Populations of *Astyanax mexicanus* (Characidae, Pisces)', *Proceedings of the Eighth International Congress of Speleology, 2,* 796-7

Erckens, W. and Martin, W. (1982a) 'Exogenous and Endogenous Control of Swimming Activity in *Astyanax mexicanus* (Characidae, Pisces) by Direct Light Response and by a Circadian Oscillator. I Analyses of the time-control systems of an Epigean River Population', *Zeitschrift für Naturforschung, 37c,* 1253-65

Erckens, W. and Martin, W. (1982b) 'Exogenous and Endogenous Control of Swimming Activity in *Astyanax mexicanus* (Characidae, Pisces) by Direct Light Response and by a Circadian Oscillator. II Features of Time-controlled Behaviour of a Cave Population and their Comparison to an Epigean Ancestral Form', *Zeitschrift für Naturforschung, 37c,* 1266-73

Erckens, W. and Weber, F. (1976) 'Rudiments of an Ability for time Measurement in the Cavernicolous Fish *Anoptichthys jordani* Hubbs and Innes (Pisces, Characidae)', *Experientia, 32,* 1297-9

Ercolini, A., Berti, R. and Cianfanelli (1981) 'Aggressive Behaviour in *Uegitglanis zammaranoi* Gianferrari (Clariidae, Siluriformes) an anophthalmic phreatic fish from Somalia', *Monitore Zoologico Italiano, 5,* 39-56

Gordon, M.S. and Rosen, D.E. (1962) 'A Cavernicolous Form of the Poeciliid Fish *Poecilia sphenops* from Tabasco, Mexico', *Copeia, 1962,* 360-8

Jankowska, M. and Thines, G. (1982) 'Etude Comparative de la Densité de Groupes de Poissons Cavernicoles et Epiges (Characidae, Cyprinidae, Clariidae)', *Behavioural Processes, 7,* 281-94

Kosswig, C. (1948) 'Genetische Beiträge zur Präadaptationstheorie', *Revue de Facultie des Science (Istambul) Series B, 5* 176-209

Kosswig, C. (1963) 'Genetische Analyse konstruktiver und degenerativer Evolutionsprozesse', *Zeitschrift für zoologische Systematik und Evolutionsforschung, 1,* 205-39

Lamprecht, G. and Weber, F. (1982) 'A Test for the Biological Significance of Circadian Clocks: Evolutionary Regression of the Time Measuring Ability in Cavernicolous Animals', in D. Mossakowski and G. Roth (eds), *Environmental Adaptation and Evolution,* Fischer Verlag, Stuttgart, pp. 151-78

MacGinitie, G.E. (1939) 'The Natural History of the Blind Goby *Typhlogobius californiensis* Steindacher', *American Midland Naturalist, 21,* 489-505

Mitchell, R.W., Russel, W.H. and Elliot, W.R.,(1977) 'Mexican Eyeless Characin Fishes, genus *Astyanax*; Environment, Distribution and Evolution', *Special Publications. The Museum of the Texas Tech. University, 12,* 1-89

Mohr, Ch. M. and Poulson, T.L. (1966): *The Life of the Cave,* McGraw-Hill, New York

Partridge, B.L. and Pitcher, T.J. (1980) 'The Sensory Basis of Fish Schools: Relative Roles of Lateral Line and Vision', *Journal of Comparative Physiology, 135,* 315-25

Parzefall, J. (1969) 'Zur vergleichenden Ethologie verschiedener *Mollienesia*-Arten einschliesslich einer Höhlenform von *M. sphenops*', *Behaviour, 33,* 1-36

Parzefall, J. (1973) 'Attraction and Sexual Cycle of Poeciliidae', in J.H. Schröder (ed.), *Genetics and Mutagenesis of Fish,* Springer Verlag, Berlin, pp. 177-83

Parzefall, J. (1974): 'Rückbildung aggressiver Verhaltensweisen bei einer Höhlenform von *Poecilia sphenops* (Pisces, Poeciliidae)', *Zeitschrift für Tierpsychologie, 35,* 66-82

Parzefall, J. (1979) 'Zur Genetik und biologischen Bedeutung des Aggressionsverhaltens von *Poecilia sphenops* (Pisces, Poeciliidae)', *Zeitschrift für Tierpsychologie, 50,* 399-422

Parzefall, J. (1983) 'Field Observation in Epigean and Cave Populations of Mexican Characid *Astyanax mexicanus*/(Pisces, Characidae)' *Mémoires de Biospéléologie, X,* 171-6

Peters, N., Peters, G., Parzefall, J. and Wilkens H. (1973) 'Über degenerative und konstruktive Merkmale bei einer phylogenetisch jungen Höhlenform von *Poecilia sphenops* (Pisces, Poeciliidae). *Internationale Revue der gesamten Hydrobiologie, 58,* 417-36

458   *Behavioural Ecology of Cave-dwelling Fishes*

Pfeiffer, W. (1963) 'Vergleichende Untersuchung über die Schreckreaktion und den Schreckstoff der Ostariophysen', *Zeitschrift für vergleichende Physiologie, 47*, 111-47

Pfeiffer, W. (1966) 'Über die Vererbung der Schreckreaktion bei *Astyanax* (Characidae, Pisces)', *Zeitschrift für Vererbungslehre, 98*, 97-105

Pfeiffer, W. (1977) 'The Distribution of Fright Reaction and Alarm Substance Cells in Fishes', *Copeia, 1977*, 653-65

Pitcher, T.J., Partridge, B.L. and Wardle, C.S. (1976) 'A Blind Fish Can School', *Science, 194*, 963-5

Poulson, T.L. (1963) 'Cave Adaptation in Amblyopsid Fishes', *American Midland Naturalist, 70*, 257-90

Poulson, T.L. (1969) 'Population Size, Density and Regulation in Cave Fishes', *Actes of the Fourth International Congress of Speleology*, Ljubljana, Yugoslavia, *4-5*, 189-92

Poulson, T.L. and Jegla, (1969) 'Circadian Rhythms in Cave Animals', *Actes of the Fourth International Congress of Speleology, 4-5*, Ljubljana, Yugoslavia, 193-5

Poulson, T.L. and White, W.B. (1969) 'The Cave Environment', *Science, 165*, 971-81

Riedl, R. (1966) *Biologie der Meereshöhlen*, Parey Verlag, Hamburg, Berlin

Romero, A. (1983) 'Behavior in an 'Intermediate' Population of the Subterranean-dwelling Characid *Astyanax fasciatus*', *Environmental Biology of Fishes, 10*, 203-8

Schemmel, Ch. (1977) 'Zur Morphologie und Funktion der Sinnesorgane von *Typhliasina pearsei* (Hubbs) (Ophidioidea, Teleostei)', *Zoomorphologie, 87*, 191-202

Schemmel, Ch. (1980) 'Studies on the Genetics of Feeding Behaviour in the Cave Fish *Astyanax mexicanus* f. *anoptichthys*', *Zeitschrift für Tierpsychologie, 53*, 9-22

Schultz, R.J. and Miller, R.R. (1971) 'Species of the *Poecilia sphenops* Complex in Mexico', *Copeia, 1971*, 282-90

Senkel, S. (1983) 'Zum Schwarmverhalten von Bastarden zwischen Fluss-und Höhlenpopulationen bei *Astyanax mexicanus* (Pisces, Characidae)', Unveröff, Staatsexamensarbeit der Universität Hamburg

Thines, G. (1955) 'Les Poissons Aveugles (I). Origine, Taxonomie, Répartition Géographique, Comportment', *Annales de la Société Royale Zoologique de Belgique*

Thines, G. (1969) *L'Évolution Regressive des Poissons Cavernicoles et Abyssaux*, Masson, Paris

Thines, G. and Legrain, J.M. (1973) 'Effets de la Substance d'Alarme sur le Compartement des Poissons Cavernicoles *Anoptichthys jordani* (Characidae) et *Caecobarbus geertsi* (Cyprinidae)', *Annales de Spéléologie, 28*, 291-7

Thines, G. and Weyers, M. (1978) 'Réponses Locomotrices du Poisson Cavernicole *Astyanax mexicanus* (Pisces, Characidae) à des Signaux Périodiques et Apériodiques de Lumière et de Température', *International Journal of Speleology, 10*, 35-55

Vandel, A. (1965) *Biospeleology*, Pergamon Press, London

Wilkens, H. (1972) 'Über Präadaptationen für das Höhlenleben, untersucht am Laichverhalten ober- und unterirdischer Populationen des *Astyanax mexicanus* (Pisces)', *Zoologischer Anzeiger. 188*, 1-11

Wilkens, H. (1976) 'Genotypic and Phenotypic Variability in Cave Animals. Studies on a Phylogenetically Young Cave Population of *Astyanax mexicanus* (Filippi) (Characidae, Pisces)', *Annales de Spéléologie, 31*, 137-48

Wilkens, H. (1982) 'Regressive Evolution and Phylogenetic Age: the History of Colonization of Freshwaters of Yukatan by Fish and Crustacea' *Texas Memorial Museum Bulletin, 28*, 237-43

Wilkens, H. and Burns, R.J. (1972) 'A New *Anoptichthys* Cave Population (Characidae, Pisces)', *Annales de Spéléologie, 27*, 263-70

Zeiske, E. (1971): 'Ethologische Mechanismen als Voraussetzung für einen Übergang zum Höhlenleben. Untersuchungen an Kaspar-Hauser-Männchen von *Poecilia sphenops* (Pisces, Poeciliidae)', *Forma et Functio, 4*, 387-93

PART FOUR: APPLIED FISH BEHAVIOUR

INTRODUCTION

T.J. Pitcher

The final section of the book contains two chapters on applied aspects of fish behaviour, the capture of fishes by fishing gear, and the management of freshwater fisheries. One topic omitted is the behaviour of fish in aquaculture (see Bardach, Magnuson, May and Reinhart 1980; Muir and Roberts 1985).

In Chapter 18, Clem Wardle gives a detailed description of how fishing gear exploits the behaviour of fishes, with the aid of some superb underwater photographs of fish interacting with commercial fishing gear taken from film recently made by his team from Aberdeen. After a brief review of the main types of fishing gear and their general relation to fish behaviours, Wardle concentrates upon how fish respond to the various elements of a towed trawl. The chapter includes a new explanation of the often-reported fountain manoeuvre in fish schools. Wardle ends the chapter by describing modifications to trawls that exploit these findings: for example, a prototype trawl which separates different species has been developed as a result of this research.

Ken O'Hara, in the last chapter in the book, considers how fish behaviour impinges upon the management of freshwater fisheries for recreation and food. By and large, the author thinks that the major impact of behavioural studies is yet to come, partly because freshwater fishery management is often a highly empirical art. Practical fishery managers tend only to believe field-based studies, and O'Hara reviews techniques of tracking and telemetry. He describes how management may take advantage of information about behaviour in feeding, migration, fish passes, vegetation, water quality, and stocking, and in habitat improvement and mitigation. O'Hara concludes with a plea for the genetic conservation of wild fish stocks.

References

Bardach, J.E., Magnuson, J.J., May R.C. and Reinhart, J.M. (1980) (eds), 'Fish Behavior and its Use in the Capture and Culture of Fishes', *ICLARM Conference Proceedings, 5*, ICLARM, Manila, Philippines, 512 pp.
Muir, J.F. and Roberts, R.J. (1985) (eds), *Recent Advances in Aquaculture*, vol. 2, Croom Helm, London, 300 pp.

18 FISH BEHAVIOUR AND FISHING GEAR
C.S. Wardle

Throughout history, human hunters for fish have made use of their knowledge of fish behaviour in order to make catches. There are more than 22000 different species of teleost fishes, each with its own characteristic world of reaction and behaviour, so that numerous appropriate fish capture systems have been invented. Outlines of our knowledge of the sensory ability and behaviour of fishes have been summarised in the earlier chapters of this book. Fish behaviour is involved in catching fish, both on the oceanic scale, where the annual cycles of maturity cause migrations so that fish are found in different locations that become known to the fisherman by observation of their availability, and on a smaller scale, where the reactions of a fish to each part of an approaching trawl can cause the fish to swim into the codend. In order to be successful, the fisherman must have local knowledge of the day-to-day movements of the fish and of their likely distribution. In all fisheries, one of the most important of the fisherman's skills is to use the appropriate gear at the right time in the right place.

Review of Fishing Methods

Although the main discussion of this chapter will be concerned with the reaction and behaviour of fish in towed trawls, it is worth considering other fishing methods briefly to realise that each system has its own complex interaction of technology and fish behaviour. The purse seine, because of its huge size, is one of the most complicated nets to operate at sea, yet it is extremely simple in its concept of, first, rapidly sinking an inpenetrable wall of netting around a group of fish, then closing the bottom edge with purse strings, and finally pumping or scooping the concentrated fish aboard the vessel. Some examples of the way in which the physiology and the behaviour of tuna affect their availability to this fishing method at the oceanic scale in the Pacific are described by Sharp (1978). Tuna, herring, mackerel and other pelagic fishes of the ocean are caught in huge numbers by purse seine nets when they aggregate at certain times of the year, usually near coasts, for spawning or feeding.

Fish aggregation devices (FADs) make easier the catching of fish in warmer waters where natural aggregations do not occur. FADs are made by hanging tree branches or tent-shaped sheets of plastic beneath rafts, and they are very effective in certain areas of the sea in aggregating fish. The fish are then gathered by purse seine, baited hooks or gill nets (Hunter and Mitchell 1977). FADs have been found, by their stimulation of aggregation behaviour, to increase the productivity of local fisheries for very little expenditure of energy (Preston 1982). Natural reefs and artificial reefs such as shipwrecks, oil rigs and pipelines all influence the behaviour of those fish that are in or moving through an area by aggregating certain species and making local fishing easier.

A length of line ending with a hook embedded in a tasty morsel of bait forms another simple fishing gear that has led to numerous variations. Fish hooks have been discovered that date from the Stone Age, and they were probably used to catch fish from long before that time. Many more types of hook are available than there are species of fish, and each type of hook has been evolved by humans matching the mechanics of the device to the behaviour, shape and size of the fish (Hurum 1977). As in all other forms of fishing, using a line and baited hook involves a degree of luck, but the odds are shortened by careful study of the behaviour and likely distribution of the species to be caught. The intricacies of angling are the source of many well-known anecdotes and wise sayings on ways to hook the elusive fish. There is a world-wide research effort into finding simple reproducible artificial baits for commercial long-line fishing. The research relies heavily on experiments investigating the feeding behaviour of fish (Mackie, Adron and Grant 1980; Johnstone and Hawkins 1981).

Fish traps are constructed by humans in great variety, often unbaited, and they operate on some seasonal facet of the animal's behaviour that causes a fish to move through a local area of coast or river. Salmon are caught around the coast of the UK by so-called beach engines or stake nets in which walls of netting supported on stakes driven into the beach lead the fish through a series of non-return funnels into a holding chamber that can be lifted and emptied at the surface. Studies of the behaviour of fish carrying ultrasonic transmitters have shown that many migrating salmon can find their way along a breach through extensive lines of these commercial stake-net traps without being caught (Hawkins, Urquhart and Shearer 1979). Larger and more elaborate traps of this type are used to catch blue-fin tuna around the coasts of the Mediterranean Sea and its islands, and off the Atlantic coast of Spain (Rodrigues-Roda 1964). Migratory fish species are caught by various ingenious traps in all parts of the world.

Spears, hand nets, snares and invisible nets that gill or tangle the fish are skilfully used in many parts of the world. These types of fishing gear, as well as poisons and explosives, work only when they stimulate no behavioural response of the fish. Gill nets, drift nets and trammel nets rely on the fish blundering into the mesh and being unaware of its presence until too late.

Trawls and Danish seine nets have developed from the historical techniques of beach seining and beam trawling. Beach seines and beam trawls rely on a minimum of reaction from the fish. In the beach seine, like the purse seine which was probably developed from it, fish are herded and surrounded by a wall of netting guiding them into a short funnel and codend bag. In a beam trawl the fish are overrun by the rigid mouth of the net and trapped in a much longer funnel and codend. During the evolution of modern trawls and the Danish seine net, it is apparent, as we shall see in the rest of this chapter, that more and more of the fishes' repertoire of behaviour has been involved as these gears have become larger, more versatile and more effective.

Evolution of Gear in the North Sea

The evolution of each of these gears is outlined in Figure 18.1. This evolution demonstrates a close link between technological change and the knowledge of humans of fish and their reactions in fishing gear. For example, in 1848 Jan Vaevers found that by adding longer ropes and a large anchor to his beach-seine net and fishing it on the sea bed in deeper water, he could catch a previously unexploited abundant stock of flatfish. Danish anchor seining was born, and developed into the modern, man- and energy-efficient, Danish seine-netting technique. It was natural that beach seines were also towed to sea, the wings held open by two sailing boats, and so a beach seine was developed to create a light pair trawl that could be fished on the surface or the sea bed. Single boat sea-bed fishing in sailing vessels, with their unreliable towing power, used beam trawls, their size limited by the length of beam that could be stowed along the side of the vessel. Once the steady, reliable pulling power of the steam engine was built into fishing vessels, otter boards were invented to replace the beam. This allowed the nets to be made much wider, and to be folded and stowed.

Humans have a natural curiosity to know how their devices work and an urge to make them better if they do not seem quite right. However, once they work, great care is taken to avoid any change. One can imagine that the beach seine and beam trawls were often used in shallow clear water where their effect in catching fish could be watched by the fisherman. As soon as gears are fished in deeper water, they are out of sight and the operator must judge his net by indirect evidence such as the presence of polished areas on chains or the size of the catch for the effort made. The Danish seine net invented by Jan Vaevers demonstrated a quite dramatic advance on his countrymen's efforts with the beach seine because he caught many more fish. Similarly, the long-wing haddock seine used between 1930 and the 1950s was replaced 'overnight' in 1953 by the commercial introduction of the so-called Vinge trawl as a seine net. When this net was fished, it had twice the headline height and it caught twice as many haddock for the same effort (Thomson 1969).

As can be seen from Figure 18.1, gears continued to evolve as technology made new materials, new ship designs, new navigation aids, new charts, echo sounders and radar available to the fisherman. It is inevitable that humans as hunters continue to apply their knowledge of fish behaviour to the design and development of fishing techniques, while at the same time they make use of new technology. The growth of railways and the introduction of steam power to fishing caused a general mobilisation of the fish-

Figure 18.1: Chronological Chart Showing an Outline of the Evolution of the Beach Seine and the Beam Trawl. The chart reflects a diversification of fishing methods increasing the exploitation of fish behaviour as technology changes. Some of the technology that has influenced the gears is shown in the left-hand margin. For more details of the gears named, see Le Gall (1931), Davis (1958), Thomson (1969) and Wardle (1983)

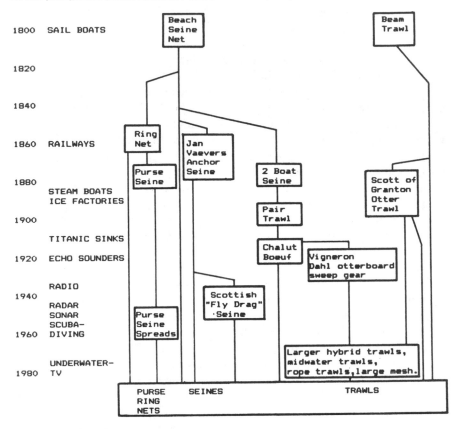

ing effort away from local traditional grounds to intrude on the fishing grounds of others. It also meant fishing quite new areas. For example, with the introduction of the otter trawl and the steam trawler, areas of the North Sea deeper than 100 m were made fishable; areas where, before that time, no fish had ever been taken by trawler (Figure 18.2). The increased range and power of the fishing fleet had generated local disputes all around the coast of the UK by 1880, when a Royal Commission was set up to investigate complaints of damage to gear and grounds by these new trawlers. The Royal Commission on Trawling (the Dalhousie Commission, 1883-85) recommended, among other things, that money should be set aside for the first time by government to support scientific investigation into how these new gears worked and to find out how they affected stocks and the sea bed. One hundred years later we are still asking similar questions for similar reasons.

Figure 18.2: A Major Change from Fishing Shallow Areas of the North Sea to Fishing Areas Deeper than 100 m was Made Possible by the Introduction of the Otter Trawl patented by Scott of Granton in 1894. The chart shows those areas from which fish were landed by steam trawlers at Aberdeen market for the first three months of the years 1891 (cross hatched) and 1901 (stippled); principal areas are darker. In 1891 using only beam trawls there were 760 landings averaging 3328 kg, and in 1901 using only otter trawls 644 landings of 6344 kg. Note the depth contours of 100 m (continuous line) and 200 m (dashed line). (Figure redrawn from Plates II and III and p. 140 of Wemys Fulton, 1902)

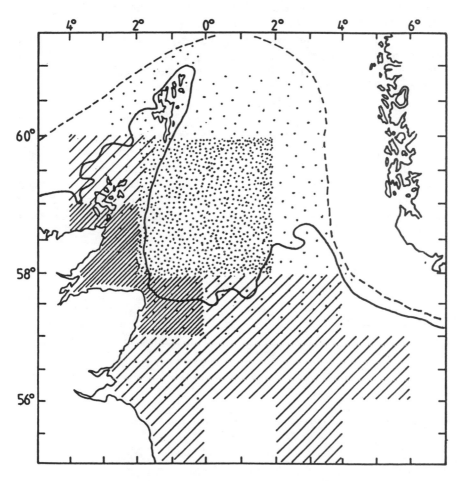

Modern Trawls

Modern trawls are made in many sizes to allow different types of fishing boat to tow them at a maximum speed: between 3 and 4 knots when using full power. A typical otter trawl, shown as a scale diagram in Figure 18.3 and as a sketch in Figure 18.4, can be towed by a 600 horse power vessel at 2 m s^{-1} (4 knots) and maintains relatively constant geometry as indicated in Figure 18.3. The 2.5 m × 1.5 m otter boards are made of wood or steel and are towed on the trawl warps (= cable or wire, Ed.) with an angle of attack that causes them to spread away from each other opening the front of the net. Otter boards remain about 30 m apart when the trawl is towed. Each wing of the net is linked to an otter board by sweeps (steel cables) 55 m long. The tips of the wings of the net are 10 m apart and so the sweeps run at an angle to the forward motion of about 10°. Sand clouds, thrown up by the turbulent swirls of water sucked in behind the two otter boards, are left as visually opaque trails spreading inwards and forming walls along the line of the sweeps (Plate 18.1). In this way a walled passage is formed on the seabed, 30 m wide at the otter boards and narrowing to 10 m wide at the net mouth. The upper lip of the trawl mouth, known as the headline, is held up by fifty 200 cm floats. It arches 6 m above the sea bed and the net tapers to the codend some 40 m behind. When a trawl is fished on smooth sand, the lower lip of the mouth of the net, called the fishing line, is weighted with chain and holds the edge of the net close to the sea bed. When fishing stony bottoms, various heavy ground gears are attached below this fishing line and can add considerably to both the visual contrast and intensity of the noise stimulus of this zone of the net. The heaviest ground gears are made up of rubber discs and wheels threaded on to chains and wires, all chosen to be extremely tough and to ease the relatively fragile advancing net over any stones and boulders that may be in its track.

Figure 18.3: A Scale Plan (Lower Part of Figure) and Elevation (Upper Part of Figure) of a Trawl that Can Be Towed by a 600 hp Vessel. The parts of the gear can be identified from Figure 18.4. The dotted lines indicate the various ranges of visibility discussed in the text in relation to the reaction behaviour of fish

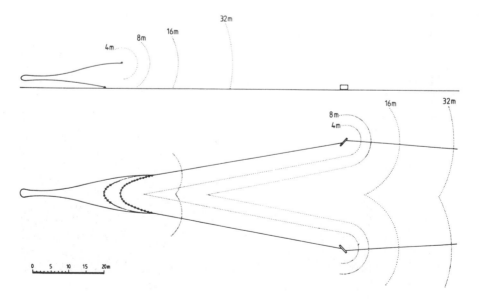

Figure 18.4: An Artist's Sketch of a Trawl in Action. A, Otter board; B, sweep wire; C, wing; D, headline; E, ground gear and fishing line; F, funnel of the net; G, codend

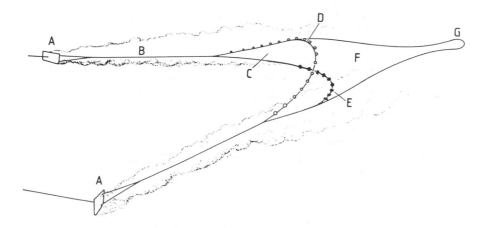

Plate 18.1: A Polyvalent Otter Board is Towed by the Warp (Bottom Left) towards the Camera. The trawl mouth (not visible) is to the left of the sand cloud some 50 m behind the otter board. Note how the sand cloud spreads in along the sweep forming a wall of turbulent, opaque water all the way to the wing end of the net

Source: Main and Sangster (1981a). (Crown Copyright.)

Fish Behaviour in Trawls

Observations of the reactions of fish to trawls have been made by unique diver-operated and remote-vehicle techniques developed and used by scientists at the Marine Laboratory at Aberdeen since 1975. These observations are recorded using a TV camera and video tape, and over this period a general pattern of reactions has been built up. A series of nine reports describe the techniques and many of the observations discussed in this chapter (see Main and Sangster 1978a-1983b). The development of the techniques, and the main conclusions of the work, are outlined in Wardle (1983). The fish behaviour patterns in trawls recorded during this period confirmed earlier observations on the Danish seine net made between 1965 and 1975 by diving (Hemmings 1973). The various parts of the trawl that will be discussed as stimulators of fish reaction are identified in Figure 18.4.

The present account attempts to interpret the reactions of fish observed during the process of capture by a trawl by applying knowledge of fish behaviour. To do this, fish will be followed through a trawl and their behaviour will be analysed at each stage. The stages of capture are outlined in Figure 18.5. In the figure the arrows indicate alternative reactions of fish at each of the positions in the gear: they do not show the relative importance of these pathways.

At the approach of a fishing boat towing a trawl, the first indication to the fish of an intrusion will inevitably be the sound of the engines and propeller of the boat. The hearing ability of fish has been discussed by Hawkins in Chapter 5 of this volume. Noise from the engines of motorised vessels towing at near full power will be heard by fish well beyond the range at which the vessel could be seen. In clear water when fish are no deeper than 40 m, a ship may be seen as a silhouette on the surface by fish beneath it. At greater distances the ship will certainly be heard but not seen. It is well known that vessel noise can cause fish to change their depth. Fish have been observed to move deeper, and even down to the sea bed, as a vessel passes. Unnecessary noise is carefully avoided on purse-seining vessels for this reason, and the skipper knows that sudden noises made on or by the vessel will frighten those fish he is pursuing and make them dive away from the ship and net.

In general, vertebrates use sounds for communication and they are also listening out for unfamiliar sounds that warn of dangers. Sound sources arouse curiosity or enhance sensitivity such that a visual explanation is expected. An animal is alerted by the sound and the source is watched out for and reacted to more effectively if and when it arrives. Sounds from ships and fishing gears will often be heard, but rarely seen due to the differences in range of these two senses under water (Hawkins 1973).

As the sounds of the towing ship begin to fade, a new sound from the otter boards will start to grow. This sound is generated by the otter boards

Figure 18.5: Outline of the Points of Fish Behaviour during Capture in a Trawl. The letters indicate the identify of the position in Figure 18.4

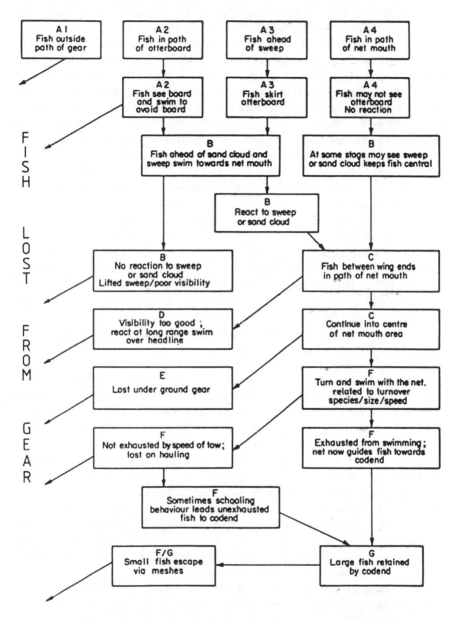

scraping and knocking on the sea bed, and will vary according to the nature of the sea bed. Observations of fish in the region of otter boards suggest that the visible range determines the distance at which a fish first reacts to the otter board. For example, in conditions of poor visibility, fish can be observed close to the board, only just avoiding collision, whereas in clear water, fish are seen reacting well away from the board, skirting the area at about the distance one would estimate they first see the board. Unfortunately it is not yet possible to observe at high enough resolution the reactions of fish outside the visible range.

When the otter board is viewed from the sea bed directly ahead in its track, the sand cloud which is illuminated by down-welling light from above is seen as a bright halo-like margin around the darker board (see Plate 18.2). The otter board is a good example of a high-contrast image, with added sound attracting attention to its approach. Four different reactions to the board, depending on the position of the fish in relation to the track of the board, are shown as A1 to A4 in Figure 18.5. It is assumed from the observations that the reactions of fish, although affected by sound, are primarily in response to the presence of the visual stimulus. The reactions to the otter board will now be analysed on this basis.

In the majority of roundfish species, each eye is able to see a field of vision of 170-180°. The two fields overlap in front of the fish, giving binocular vision, and to the rear of the fish there is a blind zone of about 20-30° on either side of the fish's tail (Walls 1942). A plan indicating the fields of view is shown in Figure 18.6. A fish positioned on the sea bed exactly in the path of the otter board reacts to the board as soon as the presence of the board is sensed visually. Observations have shown us that fish swim around the otter board at a distance just within the range of visibility. With a first limit set by the visual field of the fish, a second limit set by the visual range of the approaching otter board, and a third limit setting a swimming speed, the most direct escape or avoidance route can be predicted as in Figure 18.7a. It is assumed that the blind zone extends to 25° on either side of the fish's swimming track, and the most direct escape route, while keeping the otter board just in the rear edge of the field of view, is a course whereby the otter board (Figure 18.7a, horizontal line) is kept at 155° to the swimming direction. In the scale plan of the reaction of a fish to an otter board (Figure 18.8a) the figures indicate time in seconds for coincident positions of the otter board and the fish. The fish chooses a swimming speed in this example which matches that of the otter board. Underwater visibility can vary greatly, and in Figure 18.7b are shown the escape routes assuming first reaction of the fish at maximum visible ranges of between 2 and 8 m from the board. Figures 18.7b and c compare the resulting track of a fish swimming at a speed equal to, and at half the speed of, the approaching otter board, respectively.

This model can be extended to describe the reaction of a group of fish to the otter board. In Figure 18.a the otter board moves in a straight line from left to right and each fish shown in the figure is keeping one eye on the board while maintaining a swimming track close to an angle of 155° to the current position of the board. In order to maintain visual contact with the board, each fish decreases this angle as it approaches the maximum visible range (the circle drawn around the otter board in Figure 18.8a). In this way

Plate 18.2: Three Photographs made as the Otter Board Approached the Cameraman on the Sea Bed. The most distant view shows the otter board just visible with the edge outlined by the sand cloud catching the light. (Crown Copyright.)

Figure 18.6: Plan Showing the Two Hemispherical Fields of Vision of a Roundfish like a Cod. The line forming a tangent to each eye delimits the outer edge of the field of view of each eye as the fish passes through the centre of the circle. The sector immediately behind the fish is the blind zone and the sector ahead of the fish is a zone of binocular vision. The dotted circle represents a limit to range of visibility. (See text for discussion.)

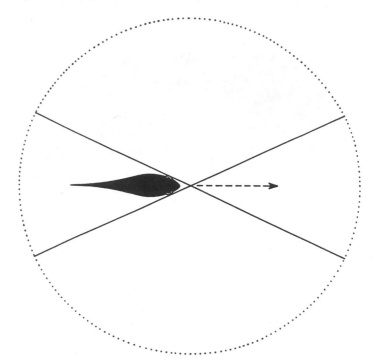

the group of fish gather together again automatically behind the board. The same behaviour is seen to any similar moving object. [This is a novel explanation of the 'fountain effect' seen in fish schools — see also Chapter 12 by Pitcher, this volume. Ed.] A human observer moving through a group of fish can observe this zigzag orientation of those fish seen directly ahead, and as they split off to the left and the right, one eye of each fish is always just visible (Wardle 1983, Figure 4.7). The fish pass on either side and rejoin as a group behind. However, the sweep and the sand cloud following the otter board come between the resulting two groups of fish (Figure 18.8b). The spreading sand cloud (Plate 18.1) takes over as the next stimulus, herding the inside fish towards the track of the mouth of the trawl and the outside fish to freedom.

The observed swimming speed used by the fish when skirting the otter board is slow and close to that of the approaching board: why is this? Studies of fish swimming have shown that fish are able and willing to cruise for long periods at slow speeds. Their endurance is virtually unlimited as long as the muscular contractions involve aerobic respiration releasing

Figure 18.7: The Predicted Tracks of a Roundfish Reacting to an Approaching Object Such as an Otter Board. The horizontal line is the track of the otter board where the numbers indicate time in seconds. The curve of the fish track is set by keeping the otter board at each position just at the edge of the visual field (dashed lines shown in Figure 18.7a). (a) shows how the track is determined; (b) shows the swimming speed of the fish set at half the approach speed of the otter board (the fish tracks shown start from points 3 to 8 ahead of the otter board); (c) shows a similar set of tracks when swimming speed is equal to the otter board approach speed. The numbers marked on the curves identify coincident positions of fish and otter board

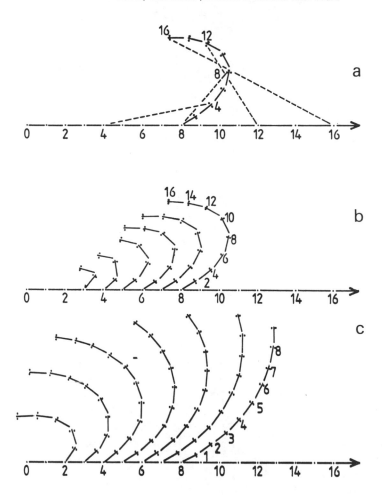

energy using continuously replaceable metabolites and oxygen (Wardle 1977; Bone 1978). On the other hand, fish can reach high swimming speeds for limited periods by contracting the large white lateral muscles which are fuelled by the rapid conversion of a limited muscle-glycogen store to lactic acid, for which no oxygen is required (Black, Robertson and Parker 1961; Wardle 1975). However, if the fish uses its high-speed swimming, it is obliged after only minutes to seek shelter and rest for as

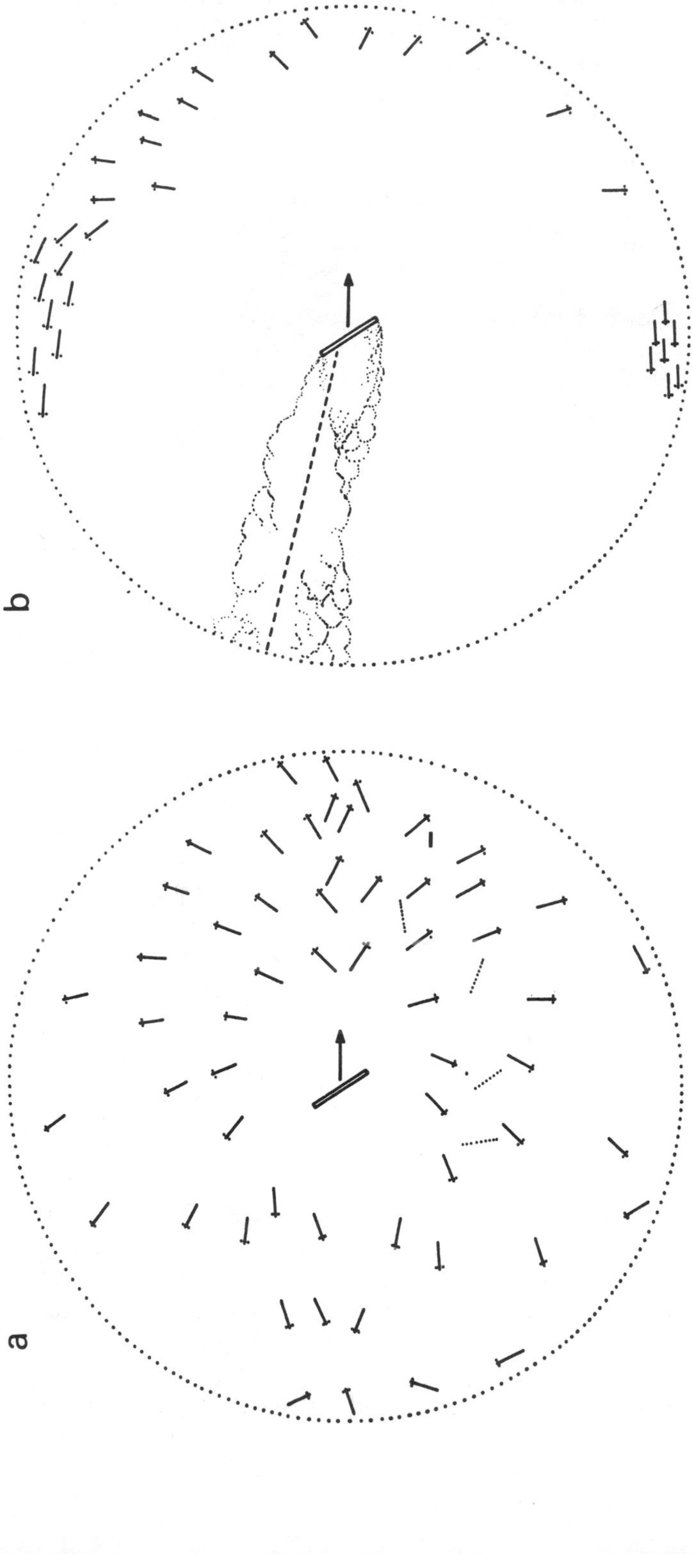

Figure 18.8: Plan View of the Predicted Reaction of Fish due to Visual Stimulus from (a) an Isolated Moving Object Like an Otter Board or Human Swimmer; and (b) an Otter Board with its Trailing Wall of Sand Cloud and Sweep Wire. The dotted circle indicates a maximum visible range of 8m

long as 24 h while its glycogen store is rebuilt. During much of this period the fish is without adequate escape ability and vulnerable to attack (Batty and Wardle 1979). Fish in general are observed to use the minimum swimming speed in order to maintain a safe distance between themselves and an identified threatening object. In this way it can be argued that they maximise their endurance for swimming and maintain their aerobic reserves for real emergencies. This limit to the fishes' swimming response, together with the limit imposed by their visual field, seems to result in a characteristic but effective avoidance strategy when an otter board threatens to overrun them.

The dimensions of a typical trawl can be related to the visible range of objects under water, which can vary from zero to perhaps a maximum of 40 m (see Chapter 4 by Guthrie, this volume). In Figure 18.3 visibility ranges of 2, 4, 8 and 16 m are indicated by the dashed contour lines. Consider, then, a group of fish spread across the track of the gear, in a line some distance ahead of the otter board (see Figure 18.9). The first fish to react visually to the otter board will be in a position directly ahead of the board at a point where, on Figure 18.9, the circumference of visible range (16 m) meets the band of fish. The model (Figure 18.7) describing the escape reaction route is now applied and these fish are drawn in Figure 18.9 moving away at ± 155°. If each fish uses the minimum speed needed to maintain the otter board just within visible range, it should be expected to follow a course similar to that plotted in Figure 18.7a, but starting at a first visual reaction distance of 16 m. For the purpose of illustrating this point on Figure 18.9, very clear water with a visible range of 16 m is chosen. However, by referring to Figure 18.3, it is clear that with only 4 m or 8 m visible range, those fish situated further than 4 or 8 m from the board as it passes will not have identified the visual stimulus of the board. They might to some extent react to the movements of neighbours that move in their direction, aided by the sound, but might otherwise show no reaction until later, when some other part of the gear approaches near enough for them to see it (see Figure 18.5).

During the 20 s after the otter board has passed, those fish swimming gently towards the as yet invisible mouth of the net are guided in by the bridles and sand cloud, but they are then quite suddenly surrounded by the visible array of netting, ropes, floats and bobbins of the trawl gear (see C, D and E, Figures 18.4 and 18.5). In the 600 h.p. net (Figure 18.3), with 8 m visibility both of the wings will be visible from the centre of the track of the net. The buoyed headline will be seen as a high-contrast silhouette as it passes overhead but not much before it does so. At about the same time the ground-gear bobbins, black against the white sand and outlined by the wisps of sand thrown up between them, form a further strong visual stimulus coming directly towards the fish across the sea bed. The fish usually turn abruptly at this point in the mouth of the trawl and swim forwards, just matching the towing speed of the gear (Figure 18.9 and Plate

Figure 18.9: Plan Showing the Reaction of Fish Predicted by Visual Reaction with a Visible Range of 16m (dotted lines). For discussion see text

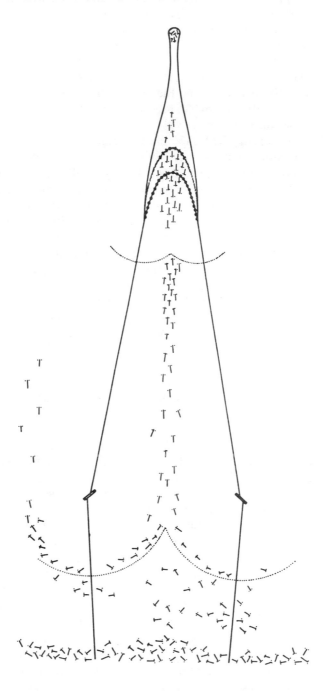

18.3). The visual field of those fish swimming forwards at this point contains monocular images of netting on either side, and chains and bobbins intruding into the rear view from both sides (Figure 18.10). It is not surprising that the fish now try to hold, for as long as possible, a stable unchanging position relative to the fast-moving visual stimulus of the gear. This behaviour appears to be an optomotor response, and Hemmings (1973) demonstrated that haddock swim holding station with the mouth of a net even when the netting behind the mouth is completely removed. For further discussion of the optomotor reflex and its relation to rheotaxis and the behaviour of fish in trawls, see Harden Jones (1963), Arnold (1974) and Wardle (1983). (See also Chapter 12 by Pitcher, this volume.)

Direct observations have repeatedly shown that larger fish such as saithe, cod and haddock swim for very long periods in the mouth of trawls and Danish seine nets, whereas the smaller species such as sprats and sandeels try hard to maintain their position but after only a few minutes give up and allow the net to pass them. These observations illustrate very directly the scale effect in fish swimming performance (Wardle 1977). At this position in the trawl mouth, all sizes of fish are being stimulated visually to swim at exactly the towing speed, say of $2\,\mathrm{m\,s^{-1}}$. It is well established that a typical teleost fish moves forwards 0.7 of its body length for each completed tailbeat cycle (Wardle and Videler 1980). A cod 1 m long therefore needs 2.9 tailbeats per second to maintain its position in this situation, whereas a 10 cm cod or sprat is struggling to swim at the same speed by beating its tail at a frequency of 28.6 Hz (see Figure 18.11). The large fish is easily cruising aerobically and could continue for hours, whereas the small fish is at or near its maximum speed using its anaerobic muscle power output, and is soon exhausted. The effect of fish size on the tailbeat frequency needed to swim at 1 and 4 knots is shown in Figure 18.12. The maximum tailbeat frequency of each length of fish at three temperatures is also shown in this figure. These maximum performance figures are calculated from measurements of the contraction time of the swimming muscle (after Wardle 1975).

Whatever the size of the fish, there is an enforced change in the behaviour response to the net if the fish becomes exhausted. This change in behaviour may be stimulated by the accumulation of lactic acid in the muscle tissue, or the corresponding depletion of glycogen, or it may simply be the result of loss of muscle power. The effect is an enforced change of tactics where the behaviour of maintaining station with the gear gives way to turning or dropping back. The fish now takes an increased risk of entering an unknown area and dodging through the threatening array of visual patterns.

Looking again at the scale plan of the net (Figure 18.3) it is seen that the distance from the mouth area to the codend is some 30 m. With a visible range of 8 m, the fish in the mouth area will see, as it turns what looks like a clear passage surrounded by a circle of netting, and the vanishing lines of

Plate 18.3: Fish Typically Hold Station with the Trawl, Swimming for Long Periods in the Mouth Area. The mackerel (top) were photographed at 100m deep with a remote-controlled vehicle. Flash was used for the photograph; the TV camera could just see the fish by natural light at this depth. The saithe (bottom) were photographed by natural light by a diver. (Photo from Main and Sangster 1981b) (Both Crown Copyright.)

Figure 18.10: Plan Suggesting the Contents of the Visual Fields of the Left and Right Eyes of a Roundfish Swimming Forwards at the Same Speed as the Net inside the Mouth of a Trawl. Note the chain in the blind zone (shown dotted) is not seen by the fish. The net panels on the opposite sides of the gear disappear outside the visible range indicated by the dotted circle

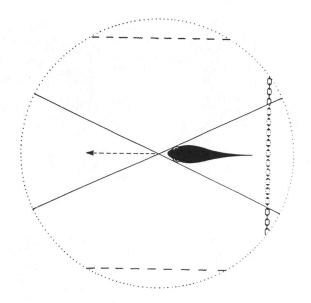

Figure 18.11: Tracks of the Tail Tip of a 100 cm and a 10 cm Fish. These tracks illustrate why a different frequency of tailbeat is required of a 10 cm and 100 cm fish swimming at 2 m s^{-1} (4 knots). For discussion see text

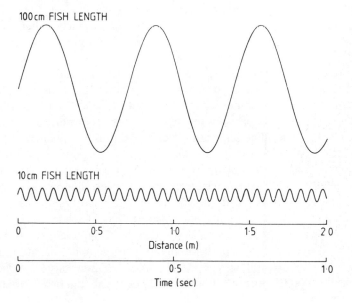

Figure 18.12: Tailbeat Frequency of Fish of Different Sizes Swimming at 0.5 and 2ms⁻¹ (1 knot and 4 knots, Firm Lines). Dotted lines show the maximum tailbeat frequencies of fish at the indicated temperatures. Dashed line shows the tailbeat frequencies at maximum aerobic cruising speed. See text for discussion

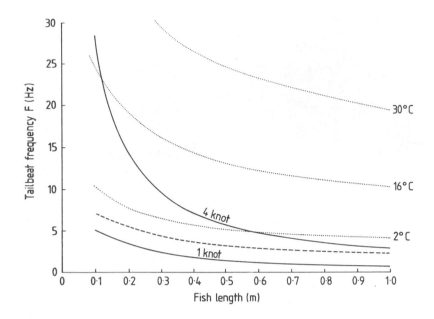

the thicker seams of the netting where the panels are joined together (Plate 18.4, top). In practice the larger fish as they become exhausted turn and swim keeping clear of the net walls and are guided down the funnel of the net. They are sometimes seen to delay their passage towards the codend by turning and attempting to swim forwards, particularly where the net narrows or is obstructed, but these fish eventually end up exhausted and trapped in the codend. Smaller fish swim only briefly, keeping station in the mouth area, and then turn back into the funnel. They are often seen to swim directly towards the meshes at various points along the net walls and particularly the top of the net funnel. Many of the smaller fish do reach the codend where they also pass out through the meshes just ahead of the catch (Plate 18.4, bottom). The scale effect described here results in a large turn-over of small fish passing into, through and out of the trawl, a very small proportion of these fish being represented in the catch. At the other extreme, groups of large fish have been observed swimming in the net mouth for long periods but do not become exhausted and are not caught; they swim away when the net is hauled from the sea bed.

Fish vision has been discussed by Guthrie in Chapter 4 of this volume, and we can now consider parts of the fishing gear as visual stimuli. In general, fish are specialists at seeing low-contrast images, and it was suggested earlier that the role of sound might be to attract the fishes' attention to

Plate 18.4: Top, Saithe are Seen Swimming Towards the Codend. The Codend is not visible but is the darker zone (middle right) where the lines of the funnel of the net vanish. The impression is of a clear passage through the net due to the underwater visibility. Some of the saithe appear to be feeding on the sea bed. (Photo from Main and Sangster 1983a). Bottom, Mesh Selection at the Codend. Small fish escaping from among the larger trapped fish. (Crown Copyright.)

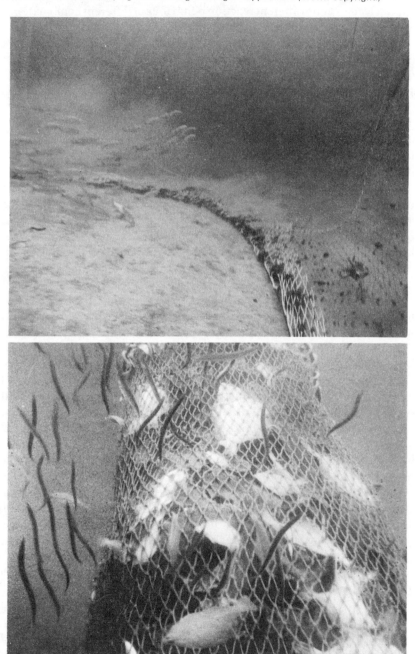

these images. Images formed by the approach of fishing gear will nearly always be low contrast when they are first presented to the fish at maximum visible range, but some, like the otter boards and ground bobbins, rapidly become of high contrast. The distance at which the image is first seen varies, and can lead to quite different reaction behaviour in different combinations of visible range and swimming performance of the fish.

Water depth has a large effect on the colour and contrast of materials and structures used to make up fishing gears, and so modifies the gear as a visible stimulus. For example, the bright-orange twines, often used to construct the panels of trawls, look to the human eye a shade of grey-green when seen below 20 m depth. When viewed horizontally against the grey-green water background, even when close up in good visibility, it has a very low contrast. Images of low contrast are detected at much shorter range than images of high contrast. Green netting becomes nearly as bright in this light as white, whereas red netting looks quite black when viewed at the same depths. These appearances of the various colours of materials can be used in appropriate ways to make the various parts of a trawl have well-defined functions as a stimulus to the fish. From these observations the net builder should aim to make the visible stimulus of the net less vulnerable to the variations that occur in the optical properties of the water background. The maximum visible range is achieved by putting highly reflective white material next to black. Observations of fish reactions indicate that there are areas of the fishing gear where application of high-contrast patterns will make the stimulus clearer to the fish. For example, the otter boards should be black with white reflective borders so that they are always seen at maximum range.

In good visibility, fish on the ground ahead of a trawl mouth have been observed to rise over the headline. To avoid fish reacting in this way, the range at which the headline is first seen can be dramatically reduced by the use of grey netting and countershaded floats. However, when this headline passes overhead, the floats and netting become silhouetted against the downwelling light and so change from least to most visible and help to keep the fish near the sea bed. The net viewed from just inside the mouth region should have maximum contrast in order to define an area where the fish will swim forwards. Strong vertical black-and-white-striped patterns would create the maximum stimulus in all visibility conditions. Behind this region of collection and exhaustion, the netting should have the least visibility when the fish turns and looks back towards the codend. Grey or orange netting used here would encourage fish to swim down this tunnel, even in clear-water conditions. A net having some of the proposed properties mentioned in this section is sketched in Figure 18.13.

The colour of the sea bed can vary and will alter the contrast and so the appearance of the lower panels of a bottom trawl. A midwater or pelagic trawl is usually towed out of visible range of the sea bed. From inside the mouth of a pelagic trawl, the top netting panels are seen by the fish against

Figure 18.13: Artist's Impression of a Trawl Design Suggested by the Fish Behaviour Observations Discussed in the Text. An enlarged view of the countershaded float is shown above the headline

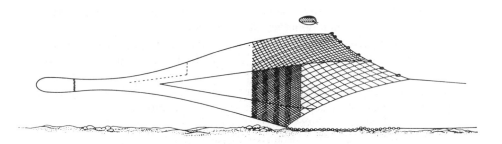

the light from above, and the lower panels are seen against a black background. In order to create a surrounding pattern of maximum contrast when seen from inside the mouth, the top of the pelagic net must be made in black, the base in white, and the sides in black-and-white stripes. To make the rear parts invisible, so that exhausted fish are stimulated to turn towards the codend, the top should be white, the base black and the sides grey.

Direct observations of nets show us that the angular size of images at different ranges has important effects. For example, when seen with one eye it is difficult to distinguish a mesh 7 m across made from 23 mm twine seen at a range of 40 m from a 1.7 m mesh of 6 mm rope seen at 10 m. Both give a 10° image at the eye. In many trawls the small mesh used in the main net panels contributes nothing to the distant view, and the image seen is the pattern of strengthening ropes that form the seams and framework of the net (for examples see Plate 18.3, bottom, and Plate 18.4, top). Large meshes made from thick ropes have been found extremely effective in mid-water trawls of large size. These large meshes have important properties of creating panels of large area with low hydrodynamic drag, allowing the same ship to tow a bigger net. The image of the large mesh made form thick rope will only stimulate from maximum visible range when attention is paid to using the appropriate colour or shade in relation to the angle of viewing. To preserve the appropriate stimulus when new designs are considered, attention must be given to contrast and dimensions in relation to variable water conditions. In this way the fish reactions may be controlled more precisely.

The stimulus to behaviour generated by the sense of vision is limited by the light level, and it is important to consider how the relation between the stimulus and the reaction may change as the light level falls. Very few animals have been studied in sufficient detail, and our main frame of reference for considering the behaviour of fish is set by knowledge of owls, cats, humans and pigeons (Martin 1983). Martin points out the behavioural threshold is at extremely low light levels of 10^{-6} to 10^{-7} lux for cats, owls and man. The low-light TV camera, used to observe fish reactions to fishing gear, ceases to form images at levels below 10^{-3} lux, and bioluminescence may then take over and be seen as flashes. The milky-green light of disturbed dinoflagellates can show up moving net panels (Wardle 1983).

As light level falls, white objects become more and more like black and they are eventually not seen against a black background. All objects become less visible due to this reduction in contrast. Reaction distance is reduced as the contrast is reduced (Anthony 1981). The loss of behavioural response to visual stimulus is related to the inability of the eye to pick up these differences in contrast. Of the few animals carefully studied, pigeons show a behaviour threshold at 10^{-4} lux, humans and owls at 10^{-6} lux, and cats at 10^{-7} lux (Martin 1983, Figure 7). Not much is known about fish, but preliminary experiments with mackerel have shown that they cease to school at levels near 10^{-7} lux (C.S. Wardle, unpublished). It is argued here that the main reactions of fish in the trawl are due to the stimulus of behaviour by visible images. Convincing proof of this argument would be the observation of no reaction when light was below the threshold for stimulating the behaviour of the fish. There are major problems in obtaining such observations; first, how to observe the fish, secondly how to know the light level when the observations are made, and thirdly to find conditions where there is no bioluminescence and low-enough ambient light levels. Flash photographs taken from a remote vehicle positioned inside the mouth of a trawl at midnight in February off Orkney, UK, have shown a few examples of fish showing no orientation to large bobbin wheels moving at $1.5 \, s^{-1}$ only centimetres before collision with the fish (see Plate 18.5). This sort of evidence, if consolidated, tells us that none of the other senses attributed to the fish can be used to orientate or react to fishing gears within the short time period available at these towing speeds.

A shortening of the reaction distance as light level drops can be observed by filming fish and by careful use of low levels of artificial light. In brightest daylight, in clear water with 40 m visible range, reaction distance may be 40 m, whereas at night, in the same water conditions, fish are seen to approach within 1 or 2 m of the gear before showing any signs that they are aware of its presence. Fishermen know that in daylight they do not catch fish in certain areas whereas at night at the same position they are successful. Recent filming at 100 m depth in daylight in such an area off Fair Isle, UK, showed large numbers of haddock swimming over the head-line of a 600 hp trawl similar in scale to that shown in Figure 18.3: none was caught during a 6 h tow. At night, however, large numbers were seen in the net mouth close to the bobbin region and a large catch was accumulated. Referring to Figure 18.3, the effect of darkness was to reduce the visible range from 40 m to less than 2 m. The effect on a fish of the reaction distance changing can be considered by estimating the time from

Plate 18.5: Are these Photographs Evidence of No Reaction when Fish are in Darkness? (Three flash photos taken inside the mouth of a trawl towing at 1.5ms^{-1} in February 1983.) The bobbin rig of the trawl ground gear is just on the right-hand edge of each photo, moving to the left. In the top photo taken at 13.50 GMT, the sandeels are oriented and swimming ahead of the ground gear. In the middle photo taken at 22.43 GMT, sandeels and other fish are pointing in all directions. In the lowest photo at 00.05 GMT, three roundfish are pointing in three directions, as are the less visible sandeels. (Crown Copyright.)

first seeing the object to collision with it, (see Figure 18.14) and then noting the distance that must be swum in this time, that is, the speed needed to pass around the object (see Figure 18.15). For example, the headline of a net first seen at 40 m range and approaching at $2 \, \mathrm{ms^{-1}}$ takes 20 s to collision. To avoid the 5 m-high headline the fish needs to swim up from the sea bed at only $0.25 \, \mathrm{ms^{-1}}$, but if the headline is first seen at 5 m, a swimming speed of $2 \, \mathrm{ms^{-1}}$ will be needed: a speed close to the maximum speed of many smaller fish.

Observations of fish both in the wild and in large tanks indicate that many species develop home grounds which are areas where the fish spends most of its time. Fish are extremely cautious when new strange objects enter their home ground. Cod, haddock and saithe can be trained to race between feeding lights through an area with which they are familiar. The same fish will not race into an area that has not been previously explored. Tank experiments in which these species were trained to race 8 m between feeding lights demonstrated a timidity of fish to pass a new object, such as a rope lain across the tank floor while the fish were feeding at one of the

Figure 18.14: Time to Collision with an Object First Seen at Visible Range Approaching at the Speeds Indicated on the Firm Lines. (One knot is approximately $0.52 \, \mathrm{ms^{-1}}$.)

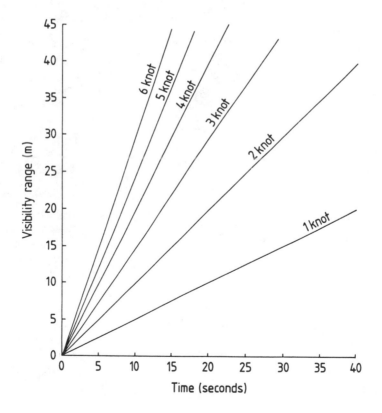

Figure 18.15: Swimming Speed Required to Clear the Headline of a Trawl Approaching at 2m s⁻¹ (4 knots). The headline heights drawn are 5m (lower line) and 10m (upper line). The fish starts to swim upwards from the sea bed at the different visible ranges indicated

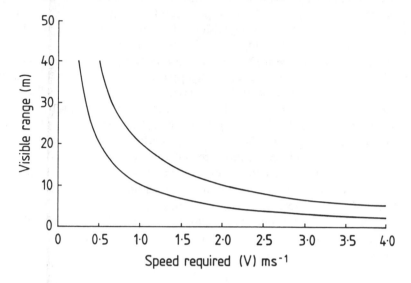

lights. When the other light was flashed, the fish started rapidly in its direction but swerved aside when they came up to the rope. Several minutes were spent patrolling before they cautiously crossed the rope and raced to the calling light and food. Replacing the rope with a large mesh net through which these fish could easily swim caused longer delays. It is equally interesting that if the large mesh net were left in position, the fish would race through it after a day or so without hesitation. The acceptance by the fish of the intruding object (that is, when timidity is lost) might be considered a process of habituation. These experiments indicate a relatively long period of timidity stimulated by quite simple objects intruding into the fishes' home ground, and have some significance in the context of a fast-moving gear where required reaction times are very short. Moving away from or holding station with the approaching fishing gear indicates a timidity to the approaching object and reluctance to allow the device to pass, and this leads to the device herding the fish. Observations have repeatedly shown that a rope towed across the sea bed will herd fish ahead of it. If the rope is angled to the forward motion, the fish slide along the rope and will be concentrated at one end. This action leads fish along the trawl sweep or the Danish seine-net rope into the mouth of the net from a wide-swept area of sea bed.

Those fish that school tend to behave *en masse* when responding to such stimuli (see Chapter 12 by Pitcher, this volume). The erratic decision-making of individuals responding to stimuli slightly differently from their neighbours is smoothed out by the general reaction of the school. The erratic individual returns to the school after a brief lonely excursion. Schooling sandeels have been observed to be herded by rope arrays, and as they become exhausted trying to swim ahead of the ropes, the whole group flows back, keeping clear of the ropes. Sandeels, sprats, mackerel and saithe have all shown this mass decision when in the mouth of trawls. A large group of saithe swimming quite easily in the mouth of a trawl and predictably able to outswim the net due to their large size all turned and swam from the mouth to the codend, apparently due to this *en masse* decision generated by their schooling behaviour.

Mackerel is a species that spends a lot of time in travelling schools (see Chapter 12 by Pitcher, this volume), and when observed in trawls they are seen to be caught just like haddock and saithe because they become exhausted swimming in the mouth area of the net (see Plate 18.3); the speed, however, must be greater than $1.5\text{-}2\,\text{ms}^{-1}$. At slower towing speeds mackerel explore all areas of the net in small schools, freely moving in all directions. Mackerel have been seen swimming forwards steadily from well back in the funnel of the net, overtaking the net and leaving through the mouth. They have also been observed feeding on sand eels swimming in the trawl mouth.

Conclusions

The more information we can gather of this sort, the more we can think in terms of making use of these patterns to make a gear more precise in its action. This does not necessarily mean more efficient in a crude sense of simply catching more fish. Although a sequence of general reaction patterns has been described here, we are beginning to notice that each species and each size group of fish have some distinct and specific reactions when passing through a trawl. Sandeels are often seen in trawls, and because of their short endurance, small size and schooling behaviour are often seen to separate from other species as they respond to the fishing gear (see Plate 18.6). Such separation of species can be exploited. It has, for example, become clear that haddock and cod with their different behaviour can be separated (Main and Sangster 1982b). When haddock become exhausted in the mouth of the net, they drop back into the net, rising high over the ground line into the top part of the net mouth (see Plate 18.7), whereas cod remain low near the sea bed. A horizontal net panel, with its leading edge dividing the net mouth, can lead fish to different codends, and such a design has been found quite practical in separating cod and haddock. Each of the codends can have different mesh sizes to select a different size range. *Nephrops norvegicus,* the scampi or Norwegian prawn, stays low in a trawl and can be separated from most fish, except cod and flatfish, by a similar approach (Main and Sangster 1982a). At present *Nephrops* trawls in Scotland are allowed to have

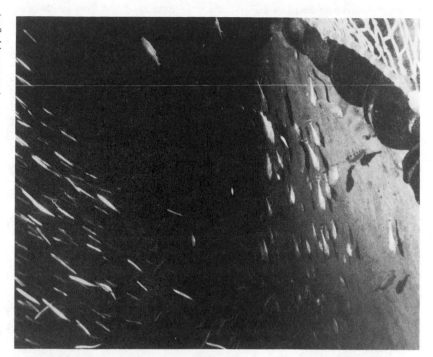

Plate 18.6: The Sandeels Separate Completely from the Other Species Still Swimming Ahead of the Ground Gear in the Mouth of a Trawl. (Crown Copyright.)

Plate 18.7: Top, the Haddock (Black Spots, Middle, Top, Left) Are Just Starting to Rise as they Tire and Stop Swimming in the Mouth of a Trawl. Cod, flatfish and saithe are seen nearer the sea bed. From Main and Sangster 1983a. Bottom: Haddock and sandeels rise up above the level of a separating panel away from cod swimming close to the sea bed. When the cod tire, they fall back beneath the separating panel.

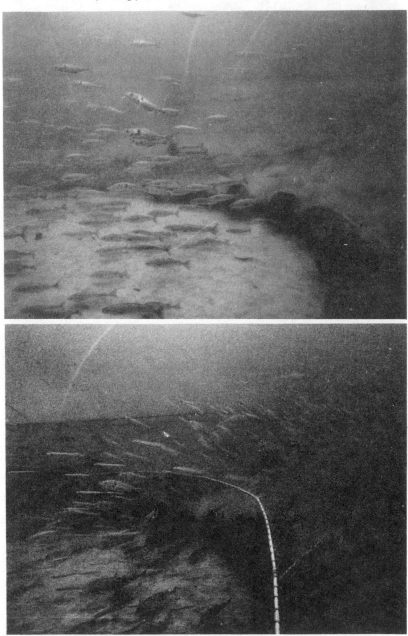

Source: Main and Sangster (1982a). (Crown Copyright.)

meshes as small as 70 mm and these nets sometimes catch a great number of young roundfish. The *Nephrops* never rise more than 70 cm from the sea bed whereas the small roundfish will tend to rise as they tire, and pass into the upper zone of a separator net. The upper codend can have a large mesh, only retaining the marketable sizes of these fish.

The possibilities for increasing the precision of fishing grow as we begin to see these patterns of behaviour, as we develop explanations, and as we apply appropriate stimuli in new designs of fishing gear.

Summary

The fishing methods devised by humans to hunt the world's stocks of fish involve many aspects of the behaviour of the different species caught. This chapter considers modern trawls and the sequence of stimuli they present to fish. At the start, long-range sound is followed by visual stimuli at a range dependent upon water conditions. Direct observation of fish in the trawl has shown a sequence of typical responses as fish progress towards the codend. First, an avoidance reaction observed at the otter board may be modelled using visual distance, swimming performance and limits of the visual field, explaining how fish are funnelled towards the mouth of the trawl. Next, the optomotor reflex, timidness, swimming performance and size explain how fish that reach the mouth of the trawl turn forwards within the net. Finally, following a period of swimming at the mouth, exhaustion leads to a change in behaviour, and the fish allow the net to overtake them, turn and seek a clear visual path back through the centre of the long funnel that leads to the codend. Although sound is important in drawing attention to the approaching gear, observations suggest that vision is the predominant sense driving precise, high-speed reactions at close range within the trawl. Selecting net construction materials to form appropriate visual stimuli can make the performance of gear more consistent in variable water conditions. Furthermore, trawls that can separate their catch into species have been designed to exploit detailed behavioural differences in the responses of some species to components of the fishing gear.

References

Anthony, P.D. (1981) 'Visual Contrast Thresholds in the Cod *Gadus morhua* L.', *Journal of Fish, Biology*, *19*, 87-103

Arnold, G.P. (1974) 'Rheotropism in Fishes', *Biological Reviews*, *49*, 515-76

Batty, R.S. and Wardle, C.S. (1979) 'Restoration of Glycogen from Lactic Acid in the Anaerobic Swimming Muscle of Plaice *Pleuronectes platessa* L.', *Journal of Fish Biology*, *15*, 509-19

Black, E.C., Robertson, A.C. and Parker, R.R. (1961) 'Some Aspects of Carbohydrate Metabolish in Fish', in *Comparative Physiology of Carbohydrate Metabolism in*

Heterothermic Animals, A.W. Martin (ed.), University of Washington Press, Seattle, pp. 89-122

Bone, Q. (1978) 'Locomotor Muscle', in W.S. Hoar and J. Randall (eds), *Fish Physiology*, vol. VII, *Locomotion*, Academic Press, New York, pp. 361-424

Davis, F.M. (1958) 'An Account of the Fishing Gear of England and Wales', *MAFF Fisheries Investigations, Series II, 21*(8), 1-165

Harden, Jones, F.R. (1963) 'The Reaction of Fish to Moving Backgrounds', *Journal of Experimental Biology, 40,* 437-46

Hawkins, A.D. (1973) 'The Sensitivity of Fish to Sounds', *Oceanography and Marine Biology Annual Reviews, 11,* 291-340

Hawkins, A.D., Urquhart, G.G. and Shearer, W.M. (1979) 'The Coastal Movements of Returning Atlantic Salmon, *Salmo salar* L.', *Scottish Fisheries Research Report, 15,* 1-14

Hemmings, C.C. (1973) 'Direct Observation of the Behaviour of Fish in Relation to Fishing Gear', *Helgolander Wissenschaftliche Meeresuntersuchungen, 24,* 348-60

Hunter, J.R. and Mitchell, C.T. (1968) 'Field Experiments on the Attraction of Pelagic Fish to Floating Objects', *Journal du Conseil permanent pour Exploration Internationale de la Mer, 31,* 427-34

Hurum, H.J. (1977) *A History of the Fish Hook*, Adam and Charles Black, London

Johnstone, A.D.F. and Hawkins, A.D. (1981) 'A Method for Testing the Effectiveness of Different Fishing Baits in the Sea', *Scottish Fisheries Information Pamphlet, 3,* 1-7

Le Gall, J. (1931) 'Les Principales Pêches Maritime de la France: Les Filets et Engins qui y Sont Employés', *La Peche Maritime, 666,* 181-200

Mackie, A.M., Adron, J.W. and Grant, P.T. (1980) 'Chemical Nature of Feeding Stimulants for the Juvenile Dover Sole, *Solea solea* (L.)', *Journal of Fish Biology, 16,* 701-8

Main, J. and Sangster, G.I. (1978a) 'The Value of Direct Observation Techniques by Divers in Fishing Gear Research', *Scottish Fisheries Research Report, 12,* 1-15

Main, J. and Sangster, G.I., (1978b) 'A New Method of Observing Fishing Gear Using a Towed Wet Submersible', *Progress in Underwater Science, 3,* 259-67

Main, J. and Sangster, G.I. (1979) 'A Study of Bottom Trawling Gear on Both Sand and Hard Ground', *Scottish Fisheries Research Report, 14,* 1-15

Main, J. and Sangster, G.I. (1981a) 'A Study on the Sand Clouds Produced by Trawl Boards and their Possible Effect on Fish Capture', *Scottish Fisheries Research Report, 20,* 1-20

Main, J. and Sangster, G.I. (1981b) 'A Study on the Fish Capture Process in a Bottom Trawl by Direct Observations from a Towed Underwater Vehicle', *Scottish Fisheries Research Report, 23,* 1-24

Main, J. and Sangster, G.I. (1982a) 'A Study of Separating Fish from *Nephrops norvegicus* L. in a Bottom Trawl', *Scottish Fisheries Research Report, 24,* 1-9

Main, J. and Sangster, G.I. (1982b) 'A Study of a Multi-level Bottom Trawl for Species Separation Using Direct Observation Techniques', *Scottish Fisheries Research Report, 26,* 1-17

Main, J. and Sangster, G.I. (1983a) 'Fish Reactions to Trawl Gear — a Study Comparing Light and Heavy Ground Gear', *Scottish Fisheries Research Report, 27,* 1-17

Main, J. and Sangster, G.I. (1983b) 'TUV II — a Towed Wet Submersible for Use in Fishing Gear Research', *Scottish Fisheries Research Report, 29,* 1-19

Martin, G.R. (1983) 'Schematic Eye Models in Vertebrates', *Progress in Sensory Physiology, 4,* 43-81

Preston, G. (1982) 'The Fijian Experience in the Utilisation of Fish Aggregation Devices', in *The 14th Regional Technical Meeting on Fisheries of the South Pacific Commission*, pp. 1-61

Rodriguez-Roda, J. (1964) 'Biologia del Atun, *Thunnus thunnus* (L.), de la Costa Sudatlantica de Espana', *Investigacion Pesquera, 25,* 33-146

Sharp, G.D. (1978) 'Behavioural and Physiological Properties of Tunas and their Effects on Vulnerability to Fishing Gear', in *The Physiological Ecology of Tunas*, G.D. Sharp and A.E. Dizon (eds), Academic Press, New York, pp. 397-449

Thomson, D. (1969) 'The Seine Net: its Origin, Evolution and Use', Fishing News (Books) Ltd, London, pp 192

Walls, G.L. (1942) *The Vertebrate Eye and its Adaptive Radiation*, Cranbrook Institute of Sciences, Michigan

Wardle, C.S. (1975) 'Limit of Fish Swimming Speed', *Nature, London, 225,* 725-7

Wardle, C.S. (1977) 'Effects of Size on Swimming Speeds of Fish', in *Scale Effects in Animal*

Locomotion, T.J. Pedley (ed.), Academic Press, New York, pp. 299-313

Wardle, C.S. (1978) 'Non-release of Lactic Acid from Anaerobic Swimming Muscle of Plaice *Pleuronectes platessa L.*: a Stress Reaction', *Journal of Experimental Biology*, 77, 141-55

Wardle, C.S. (1983) 'Fish Reactions to Towed Fishing Gears', in *Experimental Biology at Sea*, A. Macdonald and I.G. Priede (eds), Academic Press, New York, pp. 167-95

Wardle, C.S. and Videler, J.J. (1980) 'Fish Swimming' in *Aspects of Animal Movement*, H.Y. Elder and E.R. Trueman (eds), Cambridge University Press, pp. 125-50

Wemys Fulton, T. (1902) 'North Sea Investigations', in *The 20th Annual Report of the Fishery Board for Scotland, Part III*, HMSO, London pp. 73-227

19 FISH BEHAVIOUR AND THE MANAGEMENT OF FRESHWATER FISHERIES

K. O'Hara

The management of freshwater fish for human benefit is undertaken on differing levels throughout the world. In the industrialised countries, freshwater fish are often principally exploited for recreational purposes, whereas in developing countries food production is paramount. In this latter case the detail of what constitutes a fishery, and what are the requirements of the user (Larkin 1980), may be secondary to the more formal management objectives of assessing the fish yield of a water body. Nevertheless, all managers must be aware of the behaviour of their target species, whether this relates to, for example, how vulnerable the fish are to angling, or the success of an artificial reef in acting as a fish attractor.

The investigation of animal behaviour, particularly from a theoretical viewpoint, has recently been one of the principal areas of biological research, and the subject has a wider connotation than the traditional ethological approach (Barnard 1983; Huntingford 1984). Such developments may not have been wholeheartedly embraced by all fisheries managers, but whether such information is a necessity to the manager could be a similar debating point to that of the relevance of the more theoretical aspects of ecology to fisheries eclogists (Kerr 1980; Werner 1980). It would certainly be true to say that the application of some behavioural theories, such as optimal foraging, have not been warmly received by all fisheries biologists (see Regier, Paloheimo and Gallucci 1979 and discussions in Stroud and Clepper 1979).

Although fisheries management must rest ultimately on the application of ecological principles, social and financial factors are major considerations. The marrying of these often conflicting aspects may prove to be difficult, particularly where economically important species are concerned. Harris (1978), for example, notes the lack of any rigorous attempt to assess the numerous stockings of Atlantic salmon (*Salmo salar*) eggs and juveniles in the rivers of England and Wales. By comparison, it is relatively straightforward to evaluate the success or failure of supplemental stocking using the 'put and take' approach; where fish of a desired catch size are added to a water body and are caught and removed quickly by anglers so that the natural constraints of population regulations are bypassed. Although management objectives that set and meet a target of satisfying the demands of the recreational fishery consumer in terms of catch rate may totally disregard scientifically derived information on fish behaviour, such approaches may nevertheless be successful. However, in those areas such as the improvement of degraded habitats and supplemental stocking

496

of a species where natural recruitment is occurring, there is a necessity for careful consideration of population biology, and fish behaviour therefore becomes important.

There are no universal criteria for management practices, and cultural differences mean that some of the widely disseminated fisheries literature from the North American sport-fishery field, which centres on piscivorous fishes, may not be directly applicable to some European recreational fisheries. Taking just two examples to illustrate this point, the pike (*Esox lucius*) and the carp (*Cyprinus carpio*) are often managed completely differently. In Britain, pike may be subject to vigorous, if often unsuccessful, attempts to reduce their density, both in salmonid and non-salmonid waters, because they are considered by some to decrease the abundance of more preferred species. On the other hand, pike are actively and widely stocked in North America. The reverse situation pertains to carp, which are much vilified in North America and subject to active removal whereas in much of Europe the fish is widely eaten and is also an important recreational species stocked in many countries. However, if such differences are borne in mind, the transposition of knowledge can usefully be made in attempts to enhance or reduce fish numbers in the manner required by management.

This account will centre on recreational fisheries, and no major attempt will be made to discuss aquaculture in relation to food production, although there is equally a need for an appreciation of behaviour (Bardach, Magnusson, May and Reinhart 1980). Freshwater life-history stages of diadromous fishes will be considered since they are of major importance, particularly anadromous species, and because of the considerable management effort expended on stock enhancement. Given the broad scope of the subject of fisheries management, some selection of the information available has been necessary. Inevitably there are differences in depth of coverage, but four subject areas (general aspects, movements and migrations, habitat and direct stock manipulation) have been recognised in which fish behaviour can be directly related to management.

Techniques

Studies of fish behaviour can be accomplished both in the field and in the laboratory. Since the approach taken in this chapter is essentially related to applied aspects and attempts to relate behaviour to the natural environment, a field-based emphasis is provided. The assessment of behaviour in the field rested until comparatively recently on the extrapolation of results using direct capture methods, and the aquatic biologist did not have the facility to observe the animal with the ease of his terrestrial counterpart. Field techniques for investigating fish behaviour have recently been given an effective coverage by Helfman (1983) and Winter (1983). Developments in scuba diving, telemetry and improved sonar have opened whole new vistas for the manager in assessing habitat usage by fish. As recently as 1968, Harden-Jones hardly described telemetry in his list of methods available for investigating fish migration (Harden-Jones 1968).

This of course does not mean that there will not be a continued need for more traditional approaches, and the careful study of diet from stomach samples can still give information on interactions that may be undetected by other methods, particularly where laboratory studies would prove difficult (see Nilsson 1978 for examples). Two recent papers, both concerning brown trout (*Salmo trutta*), provide pleasing contrasts of approach. Tytler and Holliday (1984) used a sophisticated sonic tracking method to study trout behaviour in a lake, whereas Bachman (1984) directly observed stream-dwelling trout from a tower, identifying individual fish from their markings. The new techniques have removed some of the constraints that the very nature of the aquatic environment imposes on progress; but they have not replaced the need for careful observation.

Behaviour and General Aspects of Fisheries Management

Estimation of various stock parameters including age and growth and population size are prerequisites to successful management. Allied fisheries legislation may or may not be designed specifically to protect stocks (Everhart and Youngs 1981).

Age and Growth Determination

Age determination mostly relies on interpreting changes in the hard structures of fish and depends on abiotic factors particularly temperature. There are many ageing studies that require some understanding of fish behaviour for a correct interpretation of growth patterns. To take the most notable example, the interpretation of Atlantic salmon scales assumes a knowledge of the times of migration and periods of sea life (Harden-Jones 1968). Similarly the descent of brown trout into lakes from nursery streams is an event that leaves a record on the scales because of increased growth. The importance of accurately ageing fish is a prelude to many management manipulations of stocks, particularly exploitation rates. In tropical climates where seasonal temperature variations are too limited to be reflected in growth patterns, various behavioural events can leave a regular pattern which is used for interpretive purposes (Bagenal and Tesch 1978).

Population Estimation

Accurate assessment of animal population size presents aquatic biologists with difficulties, and, because of their mobility, problems in making quantitative estimates are often magnified for fishes. All capture of fish by man assumes some knowledge of their behaviour (see Chapter 18 by Wardle, this volume). An excellent example is the longstanding investigation of perch (*Perca fluviatilis*) population numbers in Lake Windermere, England, which takes place annually at spawning time when the fish congregate in the lake shallows and are vulnerable to fish traps (Le Cren, Kipling and McCormack 1977).

Absolute population estimates, both capture-recapture and catch-effort methods, incorporate assumptions about the behaviour of the animal (Seber 1973). Managers are aware that all fish are not equally available for capture, particularly with respect to size. This may be due to gear selectivity, or it may be caused by differential habitat selection, for example larger fish are known to often inhabit deeper areas of river, and may prove difficult to capture by electro-fishing. To overcome the problem of such behavioural differences, fish may be segregated into size groups. The division of fish populations into mobile and static components (Hunt and Jones 1974b) has obvious implications for population estimation since the two groups will have different probabilities of recapture (Roff 1973). A thorough and critical account of a population estimate that was considered inaccurate due to a number of failures to fulfil the assumptions of the models, partly as a result of the behaviour of the fish, is provided by Hunt and Jones (1974a).

Catch-effort methods can be equally prone to error because of failure to meet the underlying assumptions. Stott and Russell (1979) describe an estimate of a grass carp (*Ctenopharyngodon idella*) population that proved to be wrong since the fish were less vulnerable to capture on the second fishing because of modified behaviour after their first exposure to electro-fishing and netting. Recently Peterson and Cederholm (1984) compared a removals technique with mark-recapture for juvenile coho salmon (*Oncorhynchus kisutch*). They found that the removals technique sometimes provided unreliable estimates and this was partly due to behavioural conditioning to electro-fishing. Similarly, behavioural factors were cited by Zalewski (1983) as being responsible for inaccuracy in catch-effort estimates on small riverine fish.

It is important to know if fish are shoaling, solitary or territorial in their behaviour, since, if small sampling sections are used, it is quite possible for the complete home range of a shoaling species not to be encompassed (Hart and Pitcher 1973). Problems of sampling are compounded by the very small volume occupied by shoals in relation to the total available space (Pitcher 1980).

Legislation

Fishing regulations are usually designed to protect desired species, and are not necessarily based on fish behaviour although the timing of particular events in fish life-history stages will influence activities such as harvesting anadromous fishes which are directly dependent on a detailed knowledge of migration times (Mundy 1982). Interestingly, many non-game fish such as carp are afforded no protection by legislation in North America. In fact the opposite situation pertains and exploitation is encouraged. Game fishes, on the other hand, have close seasons and other restrictive practices imposed for their conservation (Borgeson 1979).

In England and Wales, game and non-game fisheries have close seasons resulting from statute in the Salmon and Freshwater Fisheries Act of 1975. These close seasons are designed to give protection to adults while spawning, but implicit in the legislation is the fact that fish may be more vulnerable to capture as they aggregate for spawning. Behavioural alterations as a result of handling can cause stress, and problems stemming from such treatment are well documented (Schreck 1981). Even when fish are returned after capture, handling stress (in addition to the stress imposed by reproduction) is likely to be deleterious, and protective measures are therefore probably sensible in situations where a large proportion of the stock are regularly caught.

Movements and Migrations

Precise definitions of movement and migration do not concern us here, and for a full assessment of this subject reference should be made to Baker (1978). Aspects of migration are reviewed in Gauthreaux 1980, McKeown (1984) and McLeave, Arnold, Dodson and Neil (1984). A thorough treatment of migration, including the various patterns shown by freshwater fishes with specific relation to production, is provided by Northcote (1978). The relevance of an understanding of migratory behaviour to successful management cannot be overemphasised, particularly for the important anadromous salmonids (Mundy 1982).

The migratory homing of salmon has been described in detail by Hasler and Scholz (1983); upstream migration was extensively reviewed by Banks (1969); and salmonid migration by Brannon and Salo (1982). The fact that salmon can be imprinted to return to a home river is being utilised in salmon ranching (Thorpe 1980).

Migratory behaviour of sea trout (*Salmo trutta*) is inadequately understood in comparison with the Atlantic salmon (Harris 1978). It has always been assumed that sea trout do not show the same degree of accuracy in homing as Atlantic salmon (Mills 1971), and the results of Pratten and Shearer (1983a) would tend to support this. However, Sambrook (1983) reported extremely precise returns of sea trout, and at the present time for management purposes the question remains open. The difficulties of enhancing sea-trout stocks are further compounded by the lack of understanding of the relationship of the migratory form, the sea trout,with the non-migratory form, the brown trout. In particular, the question as to whether the migratory habit is under environmental or genetic control (Pratten and Shearer 1983b) remains unresolved.

Upstream movement itself appears to be initiated by a complex of factors (Banks 1969) of which river flow appears to be of particular importance. River regulation schemes have often therefore been designed to include artificial freshets at intervals to induce salmon movement.

Catadromous fishes have not attracted the research attention devoted to anadromous species, but in Europe and Japan, eels are an important food resource. Both the juvenile (elver) and adult stages are subjected to capture by man, the former principally for stocking, and much of the elver catch of the River Severn, England, is exported to mainland Europe. An understanding of both upstream and downstream migrations is an important prerequisite for the effective management of stocks. Tesch (1977) provides an excellent account of eel biology, including good descriptions of migration. Increased interest in the eel as a resource is stimulating study, notably on species previously unfished such as *Anguilla australis australis* in Tasmania (Sloane 1984).

Migrations are also undertaken by many primary freshwater fishes, both in rivers (Hynes 1970) and lakes (Northcote 1978). However, these freshwater behaviour patterns are often ignored during river modification schemes such as dam construction, particularly for non-sport fishes. Where the species are of apparently lesser importance, such as detritivorous fishes, then little consideration may be given, although the ecological impact of destroying traditional migratory routes may be equally serious (Bowen 1983).

The movement and dispersal of fishes within lakes and rivers affect population features such as abundance and have aroused considerable interest with respect to the behaviour of new or supplemental stocked fish. Hunt and Jones (1974b) attributed the spread of barbel (*Barbus barbus*) within the River Severn, England, to the mobile component of the population. Stevens and Miller (1983) suggested that the dispersal of four species of riverine fishes was influenced by the water flow in a river system, and good recruitment years were correlated with high flows since increased dispersal reduced density-dependent mortality.

Movement of stocked fish is often a problem for management since it represents a wastage of resources. Introduced catchable-sized brown trout apparently move less than rainbow trout (*Salmo gairdneri*) or brook trout (*Salvelinus fontinalis*) (Cresswell 1981). Moring (1982) suggests that hatchery-reared rainbow trout that had an anadromous component in the stock history were more likely to move than two other strains and gave poorer angler returns. An extremely interesting finding from long-term studies on Atlantic salmon in Ireland was that hatchery fish introduced as smolts, in addition to having lower survival to adult than wild fish, also gave poorer angling returns. One reason suggested is that the fish could not migrate to their homing point and consequently were 'unsettled', spending considerable time in searching behaviour (Mills and Piggins 1983).

Some aspects of behavioural response (including movement) of prey to predators and the management implications have been considered by Stein (1979).

The development of telemetric tracking techniques has proved to be of great benefit in examining the migration patterns of anadromous salmonids, both adult and juvenile. Movement studies on non-diadromous fish are also attracting considerable interest via the use of radio and sonic-tagging studies, some of which are proving to be of great value to managers (Tyus, Burdick and McAda, 1984). Hockin, O'Hara and Eaton (in prep.) demonstrated that radio-tracked grass carp (*Ctenopharyngodon idella*) (Plate 19.1) in a linear water system showed limited movements and fed within small areas (Figure 19.1); such results have obvious implications for the use of this fish as a weed control agent.

Management of Fish Habitat

Any animal requires a place in which to live, obtain food and reproduce; considerable efforts are expended by fisheries managers in meeting these needs. The provision of living places in newly created aquatic habitats, attempts to improve existing habitats, and the restoration of degraded environments are subjects that have produced a large literature. Habitat improvement also encompasses practices such as the provision of fish passes, creation of spawning areas and the amelioration of pollution and water-quality problems. Physical improvement structures placed in lotic (Swales and O'Hara 1980) and lentic (Price and Maughan 1978) habitats are popular measures, not only because they are a directly observable attempt at management, but also because beneficial effects have been shown to accrue on populations (Hunt 1969).

Habitat Improvement — Rivers and Steams

The fact that river engineering projects associated with land drainage, flood alleviation and navigation are often detrimental to fishes is without dispute (Swales 1982a; Brookes, Gregory and Dawson 1983). Simplification and removal of habitat heterogeneity effectively reduce the places in which fish can live, the result often being a fall in species diversity, species shifts and altered abundance (Swales 1982a).

Habitat improvements are used in some developing countries as a strategy to increase fish stocks (Cooper 1980), but it is the efforts used to increase numbers in sport fisheries that have been most extensively documented. In particular there is a long history in management of the use of habitat improvement devices to enhance trout populations (Swales and O'Hara 1980). Cover within a stream can be provided by instream structures such as rocks, vegetation and tree roots, but is also afforded indirectly by the effects of overhanging vegetation, deep water and surface turbulence (Binns and Eiserman 1979; Swales and O'Hara 1980). Since it has been

Plate 19.1: Radio Tracking Grass Carp on a Canal

found that trout spend a considerable amount of time under cover, the increases in biomass of trout found after the provision of additional cover may be mediated through this behavioural response (Hartzler 1983).

Implicit in these conclusions is the supposition that trout are territorial, and the provision of increased suitable habitat increases the number of potential territories. The assumption that there is a density-dependent limit

Figure 19.1: Movements of Five Grass Carp Recorded by Radio Tracking in a 1.2 km Section of Canal (Hockin, O'Hara and Eaton, in prep.)

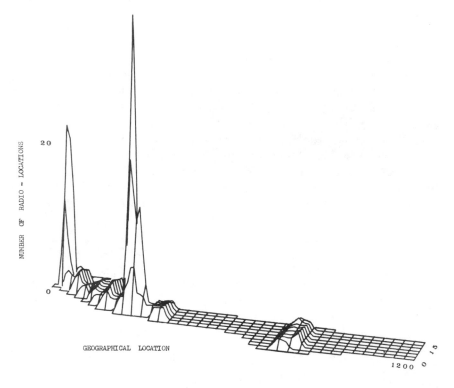

to the number of resident salmonids a steam can hold, mediated through territorial behaviour (Backiel and Le Cren 1978), is the basis of many management decisions. Recently Bachman (1984) has thoroughly reviewed the subject of territoriality in trout and concluded that, in his study stream, brown trout did not maintain a territory. Social structure was best described as 'a cost-minimising, size-dependent, linear dominance hierarchy of individuals having overlapping home ranges'. Whatever the precise definition of the behavioural interactions displayed by trout, there is nevertheless an upper limit to the number of trout a given section of river can hold on a sustained basis. Hartzler (1983) demonstrated that the inclusion of half-log covers in a stream where habitat cover was already good did not increase the number of brown trout, although biomass increased. He correctly concludes that supplemental cover is most likely to prove effective in fertile systems where existing cover is low. Similar benefits are likely to accrue where extensive drainage schemes have been performed. Kennedy (1984) has suggested that one of the factors involved in deterioration in numbers of Atlantic salmon and brown-trout juveniles following drainage schemes may be the removal of suitable-size stones from the stream bed. Again it matters little in management terms whether these changes are due to the removal of the visual basis of territory borders or whether rocks afford more energy-saving sites: the end result is the same.

If, as Shirvell and Dungey (1983) suggest, there are distinct microhabitat preferences for trout according to differing activities, and if the most limiting of these could regulate population size, then more enhancement programmes should take cognisance of microhabitat requirements. Practical strategies for enhancement of juvenile Atlantic salmon using stones to provide various microhabitat needs are described by Rimmer, Paim and Saunders (1984).

Instream flow requirements have attracted considerable attention because the altered conditions resulting from abstractions and other water-supply schemes can influence fish (Fraser 1975). In addition to the anticipated direct effect of water volume on the amount of stream-bed habitat available to fish, behavioural changes in territorial interactions have been documented. Brook trout (*Salvelinus fontinalis*), for example, have increased territory sizes at lower flow velocities (McNicol and Noakes 1984).

Whereas such interpretations might hold for species with territorial or restricted home-range behaviour, it is difficult to apply them directly to shoaling fishes such as many of the cyprinids. These have been shown to possess a home range, but they are much wider ranging in habit than salmonids (Hunt and Jones 1974b). Recently Edwards *et al* (1984) demonstrated that channelisation of a stream increased the abundance of cyprinid fish in a warm-water community to the detriment of more desired sport fish. Similarly, Karr (1981) has reported that degradation of streams will cause species shifts to give dominance to omnivorous fish such as cyprinids. However, in Britain, where the ichthyofauna is relatively species depauperate, cyprinids are the dominant fish family, and the roach (*Rutilus rutilus*) is the most popular angled species. Habitat restoration and conservation schemes must therefore of necessity include this family. Swales and O'Hara (1983) described the effects of a habitat improvement scheme on a stream community in which the most recreationally important species were dace (*Leuciscus leuciscus*) and chub (*Leuciscus cephalus*). They demonstrated that weirs (Plate 19.2) and groynes attracted both dace and chub, but overhead cover was particularly important for chub (Figure 19.2). Although these two species are closely related, chub grow to a much larger size. Recapture rates were much higher for marked chub than dace, and larger chub were apparently more solitary in behaviour than dace or young chub that shoal. The effectiveness of cover in attracting and holding chub may be due to this solitary behaviour.

Habitat Improvement — Lakes and Ponds

The addition of habitat-improvement structures to lentic habitats is well documented, but the techniques are slightly different from stream improve-

Plate 19.2: A Weir Installed in a Small Lowland Stream as Part of a Habitat Improvement Programme

ments in that submerged structures have found most favour. These are thoroughly described by Nelson, Horak and Olson (1978). Tyre reefs and related structures are primarily intended as fish attractors, providing cover usually in relatively bare environments such as newly constructed reservoirs (Prince and Maughan 1978); (see also Chapter 18 by Wardle, this volume). Their primary function is to concentrate fish for angling capture, but the structures quickly become colonised by periphyton and other organisms which act as a food source for fish. Small reefs that provide cover for only a few fish can be utilised effectively because captured fish are quickly replaced by new inhabitants (Nobel 1980). A sensible procedure that is often adopted is to leave as much natural cover material as possible on the bed of new impoundments during construction.

Oxygen levels in lakes can deteriorate to lethal conditions associated with stratification and the formation of ice cover in winter. Artificial destratification by pumps in summer, can render an anoxic hypolimnion habitable, but it also affects the thermal stratification of the lake and thus the fish distribution since fish can regulate their spatial position to remain

Figure 19.2: The Influence of Habitat Improvement Structures on the Distribution of Dace and Chub in a Small Lowland River. (Modified from Swales and O'Hara 1983.)

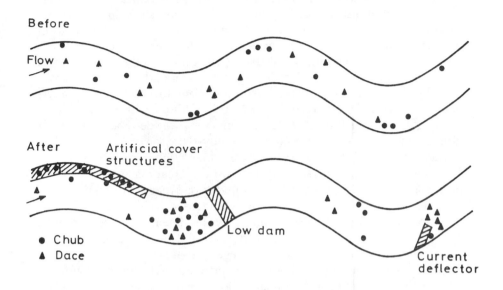

within their area of thermal preference (Magnuson, Crowder and Medvick 1979). Care must be exercised with stenothermal species such as trout which have a low upper lethal level, and 'two-tier' fisheries have been effected in this situation. Cold low-oxygen hypolimnial water is pumped to the surface to aerate it, and this is returned to the hypolimnion without disturbing the stratification, allowing trout to survive in the hypolimnion and maintaining thermally tolerant species in the epilimnion. The distribution of fishes within a thermally stratified lake was considered a contributory factor in angler yield, and artificial destratification increased the catch of channel catfish (*Ictalurus punctatus*) probably as a result of more benthic habitat becoming available (Mosher 1983). The seasonal migration of fish to wintering areas (Northcote 1978), often deeper regions, is catered for and used as a measure to protect fish from ice by including a deeper hole when constructing small impoundments (Bennett 1971).

Vegetation and Management

Weed cover and bankside vegetation have very important roles to play in the lives of many fish. The flood plains of the potamon zone of rivers are exploited by many fishes which show a behavioural seasonality in moving on to these areas to breed and feed (Welcomme, Kohler and Courtenay 1979). Failure to protect these areas from the encroachment of man has meant that river channels have been constrained, with the consequent loss of flood-plain area and thus fish production.

Similar types of movement are recorded in lakes, and are used by managers to enhance or reduce population sizes according to preference. Noble (1980) has documented the approach of raising water levels in reservoirs to enable northern pike to move into marshy spawning areas, and lowering levels to expose carp eggs. Providing cover in shoreline areas of lakes in the form of weed or other vegetation has been found to enhance abundance of juvenile largemouth bass (*Micropterus salmoides*) in these areas (McCammon and Geldern 1979). Aquatic plants can provide cover for prey fishes and may be extremely important to predator-prey interactions. Durocher, Provine and Kraai (1984) found that a reduction of submerged vegetation to below 20 per cent cover was detrimental to largemouth bass stocks. A similar result was found for the same species by Wiley, Gorden, Waite and Powless (1984), except that high weed levels were also detrimental. Grimm (1981a,b) has demonstrated that macrophyte cover is an important regulatory factor in the population dynamics of pike. One of the determining factors associated with increased vegetation cover is a reduction in intraspecific predation. A recent telemetric study on pike (Chapman and Mackay 1984) has confirmed Grimm's findings that pike smaller than 41 cm in length remain in vegetated areas, whereas large pike, fish over 54 cm in length, use both open and vegetated areas. Encouraging the growth of vegetation in barren lakes can therefore be used as a management technique to enhance sport fishes, whether these be piscivorous or non-piscivorous.

Fewer studies appear to have been conducted on the importance of instream weed to running-water fishes. Weed has been included as a physical cover factor in a number of salmonid abundance studies, and fish abundance is determined by such factors (Hermansen and Krog 1984). However, Sweales (1982b) was unable to demonstrate any measurable detrimental effect of weed cutting on non-salmonid species, although Mortensen (1977) has found a marked influence on brown trout. This may be a real difference or is possibly caused by the problems of accurately measuring population densities in shoaling species. A recent investigation on a large lowland cyprinid-dominated river with sparse instream weed showed that roach fry were present in protected lily (*Nuphar lutea*) beds (Plates 19.3 and 19.4) in much greater densities than open areas. There was no obvious attraction of these beds to the more rheophilic dace (Figure 19.3; Pearce, Eaton and O'Hara in prep.).

Fish Passes

Rheotropic responses to water current are practically utilised in a number of situations, particularly fish-guiding devices and fish passes. Mills (1971) describes the practical aspects of these structures for migrating salmonids, and Arnold (1974) has thoroughly reviewed the subject of rheotropism in fish. Where new fish passes are installed for salmonids, then shifts in the amount of time spent at sea could result if, as Schaffer and Elson (1975)

Plate 19.3: Lilies Damaged by Boat Traffic

suggest, the age structure of Atlantic salmon stocks in a river is dependent on the difficulty of the upstream migration. The provision of rheotactically attractive water flows at fish passes can be used to enable counts of upstream migrants to be made as the fish migrate through.

Water Quality

A full description of behavioural responses to water-quality parameters or particular pollutants is beyond the scope of the present chapter, and reference should be made to Alabaster and Lloyd (1983). However, the fisheries manager should be aware of behavioural changes that may be induced in fish by toxic agents. Marcucella and Abramson (1978) have used the term 'behavioural toxicology' to describe these effects, and emphasise their importance in reducing the adaptive response of fish to the environment.

One subject area that has attracted considerable attention is that of thermal influences, because of the widespread occurrence of heated power-station effluents. The effects of electricity generation on the biota,

Plate 19.4: Lilies Given Protection by Excluding Boats Using a Floating Collar

including those of thermal effluents, are comprehensively reviewed by Langford (1982). The thermal niche of fishes was described conceptually by Magnusson *et al.* (1979), and the behavioural response is of clear importance when assessing the impact of heated waters on different species.

Feeding

An awareness of feeding behaviour is often a prerequisite to a knowledge of managing stocks. Predator-prey interactions are an obvious application of such information (See Stroud and Clepper 1979). Because of the importance of fish size in many sport fisheries, growth maximisation is desirable as is production in commercially important species. Diet studies provide some elucidation of likely constraints, and recently the application of optimal foraging theory has been used to aid interpretation (Huntingford 1984).

The feeding behaviour of Atlantic salmon juveniles was found to be broadly consistent with their selecting a diet which maximises growth (Wankowsky 1981). It has been suggested that one of the reasons for the

Figure 19.3: Average Number (ñ) of 0+ Roach and Dace Caught Each Month in 20 Rising Nets Set in Shallow Open Areas of a River and within Lily Beds Dace: ○ open water; ● lily beds. Roach: △ open water; ▲ lily beds. (Pearce, Eaton and O'Hara in prep.)

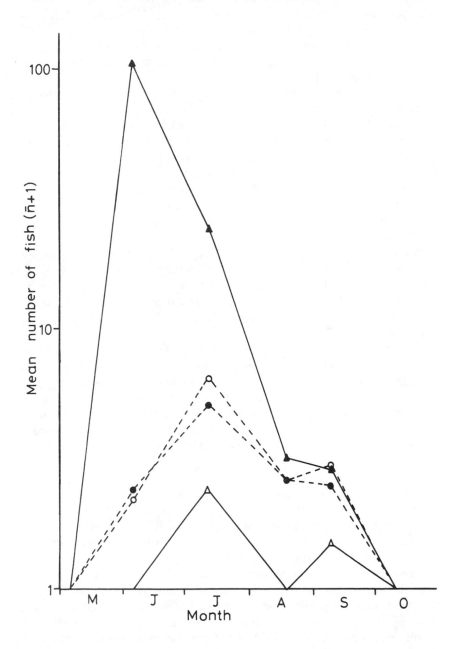

fact that stocked trout in streams show a low yield may be that they do not feed efficiently. Ersbak and Haase (1983) showed that hatchery-reared brook trout were conditioned to respond to a fish pellet and could not therefore forage effectively. Kennedy, Strange, Anderson and Johnston (1984) were unable to demonstrate any differences in feeding between hatchery-reared Atlantic salmon smolts and wild fish, and they attributed this to prior exposure to invertebrate food items because the hatchery supply was unfiltered river water.

Stein (1979) has reviewed the influence of piscivores on the behaviour of prey, and the importance of a predator on prey behaviour was shown by Werner, Gilliam, Hall and Mittelbach (1983b). The impact of predation on prey behaviour must be understood before effective manipulation can be effected. One area that has attracted considerable attention is the provision of forage fishes, the desired attributes of which were listed by Ney (1980) and include the behavioural vulnerability of the prey to the predator. An example of an ideal forage fish is the threadfin shad (*Dorosoma petenense*). This abundant species resides in open water and is vulnerable to bass predation (Heidinger 1977).

Some aspects of the feeding biology of the roach and perch (examined from both an intra- and interspecific competitive viewpoint) have recently been reported by Persson (1983a,b). He suggests that increased eutrophication is likely to encourage roach populations over perch because of their ability to switch feeding to algae/detritus when animal food is in short supply.

The feeding behaviour of the common carp has been implicated, because of its habit of rooting in the bottom mud, in encouraging water turbidity and eutrophication. Lamarra (1975) blamed digestive activities, but Nakashima and Leggett (1980) have questioned this interpretation. Selective predation by fish of large zooplankters has been suggested as a cause of increased phytoplankton levels in some lakes, and selective fish removal is a proposed amelioration technique (Leah, Moss and Forrest 1980).

Management and Direct Stock Manipulation

Considerable efforts are expended by fish managers in attempts to alter the composition of fish populations. Usually this involves the enhancement of desired species and often the reduction of perceived predators or competitors. Many such programmes are doomed to failure from the outset because of a failure to recognise the constraints that exist, even though these are often the result of behavioural traits that are well appreciated. A thorough review of European stock-enhancement practices which makes an interesting comparison with North American procedures is provided by Welcomme *et al.* (1983).

Stocking

There is considerable management justification for the creation of fisheries where none previously existed, for example in reservoir tailwaters. Other recreational practices can involve a put-and-take approach which ignores the normal constraints of carrying capacity: the fish are often removed so quickly that it matters little how they behave other than that they are vulnerable to capture by rod and line.

Among the commonest attempts at stock enhancement is the introduction of young stages of salmonids, particularly the salmons, both Pacific and Atlantic. The experience with stocking Atlantic salmon eggs, fry or parr in the British Isles is not one which suggests any great impact on the abundance of adult stocks (Harris 1978). Many failures are attributed to the addition of new individuals to areas where the stream bed is already fully occupied by naturally recruited fish. There is a considerable body of evidence to suggest that hatchery-reared salmonids are at a considerable disadvantage to indigenous fish in streams, and often these can be attributed to differences in behaviour (Ersbak and Haase 1983). Analysis of stocking policies by Egglishaw and Shackley (1980) has revealed that Atlantic salmon fry do not disperse as widely as some sources have suggested. Therefore it is unproductive to make plantings of eggs or juveniles in large numbers at a few sites, but spot stockings at short intervals may be more effective.

Attempts at establishing Pacific salmon in Europe have essentially failed (Welcomme *et al.* 1983). One possible reason for this is the lack of behavioural traits programmed to the new migratory cues. More probably, however, the introduced juveniles have been poorly suited to their new environment, and few attempts have been made to breed from the small numbers of fish that did return (D.J. Solomon, personal communication).

The influence of olfactory imprinting and homing in salmon has recently been reviewed by Hasler and Scholz (1983), and distinct management potential exists for enhancing stocks by directing the site of return and re-establishing spawning streams.

Stocking salmonids in lentic waters has often proved to be more successful than in running waters (Ersbak and Haase 1983). There is evidence that brown trout are solitary in lakes; for example, Tytler and Holliday (1984) showed that they exhibited restricted movements, but the ranges of individual fish did overlap. In lake-dwelling salmonids such as juvenile sockeye salmon (*Oncorhynchus nerka*), shoaling behaviour is typical rather than the territoriality exhibited by many stream-living juvenile salmonids (Keenleyside 1979). Donald and Anderson (1982) have noted that rainbow trout can be stocked up to a certain density in lakes without mortality occurring, but growth deteriorates and above this level density-dependent mortality occurs.

The behaviour of hatchery-reared fish can be deliberately selected to be different. In this category is the production of sunfish hybrids that have a greater vulnerability to angling capture because of their aggressiveness (Kurzawski and Heidinger 1982). The susceptibility of fish to angling capture may change as a result of learning (Beukema 1970), and this could influence the performance of a fishery where fish are returned after capture, the general practice with coarse fish in Britain. Steinmetz (1983) has suggested from results in The Netherlands that bream (*Abramis brama*) and roach have a higher probability of recapture than carp, and stocking programmes would need to take cognisance of these differences.

For non-territorial species, stocked fish may not be as subject to density-dependent constraints that result in their mortality as territorial species. However, compensatory changes must be exhibited at some level, whether this is, for example, mediated through reduced fecundity via reduced growth or survival of progeny (Goodyear 1980). The stocking of cyprinids in The Netherlands has been so successful in numbers that reduced growth rates are apparent and other management techniques are being sought (Welcomme *et al.* 1983).

Competition and Other Interactions

A common management practice is to remove undesirable fish from waters, and this has often been applied in situations where non-game fish are often regarded as the cause of reduced game-fish catches. The subject of competition is contentious, nowhere more so than in fisheries management where many species are actively removed on the flimsiest of evidence. Countless examples of removals could doubtless be located: in opposition the scientific evaluations are few but often telling where they do exist. Baltz and Moyle (1984) review this area and subscribe to the view that many declines in game-fish are often due to changed environmental conditions (see also Karr 1981). The irony of the situation is that taxonomically closely related game species have been found in some studies to be in competition rather than other fish (Marrin and Ermann 1982). Brown trout are more aggressive than Atlantic salmon juveniles and sympatrically tend to occupy deeper pool areas whereas salmon are restricted to riffles. Whether this is detrimental to salmon is open to question since they appear to be anatomically better adapted to these zones: larger pectoral fins allow the salmon to remain close to the bottom in riffles thus avoiding energetically expensive swimming into the current (Jones 1975). This division of the habitat is an example of interactive segregation (Nilsson 1978), a term which avoids the implied deleterious connotations of competition.

Nevertheless, some managers maintain that brown trout and salmon juveniles interact to the detriment of salmon, and efforts are often made to reduce densities of brown trout. This is, normally, a fruitless exercise as these are quickly replaced by new individuals (Mills 1971). This is not a surprising result, given the density-dependent nature of the population-regulatory mechanisms operating in stream salmonid communities (Backiel and Le Cren 1978).

In many non-salmonids, stunting is a feature of overstocked situations, and culling will increase growth rates (Welcomme *et al.* 1983). Werner *et al.* (1983a,b) have persuasively argued, from optimal foraging theory, optimal habitat use and the influence of a predatory fish, how the growth of bluegills (*Lepomis macrochirus*) can be restricted through the creation of intraspecific interactions. One of the usual management practices when bass/bluegill combinations 'go out of balance' (Bennett 1971) is to reduce the density of bluegills which have become stunted. This clearly would have the effect of reducing such competitive effects. An alternative has been to produce virtual monosex populations of sunfish hybrids to prevent or reduce reproduction (Kurzawski and Heidinger 1982).

Intraspecific competitive effects have often proved easier to demonstrate than interspecific changes, because changes in growth in fish are often detected. Several examples of such changes and a perceptive account of competition are provided by Weatherley (1972). There are in fact many documented shifts in feeding behaviour when new species have been added to a water body (Nilsson 1978). This is one of the criteria indicative of competition, and interestingly, as Werner, Kerr and Werner (1980) pointed out, a source of information which has generally been missed by fisheries ecologists.

Stockings of salmonids of the *Oncorhynchus* genus are often spectacularly successful, particularly for those species like pink (*Oncorhynchus gorbuscha*) and chum (*Oncorhynchus keta*), in which the fry migrate directly to sea after a very short freshwater residence period and therefore avoid the restrictions of the limited freshwater habitat in terms of space or food.

Stock Preservation

Some fisheries biologists are becoming increasingly concerned that fishing activities and stocking policies can influence the genetic composition of wild stocks. Horn and Rubenstein (1984) have discussed behavioural adaptations and life histories, including fisheries examples. As well as the likely influence of exploitation on the adjustment of sea ages of salmonid stocks (Schaffer and Elson 1975), it is possible that stocking with home river progeny fish of a specific sea age will give the same migratory pattern if this trait is genetically determined. It appears that the time of return, at least in Atlantic salmon, is both genetically and environmentally determined (Gardner 1976). However, Gardner suggests that there is sufficient genetic influence to justify using parents of specific sea ages in attempts to influence the age of return of the progeny. An excellent overview of these and related topics is provided by Saunders (1981).

Atlantic salmon show iteroparity in that a number of fish survive spawning to reproduce again. Whereas it is useless to attempt to preserve spent Pacific salmon, since they all display a semelparous life history, the conservation of Atlantic salmon kelts (spent fish returning to the sea) is one option in stock enhancement programmes (Harris 1978).

Summary

Fisheries management seeks to maximise user appreciation of the resource, whether this be in terms of catch rate, aesthetics of the experience, or food production. As such it must take cognisance of all aspects of fish biology including behaviour (and some of human behaviour!). Many fisheries management practices cannot be placed within an ethological or theoretical framework because they are often empirically derived from practical experience. However, the extensive and developing literature on fish behaviour which include a management objective suggests that the two areas are converging. It is evident that fishes differ from many other other animals in that they show indeterminate growth, and population constraints (such as competition), many mediated through behaviour, yield outcomes other than those that may be predicted from theory (Persson 1983a).

If we are to understand how factors such as food and space can influence growth and the population dynamics of fish, then ethological studies allied to ecological principles will be of great benefit. Care must be exercised to ensure that such information is not used to further increase pressures on already overfished stocks, but it should be applied constructively, as with the clear benefits that are beginning to occur in our understanding of stream-dwelling juvenile salmonids. Fisheries managers can only gain from the current upsurge in knowledge of the behaviour of their charges, even if some myths about behaviour have to be jettisoned en route.

Acknowledgements

Thanks for photographic material are due to G.S. Hockin, S. Swales and H.G. Pearce, and for preparation of the plates to B. Lewis. The patience and help of Ms A. Callaghan with typing are gratefully recorded.

References

Alabaster, J.S. and Lloyd, R. (1983) *Water Quality Criteria for Freshwater Fish*, Butterworths, London
Arnold, G.P. (1974) 'Rheotropism in Fishes', *Biological Reviews, 49*, 515-76
Bachman, R.A. (1984) 'Foraging Behavior of Free-ranging Wild and Hatchery Brown Trout in a Stream', *Transactions of the American Fisheries Society, 113*, 1-32
Backiel, T. and Le Cren, E.D. (1978) 'Some Density Relationships for Fish Population Parameters', in S.D. Gerking (ed.), *Ecology of Freshwater Fish Production*, Blackwell, Oxford, pp. 279-302

Bagenal, T.B. and Tesch, F.W. (1978) 'Age and Growth', in T.B. Bagenal (ed.), *Methods for the Assessment of Fish Production in Fresh Waters*, Blackwell, Oxford, pp. 101-36

Baker, R.R. (1978) *The Evolutionary Ecology of Animal Migration*, Hodder & Stoughton, London

Baltz, D.M. and Moyle, P.B. (1984) 'Segregation by Species and Size Classes of Rainbow Trout', *Salmo gairdneri*, and Sacramento Sucker, *Catostomus occidentalis*, in Three California Streams', *Environmental Biology of Fishes, 10*, 101-10

Banks, J.W. (1969) 'A Review of the Literature on the Upstream Migrations of Adult Salmonids', *Journal of Fish Biology, 1*, 85-136

Bardach, J.E., Magnusson, J.J., May, R.B. and Reinhart, J.N. (1980) (eds) *Fish Behavior and its Use in the Capture and Culture of Fishes*, International Center for Living Aquatic Resources Management, Manila, Philippines

Barnard, C.J. (1983) *Animal Behaviour*, Croom Helm, London

Bennett, G.W. (1971) *Management of Lakes and Ponds*, Van Nostrand Reinhold, New York

Beukema, J.J. (1970) 'Angling Experiments with Carp, 2. Decreasing Catchability through One-trial Learning', *Netherlands Journal of Zoology, 20*, 81-92

Binns, N.A. and Eiserman, F.M. (1979) 'Quantification of Fluvial Trout Habitat in Wyoming', *Transactions of the American Fisheries Society, 108*, 215-28

Borgeson, D.P. (1979) 'Controlling Predator-Prey Relationships in Streams', in R.H. Stroud and H. Clepper (eds), *Predator-Prey Systems in Fisheries Management*, Sport Fishing Institute, Washington, DC, pp. 425-30

Bowen, S.H. (1983) 'Detritivory in Neotropical Fish Communities', *Environmental Biology of Fishes, 9*, 137-44

Brannon, E.L. and Salo, E.O. (1982) (eds) *Proceedings of the Salmon and Trout Migratory Behavior Symposium*, University of Washington

Brookes, A., Gregory, K.J. and Dawson, F.H. (1983) 'An Assessment of River Channelization in England and Wales', *The Science of the Total Environment, 27*, 97-111

Chapman, C.A. and Mackay, W.C. (1984) 'Versatility in Habitat Use by a Top Aquatic Predator, *Esox lucius* L.', *Journal of Fish Biology, 25*, 109-15

Cooper, E.L. (1980) 'Fisheries Management in Streams', in R.T. Lackey and L.A. Nielsen (eds), *Fisheries Management*, Blackwell, Oxford, pp. 297-322

Cresswell, R.C. (1981) 'Post-stocking Movements and Recapture of Hatchery-Reared Trout Released into Flowing Waters — a Review', *Journal of Fish Biology, 18*, 429-42

Donald, D.B. and Anderson, R.S. (1982) 'Importance of Environment and Stocking Density for Growth of Rainbow Trout in Mountain Lakes', *Transactions of the American Fisheries Society, 111*, 675-80

Durocher, P.P., Provine, W.C. and Kraai, J.E. (1984) 'Relationship between Abundance of Largemouth Bass and Submerged Vegetation in Texas Reservoirs', *North American Journal of Fisheries Management, 4*, 84-8

Edwards, C.J., Griswold, B.L., Tubb, R.A., Weber, E.C. and Woods, L.C. (1984) 'Mitigating Effects of Artificial Riffles and Pools on the Fauna of a Channelized Warmwater Stream', *North American Journal of Fisheries Management, 4*, 194-203

Egglishaw, H.J. and Shackley, P.E. (1980) 'Survival and Growth of Salmon, *Salmo salar* (L.), Planted in a Scottish Stream', *Journal of Fish Biology, 16*, 565-84

Ersbak, K. and Haase, B.L. (1983) 'Nutritional Deprivation after Stocking as a Possible Mechanism Leading to Mortality in Stream-stocked Brook Trout', *North American Journal of Fisheries Management, 3*, 142-51

Everhart, W.H. and Youngs, W.D. (1981) *Principles of Fishery Science*, Cornell University Press

Fraser, J.C. (1975) '*Determining Discharges for Fluvial Resources*', FAO Fisheries Technical Paper No. 143

Gardner, M.L.G. (1976) 'A Review of Factors which May Influence the Sea-age and Maturation of Atlantic Salmon *Salmo salar* L.', *Journal of Fish Biology, 9*, 289-327

Gauthreaux, S.A. (1980) *Animal Migration, Orientation and Navigation*, Academic Press, New York

Goodyear, C.P. (1980) 'Compensation in Fish Populations', in C.H. Hocutt and J.R. Stauffer (eds), *Biological Monitoring of Fish*, Lexington Books, Toronto, pp. 253-80

Grimm, M.P. (1981a), 'The Composition of Northern Pike (*Esox lucius* L.) Populations in Four Shallow Waters in The Netherlands, with Special Reference to Factors Influencing O+ Pike Biomass', *Fisheries Management, 12*, 61-76

Grimm, M.P. (1981b) 'Intraspecific Predation as a Principal Factor Controlling the Biomass of Northern Pike (*Esox lucius* L.)', *Fisheries Management, 12*, 77-79

Harden-Jones, F.R. (1968) *Fish Migration*, Edward Arnold, London

Harris, G.S. (1978) (ed.) *Salmon Propagation in England and Wales*, National Water Council, 1 Queen Anne's Gate, London

Hart, P.J.B. and Pitcher, T.J. (1973) 'Population Densities and Growth of Five Species of Fish in the River Nene, Northamptonshire', *Fisheries Management, 4*, 69-86

Hartzler, J.R. (1983) 'The Effects of Half-log Covers on Angler Harvest and Standing Crop of Brown Trout in McMichaels Creek, Pennsylvania', *North American Journal of Fisheries Management, 31*, 228-38

Hasler, A.D. and Scholz, A.T. (1983) *Olfactory Imprinting and Homing in Salmon*, Springer-Verlag, Berlin

Heidinger, R.C. (1977) 'Potential of the Threadfin Shad as a Forage Fish in Midwestern Power Cooling Reservoirs', *Transactions of the Illinois State Academy of Science, 70*, 15-25

Helfman, G.S. (1983) 'Underwater Methods', in L.A. Nielsen and D.L. Johnson (eds), *Fisheries Techniques*, American Fisheries Society, Bethesda, pp. 349-69

Hermansen, H. and Krog, C. (1984) 'Influence of Physical Factors on Density of Stocked Brown Trout (*Salmo trutta fario* L.) in a Danish Lowland Stream', *Fisheries Management, 15*, 107-15

Horn, H.S. and Rubenstein, D.I. (1984) 'Behavioural Adaptations and Life History', in J.R. Krebs and N.B. Davies (eds), *Behavioural Ecology: an Evolutionary Approach*, 2nd ed, Blackwell, Oxford, pp. 279-98

Hunt, P.C. and Jones, J.W. (1974a) 'A Population Study of *Barbus barbus* (L.) in the River Severn, England. I. Densities', *Journal of Fish Biology, 6*, 255-67

Hunt, P.C. and Jones, J.W. (1974b) 'A Population Study of *Barbus barbus* (L.) in the River Severn, England. II. Movements', *Journal of Fish Biology, 6*, 269-78

Hunt, R.L. (1969) 'Effects of Habitat Alteration on Production Standing Crops and Yield of Brook Trout in Lawrence Creek, Wisconsin', in *Symposium on Salmon and Trout in Streams*, H.R. MacMillan Lectures in Fisheries, 1968. University of British Columbia, Canada, pp. 281-312

Huntingford, F. (1984) *The Study of Animal Behaviour*, Chapman & Hall, London

Hynes, H.B.N. (1970) *The Ecology of Running Waters*, Liverpool University Press, Liverpool

Jones, A.N. (1975) 'A Preliminary Study of Fish Segregation in Salmon Spawning Streams', *Journal of Fish Biology, 7*, 95-104

Karr, J.R. (1981) 'Assessment of Biotic Integrity Using Fish Communities', *Fisheries (Bethesda, Maryland), 6*, (6)21-27

Keenleyside, M.H.A. (1979) *Diversity and Adaptation in Fish Behaviour*, Springer-Verlag, Berlin

Kennedy, G.J.A. (1984) 'The Ecology of Salmonid Habitat Re-instatement Following River Drainage Schemes', *Proceedings of the Institute of Fishery Management Study Course, Northern Ireland Branch*, New University of Ulster, Coleraine, pp. 1-25

Kennedy, G.J.A., Strange, C.D., Anderson, R.J.D. and Johnston, P.M. (1984) 'Experiments on the Descent and Feeding of Hatchery-reared Salmon Smolts (*Salmo salar* L.) in the River Bush', *Fisheries Management, 15*, 15-25

Kerr, S.R. and Werner, E.E. (1980) 'Niche Theory in Fisheries Ecology', *Transactions of the American Fisheries Society, 109*, 254-60

Kurzawski, K.F. and Heidinger, R.C. (1982) 'The Cyclic Stocking of Parentals in a Farm Pond to Produce a Population of Male Bluegill × Female Green Sunfish F_1 Hybrids and Male Redear Sunfish × Female Green Sunfish F_1 Hybrids', *North American Journal of Fisheries Management, 2*, 188-92

Lamarra, V.A. (1975) 'Digestive Activities of Carp as a Major Contributor to the Nutrient Loading of Lakes', *Verhandlungen der Internationalen Vereinigung für Theoretische und Angewandte Limnologie, 19*, 2461-8

Langford, T.E. (1983) *Electricity Generation and the Ecology of Natural Waters*, Liverpool University Press, Liverpool

Larkin, P.A. (1980) 'Objectives of Management', in R.T. Lackey and L.A. Nielsen (eds), *Fisheries Management*, Blackwell, Oxford, pp. 245-62

Leah, R.T., Moss, B. and Forrest, D.E. (1980) 'The Role of Predation in Causing Major Changes in the Limnology of a Hyper-eutrophic Lake', *Internationale Revue der*

Gesamten Hydrobiologie und Hydrographie, 65, 223-47

Le Cren, E.D., Kipling, C. and McCormack, J.C. (1977) 'A Study of the Number, Biomass and Year-Classes Strengths of Perch (*Perca fluviatilis* L.) in Windermere from 1941 to 1966', *Journal of Animal Ecology, 46,* 281-307

McCammon, G.W. and Geldern, C.R. von Jr (1979) 'Predator-Prey Systems in Large Reservoirs', in R.H. Stroud and H. Clepper (eds), *Predator-Prey Systems in Fisheries Management,* Sport Fishing Institute, Washington, DC, pp. 431-42

McKeown, B.A. (1984) *Fish Migration,* Croom Helm, London

McLeave, J.D., Arnold, G.P., Dodson, J.J. and Neill, W.H. (1984) (eds) *Mechanisms of Migration in Fishes,* Plenum Press, New York

McNicol, R.F. and Noakes, D.L.G. (1984) 'Environmental Influences on Territoriality of Juvenile Brook Charr, *Salvelinus fontinalis,* in a Stream Environment', *Environmental Biology of Fishes, 10,* 29-42

Magnuson, J.J., Crowder, L.B. and Medvick, P.A. (1979) 'Temperature as an Ecological Resource', *American Zoologist, 19,* 331-43

Marcucella, H. and Abramson, C.I. (1978) 'Behavioral Toxicology and Teleost Fish', in D.I. Mostofsky (ed.), *The Behavior of Fish and other Aquatic Animals,* Academic Press, London, pp. 33-77

Marrin, D.L. and Erman, E.C. (1982) 'Evidence against Competition between Trout and Nongame Fishes in Stampede Reservoir, California', *North American Journal of Fisheries Management, 2,* 262-9

Mills, C.P.R. and Piggins, D.J. (1983) 'The Release of Reared Salmon Smolts (*Salmo salar*) into the Burrishoole River System (Western Ireland) and their Contribution to the Rod and Line Fishery', *Fisheries Management, 14,* 165-75

Mills, D. (1971) *Salmon and Trout,* Oliver & Boyd, Edinburgh

Moring, J.R. (1982) 'An Efficient Hatchery Strain of Rainbow Trout for Stocking Oregon Streams', *North American Journal of Fisheries Management, 2,* 209-15

Mortensen, E. (1977) 'Density Dependent Mortality of Trout Fry (*Salmo trutta* L.) and its Relationship to the Management of Small Streams', *Journal of Fish Biology, 11,* 613-17

Mosher, T.D. (1983) 'Effects of Artificial Circulation on Fish Distribution and Angling Success for Channel Catfish in a Small Prairie Lake', *North American Journal of Fisheries Management, 3,* 403-9

Mundy, P.R. (1982) 'Computation of Migratory Timing Statistics for Adult Chinook Salmon in the Yukon River, Alaska, and their Relevance to Fisheries Management', *North American Journal of Fisheries Management,* 2,359-70

Nakashima, B.S. and Leggett, W.C. (1980) 'The Role of Fishes in the Regulation of Phosphorus Availability in Lakes', *Canadian Journal of Fisheries and Aquatic Sciences, 37,* 1540-9

Nelson, R.W., Horak, G.C. and Olson, J.E. (1978) *Western Reservoir and Stream Habitat Improvements Handbook,* Fish and Wildlife Service, US, Department of the Interior

Ney, J.J. (1980) 'Evolution of Forage Fish Management in Lakes and Reservoirs', *Transactions of the American Fisheries Society, 110,* 725-8

Nilsson, N-A. (1978) 'The Role of Size-biased Predation in Competition and Interactive Segregation in Fish', in S.D. Gerking (ed.), *Ecology of Freshwater Fish Production,* Blackwell, Oxford, pp. 303-25

Noble, R.L. (1980) 'Management of Lakes, Reservoirs, and Ponds', in R.T. Lackey, and L.A. Nielsen (eds), *Fisheries Management,* Blackwell, Oxford, pp. 265-95

Northcote, T.G. (1978) 'Migratory Strategies and Production in Freshwater Fishes', in S.D. Gerking (ed.), *Ecology of Freshwater Fish Production,* Blackwell, Oxford, pp. 326-59

Persson, L. (1983a) 'Food Consumption and the Significance of Detritus and Algae to Intraspecific Competition in Roach *Rutilus rutilus* in a Shallow Eutrophic Lake', *Oikos, 41,* 118-25

Persson, L. (1983b) 'Effects of Intra- and Interspecific Competition on Dynamics and Size Structure of a Perch *Percha fluviatilis* and Roach *Rutilus rutilus* Population', *Oikos, 41,* 126-32

Peterson, N.C. and Cederholm, C.J. (1984) 'A Comparison of the Removal and Mark-Recapture Methods of Population Estimation for Juvenile Coho Salmon in Small Streams', *North American Journal of Fisheries Management, 4,* 99-102

Pitcher, T.J. (1980) 'Some Ecological Consequences of Fish School Volumes', *Freshwater Biology, 10,* 539-44

Pratten, D.J. and Shearer, W.M. (1983a) 'Sea Trout of the River North Esk', *Fisheries Management, 14*, 49-65

Pratten, D.J. and Shearer, W.A. (1983b) 'The Migration of North Esk Sea Trout', *Fisheries Management, 14*, 99-113

Prince, E.D. and Maughan, O.E. (1978) 'Freshwater Artificial Reefs: Biology and Economics', *Fisheries (Bethesda, Maryland), 3(1)*, 5-9

Regier, H.A., Paloheimo, J.E. and Gallucci, V.F. (1979) 'Factors that Influence the Abundance of Large Piscivorous Fish', in R.H. Stroud and H. Clepper (eds), *Predator-Prey Systems in Fisheries Management*, Sport Fishing Institute, Washington, DC, pp. 333-41

Rimmer, D.M., Paim, V. and Saunders, R.L. (1984) 'Changes in the Selection of Microhabitat by Juvenile Atlantic salmon (*Salmo salar*) at the Summer-Autumn Transition in a Small River', *Canadian Journal of Fisheries and Aquatic Sciences, 41*, 469-75

Roff, D.A. (1973) 'An Examination of Some Statistical Tests Used in the Analysis of Mark-recapture Data', *Oecologia (Berlin), 12*, 35-54

Sambrook, H. (1983) 'Homing of Sea Trout in the River Fowey Catchment, Cornwall', *Proceedings of the Third British Freshwater Fisheries Conference, Liverpool University, England*, pp. 30-40

Saunders, R.L. (1981) 'Atlantic Salmon (*Salmo salar*) Stocks and Management Implications in the Canadian Atlantic Provinces, and New England, USA', *Canadian Journal of Fisheries and Aquatic Sciences, 38*, 1612-25

Schaffer, W.M. and Elson, P.F. (1975) 'The Adaptive Significance of Variations in Life History among Local Populations of Atlantic Salmon in North America', *Ecology, 56*, 577-90

Schreck, C.B. (1981) 'Stress and Compensation in Teleostean Fishes: Response to Social and Physical Factors', in A.D. Pickering (ed.), *Stress and Fish*, Academic Press, London, pp. 295-321

Seber, G.A.F. (1973) *The Estimation of Animal Abundance and Related Parameters*, Griffin, London

Shirvell, C.S. and Dungey, R.G. (1983) 'Microhabitats Chosen by Brown Trout for Feeding and Spawning in Rivers', *Transactions of the American Fisheries Society, 112*, 355-67

Sloane, R.D. (1984) 'Preliminary Observations of the Migrating Adult Freshwater Eels (*Anguilla australis australis* Richardson) in Tasmania', *Australian Journal of Marine and Freshwater Research, 35*, 471-6

Stein, R.A. (1979) 'Behavioral Response of Prey to Fish Predators', in R.H. Stroud and H. Clepper (eds), *Predator-Prey Systems in Fisheries Management*, Sport Fishing Institute, Washington, DC, pp. 343-53

Steinmetz, B. (1983) 'Management of Fish Stocks in The Netherlands and the Need for Planning', in J.H. Grover (ed.) *Allocation of Fishery Resources*, United Nations Food and Agriculture Organization, pp. 384-95

Stevens, D.E. and Miller, L.W. (1983) 'Effects of River Flow on Abundance of Young Chinook Salmon, American Shad, Longfin, Smelt, and Delta Smelt in the Sacramento-San Joaquin, River System', *North American Journal of Fisheries Management, 3*, 425-37

Stott, B. and Russell, I.C. (1979) 'An Estimate of a Fish Population that Proved to be Wrong', *Fisheries Management, 10*, 169-71

Stroud, R.H. and Clepper, H. (1979) *Predator-Prey Systems in Fisheries Management*, Sport Fishing Institute, Washington, DC

Swales, S. (1982a) 'Environmental Effects of River Channel Works Used in Land Drainage Improvement', *Journal of Environmental Management, 14*, 103-26

Swales, S. (1982b) 'Impacts of Weed-cutting on Fisheries: an Experimental Study in a Small Lowland River', *Fisheries Management, 13*, 125-37

Swales, S. and O'Hara, K. (1980) 'Instream Habitat Improvement Devices and their Use in Freshwater Fisheries Management', *Journal of Environmental Management, 10*, 167-79

Swales, S. and O'Hara, K. (1983) 'A Short-term Study of the Effects of a Habitat Improvement Programme on the Distribution and Abundance of Fish Stocks in a Small Lowland River in Shropshire', *Fisheries Management, 14*, 135-44

Tesch, F-W. (1977) *The Eel: Biology and Management of Anguillid Eels*, Chapman & Hall, London

Thorpe, J.E. (1980) (ed) *Salmon Ranching*, Academic Press, London

Tytler, P. and Holliday, F.G.T. (1984) 'Temporal and Spatial Relationships in the Movements of Loch Dwelling Brown Trout, *Salmo trutta* L., Recorded by Ultrasonic Tracking for 24 Hours', *Journal of Fish Biology, 24,* 691-702

Tyus, H.M., Burdick, B.D. and McAda, C.W. (1984) 'Use of Radiotelemetry for Obtaining Habitat Preference Data on Colorado Squawfish', *North American Journal of Fisheries Management, 4,* 177-80

Wankowsky, J.W.J. (1981) 'Behavioural Aspects of Predation by Juvenile Atlantic Salmon (*Salmo salar* L.) on Particular Drifting Prey', *Animal Behaviour, 29,* 557-71

Weatherley, A.H. (1972) *Growth and Ecology of Fish Populations,* Academic Press, London

Welcomme, R.L. (1979) *Fisheries Ecology of Floodplain Rivers,* Longman, London

Welcomme, R.L., Kohler, C.C. and Courtenay, W.R. (1983) 'Stock Enhancement in the Management of Freshwater Fisheries: a European Perspective', *North American Journal of Fisheries Management, 31,* 265-75

Werner, E.E., Mittelback, G.G. Hall, D.J. and Gilliam, J.F. (1983a) 'Experimental Tests of Optimal Habitat Use in Fish: the Role of Relative Habitat Profitability', *Ecology,64,* 1525-39

Werner, E.E., Gilliam, J.F., Hall, D.J. and Mittelbach, G.G. (1983b) 'An Experimental Test of the Effects of Predation Risk on Habitat Use in Fish', *Ecology, 64,* 1540-8

Wiley, M.J., Gorden, R.W., Waite, S.W. and Powless, T. (1984) 'The Relationship between Aquatic Macrophytes and Sport Fish Production in Illinois Ponds: a Simple Model', *North American Journal of Fisheries Management, 4,* 111-19

Winter, J.D. (1983) 'Underwater Biotelemetry', in L.A. Nielsen and D.L. Johnson (eds), *Fisheries Techniques,* American Fisheries Society, Bethesda, pp. 371-95

Zalewski, M. (1983) 'The Influence of Fish Community Structure on the Efficiency of Electrofishing', *Fisheries Management, 14,* 177-86

AUTHOR INDEX

FISH INDEX

Fish are referred to families,
according to Wheeler (1985)

536

Reference: Wheeler, A. (1985) '*The World Encyclopedia of Fishes*' Macdonald, London. pp. 368.

SUBJECT INDEX

abyssal fish 84, 388
actin 137
active information transfer (AIT) 323-4
aggregation — floating objects 82, *see also* fish aggregation device
aggregation 1, 15, 24, 26, 31, 33, 37, 40, 54, 58, 61, 66, 98, 123, 168, 254, 259, 264, 284, 289, 290, 396, 415, 441-50
agonistic behaviour 2, 31, 33-4, 53-4, 58-9, 61, 375, 380, 396-7, 441-50, 454-5
alanine 167
alarm substance *see* fright substance
algae 118, 239, 269, 370-1, 373, 379, 388, 399, 415, 434, 512
amino acid 72, 160-1, 164, 167, 169
amphibious fish 208, 390-2
amphipod 373, 416-17, 436
anadromous *see* migration
anglers 6, 72-3, 114, 464, 496, 506, 513
anosmic males 168-9
ant 223
anti-predator behaviour 2, 53, 208, 285, 294, 296, 347-51, 411, 421-4
    cooperative 258
apparant size hypothesis 418
appetite 121
aquarium — acoustic design 72, 116, 128-9
aquarium artefact 34, 41
arcamine 167
area restricted search 30, 213
arms race 301, 306
arousal 26-7, 31, 166
artificial stream 239
Asellus 417
attack abatement effect *see* predator
attack avoidance *see* predator
audiogram 128-133
auk 303, 309, 317

backcross 8, 436, 438, 447, 451-2
bait 464
basic units of behaviour 5
bat 323, 433-4
bathypelagic fish 125, 153
beam trawl 464
bee 215, 323
behaviour genetics Ch1
    backcross 4, 8-9, 13-17
    hybrids 4, 8, 13-17
    mutation 4, 6, 8
    selection experiment 4, 6
behavioural ecology — definition 205

behavioural options 237
behavioural toxicology 509
benthic fish 84, 86-7, 225, 371, 379
benthics and limnetics 414-15, 417, 420
benthos 240, 341, 420
best of a bad job 339, 354
betaine 167
bile acids 160-1, 167, 169
binocular vision 86, 91
biological clock 405-6
bioluminescence 384, 486
bird 61, 75, 126, 133, 137, 141, 220, 232, 236, 249, 260, 262, 302, 345, 421
birth 13, 64, 409
bite 285, 442
blinded fish 192-4, 401
bluff 319
body condition 51, 289
body shape 225-7
bog 79, *see also* swamp, saltmarsh
boldness *see* timidity
bouts of behaviour 28, 442
brain 63, 438, *see also under various regions*
    fore 72, 152, 166
    hind 56
    mid 57, 103, 135
branchial cavity 140
breeding *see* reproduction, courtship, mates
brood *see* parental care
buoyancy 226, 391-2

call interception 126-7
camouflage 82-3, 108, 327
    versus conspicuousness 71, 82-3
cannibal 13, 39, 51, 262, 269, 285, 427, 441
carbon dioxide 4
carousel 51-2
cat 60, 84, 485
catastrophe theory 36
catch-effort 464, 499
causation 1, 4, 6, 41, 10-12, 17-19, 23-4
cave fish Ch1, 7, 8, 9, 59, 75, 84, 177, 185-6, 208, 366, 388
    development 59
cestode 249, 348
changeover period 207
charge 50-1, 53, 59
chase 50
chemotaxis 160
chickadee 217

543